Up and Running with AutoCAD® 2012

2D and 3D Drawing and Modeling

Up and Running with AutoCAD® 2012

2D and 3D Drawing and Modeling

Elliot Gindis

Amsterdam • Boston • Heidelberg • London
New York • Oxford • Paris • San Diego
San Francisco • Singapore • Sydney • Tokyo
Academic Press is an imprint of Elsevier

Academic Press is an imprint of Elsevier
225 Wyman Street, Waltham, MA 02451, USA
The Boulevard, Langford Lane, Kidlington, Oxford, OX5 1GB, UK

Library of Congress Cataloging-in-Publication Data
Gindis, Elliot.
  Up and running with AutoCAD 2012. 2D and 3D drawing and modeling / Elliot Gindis.
    p. cm.
  Includes index.
  Summary: "Throughout the book, the following methods are used to present material: - Explain the new concept
or command and why it is important.—Cover the command step by step (if needed), with your input and AutoCAD
responses shown so you can follow and learn them. – Give you a chance to apply just-learned knowledge to a
real-life exercise, drawing, or model.—Test yourself with end-of-chapter quizzes and drawing exercises that ask
questions about the essential knowledge"—Provided by publisher.
  ISBN 978-0-12-387029-2 (pbk.)
  1. Computer graphics. 2. AutoCAD. 3. Computer-aided design. 4. Three-dimensional display systems. I. Title.
II. Title: 2D and 3D drawing and modeling.
  T385.G5424 2011 620′.00420285536—dc23   2011014774

British Library Cataloguing-in-Publication Data
A catalogue record for this book is available from the British Library

ISBN: 978-0-12-387029-2

For information on all Academic Press publications
visit our website at www.elsevierdirect.com

Printed in the United States of America

11 12 13 14 15 9 8 7 6 5 4 3 2 1

CONTENTS

v

CONTENTS

vi

CONTENTS

ix

CONTENTS

xiii

CONTENTS

xiv

A textbook of this magnitude is rarely a product of only one person's effort. I thank all the early and ongoing reviewers of this text and Chris Ramirez of Vertical Technologies Consulting for research and ideas when most needed as well as using the text in his classroom. A big thank you also to Karen Miletsky at Pratt Institute of Design, Russell and Titu Sarder at Netcom Information Technology, and everyone at New York Institute of Technology, RoboTECH CAD solutions, and other premier training centers for their past and present support.

Extensive gratitude also goes to Joseph P. Hayton, Jeff Freeland, Michael Joyce, Maria Alonso, Becky Pease, Gnomi Gouldin and the rest of the team at Elsevier for believing in the project and for their invaluable support in getting the book out to market.

Finally, I would like to thank my friends and family, especially my parents, Boris and Tatyana Gindis, for their patience and encouragement as well as standing by me as months of work turned into years.

This book is dedicated to the hundreds of students who have passed through my classrooms and made teaching the enjoyable adventure it has become.

Elliot Gindis started out using AutoCAD professionally in a New York City area civil engineering company in September 1996, moving on to consulting work shortly afterward. He has since drafted in a wide variety of fields, ranging from all aspects of architecture and building design to electrical, mechanical, civil, structural, and rail design. These assignments, including lengthy stays with IBM and Siemens Transportation Systems, have totaled over 60 companies to date.

In 1999, Elliot began teaching part-time at Pratt Institute of Design, followed by positions at Netcom Information Technology, RoboTECH CAD solutions, and more recently at the New York Institute of Technology. In 2003, Elliot formed Vertical Technologies Consulting and Design (www.VTCDesign.com), an AutoCAD training firm that has trained numerous corporate and government clients as diverse as environmental engineering firms and the FBI in using and optimizing AutoCAD.

Elliot holds a bachelor's degree in aerospace engineering from Embry Riddle Aeronautical University. He currently resides in the Atlanta area and continues to be involved with AutoCAD education and CAD consulting. *Up and Running with AutoCAD 2012*, which carefully incorporates lessons learned from nearly 16 years of teaching and industry work, is his third textbook on the subject.

WHAT IS AUTOCAD?

AutoCAD is a drafting and design software package developed and marketed by Autodesk®, Inc. As of 2011, it has been around for approximately 29 years—several lifetimes in the software industry. It has grown from modest beginnings to an industry standard, often imitated, sometimes exceeded, but never equaled. The basic premise of its design is simple and is the main reason for AutoCAD's success. Anything you can think of, you can draw quickly and easily. For many years, AutoCAD remained a superb 2D electronic drafting board, replacing the pencil and paper for an entire generation of technical professionals. In recent releases, its 3D capabilities finally matured, and AutoCAD is now also considered an excellent 3D visualization tool, especially for architecture and interior design.

The software has a rather steep learning curve to become an expert but a surprisingly easy one to just get started. Most important, it is well worth learning. This is truly global software that has been adopted by millions of architects, designers, and engineers worldwide. Over the years, Autodesk expanded this reach by introducing add-on packages that customize AutoCAD for industry-specific tasks, such as electrical, civil, and mechanical engineering. However, underneath all these add-ons is still plain AutoCAD. This software remains hugely popular. Learn it well, as it is still one of the best things you can add to your resume and skill set.

ABOUT THIS BOOK

This book is not like most on the market. While many authors certainly view their particular text as unique and novel in its approach, I rarely reviewed one that was clear to a beginner student and distilled AutoCAD concepts down to basic, easy to understand explanations. The problem may be that many of the available books are written by either industry technical experts or teachers but rarely by someone who is actively both. One really needs to interact with the industry and the students, in equal measure, to bridge the gap between reality and the classroom.

After years of AutoCAD design work in the daytime and teaching nights and weekends, I set out to create a set of classroom notes that outlined, in an easy to understand manner, exactly how AutoCAD is used and applied, not theoretical musings or clinical descriptions of the commands. These notes eventually were expanded into the book that you now hold. The rationale was simple: I need this person to be up and running as soon as possible to do a job. How do we make this happen?

TEACHING METHODS

This book has its roots in a certain philosophy I developed while attending engineering school many years ago. While there, I had sometimes been frustrated with the complex presentation of what in retrospect amounted to rather simple topics. My favorite quote was, "Most ideas in engineering are not that hard to understand but often become so upon explanation." The moral of that quote was that concepts can usually be distilled to their essence and explained in an easy and straightforward manner. That is the job of a teacher: Not to blow away students with technical expertise but to use experience and top-level knowledge to sort out what is important and what is secondary and to explain the essentials in plain language.

Such is the approach to this AutoCAD book. I want everything here to be highly *practical* and easy to understand. There are few descriptions of procedures or commands that are rarely used in practice. If we talk about it, you will likely need it. The first thing you must learn is how to draw a line. You see this command on the first few pages of Chapter 1. It is essential to present the "core" of AutoCAD, essential knowledge common to just about *any* drafting situation, all of it meant to get you up and running quickly. This stripped down approach proved effective in the classroom and was carefully incorporated into this text.

TEXT ORGANIZATION

This book comes in three parts: Level 1, Level 2, and Level 3:

Level 1 (Chapters 1–10) is meant to give you a wide breadth of knowledge on many topics, a sort of "mile wide" approach. These ten chapters comprise, in my experience, the complete essential knowledge set of an intermediate user. You then can work on, if not necessarily set up and manage, moderate to complex drawings. If your CAD requirements are modest or if you are not required to draft full time, then this is where you stop.

Level 2 (Chapters 11–20) is meant for advanced users who are CAD managers, full-time AutoCAD draftspersons, architects, or self-employed and must do everything themselves. The goal here is depth, as many features not deemed critically important in Level 1 are revisited to explore additional advanced options. Also introduced are advanced topics necessary to set up and manage complex drawings.

Level 3 (Chapters 21–30) is all about 3D. Solid knowledge of the previous two levels is highly recommended before starting these chapters. The 3D material covers all aspects of AutoCAD solid modeling including lights and rendering.

Throughout the book, the following methods are used to present material:

- Explain the new concept or command and why it is important.
- Cover the command step by step (if needed), with your input and AutoCAD responses shown so you can follow and learn them.
- Give you a chance to apply just-learned knowledge to a real-life exercise, drawing, or model.
- Test yourself with end-of-chapter quizzes and drawing exercises that ask questions about the essential knowledge.

You will not see an extensive array of distracting "learning aids" in this text. You will, however, see some common features throughout, such as

Commands: These are presented in almost all cases in the form of a command matrix, such as the one shown here for a Line. You can choose any of the methods for entering the command.

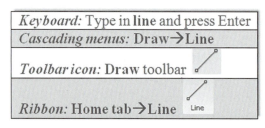

Keyboard: Type in **line** and press Enter	
Cascading menus: Draw→Line	
Toolbar icon: Draw toolbar	
Ribbon: Home tab→Line	Line

Tips and tricks: These are seen mostly in the first few chapters and one is shown here. They are very specific, deliberate suggestions to smooth out the learning experience. Do take note.

TIP 1: The Esc (Escape) key in the upper left-hand corner of your keyboard is your new best friend. It gets you out of just about any trouble you get yourself into. If something does not look right, just press the Esc key and repeat the command. Mine was worn out learning AutoCAD, so expect to use it often.

Step-by-step instructions: These are featured whenever practical and show you exactly how to execute the command, such as the example with *line* here. What you type in and what AutoCAD says are in the default font: `Courier New`. The rest of the steps are in the standard print font.

Step 1. Begin the line command via any of the previous methods.
- AutoCAD says: `Specify first point:`

Step 2. Using the mouse, left-click anywhere on the screen.
- AutoCAD says: `Specify next point or [Undo]:`

Step 3. Move the mouse elsewhere on the screen and left-click again. You can repeat Step 2 as many times as you wish. When you are done, click Enter or Esc.

Learning objectives and time for completion: Each chapter begins with this, which builds a "road map" for you to follow while progressing through the chapter, as well as sets expectations of what you will learn if you put in the time to go through the chapter. The time for completion is based on classroom teaching experience but is only an estimate. If you are learning AutoCAD in school, your instructor may choose to cover part of a chapter or more than one at a time.

Summary, review questions, exercises: Each chapter concludes with these. Be sure to not skip these pages and to review everything you learned.

In this chapter, we introduce AutoCAD and discuss the following:

- Introduction and the basic commands
- The Create Objects commands
- The Edit and Modify Objects commands
- The View Objects commands, etc.

By the end of the chapter, you will…
Estimated time for completion of chapter: 3 hours.

SUMMARY
REVIEW QUESTIONS
EXERCISES

WHAT YOUR GOAL SHOULD BE

Just learning commands is not enough; you need to see the big picture and truly understand AutoCAD and how it functions for it to become effortless and transparent. The focus after all is on your design. AutoCAD is just one of the tools to realize it.

A good analogy is ice hockey. Professional players do not think about skating; to them it is second nature. They are focused on strategy, scoring a goal, and getting by the defenders. This mentality should be yours as well. You must become proficient through study and practice, to the point where you are working with AutoCAD, not struggling against it. It then becomes "transparent" and you focus only on the design, to truly perform the best architecture or engineering work of which you are capable.

If you are in an instructor-led class, take good notes. If you are self-studying from this text, pay very close attention to every topic; nothing here is unimportant. Do not skip or cut corners, and complete every drawing assignment. Most important, you have to practice, daily if possible, as there is no substitute for sitting down and using the software. Not everyone these days has the opportunity to learn while working and getting paid; companies want ready-made experts and do not want to wait. If that is the case, you have to practice on your own in the evening or on weekends. Just taking a class or reading this book alone is not enough.

It may seem like a big mountain to climb right now, but it is completely doable. Once on top, you will find that AutoCAD is not the frustrating program it may have seemed in the early days but an intuitive software package that, with proficiency of use, becomes a natural extension of your mind when working on a new design. That, in the end, is the mark of successful software; it helps you do your job easier and faster. You can contact me at Elliot.Gindis@gmail.com. Good luck!

LEVEL 1

Chapters 1–10

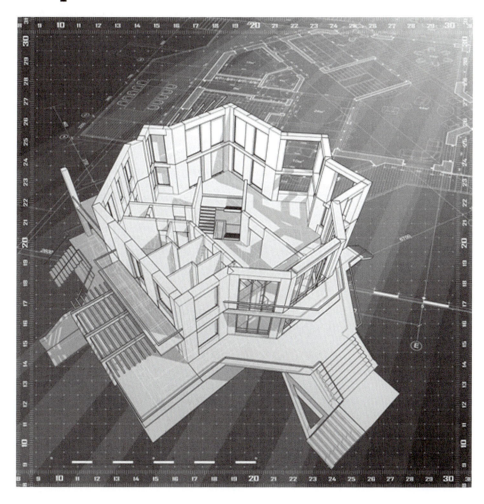

Level 1 is the very beginning of your studies. No prior knowledge of AutoCAD is assumed, only basic familiarity with computers and some technical aptitude. You are also at an advantage if you have hand drafted before, as many AutoCAD techniques flow from the old paper and pencil days, a fact alluded to later in the chapters.

We begin Chapter 1 by outlining the basic commands under Create Objects and Modify Objects followed by an introduction to the AutoCAD environment. We then introduce basic accuracy tools of Ortho and OSNAP.

Chapter 2 continues the basics by adding units and various data entry tools. These first two chapters are the most important, as success here ensures you will understand the rest and be able to function in the AutoCAD environment.

Chapter 3 continues on to layers, then each succeeding chapter continues to deal with one or more major topics per chapter: text and mtext in Chapter 4, hatching in Chapter 5, and dimensioning in Chapter 6. In these six chapters, you are asked to not only practice what you learned but apply the knowledge to a basic architectural floor plan. Chapter 7 introduces blocks and wblocks and Chapter 8 arrays. At this point, you are asked to draw another project, this time a mechanical device. Level 1 concludes with basic printing and output in Chapter 9 and finally advanced printing and output (paper space) in Chapter 10.

Be sure to dedicate as much time as possible to practicing what you learn; there really is no substitute.

CHAPTER 1

AutoCAD Fundamentals
Part I

3

LEARNING OBJECTIVES

In this chapter, we introduce AutoCAD and discuss the following:

- Introduction and the basic commands
- The Create Objects commands
- The Edit/Modify Objects commands
- The View Objects commands
- The AutoCAD environment
- Interacting with AutoCAD
- Practicing the Create Objects commands
- Practicing the Edit/Modify Objects commands
- Selection methods—Window and Crossing
- Accuracy in drafting—Ortho
- Accuracy in drafting—OSNAPs

By the end of the chapter, you will learn the essential basics of creating, modifying, and viewing objects; the AutoCAD environment; and accuracy in the form of straight lines and precise alignment of geometric objects via OSNAP points.

Estimated time for completion of the chapter: 3 hours.

1.1 INTRODUCTION AND BASIC COMMANDS

AutoCAD 2012 is a very complex program. If you are taking a class or reading this textbook, this is something you probably already know. The commands available to you, along with their submenus and various options, number in the thousands. So, how do you get a handle on them and begin using the software? Well, you have to realize two important facts.

First, you must understand that on a typical workday, *95% of your AutoCAD drafting time is spent using only 5% of the available commands, over and over again*. So getting started is easy; you need to learn only a handful of key commands; and as you progress and build confidence, you can add depth to your knowledge by learning new ones.

Second, you must understand that even the most complex drawing is essentially *made up of only a few basic fundamental objects that appear over and over again* in various combinations on the screen. Once you learn how to create and edit them, you can draw surprisingly quickly. Understanding these facts is the key to learning the software. We are going to strip away the perceived complexities of AutoCAD and reduce it to its essential core. Let us go ahead now and develop the list of the basic commands.

For a moment, view AutoCAD as a fancy electronic hand-drafting board. In the old days of pencil, eraser, and T-square, what was the simplest thing that you could draft on a blank sheet of paper? That of course is a line. Let us make a list with the following header, "Create Objects," and below it add "Line."

So, what other geometric objects can we draw? Think of basic building blocks, those that cannot be broken down any further. A *circle* qualifies and so does an *arc*. Because it is so common and useful, throw in a *rectangle* as well (even though you should note that it is a compound object, made up of four lines). Here is the final list of fundamental objects that we have just come up with:

Create Objects

- Line
- Circle
- Arc
- Rectangle

As surprising as it may sound, these four objects, in large quantities, make up the vast majority of a typical design, so already you have the basic tools. We will create these on the AutoCAD screen in a bit. For now, let us keep going and get the rest of the list down.

Now that you have the objects, what can you do with them? You can *erase* them, which is probably the most obvious. You can also *move* them around your screen and, in a similar manner, *copy* them. The objects can *rotate*, and you can also *scale* them up or down in size. With lines, if they are too long, you can *trim* them, and if they are too short, you can *extend* them. *Offset* is a sort of precise copy and one of the most useful commands in AutoCAD. *Mirror* is used, as the name implies, to make a mirror-image copy of an object. Finally, *fillet* is used to put a curve on two intersecting lines, among other things. We will learn a few more useful commands a bit later, but for now, under the header "Edit/Modify Objects," list the commands just mentioned:

Edit/Modify Objects

- Erase
- Move

- Copy
- Rotate
- Scale
- Trim
- Extend
- Offset
- Mirror
- Fillet

Once again, as surprising as it may sound, this short list represents almost the entire set of basic Edit/Modify Objects commands you need once you begin to draft. Start memorizing them.

To finish up, let us add several View Objects commands. With AutoCAD, unlike paper hand drafting, you do not always see your whole design in front of you. You may need to *zoom* in for a close-up or out to see the big picture. You may also need to *pan* around to view other parts of the drawing. With a wheeled mouse, so common on computers these days, it is very easy to do both, as we soon see. To this list we add the *regen* command. It stands for *regenerate*, and it simply refreshes your screen, something you may find useful later. So here is the list for View Objects:

View Objects

- Zoom
- Pan
- Regen

So, this is it for now, just 17 commands making up the basic set. Here is what you need to do:

1. As mentioned before, memorize them so you know what you have available.
2. Understand the basic idea, if not the details, behind each command. This should be easy to do, because (except for maybe *offset* and *fillet*) the commands are intuitive and not cryptic in any way; *erase* means erase, whether it is AutoCAD, a marker on a whiteboard, or a pencil line.

We are ready now to start AutoCAD, discuss how to interact with the program, and try all the commands out.

1.2 THE AUTOCAD ENVIRONMENT

It is assumed that your computer, whether at home, school, or training class, is loaded with AutoCAD 2012. It is also assumed that AutoCAD starts up just fine (via the AutoCAD icon or Start menu) and everything is configured right. If not, ask your instructor, as there are just too many things that can go wrong on a particular PC or laptop, and it is beyond the scope of this book to cover these situations. If all is well, start up AutoCAD. You should see the screen depicted in Figure 1.1. Your particular screen's appearance may vary slightly, which we discuss soon.

This is your basic "out of the box" AutoCAD environment for the Drafting & Annotation workspace. From both the learning and teaching points of view, this screen layout (and the layout of the last few releases) is an improvement over older versions of AutoCAD, and Autodesk has done an admirable job in continuing to keep things clean and (relatively) simple.

Note that you may also see a large panel appear upon AutoCAD startup (it is not shown here). It is called the Autodesk Exchange, an on-line tool for additional functionality and the help files. We will not discuss it right now, so just delete it via the "X" at the upper right corner. There is a "Show this window at startup" check box at the lower left of the Exchange panel. If that is left unchecked, you will not see it upon startup anymore. It may be a good idea to leave it as such until we get to talking about this tool.

5

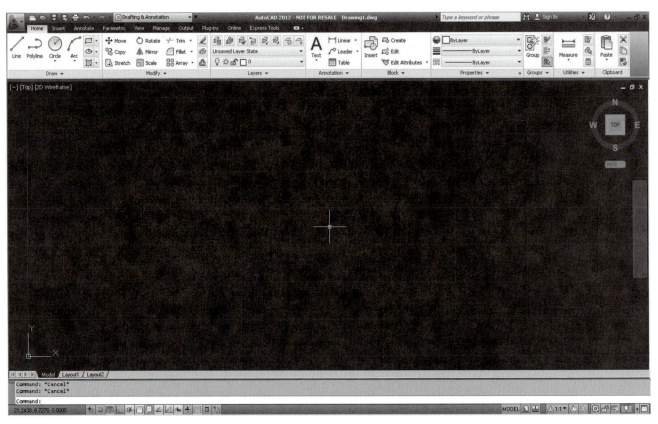

FIGURE 1.1
AutoCAD 2012.

6

AutoCAD went through a major facelift in recent years, beginning with Release 2009. If you caught a glimpse of earlier versions, you may have noticed toolbars present. They are still around, but what we have now, dominating the upper part of the screen, is called the *Ribbon*. It is a new way of interacting with AutoCAD, and we discuss it in detail soon. Other screen layouts or *workspaces* are available to you, including one with toolbars. They can be accessed through the menu seen in Figure 1.2, which is located at the top left of the screen, depending on settings.

FIGURE 1.2
Workspace switching.

If you click on the down arrow of that menu, you see the workspace switching menu, shown in Figure 1.3. Here, you can switch to AutoCAD Classic if desired, which removes the Ribbon and loads the screen with toolbars and a palette. Give it a try. So, which workspace to use? Well, for now switch back to the Drafting & Annotation workspace (seen being selected in Figure 1.3).

Next, let us take a tour of the features you are looking at on the screen, as shown in Figure 1.4. The following is a brief description of what is labeled. Be sure to read each description carefully, as the AutoCAD environment has been altered slightly out of necessity and will remain this way for the rest of the textbook. What has been added or changed (and why) is detailed in this bulleted list.

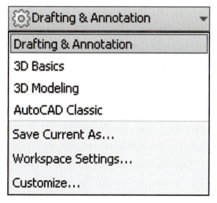

FIGURE 1.3
Workspace switching menu.

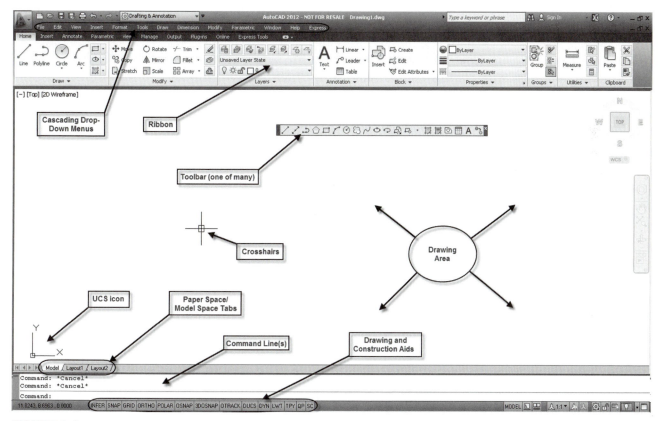

FIGURE 1.4
AutoCAD 2012 screen elements.

- **Drawing Area.** The drawing area takes up most of the screen and is colored a dark gray in the default version of the environment. This is where you work and your design appears. It is best to change the drawing area to black to ease eye strain, as less light will radiate toward you. In this textbook, however (as in most textbooks), the color will always be white to conserve ink and for clarity on a printed page. If you wish to change the color of the drawing area, you need to right-click into "Options…", choose the Display tab, then Colors… . Finally, change the color from the drop-down menu on the right. We cover all this in more detail in Chapter 14, so you should ask your instructor for assistance in the meantime, if necessary.

- **Command Line(s).** Right below the drawing area is the command line or, by default, three lines. This is where the commands may be entered and also where AutoCAD tells you what it needs to continue. Always keep an eye on what appears here, as this is one of the main ways that AutoCAD communicates with you. Although we discuss a newer "heads up" on-screen interaction method in Chapter 2, the command line remains very relevant. It is colored gray and white by default but is changed (in a manner similar to the drawing area, for the same reasons) to all white in this textbook. This is a minor detail, so you can leave your command line "as is" if you wish.

- **UCS Icon.** This is a basic X-Y-Z (Z is not visible) grid symbol. It will be important later in advanced studies and 3D. It can be turned off, as is shown via a tip in a later chapter. The significance of this icon is great, and we study it in detail later on. For now, just observe that the Y axis is "up" and the X axis is "across."

- **Paper Space/Model Space Tabs.** These Model/Layout1/Layout2 tabs, not unlike those used in Microsoft's Excel, indicate which drawing space you are in, and are important in Chapter 10 when we cover paper space. Though you can click on them to see what happens, be sure to return to the Model tab to continue further.

- **Toolbar.** Toolbars contain *icons* that can be pressed to activate commands. They are an alternative to typing and the Ribbon, and most commands can be accessed this way. AutoCAD 2012 has dozens of them. You may not have a toolbar present on your screen at the moment. If that is the case, do not worry, we activate a few toolbars shortly.

- **Crosshairs.** Crosshairs are simply the mouse cursor and move around along with the movements of your mouse. They can be full size and span the entire screen or a small (flyspeck) size. You can change the size of the crosshairs if you wish, and full screen is recommended in some cases. For now, we will leave it as is.

- **Drawing & Construction Aids.** These various settings assist you in drafting and modeling. We introduce them as necessary. By default, these aids are in graphical symbol form and some may be activated, which you can determine by observing their *color*. If they are off, they are gray; if on, then light blue. Be sure to *turn them all to off* for now. We will activate them as needed. There is also one more significant difference here. All the drawing aids have been converted from icon symbols to written symbols. The reason for this is that it is easier for new students to just read them and use them; no additional memorization of what the symbols mean is needed, only knowledge of how to use them. If you wish to do the same, *right-click* on any of them and uncheck "Use Icons." If you prefer them as icons, then leave them as is, but they are converted for the remainder of this textbook. Note that there are additional drawing and construction aids all the way to the far right, on the same band. We discuss them later in the text, as needed.

- **Ribbon.** This is a relatively new way of interacting with AutoCAD's commands, to be discussed soon. The Ribbon first appeared in AutoCAD 2009 and is somewhat similar to the approach used in Microsoft Word, Excel, and PowerPoint.

- **Cascading Drop-down Menus.** This is another way to access commands in AutoCAD. These menus, so named because they drop out like a waterfall, may be hidden initially, but you can easily make them visible via the down arrow at the very top of the screen, to the right of Drafting & Annotation (visible in Figure 1.2). A lengthy menu appears. Select "Show Menu Bar" toward the bottom, and the cascading menus appear as a band across the top of the screen, above the Ribbon. You should keep these menus in their spot from now on; they are referred to often.

Review this bulleted list carefully; it contains a lot of useful "get to know AutoCAD" information. You do not have to change your environment exactly as suggested—how you like your AutoCAD to look, after all, is personal taste to some extent. However be sure to understand how to do it if necessary and why the look of AutoCAD is the way it is in the textbook.

1.3 INTERACTING WITH AUTOCAD

OK, so you have the basic commands in hand and ideally a good understanding of what you are looking at on the AutoCAD screen. We are ready to try out the basic commands and eventually draft something. So, how do we interact with AutoCAD and tell it what we want drawn? There are *four primary ways* (Methods 1–4), which follow, roughly in the order they appeared over the years. There are also two outdated methods, called the *tablet* and the *screen side menu* (dating back to the very early days of AutoCAD), but we do not cover them.

Method 1. *Type in* the commands on the command line (AutoCAD v1.0–current).
Method 2. Select the commands from the *drop-down cascading menus* (AutoCAD v1.0–current).
Method 3. Use *toolbar* icons to activate the commands (AutoCAD 12/13–current).
Method 4. Use the *Ribbon* tabs, icons, and menus (AutoCAD 2009–current).

Details of each method including the pros and cons follow. Most commands are presented in all four primary ways, and you can experiment with each method to determine what you prefer. Eventually, you will settle on one particular way of interacting with AutoCAD or a hybrid of several.

Method 1: Type in the Commands on the Command Line

This was the original method of interacting with AutoCAD and, to this day, remains the most foolproof way to enter a command: good old-fashioned typing. AutoCAD is unique among leading CAD software in that it has retained this method while almost everyone else moved to graphic icons, toolbars, and Ribbons. If you hate typing, this will probably not be your preferred choice.

However, do not discount keyboard entry entirely; AutoCAD has kept it for a reason. When the commands are abbreviated to one or two letters (Line = L, Arc = A, etc.), input can be incredibly fast. Just watch a professional typist for proof of the speed with which one can enter data via a keyboard. Other advantages to typing are that you no longer have toolbars or a Ribbon cluttering up precious screen space (there is never enough of it) and you no longer have to take your eyes off the design to find an icon; instead, the command is literally at your fingertips. The disadvantage is of course that you have to type.

To use this method, simply type in the desired command (spelling counts!) at the command line, as seen in Figure 1.5, and press Enter. The sequence initiates and you can proceed. A number of shortcuts are built into AutoCAD (try using just the first letter or two of a command), and we learn how to make our own shortcuts in advanced chapters. This method is still preferred by many "legacy" users (a kind way to say they have been using AutoCAD forever).

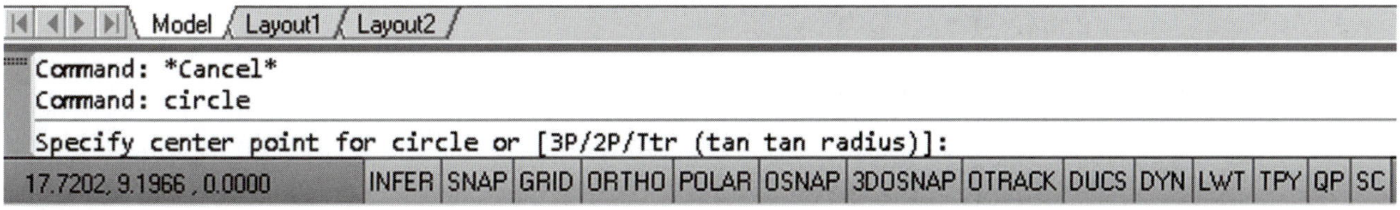

```
|◄| |◄| |►| |►|\ Model / Layout1 / Layout2 /
Command: *Cancel*
Command: circle
Specify center point for circle or [3P/2P/Ttr (tan tan radius)]:
17.7202, 9.1966 , 0.0000    INFER SNAP GRID ORTHO POLAR OSNAP 3DOSNAP OTRACK DUCS DYN LWT TPY QP SC
```

FIGURE 1.5
AutoCAD 2012 command line (`circle` typed in).

Method 2: Select the Commands from the Drop-down Cascading Menus

This method has also been around since the beginning. It presents a way to access virtually every AutoCAD command, and indeed many students start out by checking out every

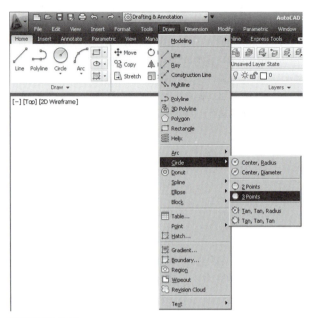

FIGURE 1.6
AutoCAD 2012 cascading drop-down menus.

one of them as a crash course on what is available—a fun but not very effective way of learning AutoCAD. Go ahead and examine the cascading menus; these are similar in basic arrangement to other software, and you should be able to navigate through them easily. We refer to them on occasion in the following format: **Menu→Command**. So, for the sequence shown in Figure 1.6, you would read **Draw→Circle→3 Points**.

Method 3: Use Toolbar Icons to Activate the Commands

This method has been around since AutoCAD switched from DOS to Windows in the mid-1990s and is a favorite of a whole generation of users; toolbars are a familiar sight with virtually any software these days. Toolbars contain sets of icons (see Figure 1.7 for an example), organized by categories (Draw toolbar, Modify toolbar, etc.). You press the icon you want, and a command is initiated. One disadvantage to toolbars, and the reason the Ribbon was developed, is that they take up a lot of space and, arguably, are not the most effective way of organizing commands on the screen.

FIGURE 1.7
AutoCAD 2012 Draw toolbar.

You can access some toolbars by activating the AutoCAD Classic Workspace or all of them at any time by simply selecting Tools→Toolbars→AutoCAD from the cascading menus. When you do that, a menu will appear (Figure 1.8) and you can simply check off the ones you want or do not want. You also can access the same menu by right-clicking on any of the toolbars themselves. Once you have a few up, you typically just dock them on top or off to the side. It is highly encouraged to bring up the Standard, Draw, and Modify toolbars for learning purposes. We need no other ones for now but bring them up as needed.

Method 4: Use the Ribbon Tabs, Icons, and Menus

This is the most recently introduced method of interacting with AutoCAD and follows a new trend in software user interface design, such as that used by Microsoft and its Office

2007 (seen in Figure 1.9 with Word®). Notice how toolbars have been displaced by "tabbed" categories (Home, Insert, Page Layout, etc.), where information is grouped together by a common theme.

So it goes with the new AutoCAD. The Ribbon was introduced with AutoCAD 2009, and it is here to stay. It is shown again by itself in Figure 1.10.

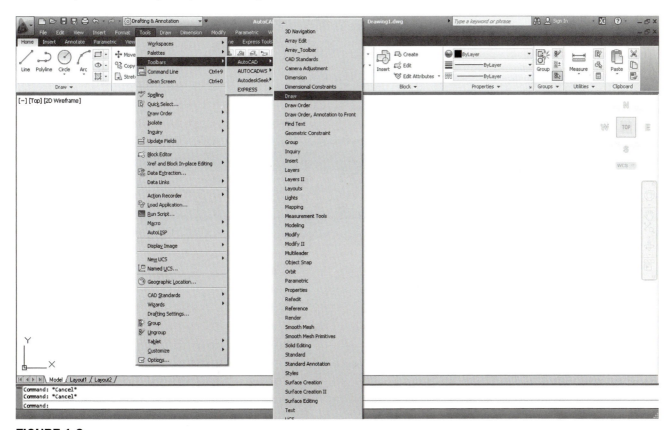

FIGURE 1.8
Toolbar menu.

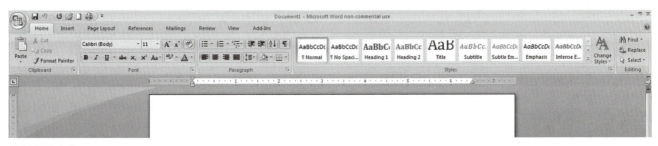

FIGURE 1.9
MS Word Ribbon.

FIGURE 1.10
AutoCAD 2012 Ribbon.

Notice what we have here. A collection of tabs, indicating a subject category, is found at the top (Home, Insert, Annotate, etc.), and each tab reveals an extensive set of tools (Draw, Modify, Annotation, etc.). At the bottom of the Ribbon, additional options can be found by using the drop arrows. In this manner the toolbars have been rearranged in what is, in principle, a more logical and space-saving manner.

Additionally, tool tips appear if you place your mouse over any particular tool for more than a second. Another second yields an even more detailed tool tip. In Figure 1.11, you see the Home tab selected, followed by additional options via the drop arrow, and finally the mouse placed over the polygon command. A few moments of waiting reveals the full tool tip; pressing the icon activates the command.

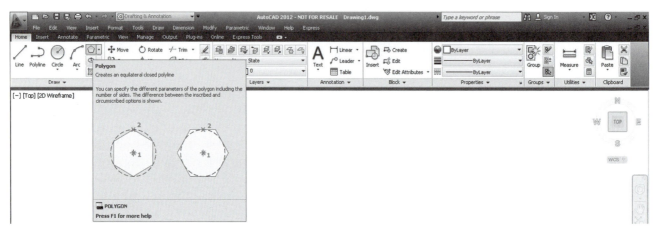

FIGURE 1.11
Polygon command and tool tip.

Familiarize yourself with the Ribbon by exploring it. It presents some layout advantages, and specialized Ribbons can be displayed via other workspaces. The disadvantage of this new method is that it is a relatively advanced tool that presents many advanced features right away and some confusion is liable to come up for a brand-new user. It is also not ideal for longtime users who type, and some veterans in my update classes turn the feature off. The Ribbon is the single biggest change to AutoCAD's user interface and represents a jump forward in user/software interaction, but the ultimate decision to use it is up to you.

TIP 1: The Esc (Escape) key in the upper left-hand corner of your keyboard is your new best friend while learning AutoCAD. It gets you out of just about any trouble you get yourself into. If something does not look right, just press the Esc key and repeat the command. Mine was worn out learning AutoCAD, so expect to use it often.

Before we try out our basic commands, let us begin a list of tips. These are a mix of good ideas, essential habits, and time-saving tricks passed along to you every once in a while (mostly in the first few chapters). Make a note of them as they are important. Here is the first, most urgently needed one.

In the interest of trying out not just the Ribbon but also toolbars, cascading menus, and typing while learning the basics, bring up the three toolbars mentioned earlier, Standard, Draw, and Modify, shutting off all the rest. This is the setup you see in all the upcoming AutoCAD screen shots. Finally, close out any floating palettes and, if you wish, remove your scroll bars (ask your instructor how). They take up room and are not an efficient way to get around the screen—we learn far more effective *pan* and *zoom* methods soon. Review everything learned thus far and proceed to the first commands.

12

1.4 PRACTICING THE CREATE OBJECTS COMMANDS

Let us try the commands now, one by one. Although it may not be all that exciting, it is extremely important that you memorize the sequence of prompts for each command, as these are really the ABCs of AutoCAD.

Remember: If a command is not working right, or you see little blue squares (they are called *grips*, and we cover them very soon), just press Esc to get back to the Command: status line, at which point you can try it again. All four methods of command entry are presented: typing, cascading menus, toolbar icons, and the Ribbon. Alternate each method until you decide which one you prefer. It is perfectly OK to use a hybrid of methods for now.

➤ Line

Keyboard: Type in **line** and press Enter
Cascading menus: **Draw→Line**
Toolbar icon: **Draw** toolbar
Ribbon: **Home tab→Line**

Step 1. Begin the line command via any of the preceding methods.
  ○ AutoCAD says: `Specify first point:`
Step 2. Using the mouse, left-click anywhere on the screen.
  ○ AutoCAD says: `Specify next point or [Undo]:`
Step 3. Move the mouse elsewhere on the screen and left-click again. You can repeat Step 2 as many times as you wish. When you are done, click Enter or Esc. You should have a bunch of lines on your screen, either separate or connected together, as shown in Figure 1.12.

13

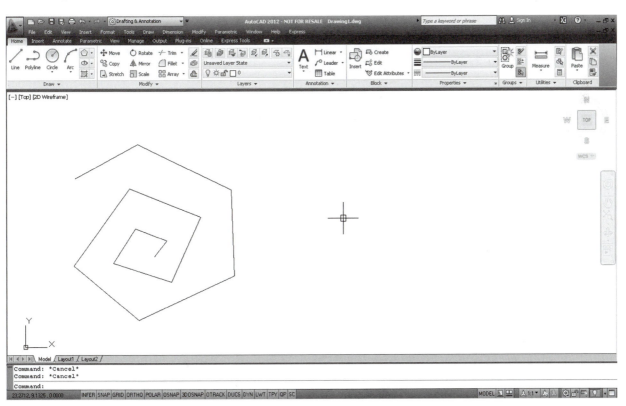

FIGURE 1.12
Lines drawn.

You need not worry about several things at this point. The first is accuracy; we introduce a tremendous amount of accuracy later in the learning process. The second is the options available in the brackets, such as [Undo]; we cover those as necessary. What is important is that you understand how you got those lines and how to do it again. In a similar manner, we move on to the other commands.

➤ Circle

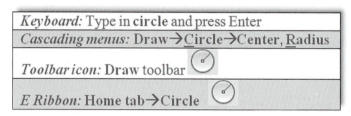

Keyboard: Type in **circle** and press Enter
Cascading menus: Draw→<u>C</u>ircle→Center, <u>R</u>adius
Toolbar icon: Draw toolbar
E Ribbon: Home tab→Circle

Step 1. Begin the circle command via any of the preceding methods.
  ○ AutoCAD says: `Specify center point for circle or [3P/2P/Ttr (tan tan radius)]:`
Step 2. Using the mouse, left-click anywhere on the screen and move the mouse out away from that point.
  ○ AutoCAD says: `Specify radius of circle or [Diameter] <1.9801>:` Notice the circle that forms; it varies in size with the movement of your mouse. The value in brackets may of course be different in your case.
Step 3. Left-click to finish the circle command. Repeat Steps 1 through 3 several times, and your screen should look like Figure 1.13.

14

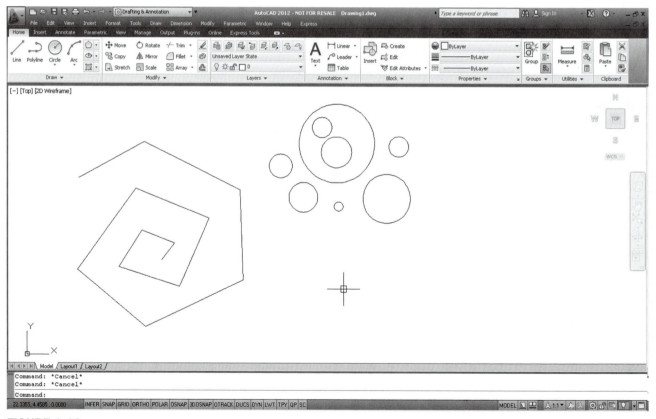

FIGURE 1.13
Circles drawn.

The method just used to create the circle was called *Center, Radius*, and you could specify an exact radius size if you wish, by just typing in a value after the first click (try it). As you may imagine, there are other ways to create circles—six ways to be precise, as seen with the Ribbon and cascading menus (Figures 1.14 and 1.15).

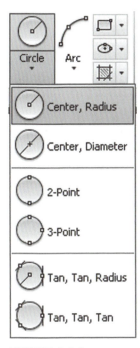

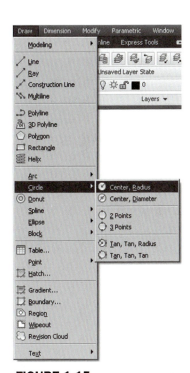

FIGURE 1.14
Additional circle options
(Ribbon).

FIGURE 1.15
Additional circle options (cascading
menus).

15

Special attention should be paid to the Center, Diameter option. Often students are asked to create a circle of a certain diameter, and they inadvertently use radius. This is less of a problem when using the Ribbon or the cascading menus, but you need to watch out if typing or using toolbars, as you will need to press d for [Diameter] before entering a value; otherwise, guess what it will be. We focus much more on these *bracketed* options as the course progresses. *You get a chance to practice the other circle options as part of the end of chapter exercises.*

➤ Arc

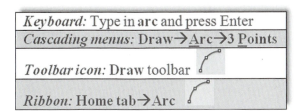

Keyboard: Type in **arc** and press Enter

Cascading menus: Draw→Arc→3 Points

Toolbar icon: Draw toolbar

Ribbon: Home tab→Arc

Step 1. Begin the arc command via any of the preceding methods.
  ○ AutoCAD says: Specify start point of arc or [Center]:
Step 2. Left-click with the mouse anywhere on the screen. This is the first of three points necessary for the arc.
  ○ AutoCAD says: Specify second point of arc or [Center/End]:

Step 3. Click somewhere else on the screen to place the second point.

 ○ AutoCAD says: `Specify end point of arc:`

Step 4. Left-click a third (final) time, somewhere else on the screen, to finish the arc.

Practice this sequence several more times to fully understand the way AutoCAD places the arc. Your screen should look something like Figure 1.16.

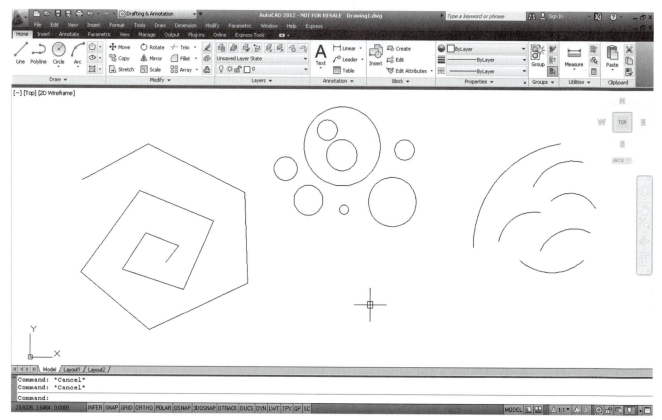

FIGURE 1.16
Arcs drawn.

The method just used to create the arcs is called *3 Point* and, without invoking other options, is a rather arbitrary (eyeball) method of creating them, which is just fine for some applications. Just as with circles, there are, of course, other ways to create arcs—11 ways to be precise, as seen with the Ribbon and cascading menus (Figures 1.17 and 1.18).

Not all of these options are used, and some you will probably never need, but it is worth going over them to know what is available. *You get a chance to practice the other arc options as part of the end of chapter exercises.*

➤ Rectangle

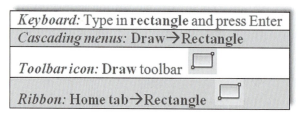

Keyboard: Type in **rectangle** and press Enter	
Cascading menus: Draw→Rectangle	
Toolbar icon: **Draw** toolbar	
Ribbon: Home tab→Rectangle	

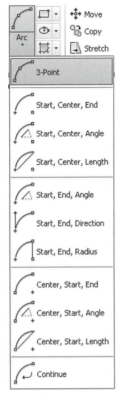

FIGURE 1.17
Additional arc options (Ribbon).

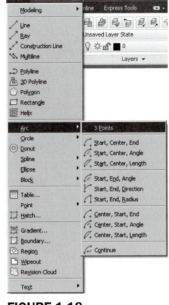

FIGURE 1.18
Additional arc options (cascading menus).

17

Step 1. Begin the rectangle command via any of the preceding methods.
- ○ AutoCAD says: `Specify first corner point or [Chamfer/Elevation/Fillet /Thickness/Width]:`

Step 2. Left-click, and move the mouse diagonally somewhere else on the screen.
- ○ AutoCAD says: `Specify other corner point or [Area/Dimensions/Rotation]:`

Step 3. Left-click one more time to finish the command. Your screen should look more or less like Figure 1.19.

There are of course more precise ways to draw a rectangle. In Step 3, you can press d for `Dimensions` and follow the prompts to assign length, width, and a corner point (where you want it) to your rectangle. Try it out, but as with all basic shapes drawn so far, do not worry too much about sizing and accuracy, just be sure to understand and memorize the sequences for the basic command—that is what is important for now.

So, now you have created the four basic shapes used in AutoCAD drawings (and ended up with a lot of junk). Let us erase all of them, and get back to a blank screen. To do this, we need to introduce Tip 2, which is quite easy to just type. Notice the use of the e abbreviation; we will see more of this as we progress.

So, let us put all of this to good use. Create the drawing shown in Figure 1.20. Use the rectangle, line, circle, and arc commands. Do not get caught up trying

TIP 2: To quickly erase everything on the screen, type in e for *erase*, press Enter, and type in all. Then press Enter twice. After the first Enter, all the objects are dashed, indicating they were selected successfully. The second Enter finishes the command by deleting everything. Needless to say, this command sequence is useful only in the early stages of learning AutoCAD, when you do not intend to save what you are drawing. You may want to conveniently forget this particular tip when you begin to work on a company project. For now it is very useful, but we will learn far more selective erase methods later.

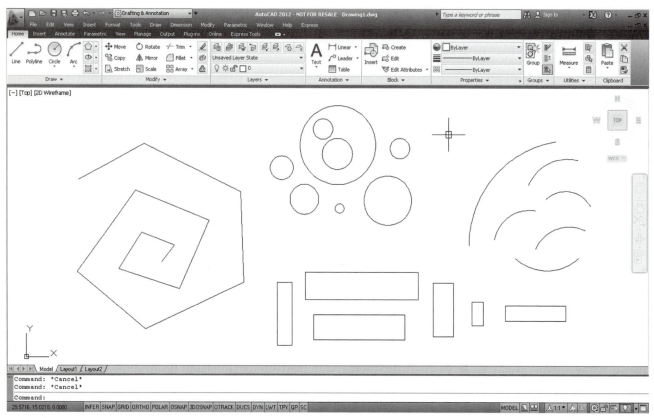

FIGURE 1.19
Rectangles drawn.

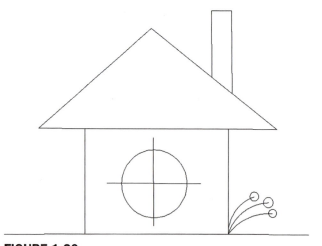

FIGURE 1.20
Simple house sketch.

to make everything line up perfectly, just do a quick sketch. Then, you can tell everyone that you learned how to draw a house in AutoCAD within the first half hour. Well, sort of.

At this point it is necessary to mention the zoom and pan commands. These are very easy if you have a mouse with a center wheel (as most PCs do these days).

1.5 VIEW OBJECTS

Zoom

Put your finger on the mouse wheel and turn it back and forth without pressing down. The house you just created gets smaller (roll back) or larger (roll forward) on the screen. Remember though, the object remains the same; your view is just zooming in and out.

Pan

Put your finger on the mouse wheel, but this time press and hold. A hand symbol appears. Now move the mouse around while keeping the wheel depressed. You are able to pan around your drawing. Remember again, the drawing is not moving; you are moving around it.

Regen

This is the easiest command you will learn. Just type in regen and press Enter. The screen refreshes. One use for this is if you are panning over and AutoCAD does not let you go further, as if you hit a wall. Simply regen and proceed.

While the mouse method is easiest, there are other ways to zoom and pan. One alternative is using toolbars, and part of the Standard toolbar (which you should have on your screen) has some Zoom (magnifying glass) and Pan (hand) icons for this purpose, as seen in Figure 1.21.

Explore the icon options on your own, but one particular zoom option is critically important: Zoom to Extents. If you managed to pan your little house right off the screen and cannot find it, this comes in handy, and a full tip on Zoom to Extents is presented next.

What if you did something that you did not want to do and wished you could go back a step or two? Well, like most programs, AutoCAD has an undo command, which brings us to the fourth tip.

Now that you have drawn your house, leave it on your screen and let us move on. We now have to go through all the Edit/Modify Objects commands, one by one. Each of them is critical to creating even the most basic drawings, so go through each carefully, paying close attention to the steps involved. Along the way, the occasional tip is added as the need arises.

Just to remind you: The Esc key gets you out of any mistakes you make and returns you to the basic command line. Be sure to perform each sequence several times to really memorize it. As you learn the first command, erase, you also are introduced

TIP 3: Type in z for *zoom*, press Enter, type in the letter e, then press Enter again. This is called *Zoom to Extents* and makes AutoCAD display everything you sketched, filling up the available screen space. If you have a middle-button wheel on your mouse, then double-click it for the same effect. This technique is very important to fit everything on your screen.

TIP 4: To perform the undo command in AutoCAD just type in u and press Enter. Do not type the entire command—if you do that, more options pop up, which we do not really need at this point, so a simple u suffices. You can do this as many times as you want to (even to the beginning of the drawing session). You can also use the familiar Undo and Redo arrows seen at the very top of the screen and also as part of the Standard toolbar. Yes, AutoCAD has a Redo; it was once called the *oops* command. Oops still exists, but its complexity has grown and we do not explore it at the moment.

FIGURE 1.21
Zoom and Pan icons.

TIP 5: Hate to keep having to type, pull a menu, or click on an icon each time you practice the same command? Well, no problem. Simply press the space bar or Enter, and this will repeat the last command you used. There, you just saved a few minutes per day for that coffee break.

to the *picking objects* selection method, which is similar throughout the rest of the commands. Also, keep a close eye on the command line, as it is where you and AutoCAD communicate for now, and it *is* a two-way conversation.

Finally, before you get started, here is another useful tip that students usually appreciate early on to save some time and effort.

1.6 PRACTICING THE EDIT/MODIFY OBJECTS COMMANDS
➤ Erase

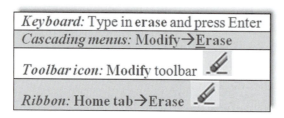

Keyboard: Type in erase and press Enter
Cascading menus: Modify→Erase
Toolbar icon: Modify toolbar
Ribbon: Home tab→Erase

Step 1. Begin the erase command via any of the preceding methods.
  ○ AutoCAD says: `Select objects:`
Step 2. Select any object from the house in the previous drawing assignment (Figure 1.20) by taking the mouse, positioning it over that object (not in the empty space), and left-clicking once.
  ○ AutoCAD says: `Select objects: 1 found`. The object becomes dashed. You saw this before when we erased everything in Tip 2. It means the objects were selected. AutoCAD continues to ask you: `Select objects:`. Watch out for this step. AutoCAD always asks this, in case you want to select more objects, so get used to it. You are done, however, so…
Step 3. Press Enter and the object will disappear.

Practice this several times, using the undo command to bring the object back. Make sure you keep it there, since you need it to practice the next few commands. You can of course select more than one object, as this is the whole point of being asked to select again. As you click on each one, it becomes dashed. You can do this until you run out of objects to select; there is no limit. To deselect any objects, hold down the Shift key and click on them. They will be removed from the selection set.

➤ Move

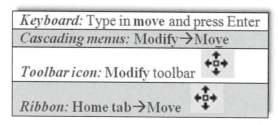

Keyboard: Type in move and press Enter
Cascading menus: Modify→Move
Toolbar icon: Modify toolbar
Ribbon: Home tab→Move

Step 1. Begin the move command via any of the preceding methods.
  ○ AutoCAD says: `Select objects:`
Step 2. Select an object by positioning the mouse over that object (not the empty space) and left-clicking once.
  ○ AutoCAD says: `Select objects: 1 found`. The object becomes dashed.
  ○ AutoCAD then asks you again: `Select objects:`

Step 3. Unless you have more than one object to move, you are done, so press Enter.
   - AutoCAD says: `Specify base point or [Displacement] <Displacement>:`
Step 4. Left-click anywhere on or near the object to "pick it up"; this is where you are moving it *from*.
   - AutoCAD says: `Specify second point or <use first point as displacement>:`
Step 5. Move the mouse somewhere else on the screen and left-click to place the object in the new location. This is where you are moving it *to*.

Undo what you just did and practice it again. Figure 1.22 shows the move command in progress. Notice how the dashed circle remains there as a "shadow" until you place your circle in its new location.

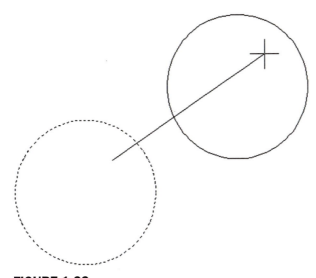

FIGURE 1.22
Move command in progress.

➤ Copy

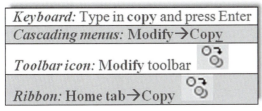

Keyboard: Type in copy and press Enter	
Cascading menus: Modify→Copy	
Toolbar icon: Modify toolbar	
Ribbon: Home tab→Copy	

Step 1. Begin the copy command via any of the preceding methods.
   - AutoCAD says: `Select objects:`
Step 2. Select an object by positioning the mouse over that object and left-clicking once.
   - AutoCAD says: `Select objects: 1 found`. The object will become dashed.
   - AutoCAD then asks you again: `Select objects:`
Step 3. Unless you have more than one object to copy, you are done, so press Enter.
   - AutoCAD says: `Specify base point or [Displacement/ mOde]<Displacement>:`
Step 4. Left-click anywhere on or near the object to "pick it up"; this is where you are copying it *from*.
   - AutoCAD says: `Specify second point or <use first point as displacement>:`

Step 5. Move the mouse somewhere else on the screen and left-click to copy the object to the new location. This is where you are copying it *to*. Notice that a dashed copy of the object remains in its original location until you complete the command. You can copy as many times as you want.

Undo what you just did and practice it again. Figure 1.23 shows the copy command in progress. The dashed circle remains there as a "shadow" until you copy your new circles into their new locations.

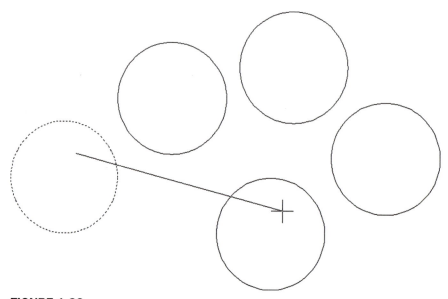

FIGURE 1.23
Copy command in progress.

➤ **Rotate**

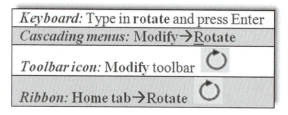

| *Keyboard:* Type in **rotate** and press Enter |
| *Cascading menus:* Modify→Rotate |
| *Toolbar icon:* Modify toolbar |
| *Ribbon:* Home tab→Rotate |

Step 1. Begin the rotate command via any of the preceding methods.
  ○ AutoCAD says: `Current positive angle in UCS: ANGDIR = counterclockwise ANGBASE = 0`, and on the next line: `Select objects:`
Step 2. Select any object as before, remembering to press Enter again after the selection. So far, it is quite similar to erase, move, and copy.
  ○ AutoCAD says: `Specify base point:` This means select the pivot point of the object's rotation (the point about which it rotates). If you select a circle for your object, try to stay away from the center point, as your rotation efforts may be less than spectacular.
Step 3. Click anywhere on or near the selected object.
  ○ AutoCAD says: `Specify rotation angle or [Copy/Reference] <0>:`
Step 4. Move the mouse around in a wide circle and the object rotates. Notice how the motion gets smoother as you move the mouse farther away. You can click anywhere for a random rotation angle or you can type in a specific numerical degree value.

Figure 1.24 shows the rotate command in progress.

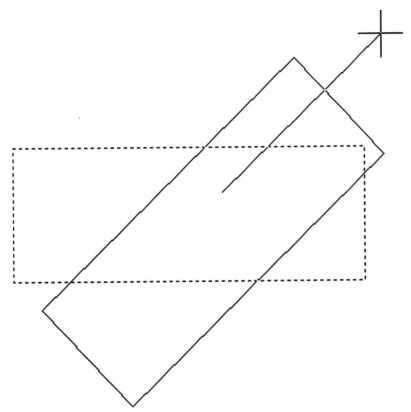

FIGURE 1.24
Rotate command in progress.

➤ Scale

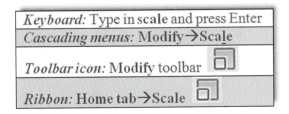

Keyboard: Type in scale and press Enter	
Cascading menus: Modify→Scale	
Toolbar icon: Modify toolbar	
Ribbon: Home tab→Scale	

Step 1. Begin the scale command via any of the preceding methods.
  ○ AutoCAD says: `Select objects:`
Step 2. Select any object as before, remembering to press Enter again after the first selection.
  ○ AutoCAD says: `Specify base point:` This means select the point from which the scaling of the object (up or down) is to occur.
Step 3. Click somewhere on or near the object or directly in the middle of it.
  ○ AutoCAD says: `Specify scale factor or [Copy/Reference] <1.000>:`
Step 4. Move the mouse around the screen. The object gets bigger or smaller. You can randomly scale it or enter a numerical value. For example, if you want it twice as big, enter 2; half size will be .5.

Figure 1.25 shows the scale command in progress.

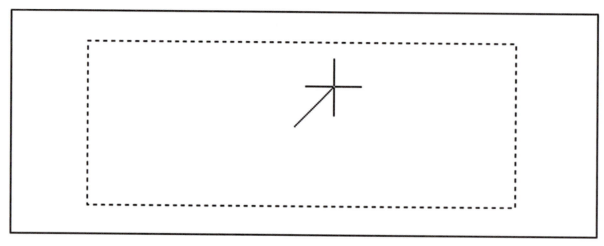

FIGURE 1.25
Scale command in progress.

➤ **Trim**

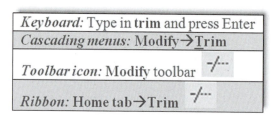

| *Keyboard:* Type in **trim** and press Enter |
| *Cascading menus:* Modify→Trim |
| *Toolbar icon:* Modify toolbar |
| *Ribbon:* Home tab→Trim |

To practice this command you first need to draw two intersecting lines, one horizontal and one vertical, similar to a plus sign. Once this is done, go ahead and perform the sequence that follows.

Step 1. Begin the trim command via any of the preceding methods.
- AutoCAD says:
```
Current settings: Projection = UCS, Edge = None
Select cutting edges...
Select objects or <select all>:
```
Step 2. Left-click on one of the lines. You can choose the vertical or horizontal, it does not matter. This is your *cutting edge*. The other line, the one you did not pick, is trimmed at the intersection point of the two lines. The cutting edge line becomes dashed.
- AutoCAD says: `Select objects or <select all>: 1 found`
Step 3. Press Enter.
- AutoCAD says: `Select object to trim or shift-select to extend or[Fence/Crossing/Project/Edge/eRase/Undo]:`
Step 4. Go ahead and pick anywhere on the line that you did *not* yet select and it is trimmed.

You can repeat Step 4 as many times as you want, but we are done, so press Enter. Practice this several times by pressing u then Enter and repeating Steps 1 through 4. Remember two things to avoid mistakes: Pick the cutting edge first, and do not forget to press Enter before picking the line to be chopped.

You can also do a trim between two or more lines, as shown in Figure 1.26. This is something we need in the first major project, so practice both types of trims. In both cases, the lines selected as cutting edges are now dashed, and the lines about to be trimmed have the cursor (small box) over them.

24

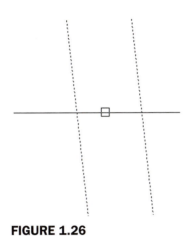

FIGURE 1.26
Trim command in progress.

➤ Extend

Keyboard: Type in **extend** and press Enter	
Cascading menus: Modify→Extend	
Toolbar icon: Modify toolbar	--·/
Ribbon: Home tab→Extend	--·/

To practice this command, we first need to draw two intersecting lines, just as for trim, but then use the move command to relocate the vertical line directly to the right (or left) of the horizontal line, a short distance away, so we can extend the horizontal line into the vertical. You can see this in Figure 1.27. Once this is done, go ahead and perform the extend command.

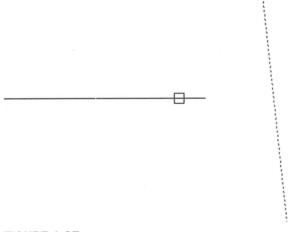

FIGURE 1.27
Extend command in progress.

Step 1. Begin the extend command via any of the preceding methods.
○ AutoCAD says:
```
Current settings: Projection = UCS, Edge = None
Select boundary edges...
Select objects or <select all>:
```

Step 2. At this point, left-click on the vertical line. This is the *target* into which you extend the horizontal line. It becomes dashed. AutoCAD says: `Select objects: 1 found.`

Step 3. Press Enter.
- ○ AutoCAD says: `Select object to extend or shift-select to trim or [Fence/Crossing/Project/Edge/Undo]:`

Step 4. Pick the end of the horizontal line that is closest to the vertical line. The *arrow* is what extends into the vertical line target.

You can repeat Step 4 as many times as you want, but we are done, so press Enter. Practice this several times by pressing u then Enter and repeating Steps 1 through 4. In Figure 1.27, the dashed vertical line is the "target" and the solid horizontal line is the "arrow." The cursor (box) is positioned to extend the arrow into the target by clicking once.

➤ Offset

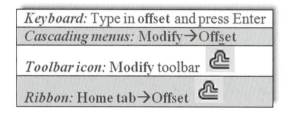

Keyboard: Type in offset and press Enter

Cascading menus: Modify→Offset

Toolbar icon: Modify toolbar

Ribbon: Home tab→Offset

This command is one of the more important ones you will learn. Its power lies in its simplicity. This command creates new lines by the directional parallel offset concept, where if you have a line on your screen, you can create a new one that is a certain distance away but parallel to the original. To illustrate it, first draw a random vertical or horizontal line anywhere on your screen.

Step 1. Begin the offset command via any of the preceding methods.
- ○ AutoCAD says: `Specify offset distance or [Through/Erase/Layer] <Through>:`

Step 2. Enter an offset value (use a small number for now, 2 or 3). Then press Enter.
- ○ AutoCAD says: `Select object to offset or [Exit/Undo] <Exit>:`

Step 3. Pick the line by left-clicking on it; the line becomes dashed.
- ○ AutoCAD says: `Specify point on side to offset or [Exit/Multiple/ Undo] <Exit>:`

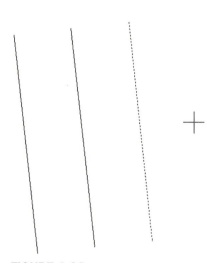

FIGURE 1.28
Offset command in progress.

26

Step 4. Pick a *direction* for the line to go, which is to one of the two sides of the original line. It does not matter exactly how far away from the original line you click, as the distance has already been specified. You can keep doing this over and over by selecting the new line then the direction (see Figure 1.28). If you want to change the offset distance, you must repeat Steps 1 and 2.

The significance of the offset command may not be readily apparent. Many students, especially ones who may have done some hand drafting, come into learning AutoCAD with the preconceived notion that it is just a computerized version of pencil and paper. While this is true in some cases, in others, AutoCAD represents a radical departure. When you begin a new drawing on paper, you literally draw line after line. With AutoCAD, surprisingly little line drawing occurs. A few are drawn, and then the offset command is employed throughout. Learn this command well, you will need it.

➤ Mirror

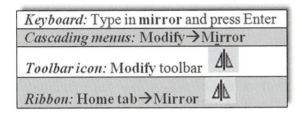

Keyboard: Type in **mirror** and press Enter
Cascading menus: Modify→Mirror
Toolbar icon: Modify toolbar
Ribbon: Home tab→Mirror

This command, much as the name implies, creates a mirror copy of an object over some plane. It is surprisingly useful because often just copying and rotating is not sufficient or takes too many steps. To learn this command, you have to think of a real-life mirror, with you standing in front of it. The original object is you, the plane is the door with the mirror attached (which can swing on a hinge), and the new object is your mirrored reflection. To practice this command, draw a triangle by joining three lines together, as seen in Figure 1.29.

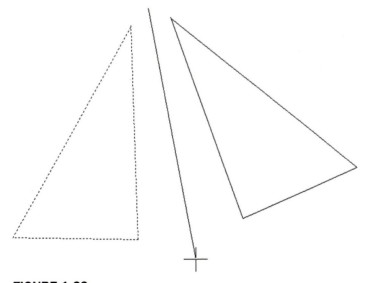

FIGURE 1.29
Mirror command in progress.

Step 1. Begin the mirror command via any of the preceding methods.
○ AutoCAD says: `Select objects:`
Step 2. Select all three lines of the triangle and press Enter.
○ AutoCAD says: `Specify first point of mirror line:`

Step 3. Click anywhere near the triangle and move the mouse around. You notice the new object appears to be anchored by one point. This is the swinging mirror reflection.
- ○ AutoCAD says: `Specify second point of mirror line:`

Step 4. Click again near the object (make the second click follow an imaginary straight line down). You notice the new object disappear. Do not panic; rather check what the command line says.
- ○ AutoCAD says: `Delete source objects? [Yes/No] <N>:`

Step 5. The response in Step 4 just asked you if you want to keep the original object. If you do (the most common response), then just press Enter and you are done. Figure 1.29 illustrates Step 3.

➤ Fillet

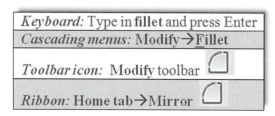

Keyboard: Type in **fillet** and press Enter
Cascading menus: Modify→<u>F</u>illet
Toolbar icon: Modify toolbar
Ribbon: Home tab→Mirror

Fillets are all around us. This is not just an AutoCAD concept but rather an engineering description for a rounded edge on a corner of an object or at an intersection of two objects. These edges are added to avoid sharp corners for safety or other technical reasons (such as a weld). By definition, a fillet needs to have a radius greater than zero; but in AutoCAD, a fillet *can* have a radius of zero, allowing for some useful editing, as we see soon. To illustrate this command, first draw two perpendicular lines that intersect, as seen in the first box of Figure 1.30.

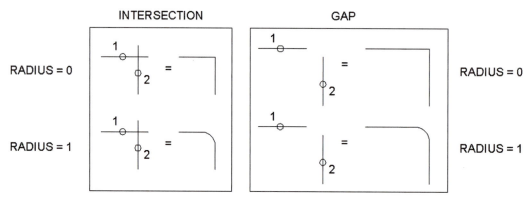

FIGURE 1.30
Fillet command summary.

Step 1. Begin the fillet command via any of the preceding methods.
- ○ AutoCAD says:
  `Current settings: Mode = TRIM, Radius = 0.0000`
  `Select first object or [Undo/Polyline/Radius/Trim/Multiple]:`

Step 2. Put a radius on the fillet. Type in r for *radius* and press Enter.
- ○ AutoCAD says: `Specify fillet radius <0.0000>:`

Step 3. Enter a small value, perhaps 0.5 or 1. Press Enter.
- ○ AutoCAD says: `Select first object or [Undo/Polyline/Radius/Trim/Multiple]:`

Step 4. Select the first (horizontal) line, somewhere near the intersection. It becomes dashed.
- ○ AutoCAD says: `Select second object or shift-select to apply corner or [Radius]:`

Step 5. As your mouse moves toward the second (vertical) line, you see a preview of the resulting fillet (if the fillet is valid). This preview capability is a new feature of AutoCAD 2012. Go ahead and click the second line somewhere near the intersection; a fillet of the radius you specified is added.

These are the steps involved in adding a fillet with a radius. You can also do a fillet with a radius of zero, so repeat Steps 1 through 5, entering 0 for the radius value. This is very useful to speed up trimming and extending. Intersecting lines can be filleted to terminate at one point, and if they are some distance apart, they can be filleted to meet together. All this is possible with a radius of zero. In these cases, the new preview feature shows a green X where the lines intersect. Try out all the variations. Figure 1.30 summarizes everything described so far. The 1 and 2 indicate a suggested order of clicking the lines for the desired effect.

Well, this is it for now. We covered the basic commands. These commands form the essentials of what we call basic CAD theory. All 2D computer-aided design programs need to be able to perform these functions in order to work, though other software may refer to them differently, such as perhaps using "displace" instead of "move," but the capability still has to be there. Be sure to memorize these commands and practice them until they become second nature. They are truly the ABCs of basic computer drafting, their importance cannot be overstated, and you need to master them to move on.

1.7 SELECTION METHODS

Before we talk about the accuracy tools available to you, we need to briefly cover a more advanced type of selection method called *Window/Crossing*. You may have already seen this by accident: When you click randomly on the screen without selecting any objects and drag your mouse to the right or left, a rectangular shape forms. It is either dashed or solid (depending on settings, these shapes may have a green or blue fill) and goes away once you click the mouse button again or press Esc. This happens to be a powerful selection tool.

29

- **Crossing.** A crossing appears when you click and drag to the *left* of an imaginary vertical line (up or down) and is a *dashed* rectangle. *Any object it touches is selected.*
- **Window.** A window appears when you click and drag to the *right* of an imaginary vertical line (up or down) and is a *solid* rectangle. *Any object falling completely inside it is selected.*

Figure 1.31 shows what both the Crossing and the Window look like. Practice selecting objects in conjunction with the erase command. Draw several random objects and start up the erase command. Then, instead of picking each one individually (or doing Erase, All), select them with either the Crossing or Window and make a note of how each functions. This is a very important and useful selection method.

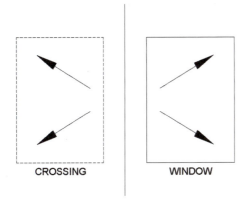

CROSSING   WINDOW

FIGURE 1.31
Crossing and Window.

1.8 DRAWING ACCURACY—PART 1
Ortho (F8)

So far, when you created lines, they were not perfectly straight. Rather, you may have lined them up by eye. This of course is not accurate enough for most designs, so we need a better method. Ortho (which stands for *orthographic*) is a new concept we now introduce that allows you to draw perfectly straight vertical or horizontal lines. It is not so much a command as a condition you impose upon your lines prior to starting to draw them. To turn on the Ortho feature just press the F8 key at the top of your keyboard, or click to depress the ORTHO button at the bottom of the screen. Then begin drawing lines. You see your mouse cursor constrained to only vertical and horizontal motion. Ortho is usually not something you just set and forget about, as your design may call for a varied mix of straight and angled lines, so do not be shy about turning the feature on and off as often as necessary, sometimes right in the middle of drawing a line.

1.9 DRAWING ACCURACY—PART 2
OSNAPs

While drawing straight lines, we generally like those lines to connect to each other in a very precise way. So far we have been eyeballing their position relative to each other, but that is not accurate and is more reminiscent of sketching, not serious drafting. Lines (or for that matter all fundamental objects in AutoCAD) connect to each other in very distinct and specific locations (two lines are joined end to end, for example). Let us develop these points for a collection of objects.

ENDPOINT

ENDpoints are the two locations at the ends of any line. Another object can be precisely attached to an endpoint. The symbol for it is a square, as illustrated in Figure 1.32.

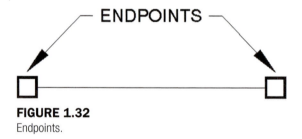

FIGURE 1.32
Endpoints.

MIDPOINT

MIDpoint is the middle of any line. Another object can be precisely attached to a midpoint. The symbol for it is a triangle, as illustrated in Figure 1.33.

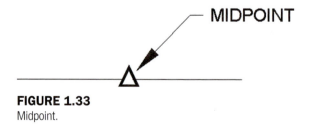

FIGURE 1.33
Midpoint.

CENTER

CENter is the center of any circle. Another object can be precisely attached to a center point. The symbol for it is a circle, as illustrated in Figure 1.34.

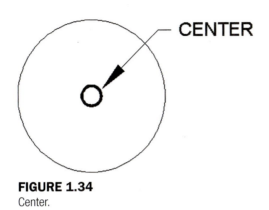

FIGURE 1.34
Center.

QUADRANT

QUADrants are the four points at the North (90°), West (180°), South (270°), and East (0°, 360°) quadrants of any circle. Another object can be precisely attached to a quadrant point. The symbol for it is a diamond, as illustrated in Figure 1.35.

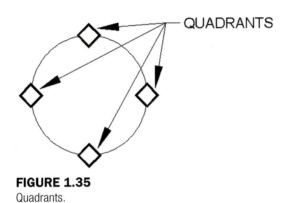

FIGURE 1.35
Quadrants.

INTERSECTION

INTersection is a point that is located at the intersection of any two objects. You may begin a line or attach any object to that point. The symbol for it is an X, as illustrated in Figure 1.36.

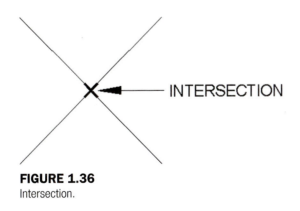

FIGURE 1.36
Intersection.

PERPENDICULAR

PERPendicular is an OSNAP point that allows you to begin or end a line perpendicular to another line. The symbol for it is the standard mathematical symbol for a perpendicular (90°) angle, as illustrated in Figure 1.37.

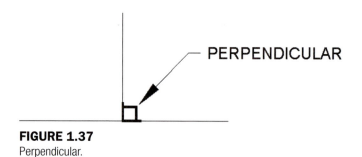

FIGURE 1.37
Perpendicular.

These are the six fundamental OSNAP points we mention for now. There are others of course, such as Node, Extension, Insertion, Tangent, *Nearest*, Apparent Intersection, and Parallel, but those are used far less often; and with the exception of Tangent, Near, and Node (covered in upcoming chapters), they are not discussed.

So how do you use the OSNAPs that we have developed so far? There are two methods. The first is to type in the OSNAP point as you need it (only the capitalized part needs to be typed in: END, MID, etc.). For example, draw a line. Then, begin drawing another line, but before clicking to place it, type in end and press Enter. AutoCAD adds the word of, at which point you need to move the cursor to roughly the end of the first line. The snap marker for endpoint (a square) appears, along with a tool tip stating what it is. Click to place the line; now your new line is precisely attached to the first line.

In the same manner, try out all the other snap points mentioned (draw a few circles to practice Quad and Center). In all cases, begin by typing line, then one of the snap points, followed by Enter. Then bring the mouse cursor to the approximate location of that point and click.

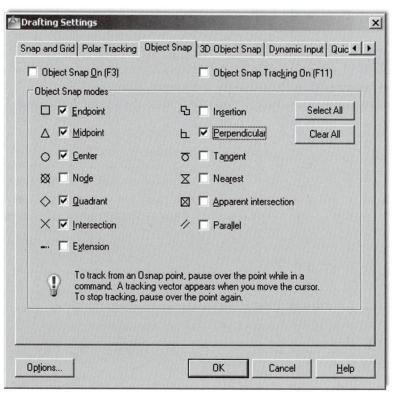

FIGURE 1.38
Drafting Settings dialog box.

1.10 OSNAP DRAFTING SETTINGS

As you may imagine, there is a faster way to use OSNAP. The idea is to have these settings always running in the background so you need not type them in anymore. To set them up, you need to type in osnap or select **Tools→Drafting Settings**… from the cascading drop-down menu. The OSNAP dialog box appears (Figure 1.38). First, make sure the Object Snap tab is selected, and then select the Clear All button. Finally, check off the six settings mentioned previously (Endpoint, Midpoint, Center, Quadrant, Intersection, Perpendicular), and press OK.

So now you have set the appropriate settings, but you still need to turn them on. For that, we need to mention another F key. If you press F3 or press the OSNAP toggle button at the bottom of the screen, the OSNAP settings become active. Go ahead and do this, then begin the same line drawing exercise; see how much easier it is now as the snap settings appear on their own. This is the preferred way to draw, with them running in the background. Feel free to press F3 anytime you do not need them anymore; OSNAP is also not something that you may want all the time. Figure 1.38 shows the Drafting Settings dialog box, with the discussed OSNAP points checked.

SUMMARY

You should understand and know how to use the following concepts and commands before moving on to Chapter 2:

- AutoCAD drawing environment
  - Drawing Area, Command Line, Cascading Menus, Toolbars, Ribbon
- Create Objects
  - Line
  - Circle
  - Arc
  - Rectangle
- Edit/Modify Objects
  - Erase
  - Move
  - Copy
  - Rotate
  - Scale
  - Trim
  - Extend
  - Offset
  - Mirror
  - Fillet
- View Objects
  - Zoom
  - Pan
  - Regen
- Selection methods
  - Crossing
  - Window
- Ortho (F8)
- OSNAP (F3)
  - ENDpoint
  - MIDpoint
  - CENter
  - QUADrant
  - INTersection
  - PERPendicular

33

Additionally, you should understand and know how to use the following tips introduced in this chapter:

TIP 1. The Esc (Escape) key in the upper left-hand corner of your keyboard is your new best friend while learning AutoCAD. It will get you out of just about any trouble you get yourself into. If something does not look right, just press the Esc key and repeat the command.

TIP 2. To quickly erase everything on the screen, type in e for *erase*, press Enter, and type in all. Then, press Enter twice. After the first Enter, all the objects are dashed, indicating they were selected successfully. The second Enter finishes the command by deleting everything.

TIP 3. Type in z for *zoom*, press Enter, type in the letter e, then press Enter again. This is called *Zoom to Extents* and makes AutoCAD display everything you sketched, filling up the available screen space. If you have a middle-button wheel on your mouse, then double-click it for the same effect.

TIP 4. To perform the undo command in AutoCAD just type in u and press Enter. Do not type in the entire command—if you do that, more options pop up, which we do not really need at this point, so a simple u suffices. You can do this as many times as you want (even to the beginning of the drawing session). You can also use the familiar Undo and Redo arrows seen at the very top of the screen and also as part of the Standard toolbar.

TIP 5. Hate to keep having to type, pull a menu, or click an icon each time you practice the same command? Well, no problem. Simply press the space bar or Enter, and this repeats the last command you used.

REVIEW QUESTIONS

Answer the following questions based on what you learned in Chapter 1. You should have the commands memorized, so copy the questions to another sheet of paper, close the book, and try to answer them without looking them up.

1. List the four Create Objects commands covered.
2. List the ten Edit/Modify Objects commands covered.
3. List the three View Objects commands covered.
4. What is a workspace?
5. What four main methods are used to issue a command in AutoCAD?
6. What is the difference between a selection Window and Crossing?
7. What is the name of the command that forces lines to be drawn perfectly straight horizontally and vertically? Which F key activates it?
8. What six main OSNAP points are covered? Draw the snap point's symbol next to your answer.
9. What command brings up the OSNAP dialog box?
10. Which F key turns the OSNAP feature on and off?
11. How do you terminate a command-in-progress in AutoCAD?
12. What sequence of commands quickly clears your screen?
13. What sequence of commands allows you to view your entire drawing?
14. How do you undo a command?
15. How do you repeat the last command you just used?

EXERCISES

1. Draw the following sets of lines on the left and practice the 2-Point, 3-Point, and Tan, Tan, Tan circle options. You may use the Ribbon, cascading menus, or typing. In all cases, use OSNAPs to connect the circle precisely to the drawn lines. The result you should get is shown in on the right. (Difficulty level: Easy; Time to completion: <5 minutes)

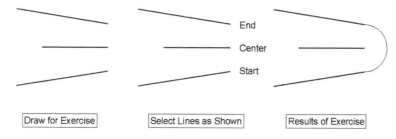

2-Point

3-Point

Tan,Tan,Tan

Draw for Exercise   Results of Exercise

2. Draw an arc using the *Start, Center, End* option (a commonly used one) as follows. First, draw the three lines shown on the left, making use of the mirror command as needed, then draw the arc itself. Be sure to select from the bottom up, as shown, and use OSNAPs. The result you should get is shown on the right. Try the same exercise by picking from the top down, noticing the difference. (Difficulty level: Easy; Time to completion: <5 minutes)

End
Center
Start

Draw for Exercise   Select Lines as Shown   Results of Exercise

35

3. Draw a simple door using the *3-Point* arc method. This requires some precise mouse clicking and we introduce an alternate method in Chapter 3, but for now this is a good exercise to try. First draw, copy, rotate, and position the lines as seen in Steps 1 through 4, then add the arc as shown in Step 5 (use Ortho and OSNAPs throughout). You may have to try it a few times to get it to look right. Finally, erase the bottom line to complete the door as seen in Step 6. (Difficulty level: Easy; Time to completion: <5 minutes)

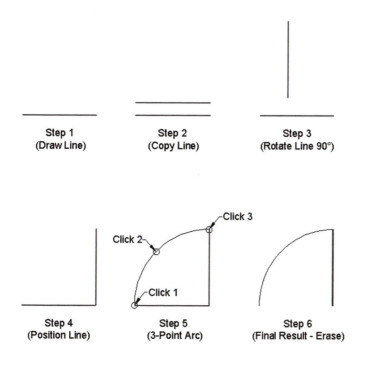

Step 1
(Draw Line)

Step 2
(Copy Line)

Step 3
(Rotate Line 90°)

Click 3
Click 2
Click 1

Step 4
(Position Line)

Step 5
(3-Point Arc)

Step 6
(Final Result - Erase)

4. Draw the following sets of arbitrary electrical connection shapes. You need only the basic Create and Modify commands, including rectangle, circle, line, offset, fillet, and mirror, as well as OSNAPs and Ortho. Specific sizes are not important at this point yet, but make your drawing look similar, and use accuracy in connecting all the pieces. (Difficulty level: Easy; Time to completion: <10 minutes)

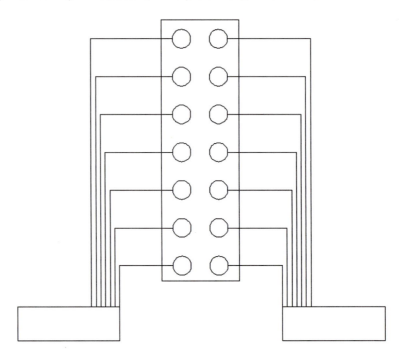

5. Draw the following house elevation using what you learned. You need only the basic Create and Modify commands. The overall sizing can be approximated; however, use Ortho (no crooked lines) and make sure all lines connect using OSNAPs. (Difficulty level: Intermediate; Time to completion: <15 minutes)

6. Draw the following sets of random shapes. You need only the basic Create and Modify commands, but you use almost every one of them, including multiple cutting edges trim. The overall sizing is completely arbitrary, and your design may not look exactly like this one, but be sure to draw all the elements you see here and use OSNAP precision throughout. (Difficulty level: Intermediate; Time to completion: <15 minutes)

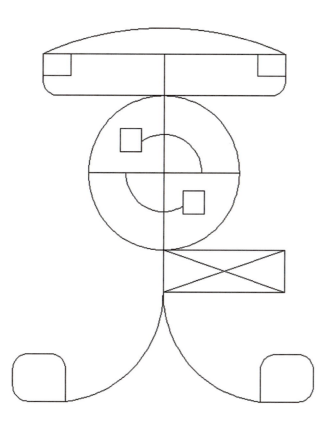

7. Draw the following side view of a car using what you learned. You need only the basic Create and Modify commands. The overall sizing (including multiple fillets needed on most corners) can be approximated; however, use accuracy and make sure all lines connect using OSNAPs. (Difficulty level: Intermediate; Time to completion: <20 minutes)

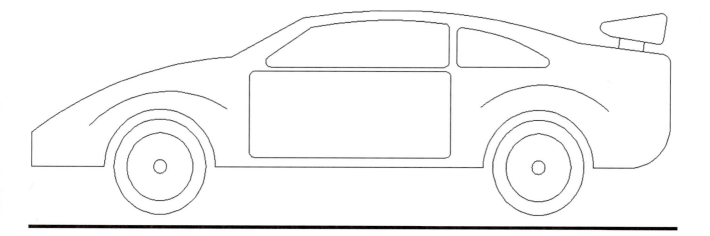

CHAPTER 2

AutoCAD Fundamentals
Part II

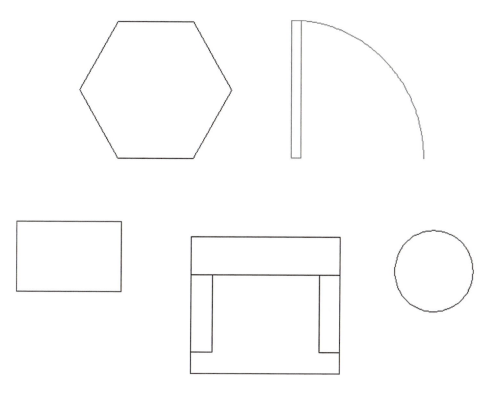

39

LEARNING OBJECTIVES

You learned a great deal in the first chapter, in fact the majority of what you need to know to get started on a design. This chapter serves as a "clean up" and concludes the basic introduction to AutoCAD, as there are just a few more topics to get to. Here, we discuss the following:

- Using grips
- Setting units
- Using Snap and Grid
- Understanding the Cartesian coordinate system
- Geometric data entry

- Inquiry commands:
  - Area
  - Distance
  - List
  - ID
- Explode command
- Polygon command
- Ellipse command
- Chamfer command
- Templates
- Setting limits
- Save
- Help files
- TANgent OSNAP

By the end of the chapter, you will have learned all the necessary basics to begin drafting your first project, including setting proper units, analyzing what you have done, and saving your work. After an introduction to layers in Chapter 3, you will begin a realistic architectural design.

Estimated time for completion of chapter: 2 hours.

2.1 GRIPS

We begin Chapter 2 with grips, a natural progression from learning the OSNAP points. These are relatively advanced editing tools, but chances are you already made them appear by accident by clicking an object without selecting a command first. That is OK; just press Esc and they go away. But, what are these mysterious blue squares? Take a close look; after learning OSNAPs, these should look familiar.

Grips are control points, locations on objects where you can modify that object's size, shape, or location. They are quite similar to the "handles" that you find in Photoshop® or Illustrator® if you are familiar with those applications.

To try out the concept, first draw some lines, arcs, circles, and rectangles. Then activate grips by clicking on any of the shapes (as seen in Figure 2.1). The objects become dashed and the grips appear. When your mouse hovers over one of them, the grip turns pink. If you click on it, it becomes red, meaning it is active (or hot). With the grip activated, move the mouse around. The object changes in one of several ways. A line or circle can be moved around if the center grip is selected. A circle can be scaled up or down in size if the outer quadrants are selected. Arcs can be made larger or their length increased. Other grips, such as the endpoints of a line or corners of a rectangle, change (or stretch) the shape of that object if activated and moved. Try them all out.

There is of course much more to grips. Formally referred to as *multifunctional grips*, they perform other construction tasks via an available vertex menu that appears if you hold your mouse over them. We explore this in depth in a later chapter.

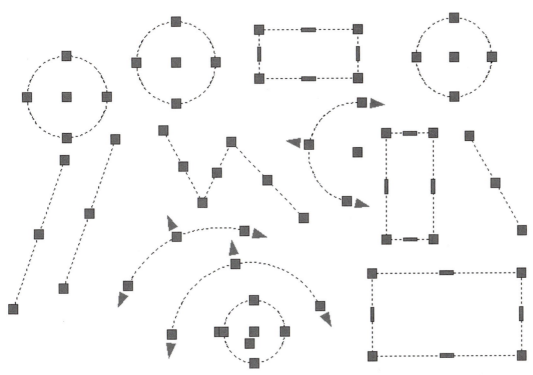

FIGURE 2.1
Grips on circles, rectangles, arcs, and lines.

41

2.2 UNITS AND SCALE

We have not yet talked about any drawing units, such as feet, inches, or centimeters. When you practiced the offset command, you entered a value, such as 2 or 3, but what did it mean? Well, as you inadvertently discovered, the default unit system in AutoCAD, until you change it, is a simple decimal system of the type we use in everyday life. However, it is unitless, meaning that the 2 or 3 you entered could have been anything—feet, inches, yards, meters, miles, or millimeters. Think about it: It is an easy and powerful idea, but it is up to you to set the frame of reference and stick to it. So, if you are designing a city and 1 is equal to a mile, then 2 is two miles and 0.5 is half a mile. AutoCAD does not care; it gives you a system and lets you adapt it to any situation or design. Then, you just draw in real-life units or, as we say in CAD, 1-to-1 units. This is the Golden Rule of AutoCAD. *Always draw everything to real-life size*. Do not scale objects up or down.

A decimal system of units is OK for many engineering applications, but what about architecture? Architects prefer their distances broken down to feet and inches, because it is always easier to understand and visualize a distance of 22'-6" than 270". To switch to architectural units, type in `units` and press Enter, or select **Format→Units**... from the cascading menus, and the dialog box in Figure 2.2 appears.

Simply select Architectural from the drop-down menu at the upper left and ignore the rest of the options in the dialog box; there is no need to change anything else at this point. Now, as you draw and enter in units, they are recognized as feet or inches. To enter in 5 feet (such as when using the offset command), type 5'. To enter in 5 inches, there is no need for an inch sign, just type in the number 5; AutoCAD understands. To mix feet and inches, type in 5'6, which is of course 5'-6"; there is no need for a hyphen either.

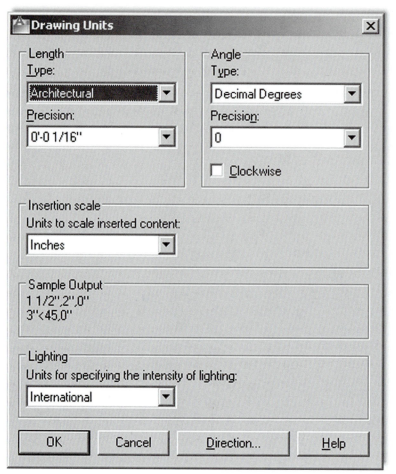

FIGURE 2.2
Drawing Units dialog box.

2.3 SNAP AND GRID

These two concepts go hand in hand. The idea here is to set up a framework pattern on the screen that restricts the movement of the crosshairs to predetermined intervals. That way, drawing simple shapes is easy because you know that each jump of the mouse on the screen is equal to some preset unit. Snap is the feature that sets that interval, but you need the grid to make it visible. One without the other does not make sense, and usually they are set equal to each other. Snap and Grid are not used that often in the industry, but it is a great learning tool, and we do some drawings at the end of this chapter using them. It also helps knowing about this feature, as it is often inadvertently activated by pressing the Snap or Grid buttons at the bottom of the screen.

To Set Snap

Type in snap.

> ○ AutoCAD says: Specify snap spacing or [ON/OFF/Aspect/Style/Type] <0' − 0 ½">:

Type in 1, setting the snap to 1 inch.

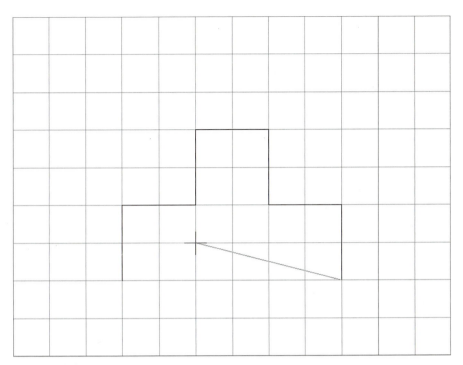

FIGURE 2.3
Figures drawn with Snap and Grid.

To Set Grid

Type in grid.

- ○ AutoCAD says: Specify grid spacing(X) or [ON/OFF/Snap/Major/ aDaptive/ Limits/Follow/Aspect] < 0' − 0 ½">:

Type in 1, setting the grid to 1 inch.

Snap and Grid are now set, although you may have to zoom to extents (Tip 3) to see everything. In previous versions of AutoCAD, the grid was hard-to-see dots. In AutoCAD 2012, it is reminiscent of graph paper, as seen in Figure 2.3. Try drawing a line; you notice right away that the cursor "jumps" according to the intervals you set. F9 turns Snap on and off, while F7 takes care of Grid, and you can always toggle back and forth using the drawing aids buttons at the bottom of the screen. Figure 2.3 is a screen shot of Snap and Grid in use while drawing a simple shape.

2.4 CARTESIAN COORDINATE SYSTEM

AutoCAD, and really all CAD programs, rely on the concepts of axes and planes to orient their drawings or models in space. In 2D drafting, you are always drawing somewhere on AutoCAD's X-Y plane, which is made up of the intersections of the X and Y axes. When you switch to 3D, the Z axis is made visible to allow drawing into the third dimension. The X, Y, and Z axes are part of the Cartesian coordinate system, and we need to briefly review this bit of basic math and illustrate why it is helpful to know it. Figure 2.4 illustrates the 2D coordinate system, made up of the X and Y axes and numbered as shown.

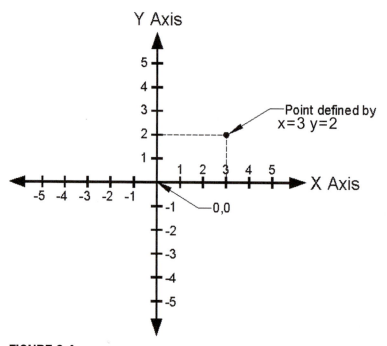

FIGURE 2.4
The Cartesian coordinate system.

This system is important not only because you may want to know where in space your drawing is but so you can create objects of a given size by entering their X and Y values. Moreover, we refer to this system often for a variety of other construction needs and distance entries, including inserting drawings into the point 0,0, so review and familiarize yourself with it again, as likely you may not have seen this in a while. Next, we introduce the various distance entry tools. As you go through Section 2.5, knowing the Cartesian coordinate system helps you understand what is going on.

2.5 GEOMETRIC DATA ENTRY

The ability to accept sizing information when creating basic objects is the "bread and butter" of any drafting or modeling software. Geometric data entry refers to the various techniques needed to convey to AutoCAD your design intent, which may involve everything from the length and angle of a line to the diameter of a circle, or the size of a rectangle.

AutoCAD has two precise methods to enter distances, angles, and anything else you may need. These techniques are the older "manual" method, where you enter data via the keyboard and it appears on your command line, and the newer "dynamic" method, where you still use the keyboard, but the data appears on your screen via what is known as a *heads-up* display. Most students probably gravitate toward the newer method, and indeed it offers many advantages, not the least of which is faster data input. We consider this the primary method in this text but present the manual method as an optional alternative. Experience has shown that some students dislike the additional on-screen clutter that Dynamic Input adds, finding the manual method less distracting. Ultimately, it is up to you to decide what to use.

Dynamic Input

Erase anything that may be on your screen and turn off all drawing aids at the bottom of the screen, including Snap, Grid, and anything else you may have on. Now press the DYN drawing aid, as seen in Figure 2.5.

FIGURE 2.5
Dynamic Input.

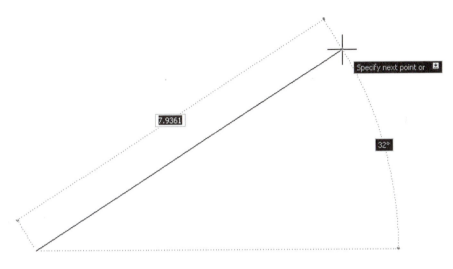

FIGURE 2.6
Drawing a line with Dynamic Input.

This type of input allows you to specify distances and angles directly on the screen. Begin drawing a line and you see an example of Dynamic Input in action. An arbitrary example is shown in Figure 2.6. Your values of course may be different, but the basic display should be the same.

Notice what AutoCAD is allowing you to do. The 7.9361 value in Figure 2.6 is surrounded by a box, meaning it is editable. All you have to do is enter a desired value, so enter a small value, such as 5, but do not press Enter yet. What if you wanted to change the angle of the line? Simply press the Tab key, tab over to the 32° value, and enter something else, perhaps a 45. Finally. press Enter. Your line 5 units long at 45° is done. The line command continues of course and you can repeat this as much as needed.

This works with Ortho on as well. In that case, you enter only distance values, no angles, as Ortho forces the lines to all be 90°. Be sure to "lead" the mouse (and in turn the lines being created) by slightly moving the mouse in the direction you want to go.

To use Dynamic Input to draw a rectangle (5" × 3" in this example), make sure DYN is on, begin the command, and click anywhere to start the rectangle. You see fields in which to enter the values for the height and width (Figure 2.7). As before, enter a new value in the first text field (5) and Tab over to the next field, entering the other value (3). As you press Enter, the 5" × 3" rectangle is created. Note of course that the first value is the *width* and the second value is the *height*.

The usefulness of Dynamic Input extends out to just about any shape with which you may work. Figure 2.8 shows DYN with an arc (left) and a circle (right). As before, you can specify on-screen data needed to create that particular shape.

You may notice also that some of the "heads up" menus show a downward pointing arrow. You can use the down arrow on your keyboard to drop down the menu and reveal other choices, which of course mimic what is available in the parentheses on the command line. Figure 2.9 is one example with the offset command. You can cycle through these options to select what you need.

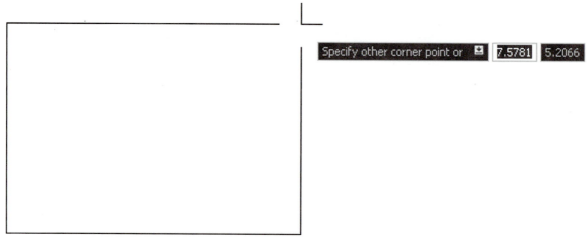

FIGURE 2.7
Drawing a rectangle with Dynamic Input.

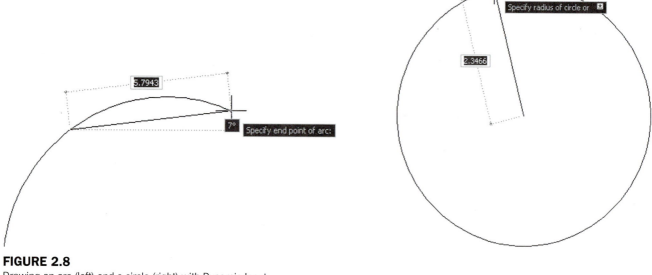

FIGURE 2.8
Drawing an arc (left) and a circle (right) with Dynamic Input.

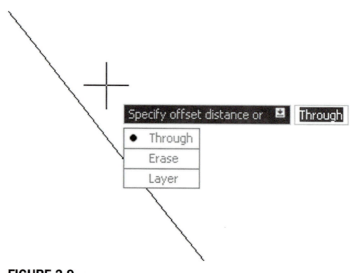

FIGURE 2.9
Dynamic Input menu, Offset command.

Manual Input

Using the manual method, you can do the same thing that can be done with the DYN input. Let us run through a few scenarios and, along the way, introduce Direct and Relative Data Entry. The simplest case just involves lines. Turn off the DYN button and begin the line command, clicking where you want to start. Now, all you have to do is point the mouse in the direction you want the line to go (moving the mouse a bit in that direction), enter a length value on the keyboard, and the line appears. You can continue this "stitching" indefinitely. Note that we have not yet explored how to draw the lines at various precise angles, except for 90°; for that just turn on Ortho. If using Ortho, be sure to once again "lead" the linework as before, by moving your mouse in the direction you want to go, so AutoCAD understands your intent.

Whether done manually or via DYN, this point-by-point stitching, called *Direct Data Entry*, is a useful technique. One example of its use occurs after you do a basic room or building perimeter survey. This may just be a wall-by-wall sequential field measurement of a space. When you return to the office you can quickly enter the wall lengths into AutoCAD one at a time via this technique. Do take a few minutes to practice it.

Direct Distance Entry can be done manually one other way. While not a common technique for day-to-day drafting, it really gets you to practice and understand the Cartesian coordinate system. What you do is enter data one click at a time by following the coordinate points. To practice this, begin the line command, then instead of clicking anymore, type in the following values, pressing Enter after each one:

0,0 0,2 2,2 2,4 4,4 4,2 6,2 6,0 then c for "close"

These are the X and Y values for the lines to follow; and if you typed correctly, then the shape seen in Figure 2.10 should emerge on your screen one segment at a time.

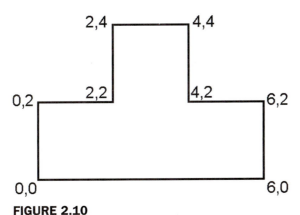

FIGURE 2.10
Shape created via Direct Distance Entry.

There is a variation on Direct Data Entry called *Relative* Data Entry. This is useful for quickly and easily creating shapes, such as rectangles, anywhere on the screen. It also allows you to enter in various angles for lines besides the Ortho-induced 90°. To do either you need to know two new symbols, @ and <.

The @ symbol frees you from 0,0 and can be interpreted as meaning "relative to." In the case of a rectangle, this means you can create it relative to wherever you are in space by simply starting the rectangle command and specifying an X and a Y value preceded by the @ symbol. So, for a 5" × 3" rectangle,

1. Start the rectangle command.
2. Click somewhere on your screen to anchor the first point.
3. Type in: @5 , 3.

The rectangle then appears.

Now that you know about the @ symbol and what it does, you can also draw lines at any angle by also employing the additional < symbol. Any numerical value after this "alligator teeth" symbol is interpreted as an angle. The format is as follows: `@Distance<Angle`. Therefore, to create a line that is 10 units long at 45°,

Start the line command,
Type in: `@10<45`.

The line (10 units long, at 45°) appears.

As we conclude Section 2.5, remember that, when it comes to creating a rectangle, the dynamic and basic manual methods are not the only ones available. A brief mention was made in Chapter 1, when the rectangle was first introduced, that you can also access length and width data as part of the command line sub-options, such a pressing d for `Dimensions`. You can vary that by specifying a total area then either the length or width. It may be worth it to briefly review all this again to have the full set of options available to you. You will need to create quite a few rectangles as you begin drafting. They are the basis of furniture, mechanical components, and even entire floor plans, as we soon see. You should be well versed in all the ways to create them.

2.6 INQUIRY COMMANDS

Our next topic in AutoCAD fundamentals is the Inquiry commands set. These commands are there to give you information about the objects in front of you on the screen, as well as other vital data like angles, distances and areas. As such, it's important for you to learn them early on; the information these commands provide is especially valuable to a beginner student, and will continue to be essential as you get more experienced. Inquiry commands were upgraded just a few releases ago and they are currently comprised of:

- *Area (gives you the area of an object or any enclosed space).*
- *Distance (gives you the distance between any two points).*
- *List (gives you essential information on any object or entity).*
- *ID (gives you the coordinates of a point).*
- *Radius (gives you the radius of a circle or arc).*
- *Angle (gives you the angle between two lines).*
- *Volume (gives you the volume of a 3D object).*
- *Mass Properties (lists various mass properties of a 3D solid).*

We cover the first four inquiry commands in detail (*Area, Distance, List,* and *ID*), and include a brief mention of *Radius* and *Angle*. *Volume* and *Mass Properties* are not discussed here, as those are for 3D work. As before, all commands can be accessed through the usual four ways: typing, cascading menus, toolbars, and the Ribbon. The toolbar used is the Inquiry toolbar, as seen in Figure 2.11, and should be brought up on screen using the methods described in Chapter 1.

FIGURE 2.11
Inquiry toolbar.

➤ Area

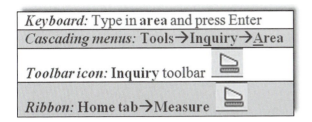

Keyboard: Type in **area** and press Enter	
Cascading menus: Tools→Inquiry→<u>A</u>rea	
Toolbar icon: **Inquiry** toolbar	
Ribbon: Home tab→Measure	

Area is a very important and useful Inquiry command. It is used often by architects to verify the area of a room or an entire floor plan. Area calculations are performed in two ways: either on an object or, more commonly, point by point (such as the area of a room). Draw a rectangle of any size. We go over both methods, although with a rectangle you would really use only the Object method, saving the point-by-point approach for a more complex floor plan.

POINT BY POINT

This is used when measuring an area defined by individual lines (not an object).

Step 1. Make sure your units are set to Architectural and you have a rectangle on your screen.
Step 2. Begin the area command via any of the preceding methods.
- AutoCAD says: `Specify first corner point or [Object/Add area/ Subtract area] <Object>:`
Step 3. Activate your OSNAPs (F3) and click your way around the corners of the rectangle until the green shading covers the entire area to be measured (five clicks total).
- AutoCAD says:
```
Specify next point or [Arc/Length/Undo]:
Specify next point or [Arc/Length/Undo]:
Specify next point or [Arc/Length/Undo/Total] <Total>:
Specify next point or [Arc/Length/Undo/Total] <Total>:
```
Step 4. When done, press Enter. You get a total for area and perimeter. Mine said:
```
Area = 67.9680, Perimeter = 34.2044
```
Yours of course will likely say something different.

OBJECT

This is used when you have an object (not an area defined by individual lines) to measure.

Step 1. Begin the area command via any of the preceding methods.
- AutoCAD says: `Specify first corner point or [Object/Add area/ Subtract area/eXit] <Object>:`
Step 2. Press the letter O for object.
- AutoCAD says: `Select objects:`
Step 3. Select the square. You get the same area and perimeter measurements.

➤ Distance

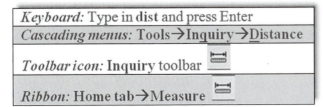

Keyboard: Type in **dist** and press Enter	
Cascading menus: Tools→Inquiry→<u>D</u>istance	
Toolbar icon: **Inquiry** toolbar	
Ribbon: Home tab→Measure	

The distance command gets you the distance between any two points. Although we practice on the corners of a rectangle, the command works on any two points, as selected by a mouse click, even an empty space.

Step 1. Make sure your units are set to Architectural, a rectangle is on your screen, and your OSNAPs are activated.
Step 2. Begin the distance command via any of the preceding methods. (Be sure to read the note after Step 4.)
  ○ AutoCAD says: `Specify first point:`
Step 3. Select a corner of the rectangle.
  ○ AutoCAD says: `Specify second point or [Multiple points]:`
Step 4. Select the other corner across the rectangle. AutoCAD then gives you some information. Mine said:
  `Distance = 13.5521, Angle in XY Plane = 0, Angle from XY Plane = 0`
  `Delta X = 13.5521, Delta Y = 0.0000, Delta Z = 0.0000`
  Yours will say something different.

Note: There is an important difference in procedures if you use the icons or the Ribbon. In those cases, DYN activates and you get a graphical representation of the distance, as seen in Figure 2.12. If you type or use cascading menus, then you get the command line results seen in Steps 1 through 4.

FIGURE 2.12
Distance command results via toolbar or Ribbon.

➤ List

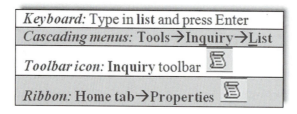

Keyboard: Type in list and press Enter
Cascading menus: Tools→Inquiry→List
Toolbar icon: **Inquiry** toolbar
Ribbon: Home tab→**Properties**

The list command gives you information about not just the geometric properties of the object but also its CAD properties, such as what layer it is on, the subject of the next chapter. It is an extremely useful command that clues you in on what you are looking at (whether a beginner or an expert, you will sometimes ask yourself this question).

Step 1. Make sure you have something on your screen to investigate. Draw a line, rectangle, or circle.

Step 2. Begin the list command via any of the preceding methods.

○ AutoCAD says: `Select objects:`

Step 3. Select your object (it becomes dashed) and press Enter. The command line in the form of a text window appears mid-screen. Look it over and close it (X, upper right, or just press F2) when done. What mine said about an arbitrary rectangle is shown in Figure 2.13.

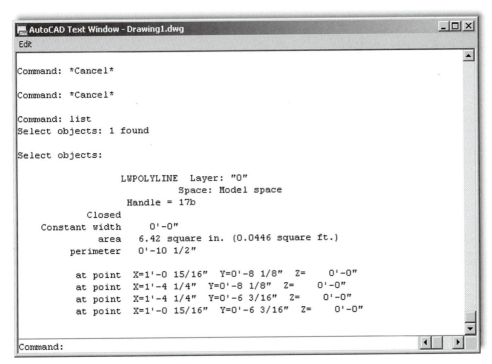

FIGURE 2.13
List command AutoCAD text window.

51

➤ ID

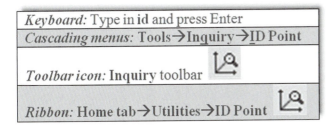

Keyboard: Type in id and press Enter	
Cascading menus: Tools→Inquiry→ID Point	
Toolbar icon: Inquiry toolbar	
Ribbon: Home tab→Utilities→ID Point	

This command is the simplest of all. It tells you the coordinates of a point in space, such as the intersection of two lines. This is useful for inserting blocks (Chapter 7) and just plain old figuring out the general location of an object, something that may be needed for Xrefs (Chapter 17).

Step 1. Make sure you have something on your screen to ID. Create a line or a rectangle.

Step 2. Begin the ID command via any of the preceding methods.

○ AutoCAD says: `Specify point:`

Step 3. Using OSNAPs, select a point, such as the corner of a rectangle or end of a line. AutoCAD gives you the X, Y, and Z coordinates. Here is what I got; you of course will likely get different values:

```
X = 11.1504  Y = 6.7753  Z = 0.0000
```

Radius and Angle

You should now have a good feel for the Inquiry toolset, so go ahead and practice the remaining two commands on your own. Radius and angle cannot be typed in; you must use the toolbars, cascading menus, or the Ribbon.

To practice *Radius*, draw an arbitrary circle, then activate the command and select it. A radius is displayed and you can just press Esc to get out. The result is shown in Figure 2.14 (left). To get an angle reading create two "alligator teeth" lines, activate the command, and select both lines. The angle is displayed and again you can press Esc to exit. The result is shown in Figure 2.14 (right). Note that these commands are "dynamic" in nature, so DYN activates when you use them.

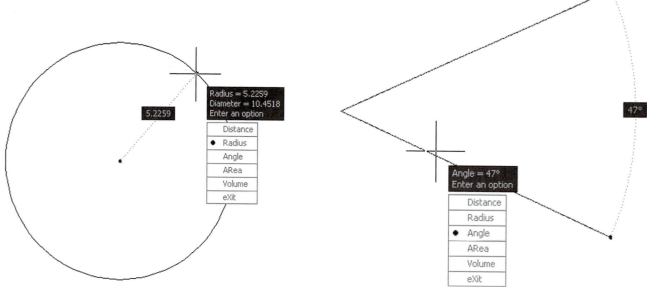

FIGURE 2.14
Radius (left) and Angle (right) measurements.

This concludes the section on Inquiry commands. We are now in the home stretch and will explore some commands and concepts that do not quite fit in neatly anywhere else, but will prove to be very useful as you start a realistic project in the next chapter, and of course everyday AutoCAD use.

2.7 ADDITIONAL DRAFTING COMMANDS

We are going to take a moment to explore four additional drafting commands: explode, polygon, ellipse, and chamfer. They are not mentioned in Chapter 1 because they are not part of the essential set of drafting commands. The need for lines and rectangles in a typical drawing is far greater than the need for ellipses or polygons. However, in the few cases you do need one of those shapes, there really is no substitute. We start off with the unusual explode command.

➤ Explode

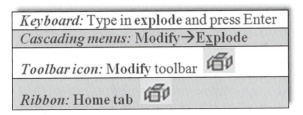

Keyboard: Type in **explode** and press Enter	
Cascading menus: Modify→Explode	
Toolbar icon: Modify toolbar	
Ribbon: Home tab	

Explode is used to make complex objects simpler. That means a rectangle becomes four separate lines, and other objects we have yet to study (blocks, text, dimensions, hatch, etc.) are simplified into their respective components. This command is mentioned in almost every chapter, as virtually all AutoCAD entities can be exploded—although many should never be! To try it out, we yet again need a rectangle of any size. Make sure you are using the actual rectangle command, not just four connected lines. Those are already in their simplest form, and explode does not work on them.

Step 1. Begin the explode command via any of the preceding methods.
  ○ AutoCAD says: `Select objects:`
Step 2. Pick the rectangle and it becomes dashed.
  ○ AutoCAD says: `1 found`
Step 3. Press Enter and the rectangle is exploded.

Click any of its lines and you notice that each line is independent, as seen in Figure 2.15. Keep in mind that you cannot explode circles, arcs, and lines (previously mentioned), as those are already the simplest objects possible. There is a way to reverse explode, in addition to an immediate undo. It is called a pedit command (discussed in Chapter 11), but that applies only to simple compound objects, such as a rectangle. If you explode a hatch pattern and save your work, that is it, no going back. This is why you see appropriate warnings when we get to those chapters.

53

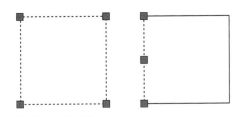

FIGURE 2.15
Unexploded rectangle (left) and exploded rectangle (right).

➤ Polygon

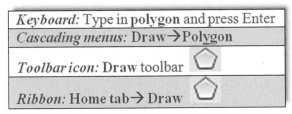

Keyboard: Type in **polygon** and press Enter	
Cascading menus: Draw→Polygon	
Toolbar icon: Draw toolbar	
Ribbon: Home tab → Draw	

The polygon command is a useful tool that we need in later chapters for a project. A polygon, strictly speaking, is any shape with three or more sides to it. Of course, in those cases when it has three sides (triangle), you can draw it with existing line techniques. Same thing with four sides (square or rectangle). However, what about five sides or more, such as a pentagon or hexagon? Here, a polygon command is needed, as it is a real chore to draw these shapes using the line command—just try figuring out all the correct angles.

Drawing polygons, however, presents a complication you do not have with simpler shapes. With a rectangle, you specify the height and width and that is it, but how do you do that with, say, a pentagon? Figure 2.16 is a picture of one. What is the height? What is the width? It is not an easily measured shape.

The solution that drafters came up with a long time ago (many of AutoCAD's techniques are just adaptations of pencil and paper board-drafting methods used for decades) is to enclose the shape in a circle of a known radius or diameter. Of course, you can also enclose the circle inside the polygon.

With either approach, it is much easier to specify a size. You just say "I need a pentagon that fits inside (or outside) a circle of 10 inches in diameter." This is where the terms *inscribed* and *circumscribed* come from. Figure 2.17 is a diagram of two pentagons, one inscribed in a circle and one circumscribed about a circle. Note the tangencies. In one case, the points of the polygon touch the circle; in the other case, its sides touch it.

Let us go through the steps in making a polygon inscribed in a circle. Begin by drawing a circle with an arbitrary radius of 5. Note that we do not need this step to create a polygon, but for your first try, it is a great help in visualizing what you are doing. Then proceed as shown next.

Step 1. Begin the polygon command via any of the preceding methods.
   ○ AutoCAD says: `Enter number of sides <4>:`
Step 2. Enter the number 5 for five sides (making it a pentagon), and press Enter.
   ○ AutoCAD says: `Specify center of polygon or [Edge]:`
Step 3. With your OSNAPs on, click to select the center of the circle.
   ○ AutoCAD says: `Enter an option [Inscribed in circle/Circumscribed about circle] <I>:`
Step 4. Now that you know what these terms mean, go ahead and press Enter to select i for "Inscribed in circle" (the default value). You can just as easily select c as well.
   ○ AutoCAD says: `Specify radius of circle:`
Step 5. Enter the value 5. A screen shot of the result is in Figure 2.18.

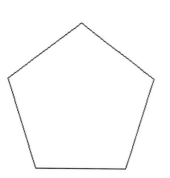

FIGURE 2.16
Basic pentagon.

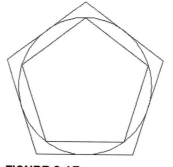

FIGURE 2.17
Inscribed and circumscribed pentagons.

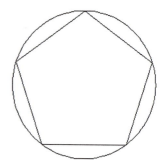

FIGURE 2.18
Results after Step 5.

You can now erase the circle, since it was just a guide. In the future you do not need it; just click to place the polygon anywhere on the screen. Go ahead and practice this command, but this time select a different number of sides (six or more) and also use the Circumscribed option. This new command leads us into the first challenge exercise.

DRAWING CHALLENGE—STAR

Try this exercise based on what you learned. Draw a perfect five-sided star, such as the one shown in Figure 2.19. Think about everything we talked about up to this point. Only four commands are needed to create the star. Take a few minutes to try to draw it. If you cannot get it, do not despair, about a third of my students through the years typically did not get it right away either. The answer is at the end of this chapter, just before the summary.

FIGURE 2.19
Drawing challenge—star.

Ellipse

Keyboard: Type in **ellipse** and press Enter	
Cascading menus: Draw→Ellipse→Center	
Toolbar icon: Draw toolbar 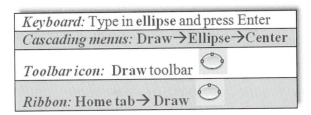	
Ribbon: Home tab→ Draw	

The ellipse command is easy to use and is needed when we draw furniture in Chapter 4, so it is included here. The command can be executed in one of two ways: Center and Axis, End. You can also create a partial ellipse by defining just a section of it.

Step 1. Begin the ellipse command via any of the preceding methods.
   ○ AutoCAD says: `Specify axis endpoint of ellipse or [Arc/Center]:`
Step 2. Click anywhere on the screen.
   ○ AutoCAD says: `Specify other endpoint of axis:`
Step 3. With Ortho on (not required but good to have), click again somewhere off to the left or right of the first point. Then, move the mouse back toward the center as the ellipse is formed, as seen in Figure 2.20.

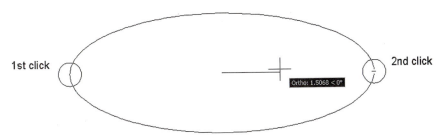

FIGURE 2.20
Ellipse after Step 3.

55

○ AutoCAD says: `Specify distance to other axis or [Rotation]:`

Step 4. Click anywhere again to complete the ellipse. You can also specify a rotation angle along the major axis (the line between clicks 1 and 2). This would be as if the ellipse were rotating away (or toward) you. For a variation on this, try the Arc submenu of the ellipse, which is essentially similar in execution until the final step, when you trace how much of the ellipse you want to see, as shown in Figure 2.21.

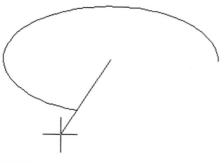

FIGURE 2.21
Ellipse, arc variation.

Chamfer

Keyboard: Type in **chamfer** and press Enter
Cascading menus: Modify→Chamfer
Toolbar icon: Modify toolbar
Ribbon: Home tab→ Draw

You may have already tried the chamfer command, as it sits right next to fillet in the cascading menus and the Modify toolbar. While normally not used as often as a fillet, it is still very much worth investigating. A chamfer is a beveled edge connecting two surfaces. Think of it as a fillet that is angled instead of round. Chamfers may have a 45° corner, but this is not required; AutoCAD allows for asymmetrical chamfers.

To try out the command, draw two perpendicular lines similar to the fillet exercise in Chapter 1 (reproduced in Figure 2.22) and then follow these steps.

Step 1. Begin the chamfer command via any of the preceding methods.
○ AutoCAD says:
`(TRIM mode) Current chamfer Dist1 = 0.0000, Dist2 = 0.0000`
`Select first line or [Undo/Polyline/Distance/Angle/Trim/mEthod/`
`Multiple]:`

Step 2. No distance has been entered yet, and any attempt at chamfer results in a chamfer radius of zero. Press d for `Distance`.
○ AutoCAD says: `Specify first chamfer distance <0.0000>:`

Step 3. Enter a small value, such as 1.
○ AutoCAD says: `Specify second chamfer distance <1.0000>:`

Step 4. Here the 1 is not needed again, so just press Enter.
○ AutoCAD says: `Select first line or [Undo/Polyline/Distance/Angle/`
`Trim/mEthod/ Multiple]:`

Step 5. Select the first line.
○ AutoCAD says: `Select second line or shift-select to apply corner or`
`[Distance/Angle/Method]:`

Step 6. As you move the mouse toward the second line, you again see a preview of the completed command (similar to fillet). Select the second line and the chamfer is created as seen in Figure 2.23. Zoom in if needed to see it.

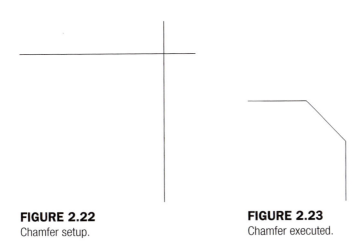

FIGURE 2.22
Chamfer setup.

FIGURE 2.23
Chamfer executed.

If the chamfer fails, it is possible that you selected values too large for the available lines. Another way to create a chamfer is using the length of one line segment together with an angle, using the `Angle` option. Experiment with both.

Templates

Templates are something of a mystery for a beginner AutoCAD student. What exactly are they? While templates can get complicated in what they contain, they are really just files with some information embedded in them, with all the proper settings but no actual design present. So it is a blank page with some intelligence built in and everything set up and ready for you to draw. These templates can be saved under a `.dwt` extension, but you really need not concern yourself with this. What is important is being able to open a new file and know what you are looking at. Using the cascading menu **File→New…**, you default to the screen in Figure 2.24. Open up the highlighted `acad.dwt` and you have a blank file, and that is it.

FIGURE 2.24
An `acad.dwt` template.

Limits

This is yet another topic that causes confusion. Limits are not technically necessary, but many textbooks persist in having you set them up. All they are is an indication of the size to which your drawing area is preset. Remember, however, that it is limitless, so all limits refer to is the area you can work on without a regen. If you draw a shape that is way beyond your limits, you simply regenerate the screen via a Zoom→Extents (see Tip 3 in Chapter 1 for a reminder) and the limits move to just beyond this new shape, so setting them is rather redundant. If you would like to set them anyway, just type in limits, press Enter, and set the lower left corner to 0,0, then the upper right corner to perhaps 100', 50', or anything else you wish.

Save

Needless to say, you have to save your work; AutoCAD will eventually lock up or crash. If not, Windows will, there will be a power failure, or you may spill coffee on your laptop, so it is just a good habit to get into. Starting in the next chapter, you need to start saving your work. To do so, you can type in save and press Enter or press the graphic of a disk at the top of the screen. Another way is to drop down the arrow located next to the big red A at the upper left of the screen (Figure 2.25) and choose Save or Save As. You then get the usual Save Drawing As dialog box (Figure 2.26) from which you should know how to proceed.

FIGURE 2.25
AutoCAD main menu with Save As selected.

Help Files

Finally we get to our almost last topic, the help files. AutoCAD, like many other applications, has extensive help files, and indeed that is one way (though not recommended) to attempt to learn the software. A better application of help is to explore it once you know how to use AutoCAD and need assistance with a particular aspect of a command. Tool tips that appear with every command can be considered to be part of the help files.

Generally, as with any software, you get to the help files via the F1 key. With AutoCAD there is a bit of a twist. By default, AutoCAD 2012's help files are bundled together with additional tutorials, features, and on-line content. Now called Autodesk Exchange, it is mentioned in Chapter 1. If you are connected to the Internet, then you see the screen in Figure 2.27. This panel can be called up via a small X symbol at the upper right of the screen. Note that you have the basic Home page, which features "What's New" and other videos. The second

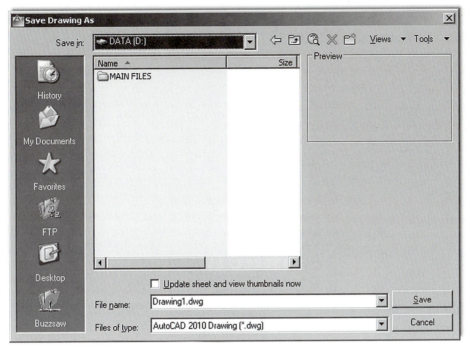

FIGURE 2.26
Save Drawing As.

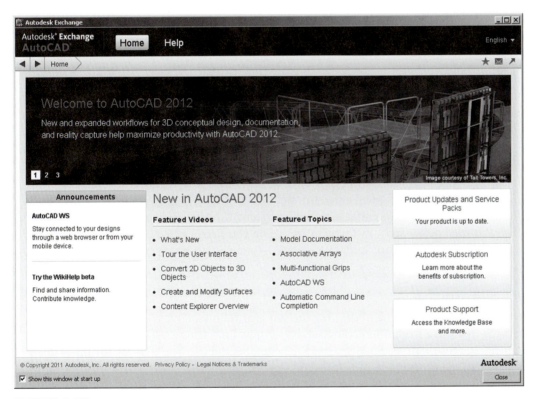

FIGURE 2.27
AutoCAD 2012 AutoCAD Exchange.

category, Help, is the actual help files. These are available whether or not you are connected to the Internet.

Click over to the Help category. Explore further by looking at the Command Reference and find ERASE, as seen in Figure 2.28.

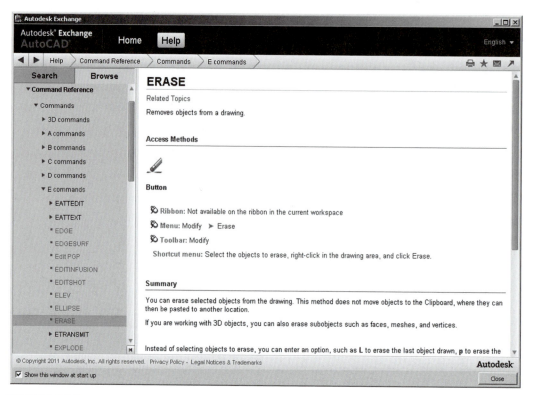

FIGURE 2.28
AutoCAD 2012 Help files.

As you can see, a tremendous amount of useful information is found here, with tutorials and reference material covering a good deal of the software, including advanced topics such as 3D and AutoLISP. You will need to occasionally explore these files for answers, so get comfortable with them.

TANgent OSNAP

The Tangent OSNAP point is necessary to draw part of Exercise 6 and is one of three additional OSNAP points mentioned throughout upcoming chapters. Tangent works by allowing you to create a line that maintains perfect tangency to a circle or an arc. Because this OSNAP is used quite rarely, just type in TAN when you need to create the tangency; you see the symbol shown in Figure 2.29. Once started, create another tangency point on another circle or arc.

DRAWING THE STAR

As promised, the technique of drawing the star is as follows. First, create a pentagon. Then, noticing that it is a five-point geometric shape, draw lines from endpoint to endpoint connecting those vertices. Finally, erase the pentagon and use trim to erase the inner lines, as shown in the progression in Figure 2.30.

The reason students do this simple exercise is to underscore a very important point. AutoCAD is used across a vast array of industries, each with its own symbols and unique objects. It would be impossible for the software to include all of them in its database. Instead AutoCAD gives you basic tools and leaves it up to your creativity and skill to create the shape you need. So, when at a loss as to how to draw something you imagined in your mind, stop and think about the basics that you learned and how you can put them together into something new.

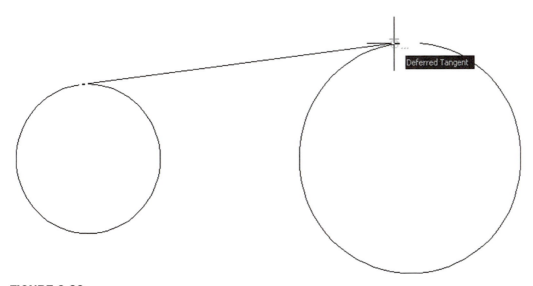

FIGURE 2.29
Using the Deferred Tangent OSNAP.

FIGURE 2.30
Drawing challenge.

61

Review everything presented so far. In Chapter 3, we introduce the layers concept and put all the preceding together to begin drawing a floor plan. Remember that learning AutoCAD is cumulative, and you need everything we have covered so far, no more and no less, to successfully deal with the upcoming material, so make sure you understand it all before moving on.

SUMMARY

You should understand and know how to use the following concepts and commands before moving on to Chapter 3:

- Using grips
  - ○ Activating them, using Esc to get out
- Setting units
  - ○ Architectural and Engineering (Decimal) units
- Using Snap and Grid
  - ○ Setting both, turning them on and off
- Understanding the Cartesian coordinate system
  - ○ X and Y axes, ordered pair (X,Y)
- Distance Entry tools
  - ○ Direct Distance Entry
  - ○ Relative Distance Entry
  - ○ Dynamic Distance Entry

- Inquiry commands
  - Area
  - Dist
  - List
  - ID
  - Radius and Angle
- Explode command
- Polygon command
  - Circumscribed and Inscribed
- Ellipse command
- Chamfer command
- Templates
  - acad.dwt
- Setting limits
- Save
- Help files (F1)
- TANgent OSNAP

REVIEW QUESTIONS

Answer the following questions based on what you learned in Chapter 1. You should have the commands memorized, so copy the questions to another sheet of paper, close the book, and try to answer them without looking them up.

1. How do you activate grips on AutoCAD's geometry?
2. How do you make grips disappear?
3. What are grips used for?
4. What is the command to bring up the units dialog box?
5. The snap command is always used with what other command?
6. Correctly draw a Cartesian coordinate system grid. Label and number each axis appropriately, with values ranging from −5 to +5.
7. On the preceding Cartesian coordinate system grid, locate and label the following points: (3,2) (4,4) (−2,5) (−5,2) (−3, −1) (−2, −4) (1, −5) (2, −2).
8. What are the two main types of geometric data entry?
9. Describe how Dynamic Input works when creating a line at an angle.
10. List the six Inquiry commands covered.
11. What command simplifies a rectangle into four separate lines?
12. What is the difference between an inscribed and a circumscribed circle in the polygon command?
13. What is a chamfer as opposed to a fillet?
14. What template should you pick if you want a new file to open?
15. What is the idea behind limits?
16. Why should you save often?
17. What is the graphic symbol for Save?
18. How do you get to the Help files in AutoCAD?

EXERCISES

1. Open a new file, set units to Architectural, and set both your snap and grid to 1". Then, draw the following shape. Keep in mind the spacing between each dot is now 1". You would be wise to use the mirror command to complete this faster. The linework in this example has been darkened slightly for clarity. (Difficulty level: Easy; Time to completion: <5 minutes)

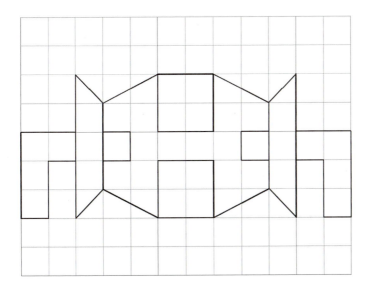

2. Open a new file and set units to Architectural. Then, draw the following figure using the DYN drawing aid. When done, use the area command to find its area. Redraw the shape again via the manual method. (Difficulty level: Easy; Time to completion: <5 minutes)

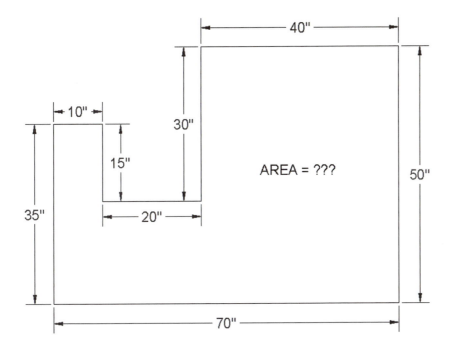

3. Open a new file and set units to Architectural. Draw the following shapes using the manual Relative Distance Entry (the @ sign) method. Repeat the exercise using the DYN method. Repeat again using the rectangle command's length and width sub-options. Finally, use the area command (Object option) to find the area and perimeter of each shape. The sizes are 24" × 36", 17.5" × 3", 12" × 12", and 3" × 40". Remember that the first number is always the X; the second is the Y as shown in the exercise figure. (Difficulty level: Easy; Time to completion: <10 minutes)

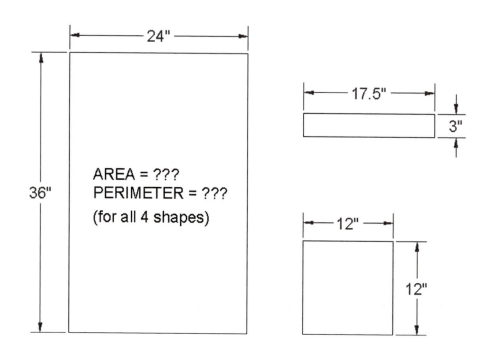

AREA = ???
PERIMETER = ???
(for all 4 shapes)

4. Open a new file and set units to Architectural. Draw the following figure using the basic circle, line, offset, trim, and fillet commands. Create the top view first using circles and project lines for the side view. (Difficulty level: Easy; Time to completion: 5–7 minutes)

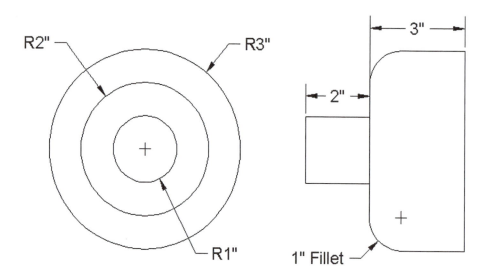

5. Open a new file and set units to Architectural. Draw the following bookshelf. Depending on your exact approach, you may need the rectangle, offset, explode, trim, rotate, copy, scale, and line commands. The spacing of the shelves and the size and rotation of the books can be anything you want. All dimensions are for reference only. (Difficulty level: Intermediate; Time to completion: 10 minutes)

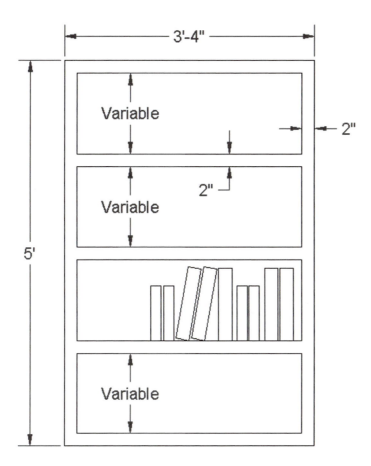

6. Open a new file and set units to Architectural. Draw the following mechanical piece. All dimensions and centerlines are for reference only; we have not yet covered how to create them. You need to use the TANgent OSNAP point. (Difficulty level: Intermediate; Time to completion: 10 minutes)

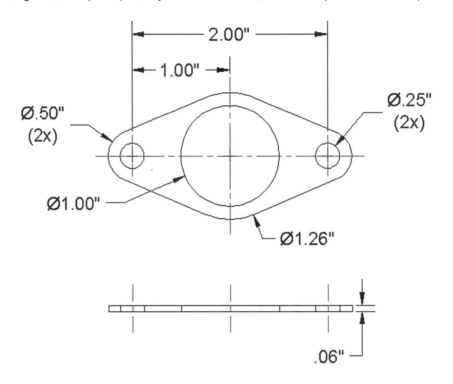

7. Open a new file and leave the units as Decimal. Draw the following simplified bolt diagram. You have all the needed information but may have to carefully project some parts of the side view. You may need the polygon, circle, line, trim, chamfer, and other commands. (Difficulty level: Intermediate; Time to completion: 5–7 minutes)

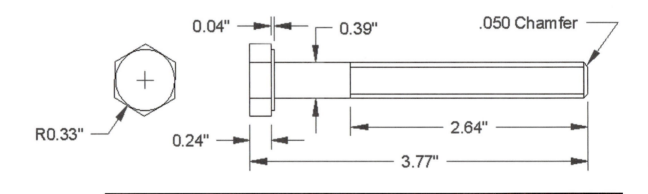

CHAPTER 3

Layers, Colors, Linetypes, and Properties

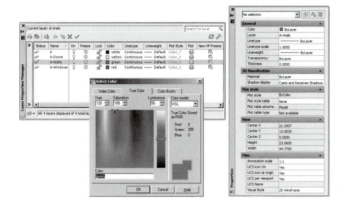

LEARNING OBJECTIVES

Layers are an essential concept and need to be introduced before a serious drawing is attempted. In this chapter, we introduce layers and discuss the following:

- What are layers?
- Why use them?
- Creating and deleting layers
- Making a layer current
- Assigning layer colors
- Index color
- True color
- Color Books
- Layer Freeze/Thaw
- Layer On/Off
- Layer Lock/Unlock
- Loading and setting linetypes
- Properties Palette
- Match Properties
- Layers toolbar

Here, you learn all the remaining tools to begin creating a floor plan, which we start at the end of the chapter.

Estimated time for completion of chapter: 3 hours (lesson and project).

3.1 INTRODUCTION TO LAYERS

Layers are our first major topic after the previous chapters of basics. It is a very important concept in AutoCAD, and no designer or draftsperson would ever consider drawing anything more complex than a few lines without using them.

What Are Layers?

"Layers" is a concept that allows you to group drawn geometry in distinct and separate categories according to similar features or a common theme. This allows you to exercise control over your drawing by, among other things, applying properties to the layers, such as assigning colors and linetypes. You can also manipulate each individual layer, making it visible or invisible for clarity, as well as being able to lock them to prevent editing.

For example, all the walls of a floor plan will exist on the "Walls" layer, all the doors on the "Doors" layer, windows on the "Windows" layer, and so on. You can think of layers as transparency sheets in an anatomy textbook, with each sheet showing a unique and distinct category of data (skeleton, muscle, circulatory system, etc.), but when viewed all at once they show the complete picture of a person. Similarly, in AutoCAD, there can be any number of layers, with all of them holding their respective data and combining into the complete design.

Why Use Them?

As just stated, the key word here is *control*. If all your geometry were lumped together on one layer, as some beginners do, then whatever you do to one set of objects, you do to all; certainly not what you want. It is very important that you learn about layers early in your AutoCAD education, and apply them right away to your first real drawing.

Let us take a look at the Layers Properties Manager. This is AutoCAD's Layers dialog box, where everything important related to layers happens. Open a new file and, if you decide to use toolbars, also bring up the Layer toolbar. We take a closer look at that toolbar in Section 3.3, but for now you need only the icon that calls up the Layers Property Manager.

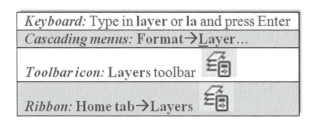

Keyboard: Type in **layer** or **la** and press Enter	
Cascading menus: Format→Layer…	
Toolbar icon: Layers toolbar	
Ribbon: Home tab→Layers	

Once you call up the layers property manager, you will see Figure 3.1

Examining the dialog box from left to right, notice the filters taking up some room on the left. We need not talk about filters until Chapter 12, and you can actually minimize that whole area by clicking the << arrows. To the right of the filters, notice the main category headers such as "Status," "Name," "On," and "Freeze." Below the categories you find the only existing layer thus far, 0, and its properties. Notice that it is white with a "continuous" linetype. Above the categories are some action buttons for creating and deleting layers, which we cover next.

Creating and Deleting Layers

Assuming you are starting from a blank file, each new layer initially needs to be typed in according to what you (or the company you work for) require. To do that, you first need to

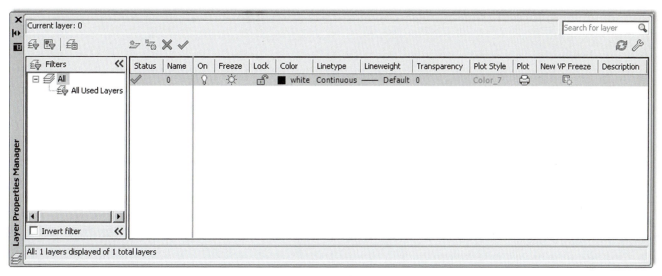

FIGURE 3.1
AutoCAD 2012 Layer Properties Manager.

click on the 0 layer to make sure it is highlighted with the blue band. Then press the New Layer icon at the top of the layer box (two icons to the left of the red X), which looks like a sheet of paper with a sparkle next to it. You can also hold down the Alt key and press N. Either way, once you do that, a new band saying Layer 1 appears. Just type in your new layer name followed by Enter, and a new layer is created. Here are some helpful tips:

- If you notice that you made a spelling mistake after you already pressed Enter, you can just click back into the layer name once or twice until the cursor enters Edit mode and retype.
- Do not attempt to change the 0 layer's name. It is the default layer and cannot be deleted or renamed.
- If you want to delete a layer (perhaps an accidentally made one), simply highlight the layer, and click the red X icon, or press Alt-D. If the layer contains nothing, it will be deleted.

Following this procedure, create the following three layers:

- A-Doors
- A-Walls
- A-Windows

What you should see is shown in Figure 3.2. The letter A in front of each layer simply means Architectural.

Making a Layer Current

The current layer is the active layer on which you can draw at any given moment. You are always on some layer (even if you are not immediately sure which one exactly). So, logically, if you are about to draw a wall, you should be on the A-Walls layer and so on. To specify what layer to have as current, either click on the green check mark button, to the right of the red X, or *double-click* on the layer name itself. At the very top left of the Layers dialog box it says "Current layer:" followed by the name of the layer. Check there if you are unsure what is what. For now, go ahead and make A-Walls the current layer.

Assigning Layer Colors

Assigning colors to layers is very important and should be done as you create them or soon after. The reasons to have color are twofold. On the most basic level, a colorful drawing is more pleasant to look at on the screen, but more important, it is easier to interpret and work with. If,

69

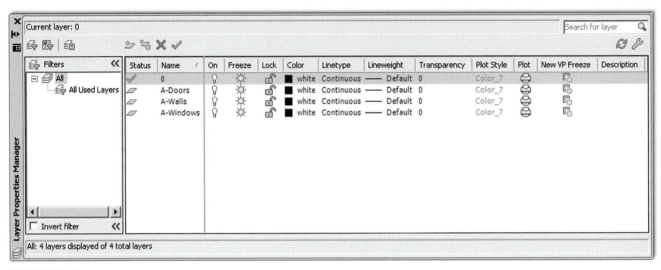

FIGURE 3.2
Layer Properties Manager with three new layers.

for example, your walls are green and you keep this in mind, then your eye is able to quickly see and assess the design. Likewise, yellow doors are easy to spot. This *color separation* greatly aids in drafting. Imagine a scenario where your building floor plan is made up of all white lines; you would be unable to tell one line from another. As that drawing grew more complex, all the lines would melt into one monochrome mess. Finally, colors are assigned pen thicknesses (an advanced topic) and contribute to more professional-looking output. We address this in Level 2.

To assign colors, let us turn our attention back to the Layers dialog box. Highlight, by clicking once, layer A-Walls. Now, follow the blue ribbon across until you see a box with the word *white* next to it. It will be under the Color column. Click on it, and the Select Color dialog box appears, as seen in Figure 3.3.

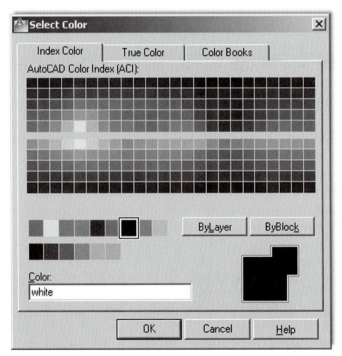

FIGURE 3.3
Select Color, Index Color.

INDEX COLOR

Let us take a tour of this Index Color tab. The two large 24 × 5 stripes of colors you immediately see at the top make up the *AutoCAD Color Index* (ACI), and it contains 240 colors. Do not be too quick in selecting one of those right away. Take a look below; there are two smaller color stripes. The rainbow-like stripes 1 through 9 are the ones in which we are most interested. If you take a close look, they are the *primary* colors and the 240 colors above them are just different shades of the primaries. You should always use up those nine colors first—they are very distinct from each other—and only then move on to the others. Also this makes setting pen weights easier later down the road, as you need not scroll through dozens of colors to find the ones you used.

The colors from left to right are Red→Yellow→Green→Cyan (not Aqua)→Blue→Magenta (not Maroon), then White and two shades of Gray, 8 and 9. The color stripe below that contains just additional shades of gray, and we do not need them for now. So, finally, pick a color for the A-Walls layer (green is what many architects use for walls), click on it, then press Enter; you are taken back to the Layers dialog box to assign more colors to more layers. Yellow is recommended for A-Doors and red for A-Windows.

TRUE COLOR

There is more to colors, of course. Go back to the Select Color dialog box. Now, click the second tab, called *True Color* (Figure 3.4). Here is where some experience in graphic design may come in handy.

The main color window you are looking at now contains over 16 million colors and should be familiar to you if you use Photoshop, Illustrator, PaintShop®, or any of those types of design programs. You can now move your mouse around, selecting colors based on the HSL method. HSL stands for *H*ue (the actual colors, moving across), *S*aturation (moving up or down), and *L*uminance (brighter or darker, moving the slider up or down next to the main window). You may also use the RGB (red, green, blue) model of color selection by using the drop-down selector at the upper right (Figure 3.5). Whichever color you settle on is shown by both color models. Designers rarely use these colors for 2D work (where you want objects to be bright and visible), but do pick these more subtle colors for realistic-looking 3D models.

71

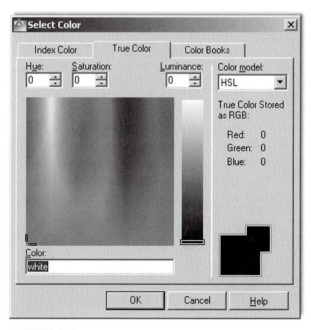

FIGURE 3.4
Select Color, HSL.

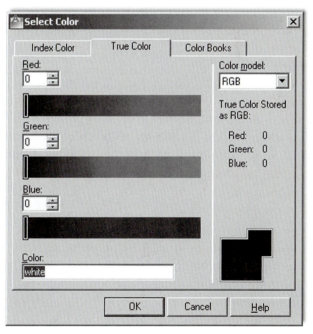

FIGURE 3.5
Select Color, RGB.

COLOR BOOKS

Here is where things get even fancier. Color Books are Pantone colors. If you are an architect or an interior designer, you already know what these are. If you are shaking your head and have never heard of Pantone, do not worry about it, chances are you will not need to use them. Just to try however, scroll through the palettes and select a color. Figure 3.6 is a screen shot of the Color Books tab.

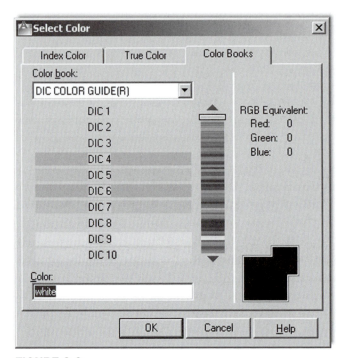

FIGURE 3.6
Select Color, Color Books.

Layer Freeze/Thaw and On/Off

Here, we come to a very important part of the Layers dialog box. Freeze and Thaw are opposites of the same command, one that makes a layer appear or disappear from view. The command freezes or thaws the layer for the entire drawing. Pick a layer, highlight it, follow the band across to the Freeze column and underneath you will find a yellow sun icon. If you click on it, that layer "freezes" and the icon becomes a blue snowflake. Clicking again undoes the action. So what does freeze mean? It makes the layer (or more precisely all the objects on that layer) disappear from view, but remember, they are *not* deleted only invisible. You will use this feature routinely in drawings to control what is displayed. The On/Off feature represented by the lightbulb (yellow to blue and back) is essentially the same as Freeze/Thaw, with some subtle (and minor) details, so we do not go into them. You can use either method; although stay with Freeze/Thaw just to keep things consistent.

Layer Lock/Unlock

This layer command, represented by the padlock, does exactly that; it locks the layer so you cannot edit or erase it. The layer remains visible but untouchable. The unlocked layer's icon is blue and looks "unlocked," while a locked layer's padlock turns yellow and changes to a "locked" position. To let you know which layers are locked, the layers fade somewhat, so you can tell them apart from non-locked geometry. You can control the level of this fading via a slider on the Ribbon. Go to the Home tab, and drop down the arrow in the Layers category.

You see a slider that says "Locked layer fading." Use it to make the locked layers darker or lighter to your liking.

Experiment with everything mentioned thus far to get a feel for layer controls. When done, make sure everything is reset (not locked or frozen) and click on the X in the upper corner of the layer box to close it.

3.2 INTRODUCTION TO LINETYPES

Linetypes are the different lines that come with AutoCAD. As a designer, architect, or engineer, you may need a variety of lines to convey different ideas in your design. Just as in hand drafting, where you create Dashed, Hidden, Phantom, and other line types to show cabinets, demolition work, or hidden geometry in part design, in AutoCAD you load the lines and use them as necessary.

The idea here is to load them all in the beginning and then assign the linetypes to either a layer or sometimes just a few items. You can also load them as needed, but it is recommended to get this out of the way so you have them at your fingertips. The procedure to do so is as follows:

Step 1. Type in `linetype` and press Enter or choose **Format→Linetype…** from the cascading menu. The Linetype Manager (Figure 3.7) appears. If it is full of linetypes, it means they have been loaded earlier. If not, and the Linetype Manager is empty, go on to Step 2.

Step 2. Press **Load…** The Load or Reload Linetypes box comes up (Figure 3.8).

73

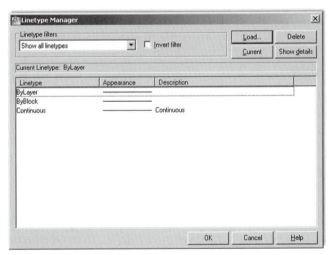

FIGURE 3.7
Linetype Manager.

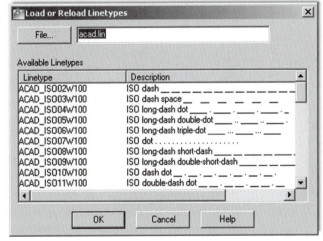

FIGURE 3.8
Load or Reload Linetypes.

Step 3. Position your mouse in the white empty space between the two columns (Linetype and Description) of the Load or Reload Linetypes box and right-click. The Select All/Clear All menu appears.

Step 4. Press Select All and every linetype in the left column turns blue (selected).

Step 5. Press OK (and possibly a Yes to All as well) and the dialog box disappears. All the linetypes are loaded now, so press OK again in the Linetype Manager and you are done.

We can now use the loaded linetypes and assign a linetype to a layer. Go back to the Layers dialog box:

Step 1. Create a new layer called Hidden and assign a linetype to it (the "Hidden" linetype, naturally).

Step 2. Make sure the Hidden layer is highlighted with the blue bar, and move your mouse to the right until it matches up with the header category Linetype. It should say Continuous right now.

Step 3. Click on the word *Continuous* and the Select Linetype (Figure 3.9) dialog box comes up with all your linetypes preloaded. If we did not load them earlier, the box would be practically empty and we would have to do the loading at that point.

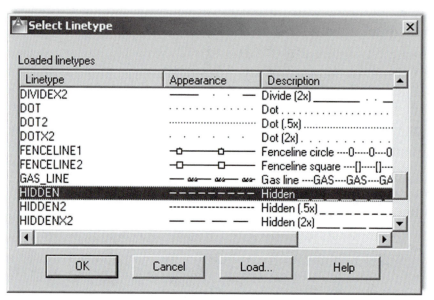

FIGURE 3.9
Select Linetype.

Step 4. So, now all we have to do is scroll up and down the list to find the appropriate linetype (Hidden in this case), select it, and press OK.

Make a note of the important linetypes for future reference: Center, Dashed, Divide, Hidden, and Phantom are some of the key ones. Ignore the ones named ACAD_ISO… for the simple reason that their names are not descriptive enough to recognize at a glance. Also, make a note of why there are three types of the same linetype (such as DASHED, DASHED2, and DASHEDX2). By checking the appearance carefully, you see that the scale is different. Finally, let us get back to the Layers dialog box. If you set your Hidden layer to Current, then everything you draw is a dashed line. Go ahead and try it out.

We conclude with one final important note on linetypes. If you cannot see the linetype, and it appears as a solid line when it should be something else, chances are the scale is too small. Type in ltscale (linetype scale) and enter a larger number. This is mentioned again as a tip in Chapter 4.

3.3 INTRODUCTION TO PROPERTIES

Sometimes, you need to change objects from one layer to another (perhaps because of an error in originally placing it on the incorrect layer or a later decision to change to another layer). Also, you may need to change the color, linetype, or other properties of objects. AutoCAD has a collection of tools to change the properties of objects, with the three most common methods being the Properties palette, the Match Properties tool, and the Layer toolbar. We explore each in order.

Properties Palette

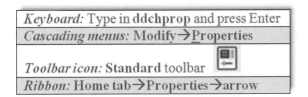

Keyboard: Type in **ddchprop** and press Enter
Cascading menus: Modify→Properties
Toolbar icon: **Standard** toolbar
Ribbon: Home tab→Properties→arrow

There are two other ways to bring up the Properties palette:

- Double-click on any geometry object (line, arc, etc.).
- Press Ctrl + 1.

Once you do any of these, the Properties box shown in Figure 3.10 appears.

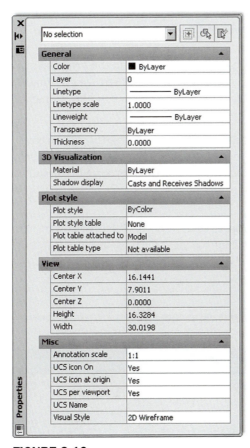

FIGURE 3.10
Properties palette.

Look at the palette in detail from top to bottom. There are a number of categories (General, View, etc.), and a number of features that can be modified in those categories (Color, Layer, Linetype, etc.). Also there are two columns, a gray one on the left and a mix of gray (inactive) and white (active) fields on the right. To use this palette first draw several lines on your screen, each of which is on a different layer. Then, follow these steps:

Step 1. First, make sure you have selected one or more items to change. They become dashed. If you selected two items, only those properties common to both that can be changed are available.

Step 2. In the Properties box, find the category you want changed, color, for example.

Step 3. Click on the right-hand column directly across from that property.

Step 4. An arrow drops down. Click on it and view the choices within that category.

Step 5. Select the different property that you want the new objects to have, such as a new color.

Step 6. Close the Properties box (the X in the upper corner) and press Esc once or twice; the objects have that property. The Properties toolbar is a quick, easy way to change a number of properties but there is another way, shown next.

Match Properties

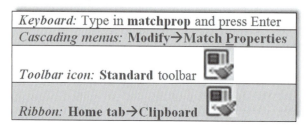

Keyboard: Type in **matchprop** and press Enter

Cascading menus: **Modify→Match Properties**

Toolbar icon: **Standard** toolbar

Ribbon: **Home tab→Clipboard**

This is a really great way to very quickly match up the properties of one object with the properties of another. The procedure is simple. Start the command via the preceding methods, select the first object whose properties you like, then click on another object that you want to have those properties.

Match Properties matches almost anything: layers, colors, linetypes, linetype scales, text sizes, hatch patterns, and many more factors. Finally, we have the Layers toolbar, which has its own method of layer changes and deserves its own separate discussion.

Layers Toolbar

The Layers toolbar is shown in Figure 3.11 with the three layers added. If you want to use the toolbars, then add it to your on-screen collection, and use it throughout the remainder of the chapters.

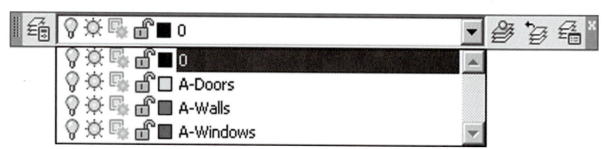

FIGURE 3.11
Layers toolbar.

To change layers using this toolbar, simply select whatever you would like to change to another layer (what you selected becomes dashed), then drop down the Layer menu in the toolbar and pick the new layer on which you want the highlighted geometry to go, and it changes. You may have to press Esc to deselect the layer(s) afterward.

This toolbar is also useful for freezing, turning off, and locking layers. It is also a great visual reference to determine what layer is current, which of course is the layer being shown. The toolbar has some useful icons, although the only one we covered so far was the one all the way to the left that brings up the Layer Properties Manager.

This is it for basic layers and associated topics. You should know the following:

- What layers are and why they are needed
- How to open the Layers dialog box
- How to enter a new layer, delete it, and change its name
- How to make a layer current
- How to assign a color to a layer and all the color options
- How to Freeze/Thaw, turn Off/On, and Lock/Unlock the layers
- How to select and set linetypes
- How to use the Properties dialog box
- How to use Match Properties
- How to use the Layers toolbar

There are a few items we did not yet cover, such as the previously mentioned Filters and Layer States, Plot, Lineweights, and a few other features of the Layers dialog box. We get to them in due time in advanced chapters. For now, it is time to put everything to good use and draw a full apartment floor plan. Review everything thus far and begin the assignment in Section 3.4.

3.4 IN-CLASS DRAWING PROJECT: FLOOR PLAN LAYOUT

This is your first major drawing assignment. You need to put together everything you learned so far and apply it to a realistic drawing situation. Although this floor plan is very simple compared to major architectural or engineering projects created in the industry, it still features components of a professional drawing, just fewer of them, so you do not get overwhelmed right away. Follow all instructions carefully and proceed slowly. Speed comes later, when all the commands become second nature. Our goal is to draw the floor plan shown in Figure 3.12.

This is a basic one-bedroom apartment. The kitchen is on the upper left, bedroom on the upper right, bathroom on the lower right, living room right in the middle, and entrance foyer in the lower left corner. There are two closets, six windows, and six doors (kitchen has no door). The dimensions are only for you to look at and use, not draw in yet, as we do not cover them until Chapter 6. The numbers inside the rectangles are the door opening sizes. All walls are 5" thick.

Outlined next is a complete discussion on how to start drawing the floor plan, intended for those students studying AutoCAD by themselves without an instructor. If you are taking a class, your instructor will assist you but should essentially follow the same format. Once you learn all the steps involved and build confidence, aim to draw it again in less than 60, then 30, and finally less than 10 minutes, starting from a blank screen. This may sound unreasonable now, but that is the required industry speed and you will be able to get to that quite soon with some practice. An outline of steps with additional information (if needed) follows.

Basic File Preparation

1. Open a brand-new AutoCAD file using the `acad.dwt` template. Save it as `YourName_ AptProject1.dwg`. We start from the very beginning for practice purposes.
2. Set your units to Architectural using the units command.
3. Set up layers: A-Walls (green), A-Windows (red), and A-Doors (yellow).

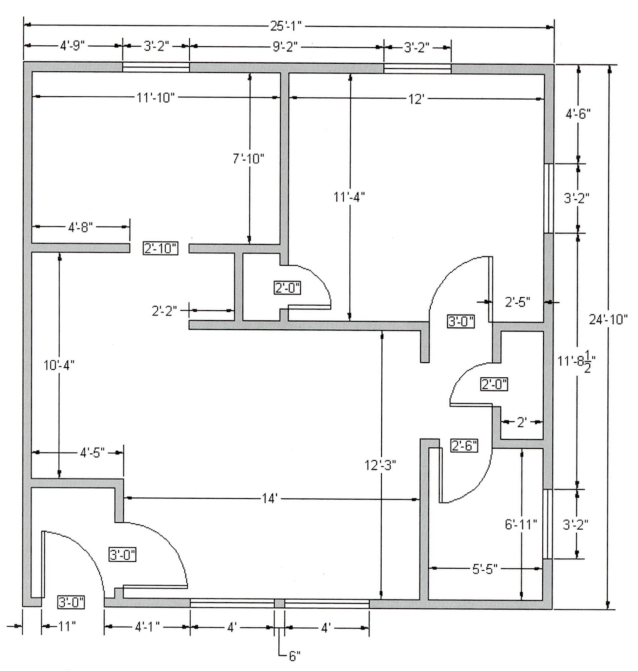

FIGURE 3.12
Floor plan layout.

4. Load all your linetypes (we need them later). If you were to save and close this file right now and use it later for another project, then it would be called a *template*, as first explained in Chapter 2.

Starting the Floor Plan

Now that the basic file is set up, how would you start this design from the very beginning? The exterior walls would be the first step, and there are a few ways to go about it. We could certainly draw line by line, but that is not too efficient. Think back to your set of tools learned thus far. The best approach is to draw a rectangle of given dimensions

(25'-1" × 24'-10") and offset it 5" to the inside. This first rectangle represents the outer walls and the offset one represents the inner walls.

5. Draw a rectangle that is 25'-1" wide and 24'-10" tall. Be careful and use the correct method of manual Relative Distance Entry or Dynamic Input, your choice. You can also just use the rectangle command, indicating the dimensions as allowed by one of the suboptions. If you do it manually, you start a rectangle, press Enter, left-click anywhere in the screen, and then type in @25'1,24'10 exactly as written. If you use DYN, start the rectangle and enter the values on screen, using the Tab key to hop between the first and second values. If you use the Rectangle Dimension option, start the rectangle, press d, and enter its length and width when prompted. Once again, make sure the units are Architectural or you will immediately run into problems.

6. Offset the rectangle 5" to the inside to create the inner walls (offset command).

7. Explode both wall rectangles. This is needed to break apart the rectangles for use as the basis for the interior walls as well as doorways and windows. You should now have a total of eight individual lines, as seen in Figure 3.13.

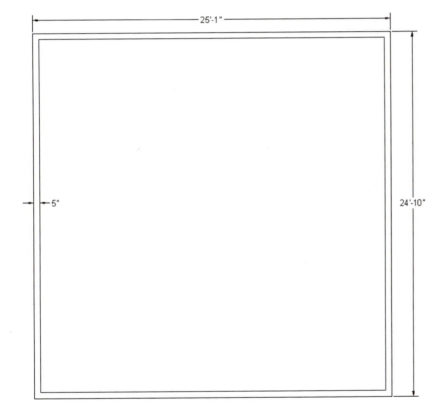

FIGURE 3.13
Floor plan, Step 7.

Drawing the Inner Wall Geometry

As mentioned in Chapter 1, one of the biggest misconceptions beginners have about drafting in AutoCAD is that they will spend lots of time drawing line after line. That expectation is natural; after all, in hand drafting that is exactly what you would do—pick up the pencil and T-square and draw lines connecting them in intricate ways to create a design. Not so in AutoCAD. You still draw lines here and there, but a lot of the time you use the offset command to create new geometry, such as in this case. For example, we use the inner main walls as a basis for the room walls and offset as needed.

8. Offset the inner left vertical wall 11'-10" to the right to create part of the kitchen wall. Then, right away offset 5" for the thickness of the wall. Repeat with the inner top horizontal wall (offset 7'-10") to get the lower part of the kitchen. Offset 5" to the right for thickness. Figure 3.14 illustrates this.

9. Fillet the two sets of walls to create a sharp corner. This is where the fillet command's advantage really shows. Follow it carefully, and select the two inner wall lines first, then the outer.

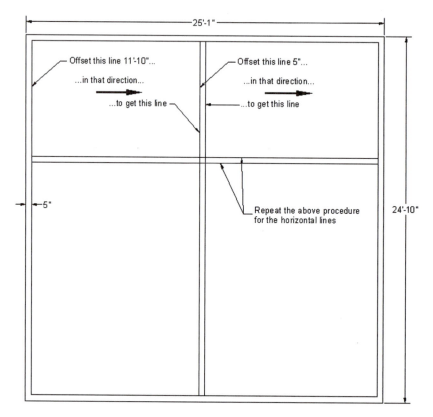

FIGURE 3.14
Floor plan, Step 8.

10. Create a 2'-10" opening for the kitchen door. Once again use the offset command. Offset the inner wall 4'-8" to the right as shown on the plan. Then, offset again 2'-10" to frame the opening. Next trim it out by selecting all the lines using a crossing, press Enter, and click on the lines you do not need. You use this technique to create the windows the same way.

11. Trim out the extra line where each inner wall meets the main wall. Figure 3.15 shows you what you should see so far (be sure to ask your instructor if something is not working out).

If you understood the previous steps, then you should have a good idea of what is needed to draw the rest of the floor plan. It really is mostly similar. You of course have to get creative on occasion, but just stop and think of the basic tools you have available. If it seems overly complicated, you are doing it wrong. After the kitchen, go on to do the bathroom in the lower right corner, then the bedroom and entrance foyer.

Drawing the Doors and Windows

Create the openings for the doors and windows early in the process, but do the actual doors and windows last. Be sure to use the correct layers. Both a door and a window are shown in Figure 3.16.

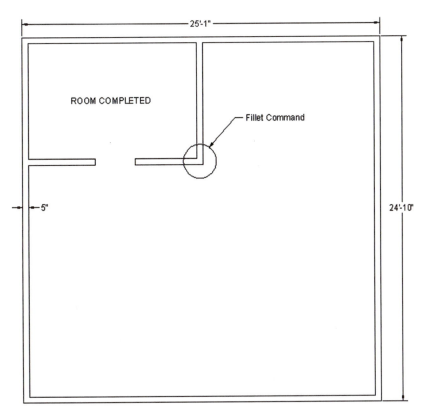

FIGURE 3.15
Floor plan, Step 11.

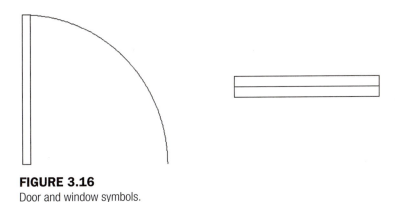

FIGURE 3.16
Door and window symbols.

Here are some additional tips to help you along.

DOORS

Once you have an opening in a wall of a certain size, adding the door involves creating a similarly sized rectangle, oriented in the proper direction. Then, you will need to add the door swing to complete the door. You already know how to create rectangles in one of several ways from Chapter 1. You can even use what you did in Exercise 3 of Chapter 1. However, if you want the doors to look more authentic, you need to give them a thickness. We use a thickness of 2" for all doors in this floor plan. Use DYN or any preferred method to create the door rectangle. A sample result is show in Figure 3.17.

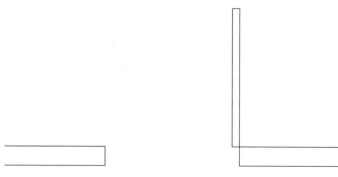

FIGURE 3.17
Creating the door symbol.

To create the door swing, you can carefully position a three-point arc (as in the previously mentioned Chapter 1 exercise) or, as a more accurate alternative, create a circle based at the hinge point and with a radius equal to the door width, as seen in Figure 3.18. Then, just trim the unneeded three quarters of the circle by using the top of the door and the left wall as cutting edges.

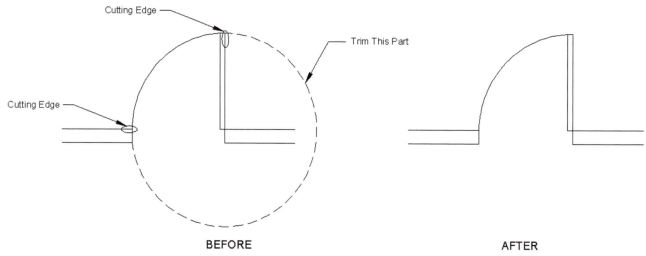

FIGURE 3.18
Creating the door swing.

WINDOWS

To create windows, draw a rectangle of the proper size to fit into the opening and add a line through it. Then, position it into the window opening as seen in the four-step process of Figure 3.19. These are of course very simplified windows; expect to see much more complex contraptions if you do architectural drafting. Finally, it bears repeating, for both the doors and the windows, be sure to be on the correct layer for each.

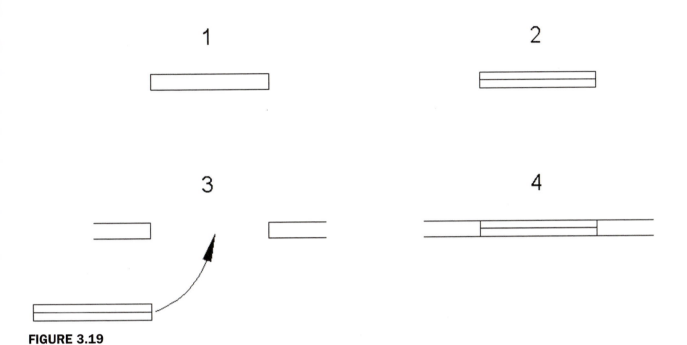

FIGURE 3.19
Creating a simple window.

SUMMARY

You should understand and know how to use the following concepts and commands before moving on to Chapter 4:

- Layers
  - Basic concept and purpose
  - Creating and deleting
  - Current layer
- Layer colors
  - Setting and changing colors
  - Index color
  - True color
  - Color Books
- Layer Freeze/Thaw
- Layer On/Off
- Layer Lock/Unlock
- Linetypes
  - Loading and selecting
- Properties palette
- Match Properties
- Layers toolbar

REVIEW QUESTIONS

Answer the following based on what you learned in Chapter 3.

1. What is a layer?
2. What is the main reason for using layers in AutoCAD?
3. What typed command brings up the Layers dialog box?
4. List two ways to create a new layer.

5. How do you delete a layer?

6. How do you make a layer current?

7. What three Color tabs are available?

8. What does freezing a layer do?

9. What does locking a layer do?

10. What are some of the linetypes mentioned in the text?

11. What are three ways to bring up the Properties toolbar?

12. What two other ways to change properties or layers are discussed?

EXERCISES

1. In a new file, load all linetypes, then open the Layers dialog box, and set up the following layers, colors, and linetypes. Then, lock all E-layers, make A-Walls-New current, and freeze A-Text. (Difficulty level: Easy; Time to completion: 5–7 minutes)
- A-Walls-Demo, Gray, Hidden
- A-Walls-New, Green, Continuous
- A-Windows-New, Red, Continuous
- A-Doors-Demo, Gray, Hidden
- A-Doors-New, Yellow, Continuous
- A-Shelves, Blue, Hidden
- A-Furniture, Magenta, Continuous
- A-Text, Cyan, Continuous
- A-Dims, Cyan, Continuous
- A-Carpeting, Gray, Continuous
- E-Outlets-Duplex, Red, Continuous
- E-Switches, Red, Continuous
- E-Switches-3way, Red, Continuous

2. In a new file, set units as Decimal; load all linetypes; set up layers M-Part, Hidden, and Center; and assign them proper linetypes and colors of your choosing. Then, draw the following mechanical object, including all hidden and centerlines. All text (CL stands for centerline) and dimensions are for drawing purposes only; we learn how to create them in Chapters 4 and 6, respectively, and you need not attempt either for now. (Difficulty level: Easy; Time to completion: 12–15 minutes)

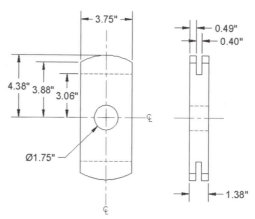

3. In a new file, set units as Decimal; load all linetypes; set up layers M-Part, Hidden, and Center; and assign them proper linetypes and colors of your choosing. Then, draw the following mechanical object, including all hidden and centerlines. All dimensions are for your reference only. (Difficulty level: Easy; Time to completion: 15–20 minutes)

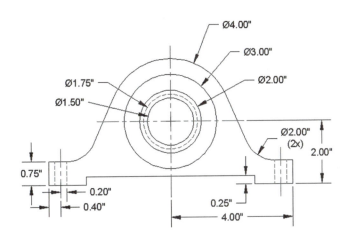

4. In a new file, set units as Decimal; load all linetypes; set up layers M-Part, Hidden, and Center; and assign them proper linetypes and colors of your choosing. Then, draw the following mechanical object, including all hidden and centerlines. All dimensions are for your reference only. (Difficulty level: Intermediate; Time to completion: 20–25 minutes)

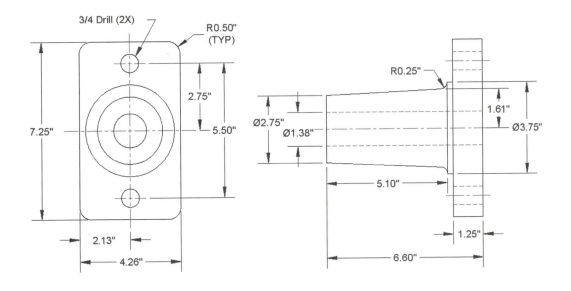

5. Often, in design work, you need to draw standard parts, because no CAD files are available. Such is the case with the following electrical termination blocks, breakers, and thermostat equipment. You are provided minimal dimensions, and your job is not to design the part but simply to document its appearance so an engineer can insert it into a larger modular design. In a new file, leave units as Decimal and set up layer E-Part, assigning it any color you wish. Then draw the electrical hardware, sizing everything relative to the overall provided dimensions and estimating most of the internal geometry. This is an excellent exercise in basic shape drafting. Be sure to use the copy command wisely to avoid significant duplication of drawing effort. (Difficulty level: Intermediate; Time to completion: 45 minutes)

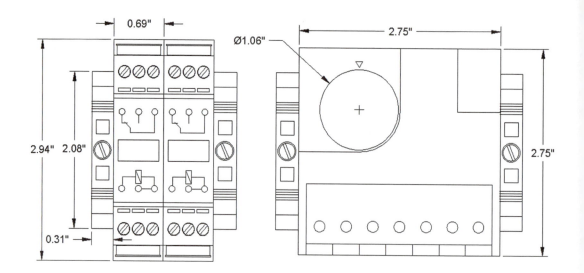

6. In a new file, set up Architectural units, the correct layers (only A-Walls and A-Doors are needed), and draw the following floor plan. Dimensions, text, and the solid hatch pattern are for clarity and construction purposes only. We learn them in upcoming chapters. (Difficulty level: Intermediate; Time to completion: 30–45 minutes)

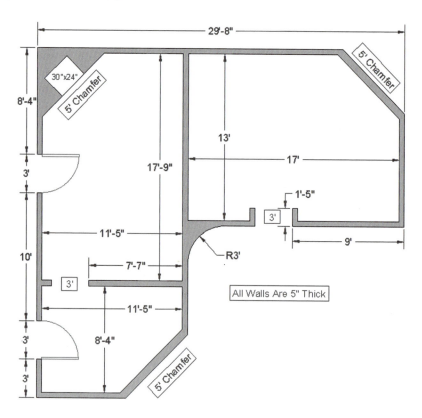

Spotlight On: Architecture

AutoCAD is the undisputed CAD software of choice among most architects, and indeed the software finds its widest application in architecture and related fields. Therefore, it makes sense to feature architecture in our first Spotlight On feature.

Architecture is the art and science of designing and constructing buildings and other physical structures for human shelter or use. It is a far-reaching field that encompasses not only construction and layout of structures but also some civil engineering, landscape design, interior design, and project management.

FIGURE 1

Architects are often described as a cross between engineers and artists. They not only need extensive knowledge of statics, strength of materials, construction methods, and codes but also artistic flair and well-developed aesthetics. Although architects often work with specialists to complete projects, many are well versed in electrical, plumbing, HVAC, landscape, and lighting design. Add to that the necessary ability to manage projects and deal with clients as well as financial and legal matters, and one can see that architecture is a demanding and multifaceted profession.

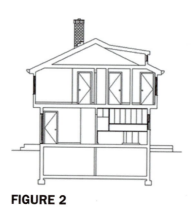

FIGURE 2

It is also not an easy profession in which to get started. In the United States, a student needs to complete a Bachelor of Architecture followed by a Masters of Architecture degree. Depending on individual programs and how they are structured, this can take anywhere from five to seven years. The student then has to gain work experience, often while earning a relatively low salary, before being eligible to sit for the Architect Registration Examination (ARE), to be licensed in the state of his or her residence. While earnings for partners and senior architects can be substantial, it does take quite a while to get established and build a reputation and client base.

Virtually all architects today have switched to computer-aided drafting, and a group of CAD drafters is a common site in most offices. Drafting is time consuming, so most senior architects focus on bringing in new business and creating top level concepts and designs, leaving the details and drafting to junior architects and CAD designers (sometimes the same person). This works out well, as junior architects have an opportunity to learn by doing actual, useful work while "learning the ropes" of the business. Acquiring cutting-edge AutoCAD skills is therefore a great way to get your foot in the door of the profession.

As of early 2011, the architecture profession is still in a decline due to the overall downturn in the housing market of recent years. Because of lack of demand, construction, and, in turn, new design work, architects have seen many of their clients cancel or postpone nonessential work. With an upturn in the economy and the eventual rebound of the housing and real estate market, this should all improve in the coming years.

Despite the downturn, compensation for architects has remained steady or risen. According to U.S. Department of Labor statistics, in 2008, architects in the United States earned between $41,320 and $119,220 a year. The mean annual wage was $76,750 per year, and the mean hourly rate was $36.90. In 2009, the mean salary for a graduate with a Bachelor of Architecture degree was $41,012, and for a graduate with a Master of Architecture degree it was $47,263. Professional architects with 20 or more years of experience earn an average income of $100,723. Twenty percent of the experienced architects earn up to $142,200, according to industry surveys.

So how do architects use AutoCAD, and what can you expect? A typical medium-sized project, such as a residential home, features a set of plans that may include

- A-1 Site Plan: The property surrounding the structure.
- A-2 Demo Plan: What needs to be demolished to make room for the new design, if applicable.
- A-3 Construction Plan: The actual design, which may have multiple floors.
- A-4 RCP: The reflected ceiling plan, which may include lighting.
- A-5 Electrical Plan: The outlets, switches, panels, and lighting, if not on the RCP or a separate plan.
- A-6 Kitchen Elevations: Shows cabinets and appliances.
- A-7 Bathroom Elevations: Shows bath, shower, sink, and other features.
- A-8 Window and Door Schedule: Lists all doors and windows, as well as sizing, quantity, and the like.
- A-9 Cellar: Shows basement or cellar plan, if needed.
- A-10 Roof: Shows construction of roof beams, rafters, braces, joists, and other elements.
- A-11 Wall Sections: Shows the various walls used, and their internal structure.
- GN-1 General Notes: Describes the overall job and any special instructions to the contractor.
- GN-2 Plumbing Riser Diagram: Describes flow of pipes and plumbing features.
- S-1 Framing Plan: Describes the construction of the house frame.
- S-2 Structural Details: Additional details of design; may run multiple pages.

Extensive use of AIA layering, or an in-house system based on it, is employed, as seen in Figure 3. Figures 4 and 5 show some architectural details.

You should also expect to see extensive use of advanced Level 1 and Level 2 concepts, such as

- Xrefs (Chapter 17): The main plan is externally referenced to the electrical and RCP plans.
- Paper Space (Chapter 10): Each of the plans is a separate layout tab.
- Attributes (Chapter 18): Found in the title block, door and window schedules, and other elements.

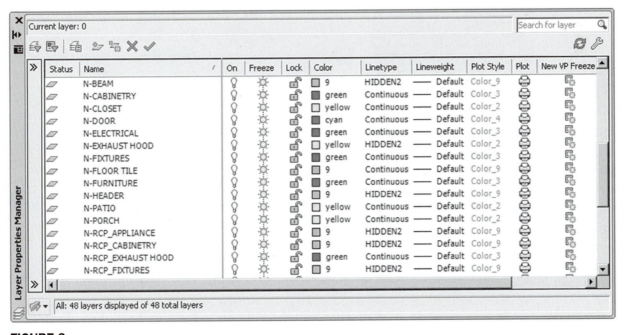

FIGURE 3
Typical AIA layers.

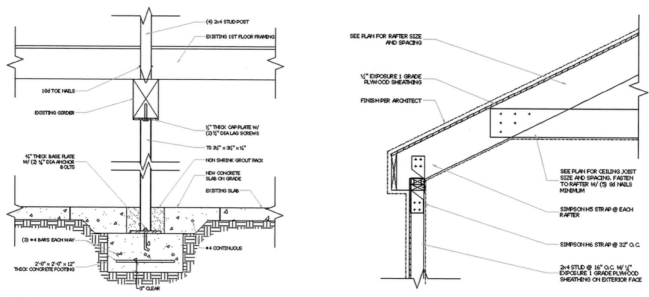

FIGURE 4
Architectural details. (*Image courtesy of RM Architect, PC, www.rma-nyc.com*)

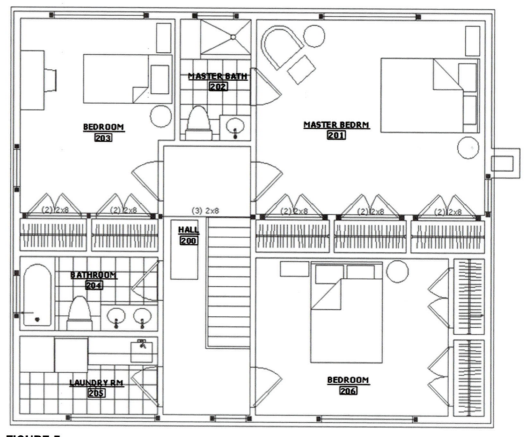

FIGURE 5
Residential architectural floor plan. (*Image courtesy of RM Architect, PC, www.rma-nyc.com*)

Architectural drafting is a demanding and skill-intensive endeavor. Many jobs are started with a survey of existing structures, if any exist. If the project calls for a brand-new building, not a renovation, then only a site plan is needed and the design may be created right away. Typically, a building core is laid out (which then becomes the xref), and the architect then works on designing the internal details of the building. When the overall design is set, other systems are added, such as electrical, plumbing, HVAC, security, fire protection, mechanical, and much more. Corporate and commercial jobs are typically more involved than residential ones, and some of the previously mentioned subsystems are applicable mostly to those. Finally, extensive construction notes, details, and schedules need to be developed. The final document set may feature 20–30 sheets or more.

Several factors make architectural drafting challenging:

- A complex building design may use virtually all of AutoCAD's tools and concepts; you need to be well versed in all of them. This means mastering every topic in Levels 1 and 2 of this book.
- There is a need for extreme accuracy, as the original design plans of the building structure are the anchor for the rest of the design. Mistakes made early on come back to haunt you.
- There is rarely a luxury of excess time. Architects bid on a job and must work within a set budget. A slow draftsperson who is not proficient (or efficient) takes too long to create the drawings and diminishes profits. Working quickly and accurately is a must. Rarely do you learn the basics and are you allowed to mature "on the job" anymore; it is just too expensive for a small architecture office.

In spite of these challenges, or maybe because of them, architectural drafting is enjoyable and rewarding, especially if you are a new graduate and picking up knowledge in your profession along the way. Even if you just do non-design drafting, the pay is good (salaries vary from $35,000 to $55,000 and up), and you acquire some frightening AutoCAD skills very quickly in this fast-paced environment. Most of today's AutoCAD experts (including the author) can trace their roots to using AutoCAD in architecture; it is the cutting edge of computer-aided drafting.

91

FIGURE 6

CHAPTER 4

Text, Mtext, Editing, and Style

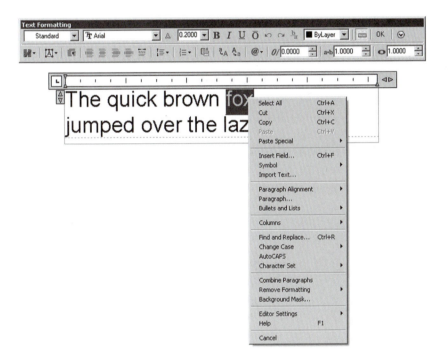

LEARNING OBJECTIVES

Text allows your design to communicate beyond what just a line drawing can. In this chapter, we introduce Text, Mtext, Style, and Editing and discuss the following topics:

- Text
- Properties and applications of text
- Editing all types of text
- Mtext
- Properties and applications of mtext
- Mtext formatting
- Mtext symbols
- Style
- Spell check
- *Nearest* OSNAP

At the end of this chapter you will be able to annotate your floor plan as well as add additional features, such as furniture and stairs.

Estimated time for completion of chapter: 2 hours (lesson and project).

4.1 INTRODUCTION TO TEXT AND MTEXT

Adding text to your AutoCAD drawing is the next logical step once you learn how to create and edit basic designs. After all, the purpose of most drawings is to describe how to build something or show what a design looks like. Text goes a long way in assisting in this and, along with dimensions, is usually the next item to be added to a new design.

Text in AutoCAD comes in two versions: regular text (single line text) and mtext (multiline text). There is some overlap and similarities between the two, and we look at how to create and edit both types. We then look at how to choose and set fonts and conclude the chapter by adding text to the previously designed floor plan.

4.2 TEXT

This is your basic text creation command, and it creates a field anywhere you click on the screen, into which you can type whatever text you need. It does have a carriage return that goes to another line upon pressing Enter, so you need to press Enter twice to get out of the field. While an unlimited number of lines of text can be typed, typically regular text is used for only a single line, as multiple lines of text are not joined together in paragraph form and cannot be formatted to any significant extent. Let us summarize:

- Text is used mostly when only one or a few lines are needed.
- The text cannot be formatted nor any effects added, beyond underlining and a few other minor effects.
- Multiple lines of text are not in paragraph form, rather they remain individual lines.

Open a new file, bring up the Text toolbar (Figure 4.1), and let us create a few sample lines.

FIGURE 4.1
Text toolbar.

Keyboard: Type in **text** and press Enter
Cascading menus: **Draw→Text→Single Line Text**
Toolbar icon: **Text** toolbar 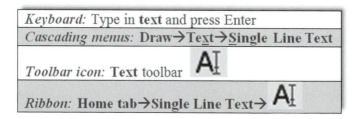
Ribbon: **Home tab→Single Line Text→**

Step 1. Begin the text command via any of the previous methods.
- ○ AutoCAD says:
  ```
  Current text style: "Standard" Text height: 0.2000
  Annotative: No
  Specify start point of text or [Justify/Style]:
  ```

Step 2. Left-click anywhere on the screen.
- ○ AutoCAD says: `Specify height <0.2000>:`

Step 3. Enter a new height if desired, say 1.0, and press Enter.
- ○ AutoCAD says: `Specify rotation angle of text <0>:`. You can rotate the text, but there really is no need to, so just type in 0 for now, and press Enter.

Step 4. A text field opens up with a blinking cursor. Go ahead and type something, pressing Enter once to go to the next line or twice to finish. Figure 4.2 shows what the basic text looks like. If yours is a different size, you may have to zoom in or out to fit it nicely on the screen.

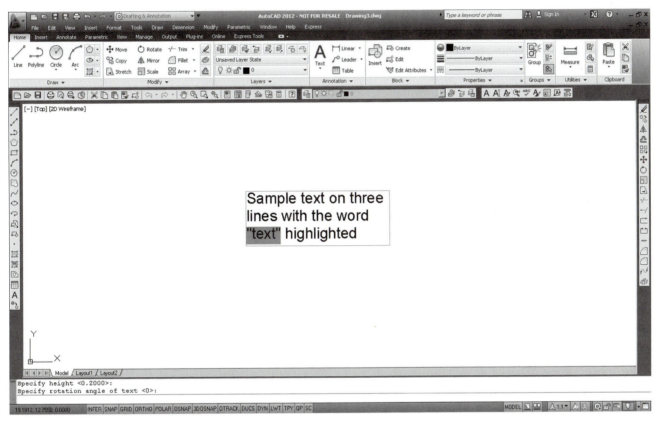

FIGURE 4.2
Basic text.

The text is still rather simple; we have not yet addressed font, sizing, or any effects. Notice that the text is on three separate lines. You can move, copy, or erase any of the two lines without affecting the other. So how would you edit this text?

Editing Text

Editing, or changing, what the text says is very simple and follows the same procedure for both text and mtext, so we go over it right away for both cases. The easiest way to edit the text is to *double-click* on it, which opens the text field. This is such an overriding and best method, preferred by most students, that in this particular case you do not see a command matrix detailing the other methods (a toolbar icon does exist for this). However it is worth noting an older typed equivalent of the double-click called `ddedit`. If you type that in, press Enter, and click the text, you can edit it. This method is rarely used, however, remember `ddedit` for Chapter 6. It is the only way to change dimension text values, so you *will* need it eventually, just not yet. Whichever method you use, when you are done editing, just press Enter again. Do not press Esc, as that cancels whatever editing you just did.

Here is a trick with regular text that you will not hear about much anymore. If you want to underline it, type in %%u just before the text string you want underlined. This old code dates back to DOS days, and it will automatically and immediately add an underline to your basic text. You can also experiment with %%c, %%p, and %%d. See what those three do. None of this is needed with mtext, as discussed next.

4.3 MTEXT

Mtext is short for *multiline* text. In other words, this is the command you use when you anticipate typing in a paragraph as opposed to one or two lines, and more important, you require some advanced formatting and effects. As of AutoCAD 2010, it even has a spell checker built in. Many designers use mtext for all their text needs, just in case formatting needs to be applied to even just one word. The only downsides are that it takes a few seconds longer to set up and the features make it a more complex command. The idea here is to define an area where your text will go. Once you do that, the new paragraph fits into that area.

Let us summarize the mtext features and try the command:

- Used primarily for writing paragraphs.
- Has extensive formatting and editing features.
- Has an extensive symbol library and spell check.
- Can accept extensive text importing.

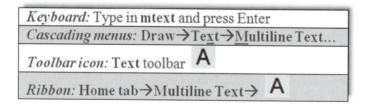

Keyboard: Type in **mtext** and press Enter
Cascading menus: Draw→Text→Multiline Text...
Toolbar icon: Text toolbar **A**
Ribbon: Home tab→Multiline Text→ **A**

Step 1. Begin the mtext command via any of the previous methods.
  ○ AutoCAD says:
    ```
    Current text style: "Standard" Text height: 0.2000
    Annotative: No
    Specify first corner
    ```
Step 2. Left-click anywhere on the screen, noticing the abc next to the crosshairs. A rectangle with an arrow appears, as shown in Figure 4.3. Continue to move your mouse down and across to the right, making the rectangle bigger. This is your text field; make it as large or small as you need it to be to hold all the text. As you do this,
  ○ AutoCAD says: `Specify opposite corner or [Height/Justify/Line spacing/Rotation/Style/Width/Columns]:`
Step 3. Click again when you have defined the field.

Here AutoCAD throws you a bit of a twist. If you are *not* using the Ribbon, you will see the text field with a toolset just above it (but not attached to it), as seen in Figure 4.4, where you can type in your text, pressing Enter to go to the next line. To finish up, press OK in the upper right of the toolset or just click anywhere outside the field. We go over some of the tools in a moment, but what if you have the Ribbon up?

If you are using the Ribbon, the formatting duties are transferred completely to the Text Editor tab, and all you see below it is the text field, no toolset, as seen in Figure 4.5 (although

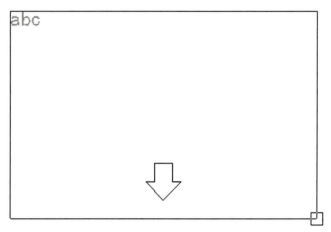

FIGURE 4.3
Mtext field.

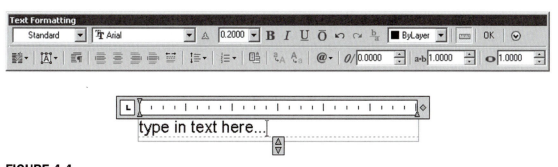

FIGURE 4.4
Text Formatting (no Ribbon).

97

you can bring it back, via the Text Editor's Options category). The functionality is essentially the same, only the presentation differs. We cover mtext's extensive formatting tools in the next section.

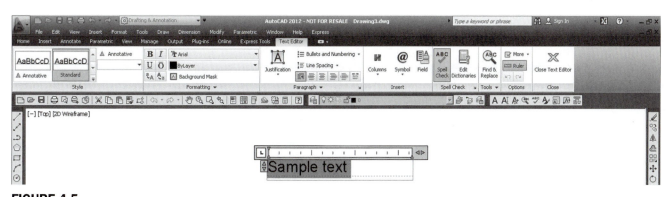

FIGURE 4.5
Text Formatting via the Ribbon's Text Editor.

Formatting Mtext

First of all, notice something: If you click on mtext, it remains a paragraph, a big difference from a few lines of regular text. To edit mtext, as already mentioned, you do the same thing as for regular text, that is, double-click on it, use the icon, or type in `ddedit`. Now, on to formatting. Mtext has a lot of additional features, and once you input the text, you can modify it significantly. There is some intent here to mimic MS Word's text editing abilities, and although AutoCAD cannot really come close to a dedicated word processing program, the available tools are still quite extensive. We take a look at the non-Ribbon toolset first and cover the tools again, this time using the Ribbon's Text Editor (in much less detail, as the tools essentially are the same).

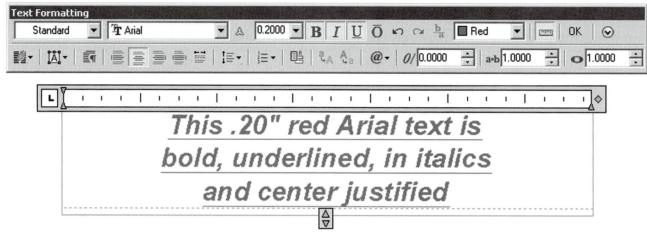

FIGURE 4.6
Text Formatting groups.

Looking at the Text Formatting toolbar from left to right, as shown in Figure 4.6, we have the following groups:

Group 1. This is where you set *name*, *font*, and *size*, although this is generally done using *style*, to be discussed later. You can also make the text annotative, to be covered in Level 2.

Group 2. These are the standard *Bold*, *Italic*, *Underline*, *Overline*, and *Undo/Redo* buttons. This is also where you set *color*, although that is usually done by layer. The Ruler icon turns off the ruler grid, the OK button closes the field, and the down arrow brings up additional *options*, to be covered soon.

Group 3. Here you find a tool for making *columns* and some of the associated settings, *paragraph justification* tools, as well as tools to change *line spacing*, *numbering*, and *fields*.

Group 4. Here you find tools to change from *upper to lower case* (and vice versa), *symbols* (to be discussed soon), *oblique*, *width factor*, and *tracking tools*.

You may be familiar with most of these if you ever typed a document in MS Word. Just as in Word, you need to highlight the text to which you want the changes to apply. Experiment with the buttons to see what they all do, and try to duplicate what you see in Figure 4.7. Notice also how Arial is the default font in AutoCAD.

FIGURE 4.7
Mtext sample.

There is more to the mtext command than just the main toolbar detailed in Figure 4.6, although it does contain most of the often used commands. Two additional menus are accessed by either right-clicking while inside the mtext box or pressing the @ symbol icon in Group 4. We cover this @ menu first; it introduces a vast array of available symbols.

Figure 4.8 shows what this menu looks like. Examine it closely; it has interesting and useful options. In architecture and engineering, one typically encounters many industry- or trade-specific symbols. AutoCAD provides a significant database of them for your use. Browse through and try a few by clicking and making them appear in your text field.

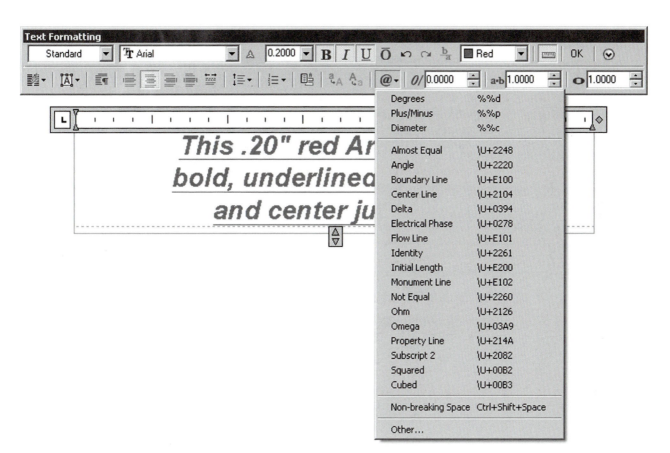

FIGURE 4.8
Mtext symbols.

In case none of those symbols is what you are looking for, there are more. At the bottom of the menu, click on Other…. A *character map* appears, as shown in Figure 4.9.

It is similar to the character map used by MS Word and functions the same way. This is basically a database of available letters (in English and several other languages) as well as symbols and characters. The choices change somewhat from font to font. To use this tool, simply click on a symbol you want (it will temporarily increase in size so you can examine it closer), and press the Select button. Then, minimize the Character Map box (or close it if you do not need it anymore), and right-click to paste back into your mtext field.

Character maps are generally for symbols not found in the main list of the drop-down table (Figure 4.8). If you see the symbol you want there first, use it. To introduce the final menu, right-click in the mtext field and you see the menu in Figure 4.10.

99

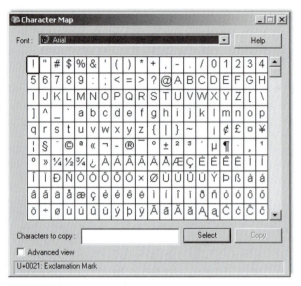

FIGURE 4.9
Character map.

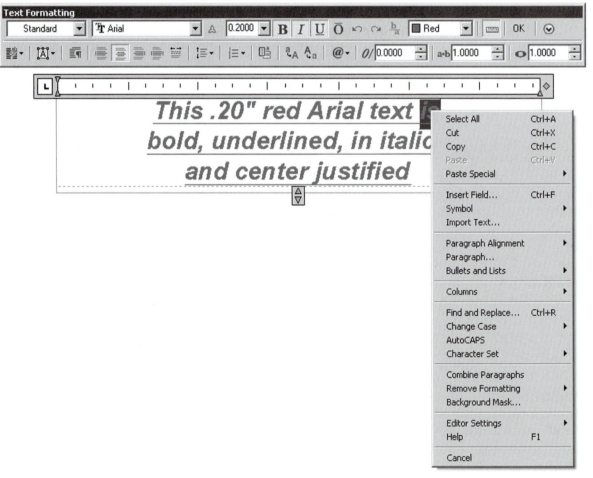

FIGURE 4.10
Mtext right-click menu.

Some of these menu choices duplicate what is already seen in the main mtext menu bar, and you can access all the symbols through here as well. Aside from some Cut and Paste choices, this menu just presents an alternate way of making option selections.

As mentioned before, Ribbon users are presented with essentially the same formatting tools but in a rearranged manner. Figure 4.11 is the Ribbon's Text Editor, again with the various drop-down menus "thumbtacked" (by clicking the tack symbol) on the drawing canvas.

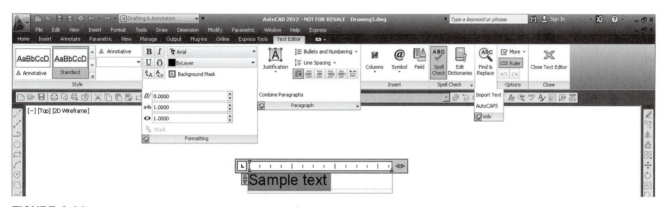

FIGURE 4.11
Ribbon Text Editor, thumbtacked.

The Ribbon's Text Editor categories are detailed next.

- **Style:** Annotation and size can be changed here.
- **Formatting:** Bold, italics, underline, overline, font, color, background mask (color), oblique, spacing, and width factor can be changed here.
- **Paragraph:** Justification, line spacing, bullets, and numbering can be changed here.
- **Insert:** Columns, symbols, and fields can be changed here.
- **Spell Check:** You can run the spell checker and set dictionaries and settings here.
- **Tools:** Find and Replace, Import Text, and AutoCAPS are found here.
- **Options:** Here you can make the editor toolbar come back (AutoCAD never really gets rid of anything), add the ruler grid, and use Undo/Redo. You can also change the background of the Editor to opaque.
- **Close:** Closes the editor, similar to the OK button.

4.4 STYLE

The idea behind the style command is very straightforward. Pick a font, give it a name and a size, and use it throughout your drawing. Drawings typically use only one font throughout the main design, and perhaps another fancier one in the title block area for logos and other designations, so it makes sense to set one style and stick to it. You have already perhaps changed the font while learning the mtext command, but it was only for *that* instance. You need to make sure that font is set globally, meaning for the entire drawing.

Prior to AutoCAD 2009, changing the font right away was a necessity, as the default font (Simplex, 0.2) was unattractive and rarely used in practice. With AutoCAD 2012 and the previous three releases, the default font is Arial (though the default size is still 0.2), so you may want to stay with this popular font and not change it. If you need to, however, here is the procedure.

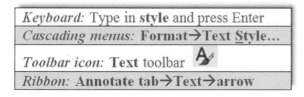

Keyboard: Type in **style** and press Enter

Cascading menus: **Format→Text Style...**

Toolbar icon: **Text** toolbar

Ribbon: **Annotate tab→Text→arrow**

101

Step 1. Begin the style command via any of the preceding methods. Whichever method you use, the Text Style dialog box of Figure 4.12 appears. Taking a look at this dialog box, let us say Arial is not what we want here; we would like to set a RomanS font that is 6" high instead.

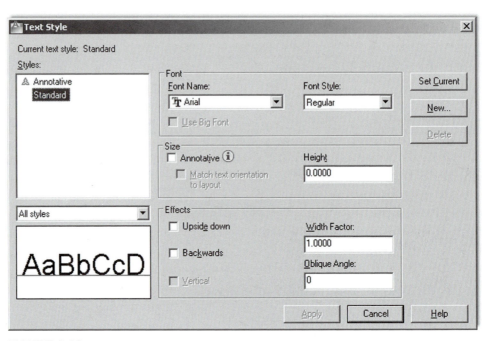

FIGURE 4.12
Text Style dialog box.

Step 2. Press the New… button and type in the name and size of the font: RomanS_6.
Step 3. Pick `romans.shx` from the Font Name: drop-down menu.
Step 4. Highlight the `0.0000` in the Height field and type in `6` (no need for the inch symbol). Press Apply and Close.

All text that you type from now on will be RomanS, 6" font. To create another font just repeat these steps. To size your font up and down, you can also create new font styles, but in this case it may be more practical to just use the scale command.

The additional options in the style dialog box are not used that often but review them just in case. A list follows. All effects can be previewed in real time in the Preview box in the lower right.

- **Upside down:** Flips the text upside down.
- **Backwards:** Flips the text backwards.
- **Vertical:** Stacks the text vertically.
- **Width Factor:** Widens the text if (>1), narrows text if (<1).
- **Oblique Angle:** Leans the text to the right if positive (+) and to the left if negative (−).
- **Annotative:** An advanced topic for Level 2.

4.5 SPELL CHECK

This section briefly describes AutoCAD's in-place, stand-alone spell checking abilities. Yes, AutoCAD does have a spell checker, and why not? For the price the software costs, it better. It is very easy to use, so let us try it out on the misspelled phrase shown in Figure 4.13.

The first thing you notice is that the misspelled word (jumped) is underlined by a dotted red line. If you right-click on the word while in editing mode, you get the in-place spell check with suggestions (Figure 4.14). Simply select the word you want to correct to and AutoCAD obliges.

Now what if you already have a significant amount of text (perhaps from an import) and need to spell check it? Then you can use the stand-alone utility, as described next.

The quick brown fox jmped over the lazy dog

FIGURE 4.13
Misspelled word.

FIGURE 4.14
In-place Spell Checker.

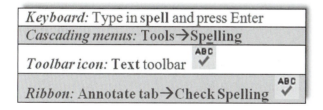

Keyboard: Type in spell and press Enter
Cascading menus: Tools→Spelling
Toolbar icon: Text toolbar ABC ✓
Ribbon: Annotate tab→Check Spelling ABC ✓

Using any of these methods, the Check Spelling dialog box appears, as seen in Figure 4.15.

Using the drop-down menu of the Check Spelling dialog box, you can select specific paragraphs or lines of text, or just press Start to initiate spell checker for all the text in the drawing. In any case, Check Spelling appears as seen in Figure 4.16.

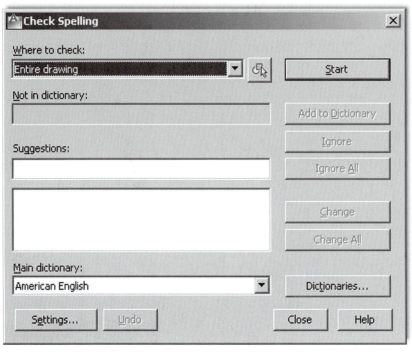

FIGURE 4.15
Check Spelling.

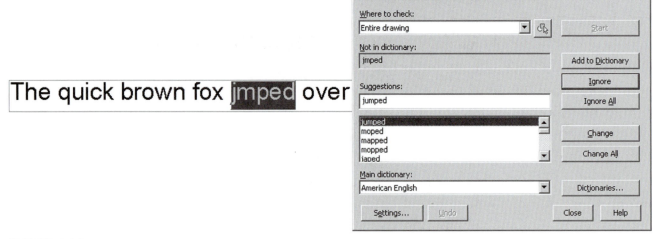

FIGURE 4.16
Check Spelling initiated.

Much like with the MS Word spell checker, you can run through some options, such as ignoring the word, changing it, or adding it to the dictionary so it is not flagged again (good for names and acronyms). There is even a variety of languages available. When done, spell checker tells you, and you can close it.

4.6 IN-CLASS DRAWING PROJECT: ADDING TEXT AND FURNITURE TO FLOOR PLAN LAYOUT

Let us now apply what you learned to our floor plan. Shown in Figure 4.17 is the same floor plan you worked on in Chapter 3, with the addition of text in the rooms, closet shelving (the dashed lines), new furniture, and appliances.

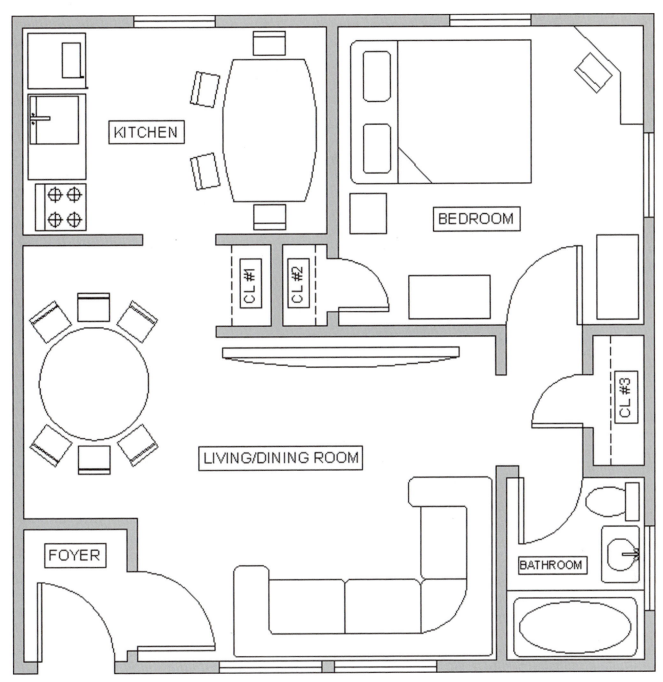

FIGURE 4.17
Floor plan layout.

Here are two general tips before you get started:

- Make your own layers and colors for each set of data, being sure the names are logical. For example, call kitchen appliances something along the lines of *A-Appliances* and so on.
- None of the furniture is drawn to scale, so approximate all the sizes. However, draft everything carefully, connecting all lines with OSNAP and using Ortho for straight lines. Most shapes are based on rectangles, with some fillets, arcs, and circles. All were done in the simplest manner possible, and you should be able to create them with basic techniques learned in previous chapters. Some hints are listed next.

Some additional specific drawing tips:

- All kitchen furniture on the left side of the kitchen is formed using basic rectangles, drawn to no particular size. At the top left there is a refrigerator with a microwave on top. Next to it is a countertop and a sink. Next to that is the range top with four burners. Figure 4.18 is a suggested procedure for creating the range top.

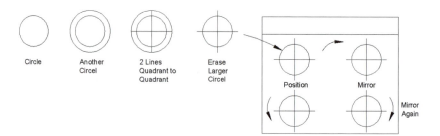

FIGURE 4.18
Constructing the range top.

- To create the kitchen table you could use a rectangle with a small straight guideline sticking out from the midpoint of the left side. That could be the anchor for the second point of the arc. Then a mirror followed by an explode and erase finishes it off, as seen in Figure 4.19.

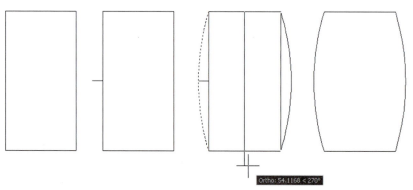

FIGURE 4.19
Constructing the table.

- Draw rectangles around each of the room text fields; we need this later for hatch.
- The bathtub and toilet are done using the basic ellipse command, introduced in Chapter 2.
- The corner desk in the bedroom, the pulled back bed cover, and all closet shelves are constructed using lines and a new OSNAP point, the second (of three) that was not initially introduced in Chapter 1; a description follows.

Nearest OSNAP

You may occasionally need a new OSNAP point called *Nearest*. It creates shapes (usually new lines) along random spots of other geometry (usually lines). In a sense *Nearest* is the exact opposite of the precise *end*, *mid*, and other points you learned. *Nearest* simply shadows a line or any object and allows you to begin a new object anywhere along that perimeter. We do not generally leave *Nearest* running along with the other OSNAP points, as it makes using them more difficult; rather, just type it in (NEA) and press Enter, or use the toolbar as needed.

An hourglass figure appears; just click on wherever you want to start a line. Figure 4.20 is a screen shot of *Nearest* in action.

Finally we have two more tips. Tip 6 has to do with `ltscale`, something briefly mentioned in Chapter 3. Tip 7 shows you how to get rid of the UCS icon, something many students ask about.

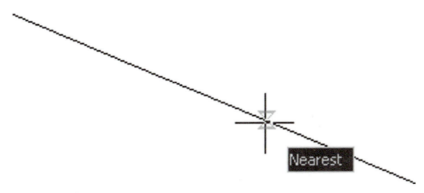

FIGURE 4.20
Nearest OSNAP.

SUMMARY

You should understand and know how to use the following concepts and commands before moving on to Chapter 5:

- Text
  - Creating basic text
- Mtext
  - Creating basic mtext
- Mtext formatting options
  - Bold
  - Underline
  - Italics
  - Justification
  - Symbols
  - Other
- Editing text
  - Double-click
  - ddedit
- Style
  - Font name
  - Font type
  - Font size
  - Width factor
  - Oblique angle
- Spell Checker
- *Nearest* OSNAP

TIP 6: If you draw your hidden lines for the closet shelving and they look solid not dashed, then the problem may be incorrect linetype scaling, a common error to watch out for. Type in `ltscale`, press Enter, and input a new value (a larger one, since your floor plan is bigger than the default original space of 8.5" × 11"). Do this a few times until you get it to look just right.

107

TIP 7: Want to get rid of that pesky Universal Coordinate System icon hanging out in the drawing area? It is somewhere near the lower left-hand corner of your screen, looks like a big *L*, and has a *Y* and an *X* on its tips. To make it go away, type `UCSICON`, press Enter, then type in `off`. Of course, if it does not bother you and you actually like it there, then ignore this whole tip.

TIPS

TIP 6: If you draw your hidden lines for the closet shelving and they look solid not dashed, then the problem may be incorrect linetype scaling, a common error to watch out for. Type in `ltscale`, press Enter, and input a new value (a larger one, since your floor plan is bigger than the default original space of 8.5" × 11"). Do this a few times until you get it to look just right.

Tip 7: Want to get rid of that pesky Universal Coordinate System icon hanging out in the drawing area? It is somewhere near the lower left-hand corner of your screen, looks like a big *L*, and has a *Y* and an *X* on its tips. To make it go away, type `UCSICON`, press Enter, then type in `off`. Of course, if it does not bother you and you actually like it there, then ignore this whole tip.

REVIEW QUESTIONS

Answer the following based on what you learned in Chapter 4:

1. What are the two types of text in AutoCAD? State some properties of each, and describe when you would likely use each type.
2. What is the easiest method to edit text in AutoCAD? What was another, typed method?
3. What feature allows you to add extensive symbols to your mtext?
4. What are the three essential features to input when using style?
5. What is the width factor?
6. What is an oblique angle?
7. How do you initiate spell check?
8. What new OSNAP is introduced in this chapter?
9. How do you get rid of the UCS icon?
10. What can you do if you cannot see the dashes in a Hidden linetype?

EXERCISES

1. Creating tables of data, such as a Bill of Materials (BOM), is an important part of many designs. AutoCAD has a table creation tool (similar to MS Word), which we cover in Level 2, but for this exercise, you need to create one from scratch with basic linework and the offset command, populating it with text data, as shown. Open a blank file and create a new style called *RomanS.2*; choose font: RomanS, height: .2, and width factor: 0.95. Then create the table shown, using the suggested sizing. Finally, fill it in with the data using your choice of text or mtext commands. Position and size the text/mtext as needed. (Difficulty level: Easy; Time to completion: 20 minutes)

9"

		LIST OF MAIN PARTS		
ITEM	QTY	DESCRIPTION	SIZE DWG	DWG. #
1	1	SKID ASSEMBLY	A1	83006
2	1	STRAINER ASSEMBLY	A1	83008
3	1	GENERATOR ASSEMBLY	A1	59470-001
4	1	GENERATOR ASSEMBLY	D1	59470-002
5	2	POWER SUPPLY (OIL IMMERSED)	B1	58717-001
6	1	POWER SUPPLY (OIL IMMERSED)	B1	58717-002
7	1	LOCAL CONTROL PANEL ASSEMBLY	D1	83017-001
8	2	LOCAL CONTROL PANEL ASSEMBLY	A1	88017-002
9	1	INSTRUMENT AIR ASSEMBLY	A1	83022
10	1	AIR PUMP	A1	86584

5" (height) · .5" (title row) · .4" (row height)

1" · 1" · 5" · 1" · 2"

2. As seen with Exercise 5 of Chapter 3 (electrical pieces), you may be asked to draft something based on a picture and use arbitrary, not strict, design dimensions. You need to be fast and efficient while maintaining strict accuracy. A good grasp of the basics is essential, and you must know exactly how you will draft something in your mind, just moments before your fingers do it via mouse and keyboard. Such is the nature of this next exercise—to build up this accuracy and speed. Draw the following set of P&ID (piping and instrumentation diagram) symbols. They are not to scale and are approximated, but you have all of the necessary tools at your disposal. Before drafting each shape, recite to yourself the approach you will use, and then execute it. Finally, add the text next to the respective symbols. Use the Arial font of appropriate size. (Difficulty level: Intermediate; Time to completion: 35–45 minutes)

SYMBOL LEGEND

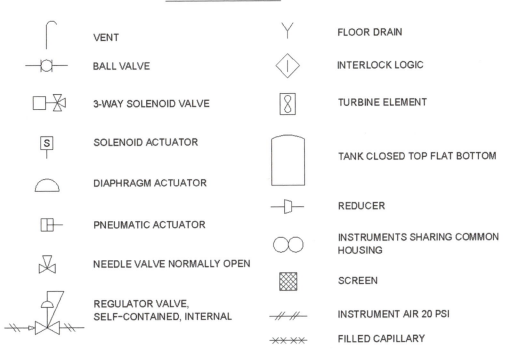

109

3. This exercise shows a simple P&ID schematic using some additional industry symbols. Draft it based on the reference dimensions given, with all internal pieces and text sizing estimated in relation to those. Use accuracy, connecting all lines with OSNAPs. (Difficulty level: Easy/Intermediate; Time to completion: 20–30 minutes)

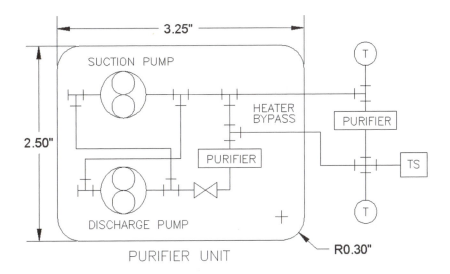

4. While we are on the subject of P&ID, let us create a simplified design of a pipe. Open a new file and set up Architectural units, creating proper layers, colors, and linetypes (Hidden and Center), as shown. Then, draw the following pipe design. Be careful in how you approach the hidden lines, noting that they are on the same plane as the solid lines. This is commonly seen on mechanical drawings. (Difficulty level: Intermediate; Time to completion: 15–20 minutes)

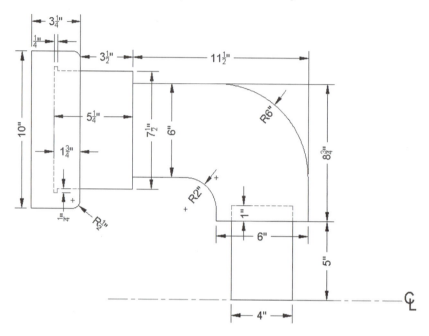

5. Valves are an important part of P&IDs. In a new file, set up the units as Architectural and set up layers M-Part, Hidden, and Center, assigning those proper linetypes and colors of your choosing. Then, draw the valve shown. Be sure to use the mirror command wisely to avoid duplication of drawing effort. The hex nuts are not specifically detailed out; you have to improvise. (Difficulty level: Intermediate/Advanced; Time to completion: 45–60 minutes)

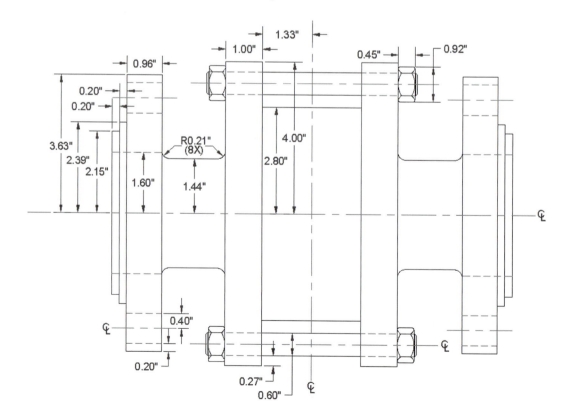

6. The last exercise in this chapter involves a floor plan. Similar to what you have already done, open a new file, set up A-Walls and A-Doors layers and colors, and leave your units as Architectural. Then, complete the following basic floor plan. All walls are 6" thick. The doors are all 3' wide and have no thickness; you may give them thickness if you like. (Difficulty level: Intermediate; Time to completion: 45–60 minutes)

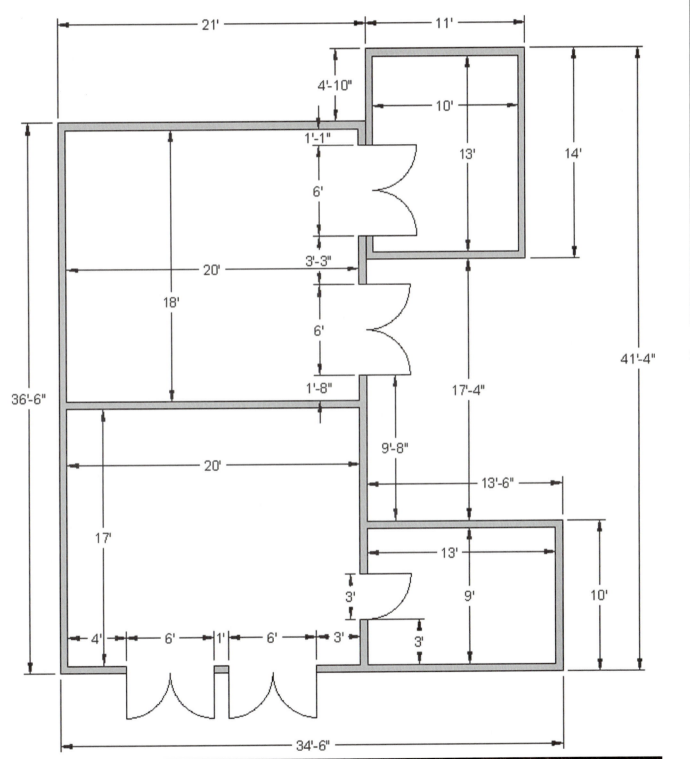

111

CHAPTER 5

Hatch Patterns

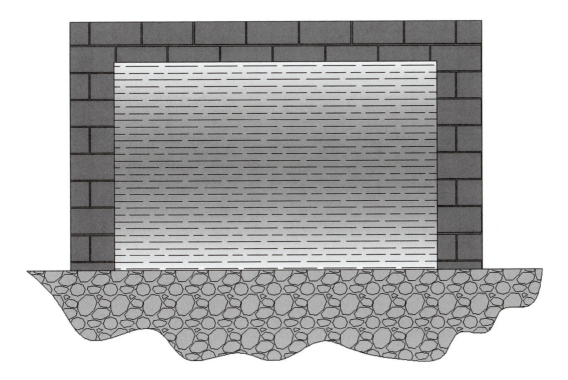

LEARNING OBJECTIVES

Hatch patterns allow you to indicate a variety of surfaces, from wood to concrete, on your design. In engineering, they are also used to indicate cross-section surfaces. In this chapter, we introduce boundary hatch and discuss the following:

- Hatch fundamentals
- Picking patterns
- Picking area or objects to hatch
- Adjusting scale and orientation
- Working with hatch patterns
- Gap tolerance
- Gradients and solid fills

At the end of this chapter, you will be able to add hatch patterns to your floor plan.

Estimated time for completion of chapter: 2 hours (lesson and project).

5.1 INTRODUCTION TO HATCH

This is a chapter on the concept of boundary hatch or just hatch for short. This is simply a tool for creating patterns and adding fills to your design. These are not merely decorative; rather they serve an important function of indicating types of materials, ground and floor coverings, and cross sections. As such, this topic is quite important and is the next logical concept to master in AutoCAD.

We cover hatch patterns in two sections. The first is with the Ribbon turned off, which makes the hatch command default to the classic dialog box seen in Figure 5.1. It is easier for a beginner to visualize hatch concepts via this method. It was also the only way to access hatch up until AutoCAD 2011, good to know if you are using an older version of AutoCAD. Starting with AutoCAD 2011 and continuing with 2012, hatch can still be accessed the old way if the Ribbon is not on, or if it is, then via the Ribbon only. The Ribbon essentially duplicates the hatch dialog box, with maybe a few more bells and whistles; we cover this approach as well.

Hatch patterns are not an AutoCAD invention. In the days of hand drafting, there was also a need to visually indicate what sort of material one was looking at. Architects and engineers created easy to understand repeating patterns that somewhat, if not closely, resembled the actual material or at least gave you a very good idea of what it was. Just a few of the uses of hatch patterns follow:

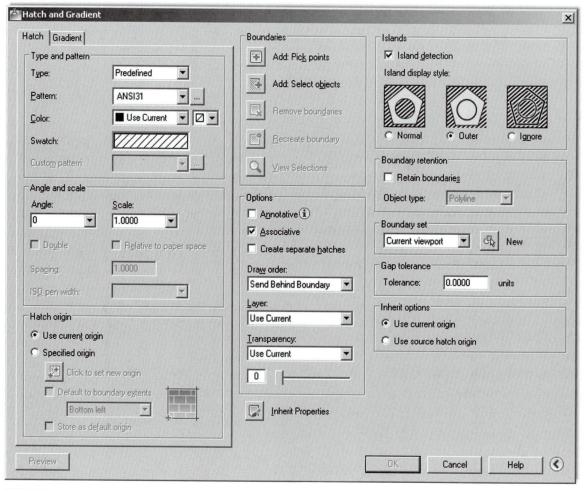

FIGURE 5.1
Hatch and Gradient dialog box.

Architecture

- **Brick:** An important and popular pattern to render the exterior and (sometimes) interior of buildings. AutoCAD has numerous brick patterns available.
- **Herringbone, parquet:** Important for flooring designations.
- **Honeycomb:** Insulation designation.

Civil Engineering

- **Concrete, sand, clay, earth, gravel:** All used in designating surfaces in civil and site plan design.

Mechanical Engineering

- **Various ANSI diagonal patterns:** Used to designate the visible inside of an object cut in cross section.

With AutoCAD, creating these patterns is easy, as they are all predrawn and saved in a library that is called up anytime you use the command. Often a designer can purchase or download additional patterns, if the basic ones are inadequate. You can even make your own and save them for future use, although this is a tedious process and only briefly covered in Appendix D.

5.2 HATCH PROCEDURES

At the most basic level, AutoCAD's hatch command requires only four steps:

Step 1. Pick the hatch pattern you want to use.
Step 2. Indicate where you want the pattern to go.
Step 3. Fine-tune the pattern by adjusting scale and angle (if necessary).
Step 4. Preview the pattern and accept it if OK.

Memorize the four steps and their sequence; this will help you when looking at the hatch dialog box.

Before we get into the details, it goes without saying that you first need something to hatch. The command will not work on a blank screen or a bunch of unconnected lines. You need a closed area (although later on we violate that rule), and the easiest to do is a circle or rectangle. Open a new file and draw either of the two shapes. Follow the steps carefully as described next. All advanced hatch functions flow from these basics.

Step 1. Pick the Hatch Pattern You Want to Use

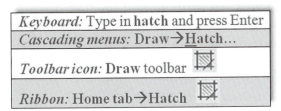

Keyboard: Type in **hatch** and press Enter
Cascading menus: Draw→Hatch...
Toolbar icon: **Draw toolbar**
Ribbon: **Home tab**→Hatch

With the Ribbon turned off (**Tools→Palettes→Ribbon**), begin the hatch command via the keyboard, cascading menus, or icons. The Hatch and Gradient dialog box appears (Figure 5.1). At the bottom right of the dialog box is an arrow that expands or collapses additional options. Make sure it is expanded, as seen in the figure.

Now select a pattern. The choices are listed in the upper left of the dialog box under the heading `Type and pattern`. Below that are four fields called `Type:`, `Pattern:`, `Color:`,

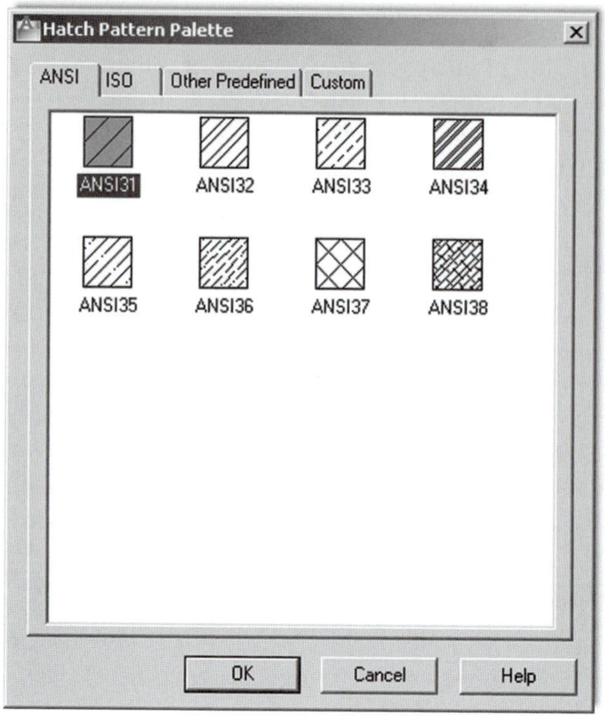

FIGURE 5.2
Hatch Pattern Palette.

and Swatch:. Leave Type as Predefined and also ignore Color for now. For the actual pattern selection, if you use Pattern, then select it *by name*; if you pick Swatch, patterns are selected *visually*, a much better option. Click on the diagonal lines to the right of Swatch: (or the little box with the three dots to the right of Pattern) and a new box, called the *Hatch Pattern Palette*, appears (Figure 5.2).

Take a look through the tabs: ANSI, ISO, Other Predefined, and Custom. We have a use for some of the ANSI patterns, not so much for the ISO ones, and Custom is where the custom defined patterns go if you are inclined to create them. We are most interested in the Other Predefined tab. Go ahead and select it and you will see what is shown in Figure 5.3.

Scroll up and down the patterns. You will notice some of the ones mentioned at the start of this chapter. Go ahead and pick one that you like, preferably one that is distinct and stands out; AR-HBONE is a good choice and is used in this example. When you click on it, the pattern is highlighted blue. Go ahead and click on OK. We are done with Step 1. The new pattern choice is reflected in the Swatch area, as seen in Figure 5.4.

Step 2. Indicate Where You Want the Pattern to Go

There are two ways to indicate where to put the pattern, either by *directly picking* the object that will contain the pattern or by *picking a point inside* that object. Which method to use is easy to determine. If the object is actually made of joined-together lines and is one piece (such as your rectangle or circle), then it can be picked directly. If not, and the "object" is really just a collection of connected lines defining an area, then the best way is to pick a point in the middle of that area; and indeed this is the more common situation. The pattern then behaves like a bucket of paint spilled in the middle of the room. It flows outward evenly and stops only when it hits a wall. Once again, no holes or gaps are allowed yet; that is addressed later.

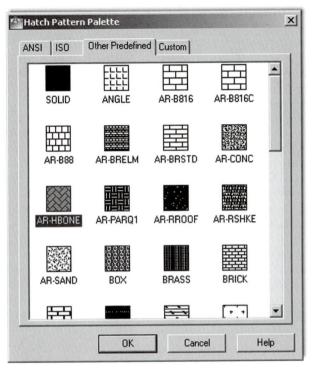

FIGURE 5.3
Other Predefined tab.

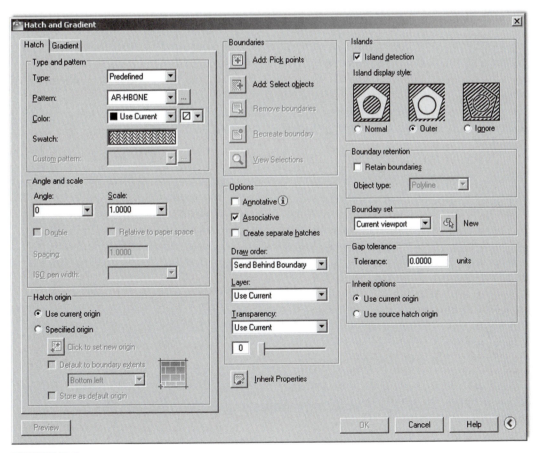

FIGURE 5.4
Step 1 completed.

In the Boundaries section of the Hatch and Gradient dialog box (top middle), notice the two choices just discussed—Add: Pick points and Add: Select objects. Choose either one to try. The AutoCAD procedure for both is outlined next.

ADD: PICK POINTS

If you choose this option, AutoCAD says:

```
Pick internal point or [Select objects/remove Boundaries]:
```

Click somewhere *inside* the object, and AutoCAD says:

```
Selecting everything...
Selecting everything visible...
Analyzing the selected data...
Analyzing internal islands...
```

Press Enter and you return to the hatch dialog box with Step 2 completed. Notice that a preview of the hatch is shown to you as you complete this step.

ADD: SELECT OBJECTS

Alternatively, if you pick this option, AutoCAD says:

```
Select objects or [picK internal point/remove Boundaries]:
```

Click on the *object* itself (not the empty space) and press Enter. You return to the hatch dialog box with Step 2 completed.

Successfully picking a boundary can sometimes be tricky business. The existence of gaps, no matter how small, can be a source of occasional frustration to designers; and until relatively recently, it used to be difficult to tell where those gaps were hiding in order to fix them. As of AutoCAD 2010, red circles now appear where the gaps are, aiding you in locating them, but this all can be avoided in the first place by not doing sloppy drafting (i.e., not connecting lines together properly). It is a sign of advanced AutoCAD skills when a complex area hatches right the first time.

In cases where boundary selection is not successful, complex areas can often be broken down to smaller ones by using lines to divide the area into smaller pieces and the non-problematic portions hatched first, although this is a last resort. A few releases ago, AutoCAD added the gap tolerance command that ignores gaps up to a preset limit. This is a great tool, which we cover shortly, along with the entire gap issue in general, but it is somewhat of a Band-Aid and still does not address the underlying sloppy drafting, but merely allows you to get away with it.

Remember, you get only the level of accuracy that you put in. It is essential to master the fundamentals of drafting early on. The hatch command is an "early warning" to students. If they are having problems using it smoothly, they need to go back and refine their basic drafting (linework and accuracy) skills.

Step 3. Fine-Tune the Pattern by Adjusting Scale and Angle (If Necessary)

You are almost done and, at this point, could probably just press OK and finish the hatch pattern; however, press Preview instead. It is found at the bottom left of the dialog box and

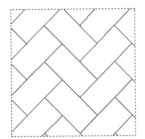

FIGURE 5.5
Hatch preview.

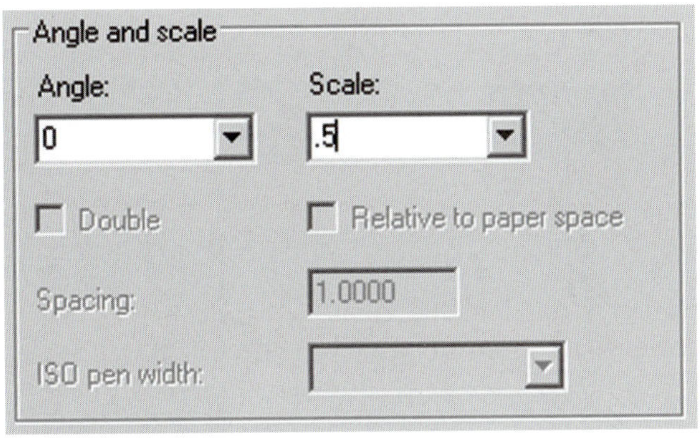

FIGURE 5.6
Angle and scale.

119

is generally a good habit to get into. What you see is your hatch pattern (if it is the right size) with the object's border in dashed lines (Figure 5.5).

If everything looks right (as it does in this example), AutoCAD says what to do next:

```
Pick or press Esc to return to dialog or <Right-click to accept hatch>
```

Go ahead and right-click or press Enter.

Sometimes, however, the pattern is too big or too small. You can usually tell it is too big by either not being able to see it, or seeing a small part of it. If it is too small the problem is worse, because you may not be able to finish the pattern at all. Instead, AutoCAD tells you the following:

```
Hatch spacing too dense, or dash size too small. Pick or press Esc to
return to dialog or <Right-click to accept hatch>:
```

Nothing appears. In either case, press Esc once and you return to the dialog box. Under Angle and scale, change the scale to a larger or smaller number (by typing it in) and preview it again. You may need to do this several times until the hatch pattern is just right. In the same manner, you may adjust the angle of the entire pattern if desired. Angle and scale is shown in Figure 5.6.

Step 4. Preview the Pattern and Accept It If OK

Finally, you are done. Do one more preview and press OK. Note again that a preview was available to you right after Step 2, as your mouse hovered over the area you were planning on hatching. It is still a good idea to do one more preview though before the final OK.

FIGURE 5.7
Final hatch pattern.

The sample hatch pattern should look like Figure 5.7. Yours may or may not be similar depending of course on what pattern and scale you selected, as well as what shape you used.

5.3 WORKING WITH HATCH PATTERNS

In this section, we mention some of the additional tools and concepts that pertain to hatch patterns. First of all, the most basic question at this point is: How do you edit a hatch pattern after you have created it? It is very easy: To edit a hatch pattern, simply double-click on it. Make sure you get one of the lines of the pattern, not just empty space. You will see grips and the standard Properties Palette (Figure 5.8); here, you can change whatever you need to. This is different from older releases of AutoCAD, where you were taken back to the Hatch dialog box. The thinking was that the hatch dialog box was too big and covered up your work, so a sleeker Properties Palette was a better option. If you are using the Ribbon, to be discussed shortly, you need to click only once to edit the pattern.

Exploding Hatch Patterns

This may be one of the worst things you can do. Never explode hatch patterns; depending on the size, they turn into hundreds of pieces that cannot be put back together again in the same editable form. All the reasons designers give for exploding hatch patterns (making them fit, trimming them, moving the containment border, etc.) can be addressed with additional hatch tools.

Hatch Pattern Layers and Colors

Hatch patterns should be on a generic A-Hatch layer or whatever layer name best describes what the pattern represents (A-Carpet, M-Cross-Sec, C-Gravel, etc.). Choose any color you want for the hatch patterns, though you should not go with anything too bright, as it may overwhelm the design. Shades of gray are common for cross sections in mechanical design, as are various shades of rusty orange for brick. As a side note, hatch patterns can be easily erased by the usual erasing methods. The patterns can also be moved, copied, rotated, and mirrored, although there is rarely a reason to do so.

As of the previous release of AutoCAD, you can now assign a color directly to a hatch pattern regardless of what layer, and associated color, it is on. You can also assign a

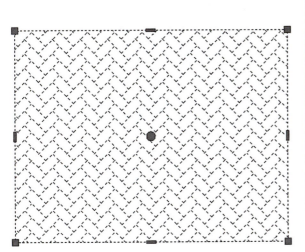

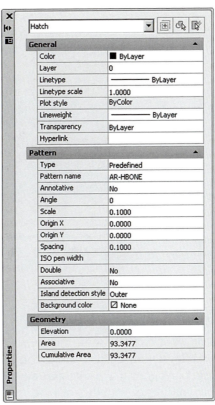

FIGURE 5.8
Editing the hatch pattern (no Ribbon).

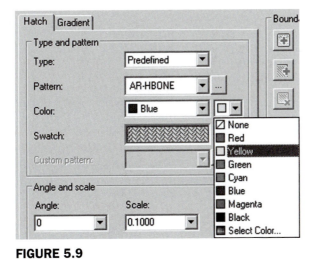

FIGURE 5.9
Hatch (blue) and background (yellow) color selection.

background color (solid, not a gradient). To do this when first creating the hatch, you can select both the hatch color and the background color (Figure 5.9) from two drop-down menus at the top left of the hatch dialog box. The same steps can be performed with the Ribbon, as shown a bit later. One sample result of assigning a hatch and background color is shown in Figure 5.10.

FIGURE 5.10
Hatch with a background color assigned.

Advanced Hatch Topics

Here, we summarize some additional advanced options available for working with hatches, including a more detailed discussion concerning the gap tolerance feature. You should explore these on your own as well; this is an important part of learning AutoCAD and greatly enhances your confidence with the software.

HATCH ORIGIN

This allows you to begin hatch patterns from a precise location (origin) as opposed to a random fill-in. AutoCAD chooses the origin on the pattern itself, and you get to choose to what point on the object that origin is aligned. Try it out.

OPTIONS

- **Annotative.** Will not be covered until Level 2.
- **Associative.** Keep this checked. This simply means that, if you move the border, the hatch moves with it; very useful in case of future updates.
- **Create separate hatches.** If there are multiple separate areas to hatch, this option keeps them separate; sometimes useful, sometimes not.
- **Draw order.** This is exactly what it sounds like; it gives you options on how you want to stack the hatch and its boundary. Check out the options in the drop-down menu.
- **Inherit properties.** This option lets you borrow the features of another hatch pattern and use them in a new pattern you are creating, saving you some effort and time in picking out a pattern from scratch.

ISLANDS

This gives AutoCAD guidance in how to interpret boundaries inside boundaries (e.g., a chair in the middle of your floor). It simply tells AutoCAD to ignore them or, if not, how to deal with them. Try drawing another rectangle inside your first one and running through the options represented by the pentagon shapes.

GAP TOLERANCE

This is a simple and popular new tool that was mentioned earlier. Simply set the tolerance to a value big enough to span anticipated holes or gaps in your design and hatch as usual. Initially, if your tolerance is set to 0 units and you try to hatch an area with a gap, you get Figure 5.11.

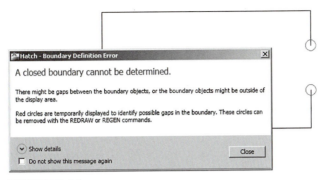

FIGURE 5.11
Boundary not closed alert.

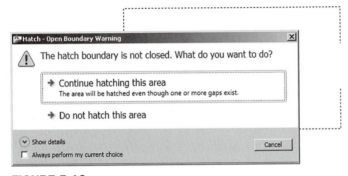

FIGURE 5.12
Open boundary warning.

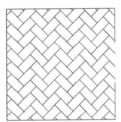

FIGURE 5.13
Hatch with gap tolerance.

Close the alert and press Esc to continue. Then, returning to the main dialog box, set a value for the gap tolerance (you can use 10 for starters) and rehatch the area. You get another warning, which can be eliminated from future occurrence by checking off `Always perform my current choice`, as seen in Figure 5.12.

Select `Continue hatching this area` and AutoCAD temporarily closes the gap to continue the hatch command. After successful completion, the gap shows again but the hatch remains, as seen in Figure 5.13. Some designers have used the command to ignore door openings when hatching carpeting in floor plans. You can try this as well on your floor plan later.

5.4 GRADIENT AND SOLID FILL

At the top left of the Hatch and Gradient dialog box, click on the Gradient tab. The left side of the box changes while the right-hand side stays essentially the same, as shown in Figure 5.14.

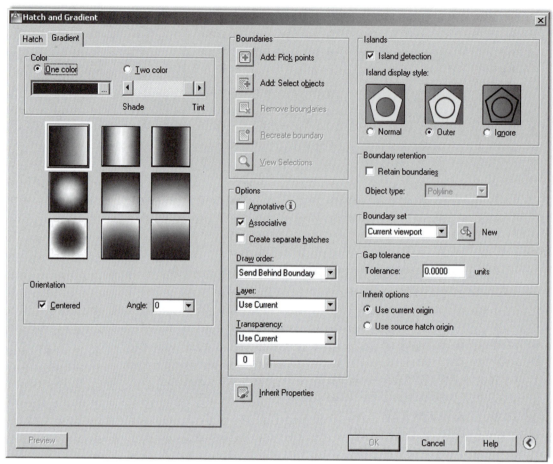

FIGURE 5.14
Gradient tab.

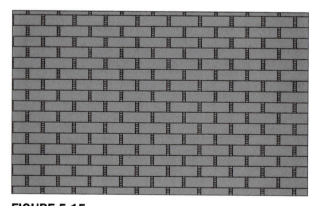

FIGURE 5.15
Brick wall created with Hatch and Gradient.

The gradient option is just a solid fill with some flair. You can choose the desired pattern, in one or two colors, and adjust the Shade, Tint, Orientation, and Angle of the gradient. The rest of the steps are similar; Pick point or Select object, Preview, and done.

Shown in Figure 5.15 is a possible use for gradient in rendering a brick wall. A brick pattern is created and a brick-colored fill tops it. Note that you may need to use **Tools→Draw Order →Bring to Front** from the cascading menu to position the brick above the gradient fill (or the gradient fill behind the brick, depending on what you pick first). Try it.

Solid Fill

This fill is the first choice (black box, top left) in the Other Predefined tab and deserves special mention as our final topic.

This solid fill is similar to a gradient but with no fancy color mixing, just one color. It is very handy as a fill for walls, some types of cross sections, or any location where you want a dark area. It becomes even more useful when paired up with a lineweight feature called *screening*, which is discussed in detail in Chapter 19. It has to do with being able to fade out (or screen) the intensity of any color, typically down to 30%, and is perfect for indicating "area of work" on key plans, some types of carpeting, and even pavement on civil engineering drawings.

A way to mimic this now without screening is to change the solid fill's color to Gray 9; a soft gray haze is then seen that is distinct and visible, yet not overwhelming. If you are wondering how the walls were filled in on the apartment drawing, this is how.

The previous steps using the hatch dialog box are essentially duplicated using the Ribbon. Bring it back by typing in Ribbon (or **Tools→Palettes→Ribbon** via the cascading menus). Draw a shape to hatch and begin the command in the same manner as before; you will see the Ribbon switch to Figure 5.16. Here, you still have to pick the pattern by dropping down the menu, as seen in Figure 5.17. Note how *all* the hatch patterns are present and you do not have tabs anymore.

You can then hover your mouse over the area and get an instant preview of the pattern to get an idea of what you have. When you move the mouse away, the pattern disappears. Selection of boundaries is all the way on the left, under Boundaries, and the fine-tuning is in the middle of the Ribbon under Properties. Here, you can select the Scale, Angle (with instant rotation

FIGURE 5.16
Ribbon—Hatch Creation.

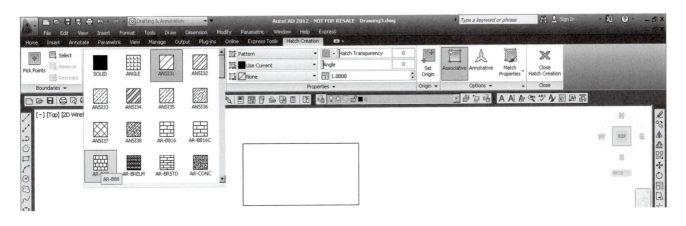

FIGURE 5.17
Selecting patterns.

via a slider), and Background Colors. The rest of the features, such as Gap Tolerance, Origin, and others, are found under Origin and Options on the right.

Which method you use is entirely up to you, although if you have the Ribbon up all the time, it makes more sense to continue using it for the hatch command.

5.5 IN-CLASS DRAWING PROJECT: ADDING HATCH TO FLOOR PLAN LAYOUT

We continue work on the floor plan and add hatch to the rooms and walls. Some tips follow.

- To add hatch to rooms, freeze the A-Doors layer and instead draw *temporary* lines closing off the door entrances. After hatching, remove the lines. You may also be able to use Gap Tolerance set to at least 3 feet, but that is not recommended.
- Make sure the hatch patterns are on the proper layers, whatever they may be (Carpeting or just Hatch_1, Hatch_2; you decide). You do not have to pick the hatch patterns shown, but make sure they are reasonable and at the proper scale.

To fill in the apartment walls, freeze everything except the A-Walls and the Hatch layer. Then, use the solid fill, color Gray 9, as described earlier. If you created everything properly, there should be no gaps or breaks in the wall islands. Check to make sure the windows are drawn correctly, meaning the wall has to have an actual closed gap, into which the window is inserted. Figure 5.18 shows what everything should look like *prior to wall hatching* (1) and *right after* (2). One possible approach to hatching the apartment is shown in Figure 5.19.

126

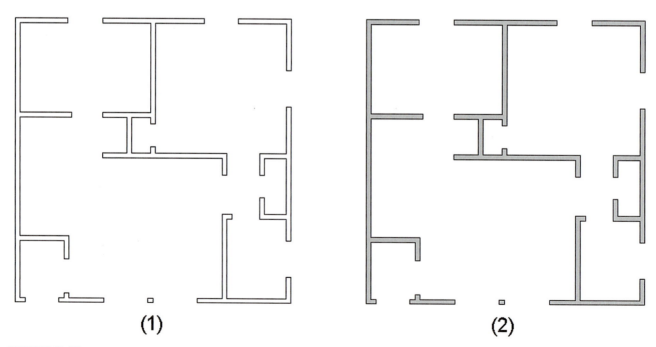

(1)   (2)

FIGURE 5.18
Wall hatching before (1) and after (2).

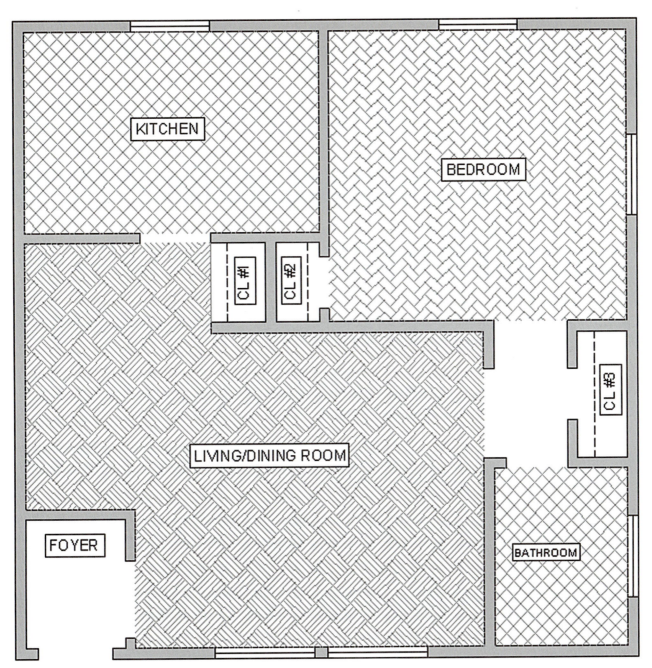

FIGURE 5.19
Floor plan with wall and floor hatching.

SUMMARY

You should understand and know how to use the following concepts and commands before moving on to Chapter 6:

- Hatch
- Picking a pattern
- Selecting area or object
- Scale and orientation adjustments
- Preview

- Editing a hatch
  - Double-click
- Additional options
  - Islands
  - Inherit properties
  - Associative
  - Draw order
- Gap tolerance
- Gradients and fills

REVIEW QUESTIONS

Answer the following based on what you learned in Chapter 5.

1. List the four steps in creating a basic hatch pattern.
2. What two methods are used to select where to put the hatch pattern?
3. What commonly occurring problem can prevent a hatch being created?
4. What is the name of the tool recently added to AutoCAD to address this problem?
5. How do you edit an existing hatch pattern?
6. What is the effect of exploding hatch patterns? Is it recommended?
7. What especially useful pattern to fill in walls is mentioned in the chapter?

EXERCISES

1. In a new file, create the following rectangle, offset to get the smaller ones, and fill them all in with the shown hatch patterns. Do try to duplicate what is shown, including hatch patterns, scale, and angles as closely as possible. (Difficulty level: Easy; Time to completion: <5 minutes)

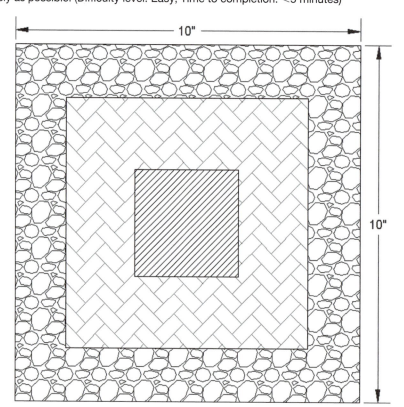

2. In a new file, create the following shapes and hatch patterns. Some overall sizing is shown, but you have to improvise the rest via the offset and line commands. Do try to duplicate what is shown, including hatch patterns, scale, and angles as closely as possible. (Difficulty level: Easy; Time to completion: <5 minutes)

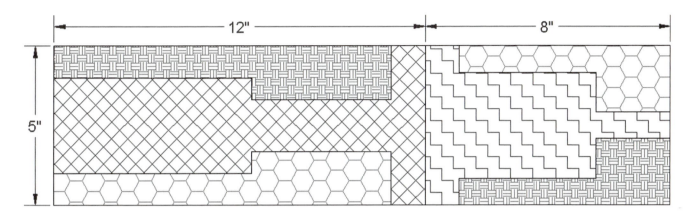

3. Open a new file, create the appropriate layers and set units to Architectural. Create the following stair layout and hatch patterns. Any sizes and dimensions that are not given can be estimated, as can the scale of the patterns. (Difficulty level: Intermediate; Time to completion: 20–30 minutes)

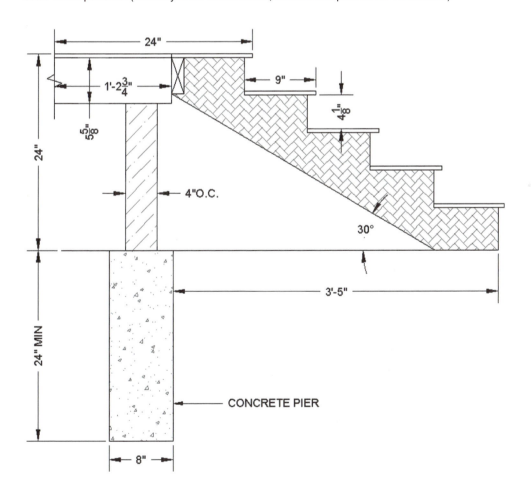

4. Open a new file and set units to Architectural. Create the following layout of a fictional set of cross-sectioned cylinders. Dimensions are provided, but sizing is of secondary importance to creating the hatch and gradient patterns. Notice how gradients are used to depict curvature inside the pipe cross sections: a good trick to add realism to 2D drawings. (Difficulty level: Intermediate; Time to completion: 20 minutes)

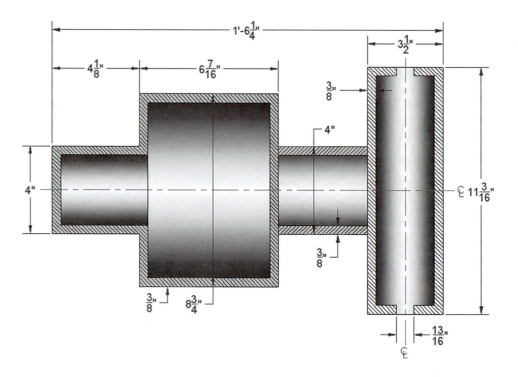

130

5. Open a new file and set units to Architectural. Create the following brick wall design, a slightly modified version of what is seen on the cover page of this chapter. Note that some of the backgrounds are gradients. Sizing and scale is up to you, but do try to duplicate what you see. (Difficulty level: Easy; Time to completion: <10 minutes)

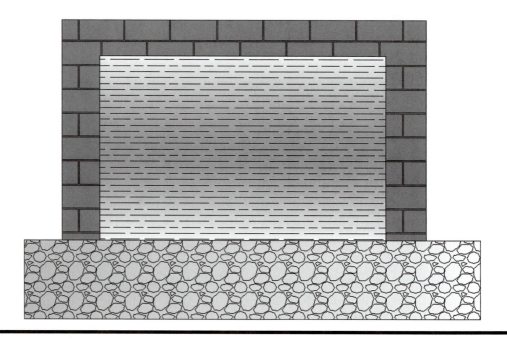

CHAPTER 6

Dimensions

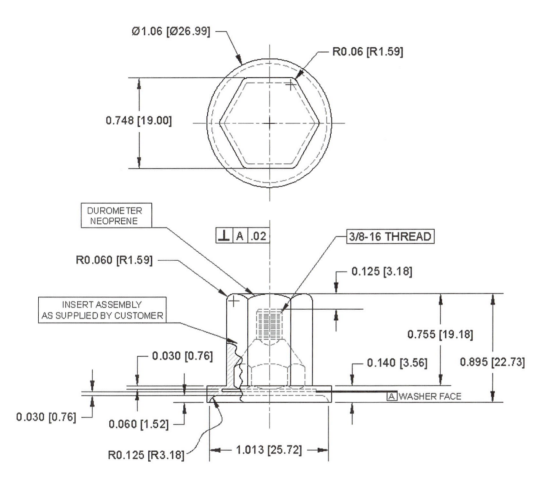

LEARNING OBJECTIVES

In this chapter, we introduce the first parts of AutoCAD's extensive dimensioning capabilities and discuss the following:

- Linear (horizontal and vertical) dimensions
- Aligned dimensions
- Diameter dimensions
- Radius dimensions

- Angular dimensions
- Continuous dimensions
- Baseline dimensions
- Leader and multileader
- Arc length dimensions
- Ordinate dimensions
- Jogged dimensions
  - ddedit
  - ddim
- Dimension units
- Dimension font
- Dimension arrowheads
- Dimension overall size

At the end of this chapter, you will be able to add dimensions to your floor plan.

Estimated time for completion of chapter: 3 hours (lesson and project).

6.1 INTRODUCTION TO DIMENSIONS

Dimensioning in AutoCAD is a major topic, one so extensive that it is split into two separate discussions. In this chapter, we address the basics of what dimensions are, which ones are available, and how to use and edit them. In Chapter 13, we look at extensive customization and additional options and explore some relatively new additions to the dimensioning family: parametric dimensioning and the concept of constraints and dimension-driven design.

So, what are dimensions? They are simply visible measurements for purposes of conveying information to the audience that will be looking at your design. It is how you describe the size of the design and where it is in relation to everything else. Dimensions can be natural, meaning they display the actual value, or forced, when they display an altered value, such as when you dimension an object with a break line.

6.2 TYPES OF DIMENSIONS

The first step in learning dimensions is to know what is available, so you can use the appropriate type in any situation. The primary and secondary dimensions available to you are shown next. Commit them to memory, as you may need most of them on a complex project.

Primary dimensions are the dimensions used most often by the typical architectural or engineering designer:

- Linear (horizontal and vertical)
- Aligned
- Diameter
- Radius
- Angular
- Continuous
- Baseline
- Leader

Secondary dimensions are used less often or are quite specialized but still need to be learned:

- Arc length
- Ordinate
- Jogged

Let us take the primary list, and discuss the dimensions one by one to understand how each is used. In each case, you need to draw the appropriate shape to practice with, and then add the new dimension(s) by following the step-by-step instructions. As before, we use typing, toolbars, cascading menus, and the Ribbon. You can of course stick to just one method if you prefer. Open a new file and, if using toolbars, bring up the Dimension toolbar (Figure 6.1), adding it to those already on your screen.

FIGURE 6.1
Dimension toolbar.

Note an important point if typing dimensions: You need to type in `dim` and press Enter. This gets you into the dimension submenu. The command line then says `Dim:` and you can type in whatever dimension you want to create, usually abbreviating by typing only the first few letters, as shown (such as *hor* for *horizontal*); then press Enter again. The command then executes normally. So there is an advantage to using other input methods here rather than typing, though the differences in effort and speed are rather small. Remember also that you need to press Esc to get out of this `Dim:` mode and back to the regular command line.

Linear Dimensions

These are any dimensions that are strictly horizontal or vertical (Figure 6.2). We go over only the horizontal dimension, as reflected in the matrix that follows. Of course, the same steps apply to the vertical dimension and you should practice it on your own.

Keyboard: Type in **dim**, press Enter, type in **hor**, press Enter
Cascading menus: **Dimension→Linear**
Toolbar icon: **Dimension** toolbar ⊢
Ribbon: **Annotate tab→Dimension→Linear** ⊢

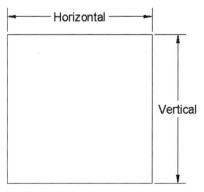

FIGURE 6.2
Linear dimensions.

133

Step 1. Create a small to medium-size rectangle or square, as seen in Figure 6.2, and begin the linear dimension command via any of the preceding methods.
  ○ AutoCAD says: `Specify first extension line origin or <select object>:`
Step 2. Turn on your OSNAPs, and using ENDpoint, select the upper left corner of the rectangle or square.
  ○ AutoCAD says: `Specify second extension line origin:`
Step 3. Again using ENDpoint, select the upper right corner of the rectangle or square (since we are doing the *horizontal* dimension first). The dimension appears, attached to the movements of your mouse.
  ○ (Keyboard) AutoCAD says: `Specify dimension line location or [Mtext/ Text/Angle]:`
  ○ (If using other three methods) AutoCAD says: `Specify dimension line location or [Mtext/Text/Angle/Horizontal/Vertical/Rotated]:`
Step 4. Move the mouse up and down and click where you want the dimension to go, usually a short distance from the object. It sets and the value is shown.
  ○ (Keyboard) AutoCAD says: `Enter Dimension text = 4.4681`
  ○ (If using other three methods) AutoCAD says: `Dimension text = 4.4681`

Your value likely is different, of course.

Note an important distinction with the keyboard entry method. Instead of setting the value in Step 4, AutoCAD throws in one more step, call it 4b, allowing you to alter the dimension on the spot (forcing it). You can enter a new value or press Enter to accept the natural one. With the other input methods you get no chance to alter the value until after you have already created it, a slight disadvantage in some cases.

The previous linear dimensions are by far the most common in most drawings and are also the basis for the continuous and baseline dimensions, to be covered later on. Be sure to also run through the vertical version of the linear dimension, selecting the upper right and lower right (or upper and lower left) corners of the rectangle or square. Next is the aligned dimension, which gives the true distance of a slanted surface.

Aligned Dimension

This is a dimension that measures a slanted line or object (Figure 6.3).

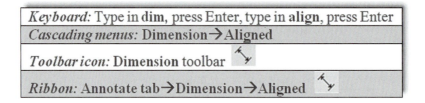

Keyboard: Type in **dim**, press Enter, type in **align**, press Enter
Cascading menus: Dimension→Aligned
Toolbar icon: Dimension toolbar
Ribbon: Annotate tab→Dimension→Aligned

Step 1. Create a small to medium-size rectangle or square and chop off any corner, or just do this to your existing shape, as seen in Figure 6.3. Then, begin the aligned dimension command via any of the preceding methods.
  ○ AutoCAD says: `Specify first extension line origin or <select object>:`
Step 2. Turn on your OSNAPs and, using ENDpoint, select the upper left start of the slanted line.
  ○ AutoCAD says: `Specify second extension line origin:`

Step 3. Again using ENDpoint, select the lower right end of the slanted line. The dimension appears.

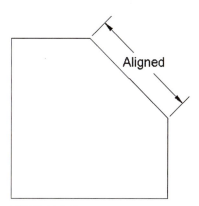

FIGURE 6.3
Aligned dimension.

- ○ AutoCAD says: `Specify dimension line location or [Mtext/Text/Angle]:`
Step 4. Move the mouse up and down and click where you want the dimension to go, usually a short distance from the object. It sets and the value is shown.
  - ○ AutoCAD says: `Dimension text = 2.9255`

Your value likely is different, of course.

Again, note an important distinction with keyboard entry. Instead of setting the value in Step 4, AutoCAD throws in one more step, allowing you to alter the dimension on the spot. You can enter a new value or press Enter to accept the natural one. We do not detail the AutoCAD responses with keyboard entry here; the previous linear dimensions should serve as a good example.

The next two dimensions, diameter and radius, are essentially similar, although we cover them separately for clarity. Note the primary differences: the symbol indicating diameter (Ø) and the symbol for radius (R) prior to the value. Although you can set this up, by default there is no line crossing the circle halfway for radius, or all the way across for diameter, so you have to read the values carefully and look for the appropriate symbol to know what you are looking at.

Also, note the location of the values. They are positioned at roughly 10, 2, 4, and 8 o'clock relative to the circle. Do not just put them anywhere; hand-drafting rules still apply in CAD, and designers have typically left them in these positions for clarity, consistency, and style.

Diameter Dimension

This is a dimension that measures the diameter of a circle or an arc (Figure 6.4).

Keyboard: Type in **dim**, press Enter, type in **dia**, press Enter
Cascading menus: Dimension→Diameter
Toolbar icon: Dimension toolbar ⊘
Ribbon: Annotate tab→Dimension→Diameter ⊘

Step 1. Create a small to medium-size circle, as seen in Figure 6.4. Then, begin the diameter dimension command via any of the preceding methods.
  - ○ AutoCAD says: `Select arc or circle:`

135

— (Ø) Diameter

FIGURE 6.4
Diameter dimension.

Step 2. Pick the circle and the dimension appears attached to your mouse.
  ○ AutoCAD says: Dimension text = 3.1357 Specify dimension line
    location or [Mtext/Text/Angle]:

Your value likely is different, of course.

Step 3. Position your mouse somewhere to the upper right of the circle (rarely inside, usually outside) at 2 o'clock or at any of the other accepted positions and click. The value and the dimension are shown.

Radius Dimension

This is a dimension that measures the radius of a circle or an arc (Figure 6.5).

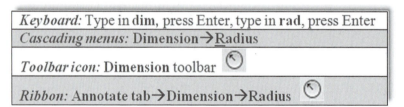

Keyboard: Type in **dim**, press Enter, type in **rad**, press Enter
Cascading menus: Dimension→Radius
Toolbar icon: Dimension toolbar
Ribbon: Annotate tab→Dimension→Radius

— (R) Radius

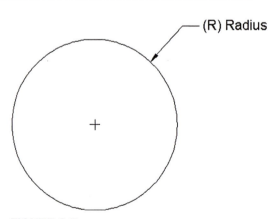

FIGURE 6.5
Radius dimension.

Step 1. Create a small to medium-size circle, as seen in Figure 6.5. Then, begin the radius dimension command via any of the preceding methods.
  ○ AutoCAD says: Select arc or circle:

Step 2. Pick the circle and the dimension appears attached to your mouse.

    ○ AutoCAD says: `Dimension text = 1.5679`
      `Specify dimension line location or [Mtext/Text/Angle]:`

Your value likely is different, of course.

Step 3. Position your mouse somewhere at the upper right of the circle (rarely inside, usually outside) at 2 o'clock or at any of the other accepted positions and click. The value and the dimension are set.

Angular Dimension

This is a dimension for angles between two lines or objects (Figure 6.6).

Keyboard: Type in **dim**, press Enter, type in **ang**, press Enter
Cascading menus: Dimension→Angular
Toolbar icon: Dimension toolbar ◿
Ribbon: Annotate tab→Dimension→Angular ◿

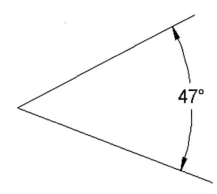

FIGURE 6.6
Angular dimension.

Step 1. Create two lines spread out at a random angle, as seen in Figure 6.6. Then, begin the angular dimension command via any of the preceding methods.

    ○ AutoCAD says: `Select arc, circle, line, or <specify vertex>:`

Step 2. Pick both lines, one after the other.

    ○ AutoCAD says: `Specify dimension arc line location or[Mtext/Text/`
      `Angle/Quadrant]:`

Step 3. Drag the mouse out and move it around, selecting the best position for the new dimension, and click when you find a spot you like. Be careful, as the supplement of the degree value shows if you move behind the lines, which of course may be desirable in some cases.

    ○ AutoCAD says: `Dimension text = 47`

Your value of course may be different.

Once again, note an important distinction with keyboard entry. Instead of setting the value in Step 3, AutoCAD throws in one more step, allowing you to alter the dimension on the spot (forcing it). You can enter a new value or press Enter to accept the natural one and position the text.

Angular completes the basic set of new fundamental dimensions. Next we have *continuous* and *baseline*, which are not really new, as we will see, but rather an extension (or automation) of the linear dimension learned earlier. We then conclude with a leader, not so much a dimension but a useful member of that family.

Continuous Dimensions

These are continuous *strings* of dimensions, as seen in Figure 6.7.

A short explanation is in order, as continuous and baseline dimensions give new students some initial problems. Continuous dimensions are nothing more than a string of familiar horizontal or vertical ones. The idea here is to create one of those two types of linear dimensions, then start up the continuous dimension where you just left off and let AutoCAD quickly fill them in as you pick contact points. It is really the same as using linear dimensions over and over again but faster and more accurate since it is partially automated.

The easiest way to practice continuous dimensions is to draw a set of squares attached to each other. These represent a simplified view of building and room walls. Draw one random-sized square and copy it, endpoint (lower left corner) to endpoint (lower right corner), until you have what is shown in Figure 6.8.

The next step is to draw one horizontal linear dimension as you already have done (it could have easily been a vertical line, too, if the rectangles were stacked vertically). Locate it about where you would like the entire string to go, as seen in Figure 6.9.

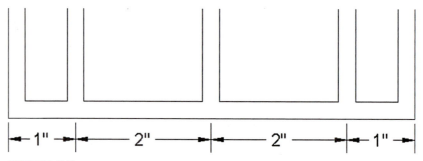

FIGURE 6.7
Continuous dimensions.

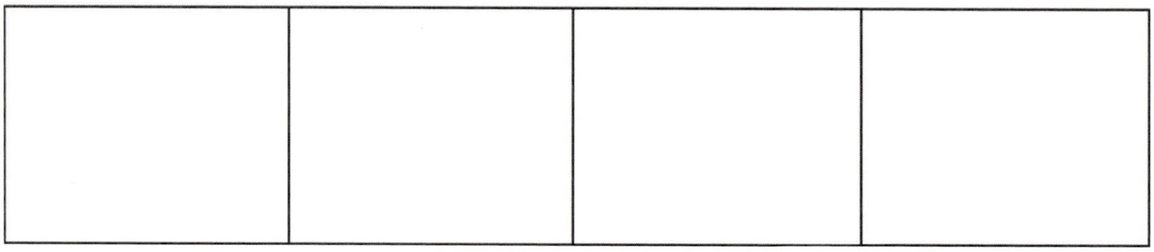

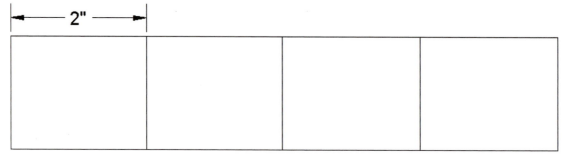

FIGURE 6.8
Setup to practice continuous dimensions.

FIGURE 6.9
First step to create continuous dimensions.

This is the essential first step in creating both continuous and baseline dimensions. Now, you can practice creating continuous dimensions as outlined next.

Keyboard: Type in **dim**, press Enter, type in **cont**, press Enter
Cascading menus: Dimension→Continue
Toolbar icon: Dimension toolbar ⊬⊣
Ribbon: Annotate tab→Continue ⊬⊣

Step 1. Once you have the squares and one linear dimension as seen in Figure 6.9, start up the continuous command via any of the preceding methods.
- AutoCAD says: `Specify a second extension line origin or [Undo/ Select]<Select>:`

Step 2. A new dimension appears. Pick the *next* point (corner) along the string of rectangles and click on it (always using OSNAP points, no eyeballing).
- AutoCAD says: `Dimension text = 1.9709`

Your value of course may be different.

Step 3. You can continue this until you run out of objects to dimension; at that point, just press Esc. Your result is shown in Figure 6.10. Notice how all the values are neatly spaced in a row.

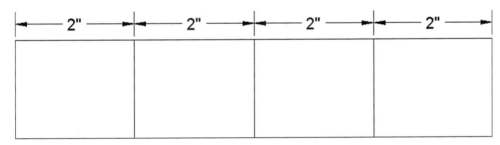

FIGURE 6.10
Continuous dimensions result.

As you are already aware, the difference between keyboard entry and the other methods is the opportunity to enter a value other than what is given (force the dimension) when typing. You then have to repeat the continuous dimension (in other words, retype `cont`) to carry on.

An interesting side note concerns what happens if you press the space bar before Enter while doing Step 3 (via the keyboard entry method). This causes the numerical value to disappear and a solid dimension line to take its place. If you press the space bar a few times, a gap appears where the dimension value should have been. This may be useful if you are unsure of dimensions on an "as built" floor plan and will be taking the plans out into the field to verify them. You can then just write in the values as you measure them.

Baseline Dimensions

These are continuous *stacks* of dimensions (Figure 6.11).

The baseline dimension, as mentioned before, is very similar in principle to the continuous dimension. The goal is to make a neat stack of evenly spaced dimensions that all start at one point (the *base* in *baseline*). To begin, erase the previous continuous dimensions (leaving the squares) and once again draw one linear (horizontal) dimension, as seen before in Figure 6.9.

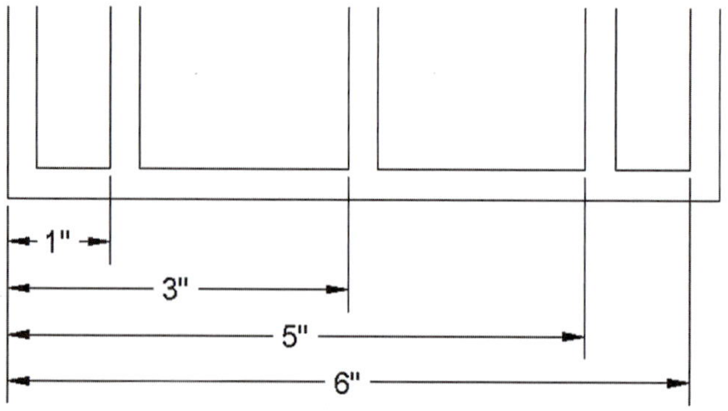

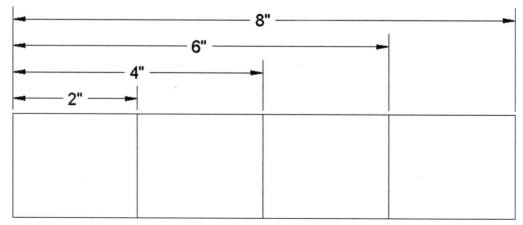

FIGURE 6.11
Baseline dimensions.

| *Keyboard:* Type in **dim**, press Enter, type in **base**, press Enter |
| *Cascading menus:* Dimension→Baseline |
| *Toolbar icon:* Dimension toolbar 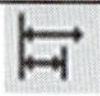 |
| *Ribbon:* Annotate tab→Baseline 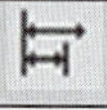 |

Step 1. Once you have the squares and one linear dimension, start up the baseline command via any of the preceding methods.
- AutoCAD says: `Specify a second extension line origin or [Undo/Select]<Select>:`

Step 2. A new dimension appears. Pick the *next* point (corner) along the string of rectangles and click on it (always using OSNAP points, no eyeballing).
- AutoCAD says: `Dimension text = 3.9417`

Your value of course may be different.

Step 3. You can continue this until you run out of objects to dimension; at that point, just press Esc. Your result is shown in Figure 6.12.

Leader and Multileader

This is an *arrow and label* combination pointing at something (Figure 6.13).

While a leader is not a true dimension by definition, it is still a very common and necessary member of the dimension family. The values shown by the leader can be not only numerical

FIGURE 6.12
Baseline dimensions result.

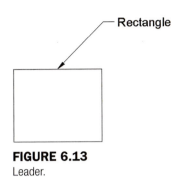

FIGURE 6.13
Leader.

but also text comments from the designer, concerning the part or object to which it is pointing.

Leaders are so important that AutoCAD has given them a major rework in recent releases. We go over the two main options in increasing order of complexity, starting with the basic leader command, followed by the more feature-rich multileader, with its Add, Remove, Align, and Collect options.

LEADER COMMAND

Step 1. Draw a small box on your screen, making sure Ortho is off. Type in the command leader and press Enter.
- AutoCAD says: `Specify leader start point:`

Step 2. Click with OSNAP precision on your shape, perhaps a midpoint on the top side, as seen in Figure 6.13.
- AutoCAD says: `Specify next point:`

Step 3. Move your mouse at a 45° angle away from the first point and click again at a reasonable distance away. You created the *leader line and arrowhead* seen in Figure 6.14.
- AutoCAD says: `Specify next point or [Annotation/Format/Undo] <Annotation>:`

Step 4. Turn Ortho back on and draw a short line to the right, clicking when done; this is your *horizontal landing*, as seen in Figure 6.15.
- AutoCAD says: `Specify next point or [Annotation/Format/Undo] <Annotation>:`

Step 5. You are done, so press Enter.
- AutoCAD says: `Enter first line of annotation text or <options>:`

Type something in, pressing Enter if you wish to do another line, and Enter again if you are done, as seen in Figure 6.16.

141

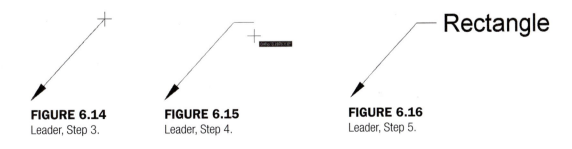

FIGURE 6.14
Leader, Step 3.

FIGURE 6.15
Leader, Step 4.

FIGURE 6.16
Leader, Step 5.

MULTILEADER

This method is relatively new and is meant to add more flexibility and usefulness (and inadvertently some complexity) to the leader command. To work with this, you need to bring up the Multileader toolbar, shown in Figure 6.17.

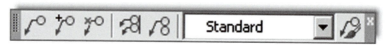

FIGURE 6.17
Multileader toolbar.

There is something new to the multileader that you did not have with the basic leader. You can set up a multileader style, so all leaders have the exact look you want. In our sample case, we give our multileader a bubble in which to add text. Press the very last button on the right (mleader with a paint brush) and the dialog box shown in Figure 6.18 appears.

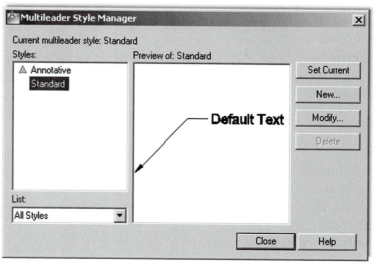

FIGURE 6.18
Multileader Style Manager.

Press New... and give the multileader a name, such as *Sample Style*, and press Continue. You are taken to the Modify Multileader Style dialog box (Figure 6.19). Examine the three tabs, Leader Format, Leader Structure, and Content, carefully. Most of the options are reasonably self-explanatory, and you will see some of them again later on. Under the Content tab, select Multileader type: as Block and Source block: as Detail Callout. Finally press OK, Set Current, and Close. Now let us try out our new multileader style.

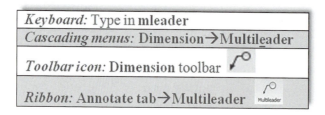

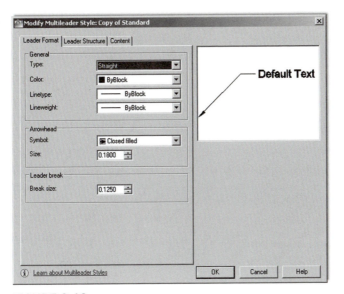

FIGURE 6.19
Modify Multileader Style.

Step 1. Start the multileader command via any of the preceding methods.
- ○ AutoCAD says: `Specify leader arrowhead location or [leader Landing first/Content first/Options] <Options>:`

Step 2. Click anywhere and extend your mouse (Ortho off) at 45° to the right.
- ○ AutoCAD says: `Specify leader landing location:`

Step 3. Click again to place the multileader and bubble. You see something like Figure 6.20.
- ○ AutoCAD says: `Enter view number <VIEWNUMBER>:`

Step 4. Enter in some value.
- ○ AutoCAD says: `Enter sheet number <SHEETNUMBER>:`

Step 5. Enter in another value.

Once you do this, you see the final result (Figure 6.21), the appearance of which can of course be fine-tuned.

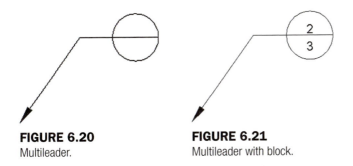

FIGURE 6.20
Multileader.

FIGURE 6.21
Multileader with block.

Now that you have one leader on the screen, it is easy to experiment with other interesting options. Move across the toolbar, starting first with Add Leader (Figure 6.22) and get rid of it via Remove Leader. Then, after adding it back in, create a new set of leaders and line them up using Align Multileaders (Figure 6.23), which aligns the leaders with one that you pick as the primary. Finally, Collect Multileaders (Figure 6.24) combines them into one string.

Secondary Dimensions

It was mentioned that there are some secondary dimensions, meaning they are not used as often. We briefly discuss them here, and you may want to also go over them in detail on your

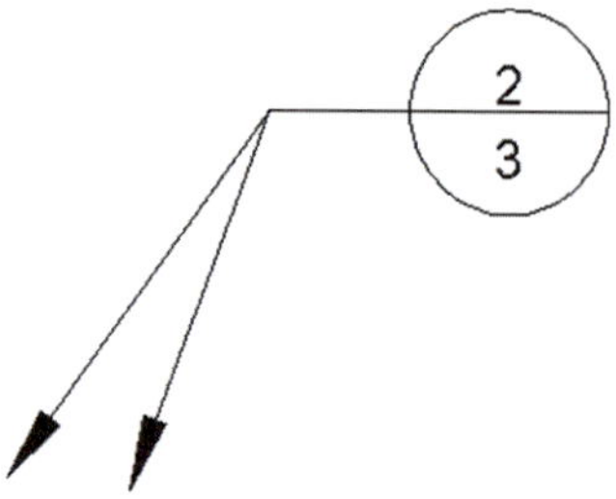

FIGURE 6.22
Add Leader.

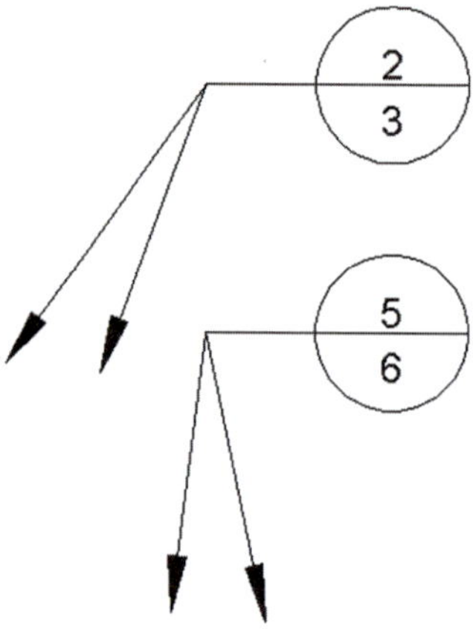

FIGURE 6.23
Align Multileaders.

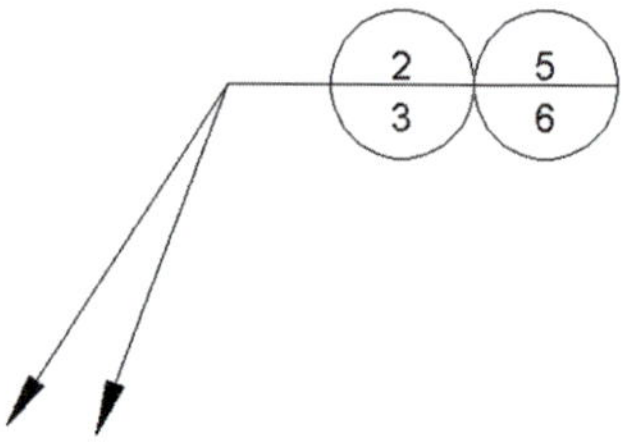

FIGURE 6.24
Collect Multileaders.

own, especially if you feel one or more of them may be extremely useful to your particular field.

- Arc length
- Ordinate
- Jogged

Arc length, as you may have guessed, measures the length of an arc. Simply select the toolbar icon and click on the arc you want to measure, as seen in Figure 6.25.

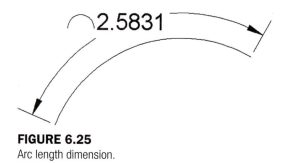

FIGURE 6.25
Arc length dimension.

Ordinate dimensions (Figure 6.26) measure horizontal or vertical distances from a datum point (0,0 in this case). They are used in manufacturing to prevent accumulation of errors that can occur using continuous measurements. Draw the shapes shown in Figure 6.26, positioning the lower left corner of the rectangle at 0,0, and begin the ordinate dimension. Then click major points along the way in either horizontal or vertical directions to get precise values at that location from the origin.

Jogged dimensions are useful when the radius or diameter dimension's center is off the sheet of paper and cannot be shown directly, so a "break line" of sorts is used. You simply select the arc or circle then the center location override and a jogged dimension appears (Figure 6.27).

There are other dimension options we have not yet explored. We come back to them in Level 2. For now we must move on and learn how to do editing and customizing.

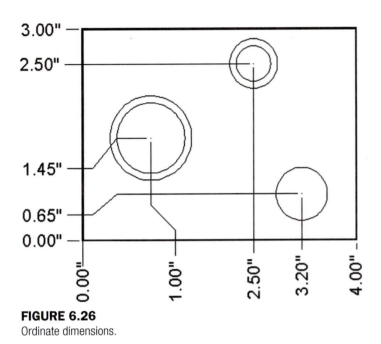

FIGURE 6.26
Ordinate dimensions.

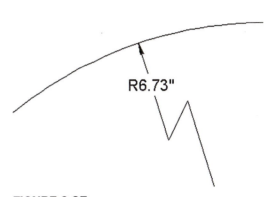

FIGURE 6.27
Jogged dimension.

6.3 EDITING DIMENSIONS

You should now be quite familiar with the types of dimensions available and how to apply them to basic shapes. The next step is to be able to edit them or change their values, if needed. This is a very simple and short topic.

You already learned that you can edit text and mtext by double-clicking on it. The text in the dimensions is essentially editable mtext, so you would be tempted to double-click on the value to edit it. Not so fast though. If you do that, then the Properties dialog box appears. You first saw it in Chapter 3, and in this case, it allows you to change virtually any property of the dimension. We have not yet gone over any of those properties (we will in the next section), and in any case, it is not what you want. So how do we edit dimension values?

Recall in Chapter 4, we covered *ddedit* and mentioned it was also useful for editing. Go ahead and create a horizontal dimension again, type in `ddedit`, and press Enter. Click on the dimension value and the mtext editing box appears (Figure 6.28). You can now type in any value you need. Edit as usual and click on OK when done.

Here is a useful tip. If you want to reset your forced dimension value back to its natural value and you forgot what that was, just type in ⟨ ⟩. These two "alligator teeth" brackets restore the natural value. Type them in instead of the forced value and click on OK. Try it yourself.

6.4 CUSTOMIZING DIMENSIONS

We have one more topic to cover on our way to basic proficiency in dimensioning. It is customization, and as mentioned at the start of the chapter, it is an extensive subject, necessitating it being split into the fundamentals here and advanced customization in Chapter 13. The goal here is to learn what experience has shown to be the most important four customization tools. This allows you to be productive in most of the drafting situations you are likely to encounter as a beginner. Later, you will add to this knowledge by learning tools used by CAD administrators and senior designers.

So what customization is necessary at this level? We need to learn how to

- Change the *units* of the dimensions.
- Change the *font* of the dimensions.

145

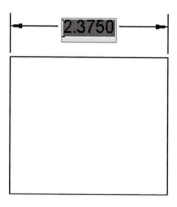

FIGURE 6.28
Mtext edit of dimensions.

- Change the *arrowheads* of the dimensions.
- Change the *fit* (size) of the dimensions.

To be able to change any of these we need to introduce a new tool, the Dimension Style Manager dialog box or *dimstyle* for short. There you will find a button to bring up the Modify Dimension Style dialog box. It is a "one-stop-shopping" dialog box for dimension modification that you will eventually need to learn backward and forward.

Dimstyle

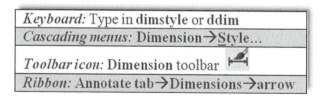

Keyboard: Type in **dimstyle** or **ddim**
Cascading menus: Dimension→Style…
Toolbar icon: Dimension toolbar
Ribbon: Annotate tab→Dimensions→arrow

Using any of the preceding methods, bring up the Dimension Style Manager, as seen in Figure 6.29.

The Dimension Style Manager dialog box has some options for creating a new dimension style or just modifying an existing one. For us at this point it really does not matter which we pick, so just press Modify….

You then are taken to the Modify Dimension Style dialog box, shown in Figure 6.30. Note that you may not necessarily see the Lines tab as the open tab as seen in Figure 6.30. If you do not, just click on it to have the same thing on your screen as what is in the text.

This rather imposing dialog box contains tools to significantly modify any dimension. We, in fact, do so in Chapter 13 as we explore nearly every feature. For now, we just want to focus on learning how to modify the handful of items shown in the bulleted list at the start of Section 6.4.

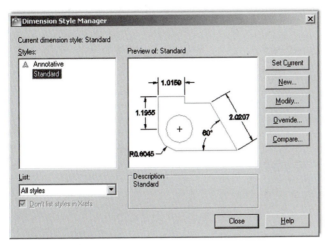

FIGURE 6.29
Dimension Style Manager.

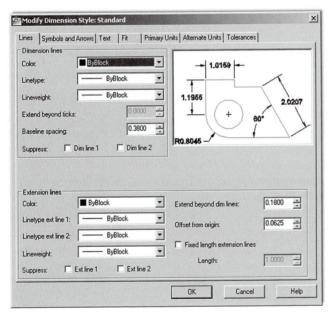

FIGURE 6.30
Modify Dimension Style (Lines tab).

STEP 1. CHANGE THE *UNITS* OF THE DIMENSIONS

Pick the Primary Units tab and simply select the drop-down menu at the very top left, where it says `Unit format:`. The default value is Decimal; change it to Architectural and select the appropriate Precision in the menu just below it (the default value is fine). That is all we need from that tab for now. Notice how the preview window on the upper right reflects your choice of units, as seen in Figure 6.31.

STEP 2. CHANGE THE *FONT* OF THE DIMENSIONS

To do this you need to click on the Text tab. You then see `Text style:` on the upper left. If you just opened a new AutoCAD file to practice dimensions, you probably do not have

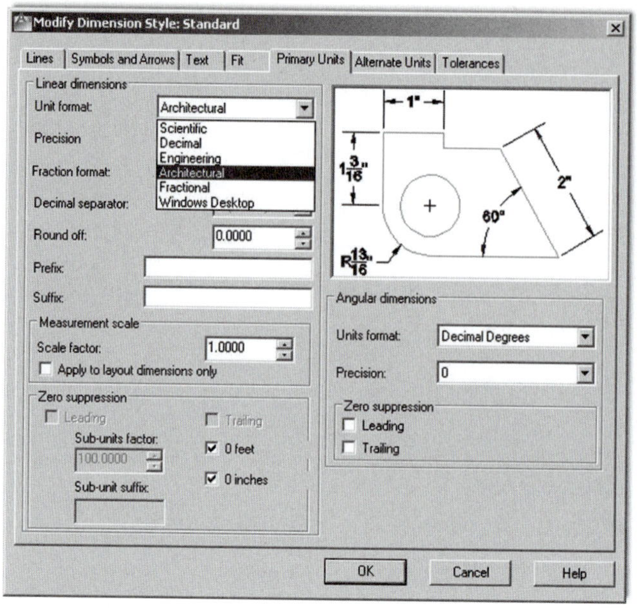

FIGURE 6.31
Selecting Architectural units.

a font set and get only Standard as your choice if you click the down arrow. Fortunately, Standard is now a more attractive Arial font, not the stick-figure Simplex it used to be, so changing fonts is optional but still useful to know. For example, the RomanS font is a popular alternative.

Something else to keep in mind is that, in a real working drawing, you would likely set your text style before you work on any dimensions (an important point, take note). Then, you would just select the font from the list. The idea here is to match up your regular font used in text and mtext with the font used in dimensioning, another important point. In general, you should stick to one font in a given drawing and just vary the size as needed. The title block is exempt from this, as that may have special fonts, logos, and the like.

AutoCAD of course anticipates that you may not have set a font style when you first set up dimensions and allows you to do this from the Text tab by bringing up the Style box when you click the button just to the right of the Text style menu (with the three dots, …). Go ahead and set a different font if you wish, RomanS 0.25" perhaps, and the new setting appears, as seen in Figure 6.32.

STEP 3. CHANGE THE ARROWHEADS OF THE DIMENSIONS

The default for all dimensioning is the standard arrowhead. You can easily change that to any other type, including the Architectural tick, popular with architects. You can even create a custom arrowhead (not something we try to do here).

Click over to the Symbols and Arrows tab. In the upper left corner under the Arrowheads section, you will see First:. That is the first arrowhead type, and if you change it, AutoCAD assumes you want the second one to be the same and changes it as well. Simply click the down arrow and select the Architectural tick. Leave everything else the same. The result is shown in Figure 6.33.

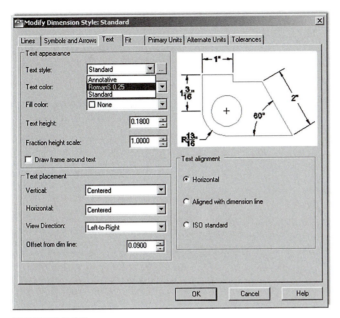

FIGURE 6.32
Setting a new font.

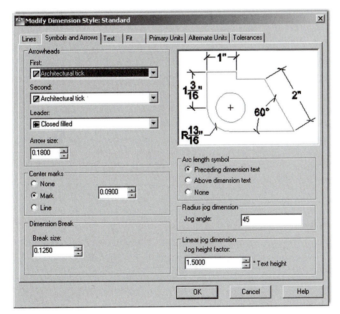

FIGURE 6.33
Changing the arrowheads.

STEP 4. CHANGE THE FIT (SIZE) OF THE DIMENSIONS

This function is quite important. Located in the Fit tab, just underneath the preview window, it features only a few lines and is shown in Figure 6.34.

The idea here is to increase the size of the dimensions proportionately so everything scales up evenly and smoothly in step with the size and scale of your overall drawing. This concept,

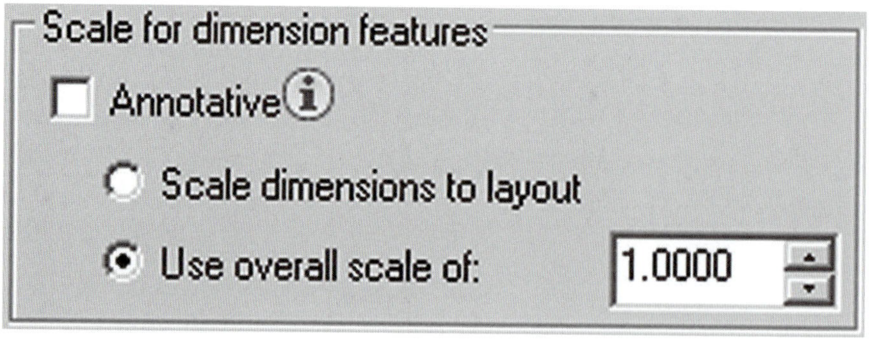

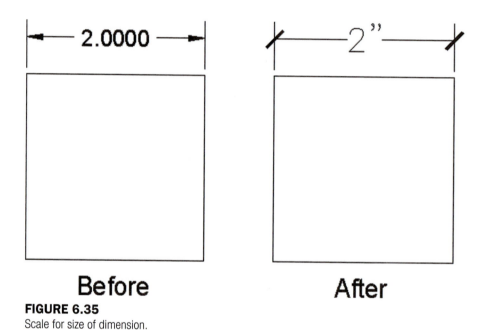

FIGURE 6.34
Scale for dimension features.

Before After

FIGURE 6.35
Scale for size of dimension.

however, opens up a number of questions. Chief among them, What value do you enter into there and why? We are not ready to discuss this yet but we do in Chapter 10 (on paper space). So for now do not enter in anything but be aware of how to do it when needed. When you get to dimensioning the floor plan in the next section, enter 15 in that space.

This is it for now; just click on OK and AutoCAD returns you to the Dimension Style Manager, where you can just press Close. Try creating a dimension and see what it looks like now. Figure 6.35 shows a set of horizontal dimensions before the changes we just went through and another one afterward. The difference is obvious. For now, this is all you need to make professional-looking dimensions. Review and memorize everything you learned and go on to the floor plan in Figure 6.36, adding dimensions to complete the drawing.

6.5 IN-CLASS DRAWING PROJECT: ADDING DIMENSIONS TO FLOOR PLAN LAYOUT

Let us now apply what we learned to the floor plan. Open the file and freeze all the layers except the walls and windows. If you really want to get fancy, put the wall solid hatch on its

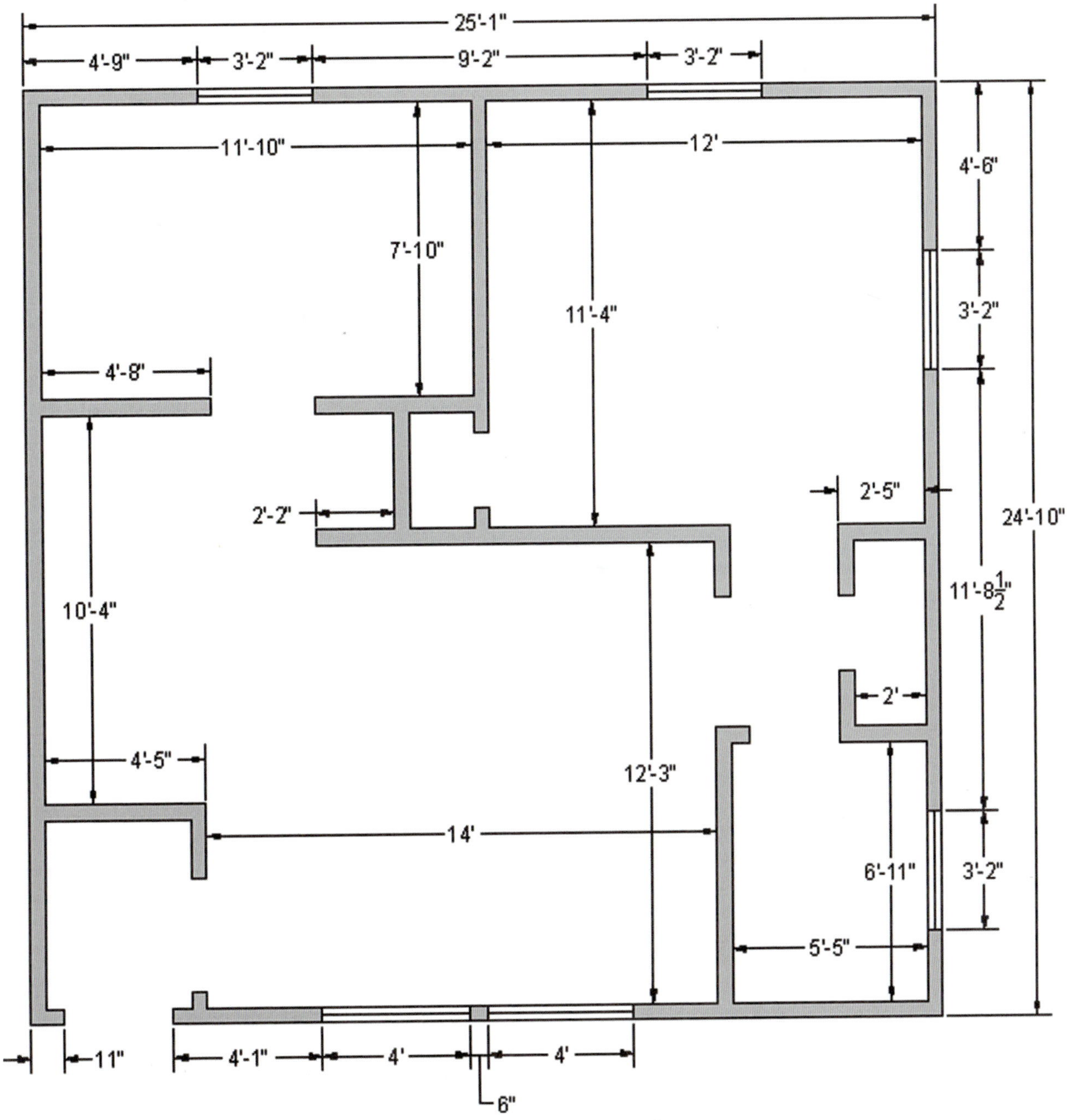

FIGURE 6.36
Adding dimensions to floor plan.

151

own (visible) layer and freeze the regular floor hatch layer so the carpeting and floors do not show. Next create a new layer for the dimensions, A-Dim. Finally, set up the dimensions: Use Architectural units and Arial 6" font, leave the arrowheads as they are, and change the scale under the Fit tab to 15. Dimension the floor plan any way you want; what is shown in Figure 6.36 is a guide but not the only way to do it. Note that the outer dimensions are continuous.

SUMMARY

You should understand and know how to use the following concepts and commands before moving on to Chapter 7:

- Dimensions
  - Linear (horizontal and vertical)

- ○ Aligned
- ○ Diameter
- ○ Radius
- ○ Angular
- ○ Continuous
- ○ Baseline
- ○ Leader and multileader
- ○ Arc length
- ○ Ordinate
- ○ Jogged
- Editing dimension values
  - ○ `ddedit`
  - ○ The effect of < >
- `ddim` command
  - ○ Dimension units
  - ○ Dimension font
  - ○ Dimension arrowheads
  - ○ Dimension overall size

REVIEW QUESTIONS

Answer the following based on what you learned in Chapter 6.

1. List the 11 types of dimensions discussed.
2. What command is best for editing dimension values?
3. What is the difference between natural and forced dimensions?
4. How do you restore a natural dimension value when you forgot what it was?
5. List the four items that needed to be customized in our dimensions.
6. What command brings up the Dimension Style Manager?
7. What type of arrowhead do architects usually prefer?

EXERCISES

1. Retrieve Exercises 2, 3, 4, and 6 from Chapter 3 and set up and add all shown dimensions. (Difficulty level: Easy; Time to completion: 10–15 minutes)
2. Retrieve Exercises 4, 5, and 6 from Chapter 4 and set up and add all shown dimensions. (Difficulty level: Easy; Time to completion: 10 minutes)
3. Retrieve Exercises 3 and 4 from Chapter 5 and set up and add all shown dimensions. (Difficulty level: Easy; Time to completion: <10 minutes)

4. In a new file, set up the appropriate layer, colors, and linetypes. Then, draw and fully dimension the following mechanical part. Select Decimal as the dimstyle, with precision of 0.00 and inch symbols (") for a suffix. (Difficulty level: Easy/Moderate; Time to completion: 35–45 minutes)

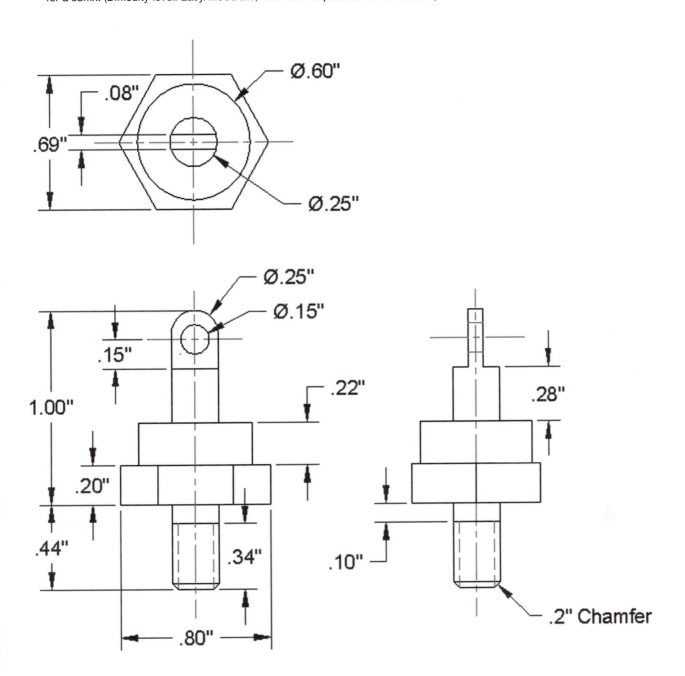

153

5. In a new file, set up the appropriate layer, colors, and linetypes. Then, draw and fully dimension the following mechanical plate. Select Decimal as the dimstyle, with precision of 0.00 and inch symbols (") for a suffix. Make careful use of mirror to minimize redundant work. All sizing is based on the smallest two circles and their position from an arbitrary centerline. (Difficulty level: Moderate; Time to completion: 35–45 minutes)

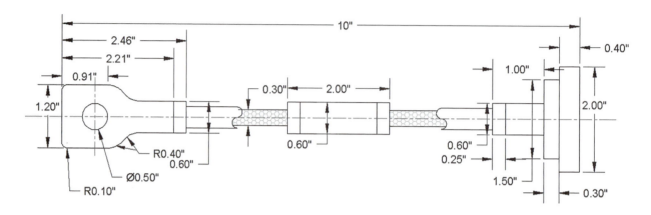

6. In a new file, set up the appropriate layer, colors, and linetypes. Then, draw and fully dimension the following mechanical part. Select Decimal as the dimstyle, with precision of 0.00 and inch symbols (") for a suffix. (Difficulty level: Moderate; Time to completion: 35–45 minutes)

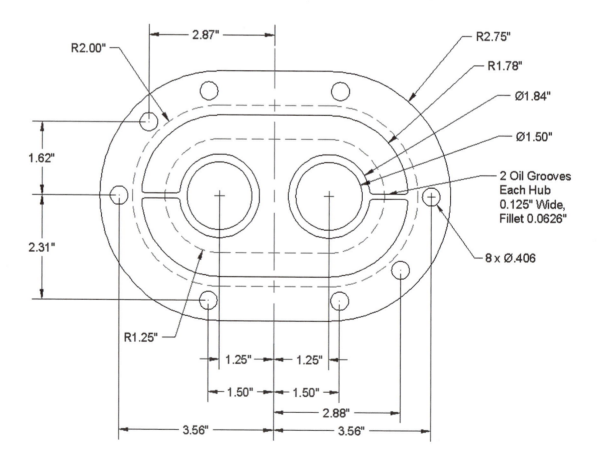

7. In a new file, set up the appropriate layer, colors, and linetypes. Then, draw and fully dimension the slightly simplified version of the bolt assembly seen on the cover of this chapter. Select Decimal as the dimstyle, with precision of 0.00 and inch symbols (") for a suffix. Note that no dimensions are available for some internal parts of the bolt; you have to improvise. (Difficulty level: Moderate/Advanced; Time to completion: 60 minutes)

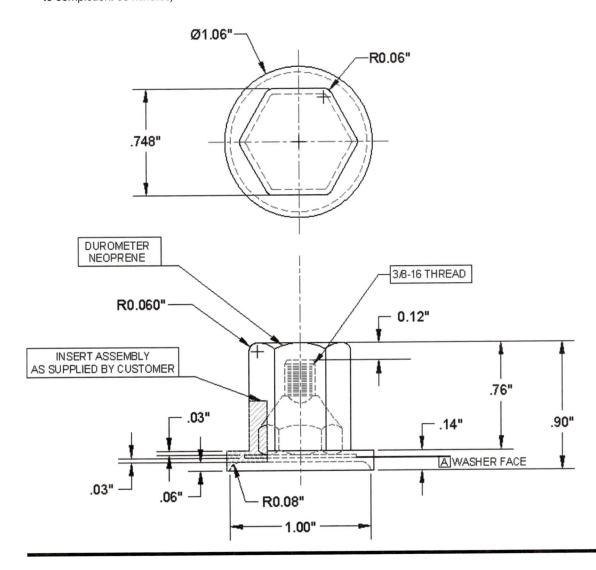

CHAPTER 7

Blocks, Wblocks, Dynamic Blocks, Groups, and Purge

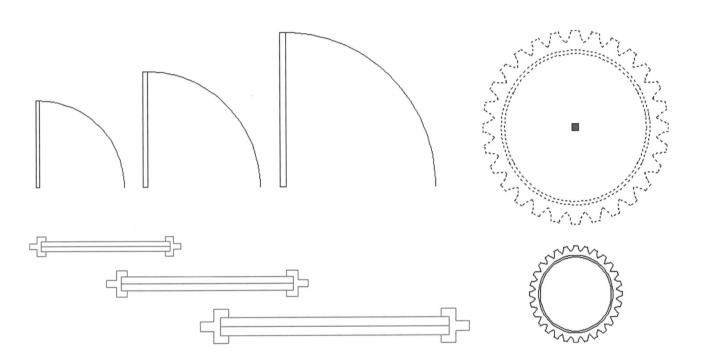

At the end of this chapter you will be able to create blocks, wblocks, dynamic blocks, and groups as well as insert symbol libraries and purge your drawing.

Estimated time for completion of chapter: 1–2 hours.

Up and Running with AutoCAD® 2012: 2D and 3D Drawing and Modeling.

7.1 INTRODUCTION TO BLOCKS

This chapter is about some very useful AutoCAD concepts that greatly simplify your workload. Blocks, wblocks, dynamic blocks, and groups are all members of the same family. They exist around the idea that objects can and should be grouped together if possible, to allow for easier handling and storage for future reuse.

Let us formally define a block. It is a collection of objects that are grouped and held together under some identifying name. These objects can then be copied, moved, erased, and just about anything else, all as one unit. Think of blocks as electronic "glue" that binds the objects they contain to each other.

There are many benefits to this. If you have an office table with eight chairs around it, and you need to move them all around many times while optimizing a room layout, it is a lot easier to click on a block once than attempt to select all nine pieces using a window or a crossing or one at a time. Imagine an engineering example involving a simple gear with all its intricately designed teeth. Dozens if not hundreds of lines and arcs may make up the gear. It would be not just wise but mandatory to form them all into a single block.

There is however another, more compelling, reason for blocks. That is the concept of *reuse and symbol libraries*. Blocks can be saved and reused for future drawings that may need the same exact object. That way you never have to draw the same thing twice, an AutoCAD Golden Rule. A collection of these objects cataloged in some orderly fashion becomes a symbol library, an indispensable part of just about any designer's toolset. Likely candidates for these libraries include doors, windows, furniture, bolts, standard parts, fasteners, and really just about anything that can be used in more than one design or drawing.

Difference between Blocks and Wblocks

Ideally, you are sold on the importance of blocks as we go through the specifics. The differences between blocks and wblocks (write blocks) are blurred; they are alike in many ways. The key difference is that blocks are *internal* to the file you are working on and wblocks are *external*. This means that blocks are saved inside whatever file you are currently working on. As such, their main use is for grouping things together and reuse in that particular AutoCAD drawing. While they can certainly be moved between files, via the clipboard Copy/Paste tool, this is not really what they are intended for.

Wblocks are intended for creating symbol libraries. Wblocks are saved externally, anywhere you want to put them, and become independent, stand-alone AutoCAD drawings, which you can open and work on if needed. Both types of blocks can of course be easily inserted back into drawings, and we cover that procedure in detail.

Creating a Block

To create a block you first need something to make a block out of. Once you draw something, only three things need to be done:

1. Select the object(s).
2. Give the block a name.
3. Select a base point.

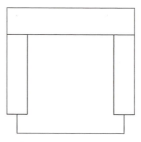

FIGURE 7.1
Basic chair symbol.

Let us try this out. Draw a chair, as shown in Figure 7.1. It is made up of three rectangles and three lines. Use accuracy tools.

We need to now group them all together into one object.

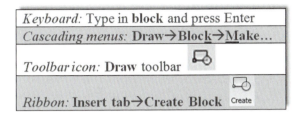

Keyboard: Type in **block** and press Enter	
Cascading menus: **Draw→Block→Make...**	
Toolbar icon: **Draw** toolbar	
Ribbon: **Insert tab→Create Block** Create	

Start the block command via any of the preceding methods. The Block Definition dialog box appears, as seen in Figure 7.2.

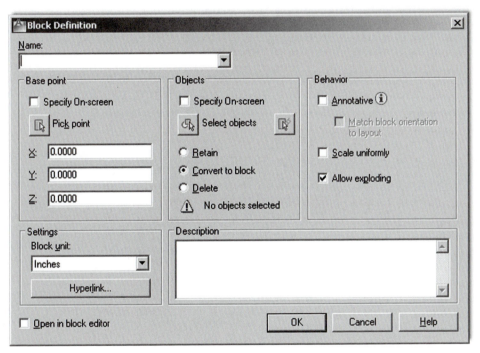

FIGURE 7.2
Block Definition dialog box.

159

Step 1. Press the Select objects button in the middle column. Select the entire chair and press Enter.
Step 2. Enter a name for this block in the top left field under Name.
Step 3. Pick a base point with the Pick point button at the upper left. Select a corner or midpoint somewhere on the object. The significance of this becomes more apparent when we insert the object back into the file.

There are of course a few other items in the dialog box under Settings, Behavior, and Description, but they are not usually needed (though feel free to enter a description). Annotative is covered in advanced chapters. The one setting of importance, however, is the Retain, Convert to block, and Delete choices under the Select objects button in the middle column. They are explained next and appear with almost similar wording in wblock as well.

RETAIN

This means that the block is created and stored in memory, but the original object (the chair you are looking at) remains in separate pieces, useful if you wish to make several blocks of the chair, one after the other, each slightly different yet based on the same original.

CONVERT TO BLOCK

This option is what you are likely to use the most. It creates the block and stores it in memory and also makes a block out of the original object. Use it when you are sure one block is all you need to make out of this one object and you have a need for it right away.

DELETE

Here the block is created and stored in memory, but the original is deleted from the drawing. This is useful when you are making a block for future use but have no need for it now.

The common theme with all three of these choices is that, whichever way you use it, the block is created and saved to memory. The only real choice is what to do with the original.

Finally you are ready to click on OK. The Block Definition dialog box should look like Figure 7.3. Notice the small preview window in the upper middle of the dialog box.

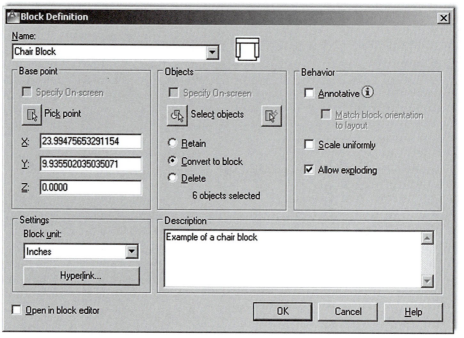

FIGURE 7.3
Completed block definition.

The chair is now a block, which can be proven by clicking on it once with the mouse. Notice that only one blue grip point appears (wherever you picked the base point; see Figure 7.4). If you now erase it, the entire chair will disappear. Go ahead and erase it; we need to do this to introduce two new commands before we discuss wblocks.

7.2 INSERT

As mentioned before, the process of making the block adds it into the memory of the file you have open. Then, you can insert it back into the drawing when needed.

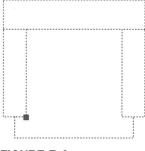

FIGURE 7.4
Chair as a block.

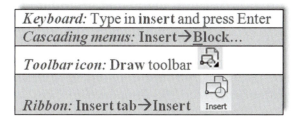

Keyboard: Type in **insert** and press Enter	
Cascading menus: Insert→Block...	
Toolbar icon: Draw toolbar	
Ribbon: Insert tab→Insert	Insert

Start up the insert command via any of the preceding methods and the Insert dialog box appears, as seen in Figure 7.5. Notice the chair is cued up and ready to go, as it is the only block created so far.

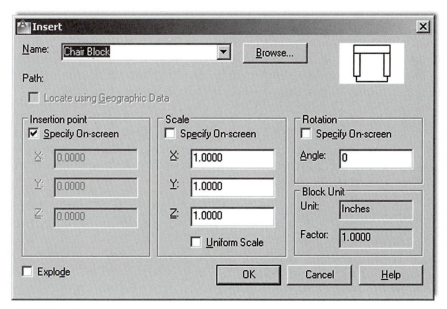

FIGURE 7.5
Insert dialog box.

Look over the various options, such as Insertion point, Scale, Rotation, and Block Unit, but at this point you need not do anything extra. Go ahead and click on OK, and the chair appears attached at the insertion point to the mouse. The point of attachment was specified when you made the block. Click anywhere on the screen and the block becomes part of the drawing. This process can be repeated as often as necessary as long as the block remains in memory. Note that you can always specify a precise point of insertion for the block by typing in the coordinates (example: 0,0,0).

There is something important to remember concerning redefining blocks. If a block is saved under a certain name and another block is created and saved under the same name, the first block is redefined, a sort of overwriting process. Sometimes this is bad, as you are destroying the old block. Sometimes, however, this is desirable, such as when you are looking to update the appearance of the block. When there are many blocks in the drawing under the same name, all are updated in this manner.

7.3 PURGE

Between introduction of the block and wblock concepts is a good time to take an intermission of sorts and discuss a command called *purge*. Purge is a very useful tool to "clean up" your drawing of unwanted and often unseen geometry and data. It is especially useful when it comes to blocks and wblocks, hence its inclusion here.

Conceptually, here is the idea. A complex AutoCAD drawing often grows in size and becomes weighed down, similar to an iceberg floating in the ocean. Much like the mountain of ice, mostly hidden under the surface, what you see on screen may only be a small fraction of what actually is present in the file.

The reason why becomes apparent when you think about the drafting process for a moment. If you create 50 layers for anticipated future use but end up needing only 40 of them, the other 10 are uselessly hanging out in your layer dialog box. Same thing with unused fonts or linetypes (remember loading *all* of them but using only a few). Layers, linetypes, and fonts really take up little space, but with blocks and wblocks, it is a whole different story.

Wblocks are miniature drawings all to themselves, and while many are just simple doors or window symbols, others can be quite large, such as entire furniture sets or appliances in an architectural layout. Worse yet, these blocks rarely come alone; they bring friends, lots of them. Complex multistory sprawling building layouts feature not dozens but *hundreds* of blocks. They all take up room in a file, ballooning it up to a rather huge size.

If all these blocks are needed, then fine; this is a necessary evil. But if not—say, a type of chair was deleted from the specs—they need to be removed. Erasing them all is easy enough, but are you done? Are they all gone for good? Unfortunately, no. They may not be visible but they are still there weighing down your file, slowing down computer performance and just about everything else. This is where purge comes into play.

Purge gets rid of items that are not actually used in the drawing file and permanently deletes them. You can use the command to selectively purge items or purge everything all at once. The two main points to remember are

- Purge will not get rid of objects that are visible and present in the drawing—it is *not* an erase command.
- Purge is permanent (unless you immediately undo it), so be careful of what you purge. If you anticipate eventually needing something, do not purge it out.

Let us give the command a try.

Keyboard: Type in **purge** and press Enter	
Cascading menus: **File→Drawing Utilities→Purge…**	
Toolbar icon: **none**	
Ribbon: **none**	

First of all, erase the chair from your screen. Then, start up the purge command via one of the two methods shown in the command matrix and the dialog box in Figure 7.6 appears.

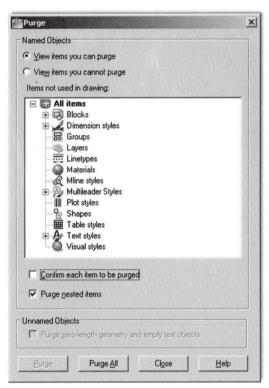

FIGURE 7.6
Purge dialog box.

As you can see from looking at the list of items in the dialog box, quite a few items are eligible for purging. Some were already mentioned, like Layers and Linetypes, while others you may not have heard of, like Materials and Shapes. Either way, what is important is whether or not the item has a plus sign next to it. If it does, you can click the plus sign and expand the folder to see what specific items are in those general categories. In our case, you see the chair block under the Blocks category. You will also see Annotative under some of the other categories. We look at Annotative in later chapters, so for now just ignore it and purge Annotative alongside the chair block.

The actual process of purging is easy. You can either select each item one at a time (usually done only when you want to selectively purge) or you can click on Purge All at the bottom. Be sure the Purge nested items box is checked and the one above it is not, as seen in Figure 7.6. When all the items are purged, there are no more plus signs and the buttons gray out as well. Press Close after you are done and that is it.

Experienced users purge often and keep little useless junk in the file unless absolutely necessary. Purging is certainly something that should be done prior to saving the file for the day, emailing it, or archiving it for storage and record keeping. Be careful though, and do not get rid of needed items. If in doubt, do not purge the items.

7.4 WBLOCKS

Pretty much everything said about blocks applies to wblocks as well, and as a result, this section is short, as you have seen this before. The central idea to keep in mind is that a wblock is external, and as such, you need to tell AutoCAD where to put the wblock as you make it. Let us try this out. Clear your screen and make sure all items from the previous exercise are purged. Now go ahead and draw another furniture object, perhaps a sofa this time.

The only way to start up this command is via typing `wblock` and pressing Enter. Once you do that, the dialog box in Figure 7.7 appears.

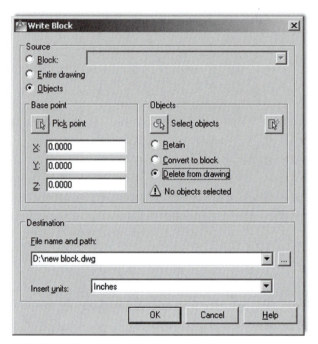

FIGURE 7.7
Write Block dialog box.

For the screen shot in this figure, a specific path has not been selected, but you can do that by clicking on the button with the three dots, …, to the right of File name and path: and browsing around until you find where you want the file to go. Be sure to also change the name of the file at the end of the string, so it does not just say New Block. Then select the sofa and pick one of the three options to Retain, Convert, or Delete the original. When done, press on OK and the wblock is created and dropped off where you indicated in the path.

Inserting Wblocks

Go ahead and insert your sofa back in using the insert command. The procedure is essentially the same, except that now you have to browse for the file. The Browse… button is just to the right of the Name: field. Look around for your block and when you find it, click on it and press OK. It inserts just as with the previous regular block command.

7.5 DYNAMIC BLOCKS

Dynamic blocks were introduced relatively recently to AutoCAD (in Release 2006). Like many new features, they are not necessarily a radically new concept. Rather, they are an evolution of the regular block or wblock. The idea here is to be able to modify a block in response to design conditions. A simple example is of a door. If you insert a block of a 3' door, rotated at 45°, and later realize that you need a 3'-6" door, rotated to a closed position, you can change that in place without redefining the block or inserting another one.

So what we are doing essentially is adding a level of intelligence and automation to the standard block concept. This is especially useful when it comes to scaling, stretching, and otherwise modifying the sizes, as that is cumbersome to do with regular blocks (as opposed to rotating them, which is easy).

To accomplish all this, the designer needs to modify a regular block by adding Parameters and Actions. Three things make up a dynamic block: the *geometry* itself, at least one *Parameter*, and at least one *Action* associated with that parameter. The result is a block that has one or more special grip points, each defined with a certain modifying property (e.g., stretch, rotate, scale). The block can then be modified in place as needed. Working with our existing chair block, follow the procedures outlined.

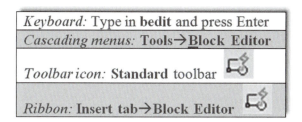

Keyboard: Type in **bedit** and press Enter

Cascading menus: **Tools→Block Editor**

Toolbar icon: **Standard** toolbar

Ribbon: **Insert tab→Block Editor**

Regardless of which method you use to start the command (you can also just double-click on the block itself), you will see the dialog box in Figure 7.8. Select the block you want to work with, which in this case is the only one available.

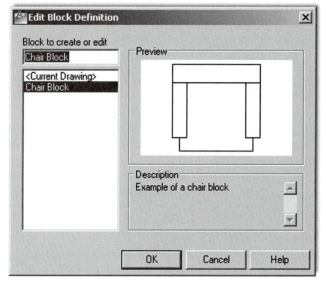

FIGURE 7.8
Edit Block Definition.

After you select Chair Block, you are taken to the main editing screen, called a *Block Editor* (recognizable by its gray background), shown in Figure 7.9. This is where the settings are applied.

A Block Authoring Palette also appears. Sets of parameters and actions are available from this palette, and the block can be set up to have any of those. When done, Close Block Editor is pressed (top right of screen) to return to the regular drawing space. So what can we do here? Well, let us add two *actions* and *parameters* to the chair that gives it some scaling and rotational abilities.

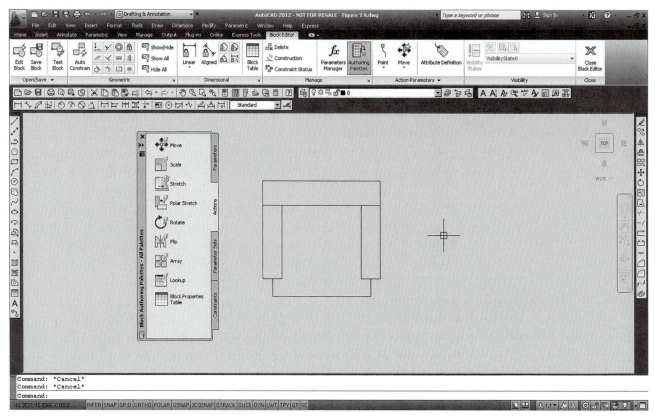

FIGURE 7.9
Dynamic Block Editor.

Step 1. Open the Parameter tab on the palette and pick the Linear parameter. That, after all, is the way scale works, by linearly scaling up an object. First pick the upper left corner start point, then pick the upper right corner endpoint of the chair (like adding a horizontal dimension). The result is shown in Figure 7.10.

Step 2. While you are there, add a rotation by running through the required Base Point (lower left), Radius (any value), and Angle (30°) prompts. The result is shown in Figure 7.11.

Step 3. Now, you can add actions associated with those parameters. Pick the Actions tab and pick scale. Then, select the Parameter and the Objects (the chair), as prompted. Repeat the same steps for the rotation. The result is shown in Figure 7.12. Notice the new (faded) scale and rotate symbols.

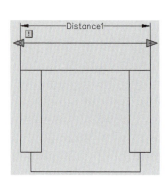

FIGURE 7.10
Dynamic block, Step 1.

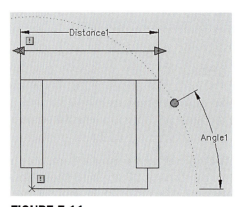

FIGURE 7.11
Dynamic block, Step 2.

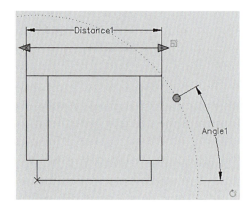

FIGURE 7.12
Dynamic block, Step 3.

Close the Block Editor (button at upper right) and you are prompted to save your work, as seen in Figure 7.13.

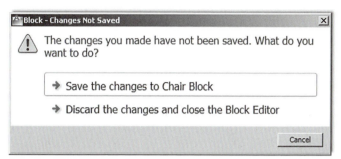

FIGURE 7.13
Save changes prompt.

Finally, you are back where you started. Click on the new block and notice the dynamic grips, as seen in Figure 7.14. You can now use these to change the size and rotation of the block, as seen in Figure 7.15 with scaling.

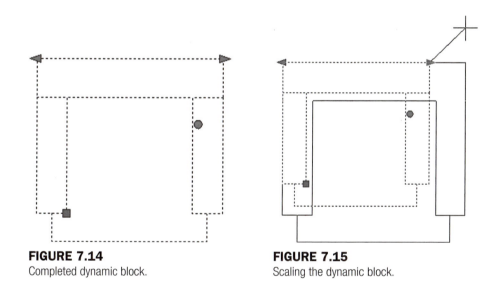

FIGURE 7.14
Completed dynamic block.

FIGURE 7.15
Scaling the dynamic block.

There are of course a few more tricks to dynamic blocks, chief among them being the new parametric tools, but we address those in the advanced level.

So, how do dynamic blocks fit into the big picture? Are they as useful as they seem? The results are mixed. The idea is certainly a good one, but the effort involved to add meaningful and extensive dynamic intelligence to a large group of different blocks is considerable. Existing libraries of blocks have sufficed for most users, who diligently built them up over the years, and experience has sometimes shown that there has been a reluctance to redefine these libraries. A lack of time and money to tinker with AutoCAD (as opposed to using AutoCAD for billable drafting work) has always been a problem. In a sense, it may be a shame, as dynamic blocks, by allowing multiple variations on a single block, may actually reduce the sizes of libraries. A door may just be a door, adjustable to any size on the go, instead of 20 doors for every conceivable situation. It is a useful, if underused, idea.

7.6 GROUPS

A group is a useful, but often overlooked, command from the block family. It is perhaps the easiest way to collect objects together and presents some advantages (and disadvantages) over the familiar block as shown next.

Similar to a block, a group can

- Be given a name.
- Bind a collection of objects together.
- Be moved around, copied and rotated.

Advantages of a group:

- A group can have objects added and subtracted from it.
- A group can have individual objects (and their properties) edited.

Disadvantage of a group:

- A group cannot be transferred between drawings (for "local" use only).
- An update to one group does not update other groups with the same name.

Group went through some changes for AutoCAD 2012. You now generally have to use the command line or the Group toolbar (Figure 7.16) to create and manage the groups, but the classic Object Grouping dialog box is still available as one of the options and is discussed.

FIGURE 7.16
Group toolbar.

To try this command, first draw a random collection of objects as seen in Figure 7.17, then follow the steps shown.

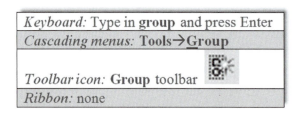

Keyboard: Type in **group** and press Enter
Cascading menus: **Tools→Group**
Toolbar icon: **Group** toolbar
Ribbon: none

Step 1. Begin the group command via any of the preceding methods.
- AutoCAD says: `Select objects or [Name/Description]:`
Step 2. Press n for `Name`.
- AutoCAD says: `Enter a Group Name or [?]:`
Step 3. Type in a descriptive name: `Sample_Group` in this case. Note that spaces in the name are not allowed.
- AutoCAD says: `Select objects or [Name/Description]:`
Step 4. Select all the objects.
- AutoCAD says: `Group "Sample_Group" has been created.`

The group is now completed, and you can see results by clicking on any object (Figure 7.18). Note how there is a rectangle around the objects and a single grip in the middle. You can now manipulate all the objects together as one unit.

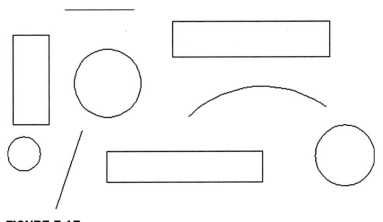

FIGURE 7.17
Random objects for grouping.

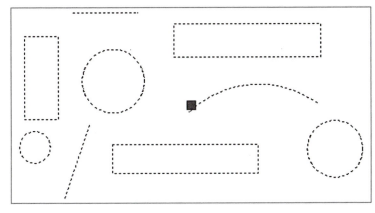

FIGURE 7.18
Completed group.

Let us take a look now at some of the features of a group and some additional functionality. One property you can take advantage of right away, and the main reason to use a group, is the ability to alter the properties of the objects in the group (or the shapes of the objects themselves). Double-click on any of the shapes and a palette called QuickProperties appears, as seen in Figure 7.19.

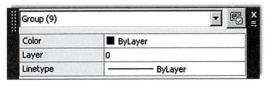

FIGURE 7.19
Group QuickProperties.

Drop down the top menu and you see a breakdown of the group, object by object. There are three rectangles (polyline), two lines, one arc, and three circles, for a total of nine objects. Select Circle (3) in the menu and the palette shows only the properties of the circles. Go ahead and change the color to red and observe the result (Figure 7.20).

169

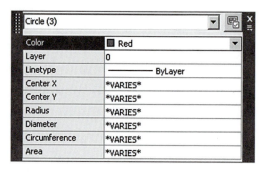

FIGURE 7.20
Color change to circles in group.

You can also work with individual objects in the group by clicking on the very last icon on the Group toolbar, called *Group Selection On/Off*. You can do the same by typing in `pickstyle` and entering a 0 when prompted. Pickstyle is a system variable, and more will be said about system variables in Chapter 15. For now, go ahead and modify any of the shapes, and when done, press the same icon or type in the same system variable (entering a 1 this time) to get back to the grouping. Figure 7.21 shows the editing in progress.

170

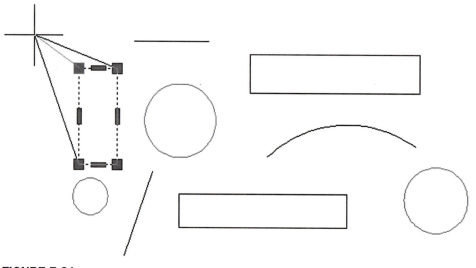

FIGURE 7.21
Group object editing.

Take a look at the other toolbar icon. The second one from left, called *Ungroup*, as you would expect, removes the grouping from the objects. Note that exploding the group does *not* delete the group itself but rather just explodes the objects inside of it—only the rectangles in this case.

The next icon to the right (with the +1/−), called *Group Edit*, allows you to add an object to or subtract and object from the group. Go ahead and try this on your own by deleting a rectangle via the `Remove objects` option and adding a new object (a polygon in this example) via the `Add objects` option. An example of the result is shown in Figure 7.22.

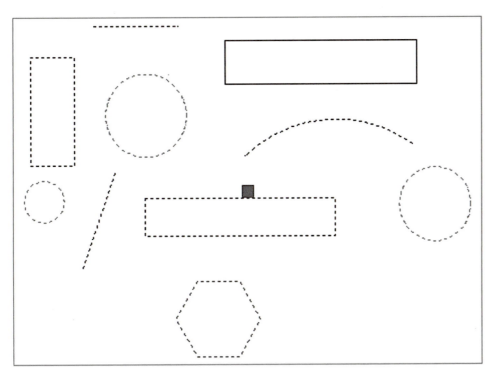

FIGURE 7.22
Object added (polygon) and another subtracted (rectangle).

Finally, it is necessary to mention the legacy Object Grouping dialog box (Figure 7.23). It serves two purposes here. One is to provide you an additional way of managing your groups if you are using AutoCAD 2012. The second is to familiarize you with the process if you are using an older version of AutoCAD, prior to the 2012 release, where this was the only way to create and manage groups.

171

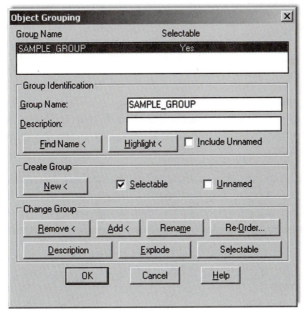

FIGURE 7.23
Object Grouping dialog box.

The Object Grouping dialog box can be called up by typing in `classicgroup` and pressing Enter. Note that, in this example, the SAMPLE_GROUP is already created and is highlighted in the dialog box to reveal a multitude of options that should generally be familiar to you, based on the preceding discussion.

From this dialog box you can create new groups, remove and add objects, and execute many other functions. Take some time to go over this dialog box on your own to complete your tour of the group command.

SUMMARY

You should understand and know how to use the following concepts and commands before moving on to Chapter 8:

- Block
  - Select objects
  - Name the block
  - Select a base point
- Wblock
  - Select objects
  - Select path and enter block name
  - Select a base point
- Options
  - Retain
  - Convert to block
  - Delete
- Insert command
- Purge command
- Dynamic blocks
  - Parameters
  - Actions
- Groups

REVIEW QUESTIONS

Answer the following based on what you learned in Chapter 7.

1. What are blocks and why are they useful?
2. What is the difference between blocks and wblocks?
3. What are the essential steps in making a block? A wblock?
4. How do you bring the block or wblock back into the drawing?
5. Explain the purpose of purge.
6. Explain the basic idea behind dynamic blocks. What steps are involved?
7. What are the advantages and disadvantages of using groups?

EXERCISES

1. In a new file, create the following architectural table, chair, keyboard, phone, and computer screen and make a block out of the individual pieces as well as the overall "assembly." Nested blocks, such as these, are quite common in AutoCAD design. Insert the block back into the file. Only general dimensions are given, so you have to improvise for the individual pieces. Also, naming the blocks is up to you, just make them descriptive. (Difficulty level: Easy; Time to completion: 15–20 minutes)

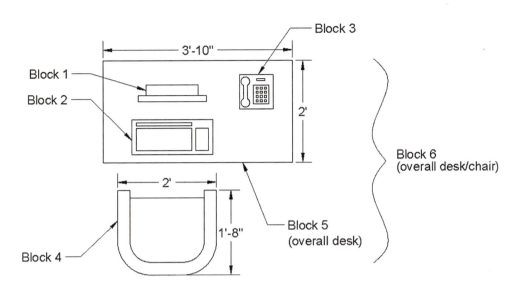

2. In a new file, create the following set of bathroom toilet and sink fixtures, doors, and other items. The overall dimensions are provided, but you need to create the details on your own. Be sure to use the appropriate layers. Create basic blocks out of the fixtures, and copy as needed. Then, create the floor plan depicted next. Insert the blocks into the floor plan as shown, adding the wall hatching as a last step. (Difficulty level: Moderate; Time to completion: 30–45 minutes for blocks, 20–30 minutes for floor plan)

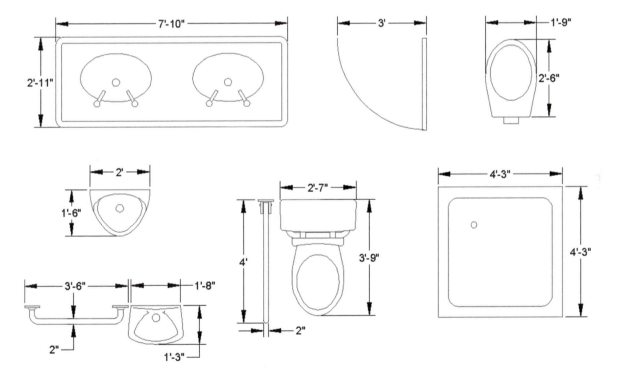

173

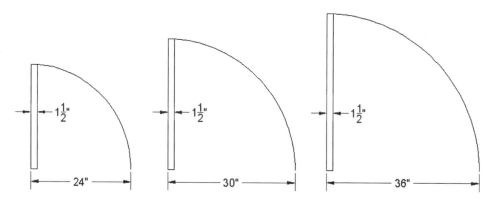

33'-10"

6"

4'-1"

2"

1'-6"

1'-3"

1'-4"

9"

16'-11"

2'-7"

12'-5"

7'-2"

28'-11"

9"

9'-1"

9'-1"

9'-8"

6'-6"

8'-1"

3'-1"

9"   5'-3"   5'-6"   4'-8"   4'-8"   4'-8"   4'-5"

1'-4"   2'-7"

3. In a new file, create the following set of doors and make a wblock out of each one (Door_24in, Door_30in, etc.). Drop them all into a premade folder called *Doors*. Insert each wblock back into the file. When all three doors are inserted, explode them all and purge the file. Then, create dynamic blocks out of all three. Add the scale action and linear parameter. (Difficulty level: Easy; Time to completion: 5–10 minutes)

$1\frac{1}{2}$" $1\frac{1}{2}$" $1\frac{1}{2}$"

24"   30"   36"

4. Electrical symbols are all usually blocks and electrical designers/draftspersons have an extensive library at their fingertips. You have already drawn a small library of P&ID symbols in a previous chapter; now let us turn our attention to electrical symbols. Draw the following symbol library and create a block out of each symbol. While you are drawing these symbols, try to also memorize them. You will see them again in almost any architecture-related field. (Difficulty level: Moderate; Time to completion: 30–45 minutes for blocks)

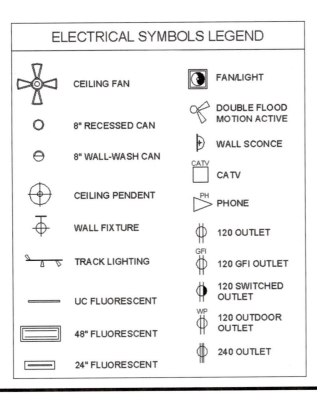

Spotlight On: Mechanical Engineering

Mechanical engineering is the science of designing and manufacturing mechanical systems or devices. It is one of the oldest and most widespread engineering specialties and is often the profession one has in mind when thinking of a typical engineer. Mechanical engineering is also by far the most diverse of the engineering specialties and "mechies" can be found working in fields as wide-ranging as aerospace, automotive, naval, rail, power, infrastructure, robotics, manufacturing, consumer products, and much more. It is truly a profession that keeps our society moving.

FIGURE 1

FIGURE 2

Mechanical engineers need a mastery of a wide variety of engineering sciences, structures, materials, vibrations, machine design, control systems, electrical engineering, and more. They also need to be well versed in business practices, cost analysis, and risk assessment, as design and production of just about anything is closely tied to these. Mechanical engineers have been the driving force behind development of high-end software tools, such as Finite Element Analysis (FEA) for structural work.

Education for mechanical engineers starts out similar to education for the other engineering disciplines. In the Unites States, all engineers typically have to attend a four-year ABET-accredited school for their entry level degree, a Bachelor of Science. While there, all students go through a somewhat similar program in their first two years, regardless of future specialization. Classes taken include extensive math, physics, and some chemistry courses, followed by statics, dynamics, mechanics, thermodynamics, fluid dynamics, and material science. In their final two years, engineers specialize by taking courses relevant to their chosen field. For mechanical engineering, this includes classes on structures, controls, vibration analysis, thermodynamics, and machine design, among others.

Upon graduation, mechanical engineers can immediately enter the workforce or go on to graduate school. Although not required, some engineers choose to pursue a Professional Engineer (P.E.) license. The process first involves passing a Fundamentals of Engineering (F.E.) exam, followed by several years of work experience under a registered P.E. and finally sitting for the P.E. exam itself. Mechanical engineering is also an excellent undergraduate degree for entry into law, medicine, or business school. Just some of the careers you can pursue with an engineering bachelor's followed by a J.D., M.D., or an MBA include patent attorney, biomedical engineer, and manager or CEO of an engineering firm. If you can handle the education requirements, the mechanical engineering undergraduate degree may truly be one of the most useful ones in existence.

Mechanical engineers can generally expect starting salaries (with a bachelor's degree) in the $54,000–$58,000/year range, which is somewhat toward the higher end among engineering specialties. This of course depends highly on market demand and location. A master's degree is highly desirable and required for many management spots. The median salary in 2008 was $74,920, with the top 25% making $94,400. The job outlook for mechanical engineers is excellent, as the profession is so diversified that there is always a strong demand in some of the sectors, even if others are in a downturn.

So, how do mechanical engineers use AutoCAD and what can you expect? Industrywide, AutoCAD enjoys significant amount of use, even as its share shrinks as companies switch to low-cost 3D solutions such as SolidWorks and Inventor.

In truly high-end design (aerospace, automotive, and naval, for example), the dominant software applications continue to be CATIA, NX, and Pro/Engineer, and AutoCAD does not compete with these. However, this still leaves many applications where AutoCAD is appropriate, such as small part design, schematics, and overall layout of systems. Many companies also do not manufacture but rather assemble modular pieces into new products; here, too, AutoCAD is the appropriate low-cost solution. Let us take a look at a few examples.

The acronym P&ID stands for piping and instrumentation diagrams, a major mechanical engineering field, commonly referred to as *process control*. Engineers use the principles of fluid flow, material science, and hydrodynamics to design piping and valves that regulate fluid flow. This can be anything from a chemical plant to an electrochlorination system to waste treatment. Usually, systems are assembled out of standard industry parts, so AutoCAD is the appropriate tool to put them together into a design. Figure 3 shows a typical piping valve profile diagram.

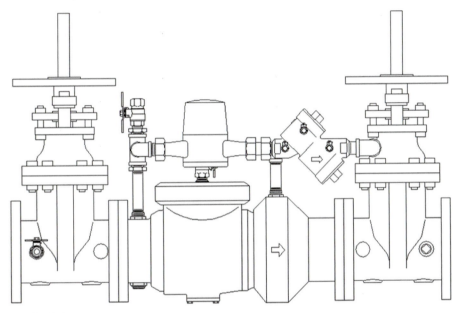

FIGURE 3
Piping valves diagram.

Small device assemblies are another example of mechanical design. If the parts are standard and no manufacturing is necessary, only assembly drawings, then AutoCAD can be used to depict the ideas. Figure 4 shows an AutoCAD drawing of an instrument tuning peg assembly.

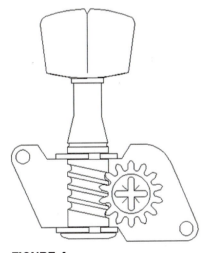

FIGURE 4
Instrument tuning peg assembly.

CHAPTER 8

Polar, Rectangular, and Path Arrays

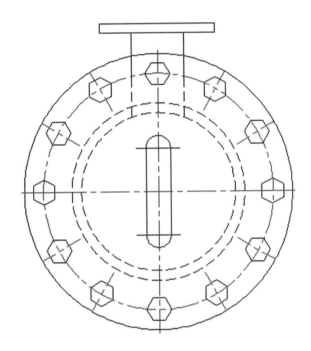

LEARNING OBJECTIVES

In this chapter we introduce the concept of an array and discuss the following:

- Creating a polar array
  - Object, center, quantity, and degrees
- Additional operations with polar arrays
- Legacy polar array (pre-AutoCAD 2012)
- Creating a rectangular array
  - Object, rows and columns, and distances
- Additional operations with rectangular arrays
- Legacy rectangular array (pre-AutoCAD 2012)
- Creating a path array
- Additional operations with path arrays

At the end of this chapter you will be able to create polar and rectangular arrays. You will then start a new in-class mechanical project.

Estimated time for completion of chapter (lesson and project): 3 hours.

Arrays are very useful tools in AutoCAD to create patterns of objects. These patterns can be of a circular, linear, or path type, hence the polar array, rectangular array, and path array discussions that follow. In a strict sense, these tools are not something entirely new but merely automate what you can do (albeit tediously) by regular commands learned in Chapter 1, and they are huge time savers.

The array command underwent a major rework in AutoCAD 2012. If you are using this textbook to refresh your skills and have used older versions of AutoCAD before, you are in for a surprise. Not only has an entirely new type of array been added (path array), but the tool is now command line activated, with the Ribbon assisting with modifications. For better or for worse, the familiar array dialog box is gone, and a whole new set of possibilities are opened up.

This chapter introduces the various arrays by assuming that the student is new to AutoCAD and has never used or seen the older, legacy, version of the command. It is, however, referred to and described, just in case you end up using a pre-2012 AutoCAD at some point in the future.

8.1 POLAR ARRAY

A polar array is a collection of objects around some common point arranged in a circle (or part thereof). An example in architecture is a group of chairs arranged around a circular table. The idea is to draw the table, then one chair, and after positioning the chair exactly where it will be in relation to the table, use a polar array to copy it around the table to create, let us say, ten of them. The array command copies and rotates them into position and spaces them out evenly, a task that would have taken a while to do one chair at a time.

In mechanical engineering, an example may be teeth of a sprocket gear. You draw the circle representing the wheel of the gear; put in one tooth properly drawn, detailed, and centered on top; and then array to get the full gear. Figure 8.1 shows illustrations of what was just discussed. There are of course many other examples.

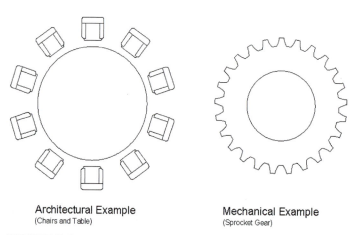

Architectural Example
(Chairs and Table)

Mechanical Example
(Sprocket Gear)

FIGURE 8.1
Polar array examples.

Steps in Creating a Polar Array

First of all, we need to create something to array. So go ahead and draw a circle of any size to represent the table just discussed, and then create a convincing-looking chair. You may want to make a block out of it after you are done for convenience. Next, position it at the top of the table, as seen in Figure 8.2. Be sure to line it up perfectly on the vertical axis, or else the array will be skewed off center; and move it back a bit from the table with Ortho on. So, what does AutoCAD need to know to do the array?

- It needs to know *what object(s)* to array (that would be the chair, *not* the table).
- It needs to know the *center point* of the array (center of the table).
- It needs to know *how many objects* to create (let us stick to ten or so).
- It needs to know if the pattern goes all the way around (that is, is it 360° or less).

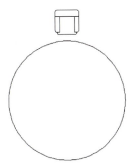

FIGURE 8.2
Chair and table, initial setup.

It is useful to memorize these steps and realize what you need to do before even starting up the array command; that way you have a clear idea of what to do and what data to enter as you get going. Once you have the table and chair drawn, follow the steps shown to create the polar array.

Keyboard: Type in **arraypolar** and press Enter
Cascading menus: **Modify→Array→Polar Array**
Toolbar icon: **Array_Toolbar**
Ribbon: **Home tab→Modify→Polar Array**

Step 1. Start up the array command via any of the preceding methods.
  ○ AutoCAD says: `Select objects:`
Step 2. Select the chair block.
  ○ AutoCAD says: `1 found`
Step 3. Press Enter.
  ○ AutoCAD says: `Type = Polar Associative = Yes`
    `Specify center point of array or [Base point/Axis of rotation]:`
Step 4. You successfully selected the *object to array.* You now need to specify the *center point of the array.* Simply select the center of the large circle (the table) using the CENter OSNAP.
  ○ AutoCAD says: `Enter number of items or [Angle between/Expression]`
    `<4>:`
Step 5. Here you need to state *how many objects to create.* Enter 10, noting that you can *dynamically* do this as well by moving the mouse around the circle and watching new chairs appear.

183

○ AutoCAD says: `Specify the angle to fill (+=ccw, -=cw) or [EXpression] <360>:`

Step 6. Here, you need to state if you want the chair pattern to go all the way around the circle or part thereof. In this case, we fill up the circle, so just press Enter. You could have just as easily entered a numerical degree value for any part of a circle.

○ AutoCAD says: `Press Enter to accept or [ASsociative/Base point/ Items/Angle between/Fill angle/ROWs/Levels/ROTate items/ eXit]<eXit>`

Step 7. Your array is done, so just press Enter to accept. We explore some of the options in Step 7's menu in the next section. The completed polar array is shown in Figure 8.3.

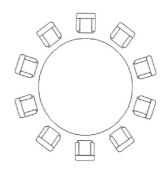

FIGURE 8.3
Polar array complete.

Additional Operations with Polar Array

Your new array behaves like a block, a major difference between AutoCAD 2012 and all earlier releases, where the new pieces of the array were not connected in any way. Click on one of the chairs with your mouse and it becomes selected, with several varieties of grips visible. If you hover your mouse over the chair grip, you see a menu of additional options. Finally, a new Ribbon Array tab appears. All are seen in Figure 8.4. Let us take a closer look at the Ribbon and the Grip menu.

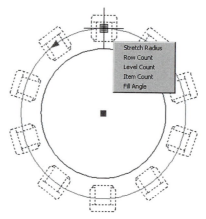

FIGURE 8.4
Polar array editing.

The Ribbon has two areas of interest, the Items category and the Options category, as seen in Figure 8.5. The Items category is relatively self-explanatory. Top to bottom, the first value (10) is the number of items in the array, and you can change that for an instant update. Below that is the angle between all the objects (36°), and finally, below that, the total angle filled (360°, or a full circle). This should make sense as 36 × 10 = 360.

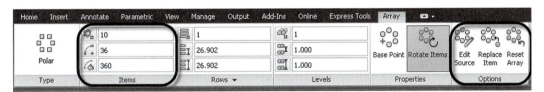

FIGURE 8.5
Polar Array tab.

The Options category's tools are a bit more involved. They are described next:

- **Edit Source.** This option allows you to edit and modify the source object. These changes are then reflected in the overall array. If you forget which object was the original source, it does not matter; you can pick any of the chairs and work on it. Execute the command, select the object, modify it, and when done, type in ARRAYCLOSE. Figure 8.6 shows Edit Source in progress.

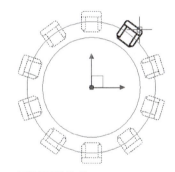

FIGURE 8.6
Edit Source.

- **Replace Item.** If editing the source is too time consuming, then just replace it with another design via this option. AutoCAD asks you to select the replacement object and a base point. Use the Key Point option to pick a point on the replacement object. Finally, select an item (or several items) in the array to replace. Figure 8.7 shows Replace Item in action as squares replace the chair symbol.

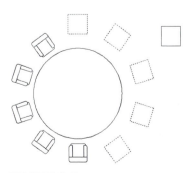

FIGURE 8.7
Replace Item in progress.

- **Reset Array.** This is an easy one. Autodesk figured you will mess up a few arrays while learning this tool (speaking from experience of course) and included a Reset button. Its function is apparent immediately on use.

Next, we have the menu seen when the mouse is held over the chair grip. Run through all the options as you read the descriptions. It contains the following tools:

- **Stretch Radius.** This option simply expands the radius of the overall array. The effect is the same as if you just activate the grip and move the mouse. Stretch Radius in action is seen in Figure 8.8.
- **Row Count.** This option adds another set of row(s) of items in front or behind the main row, as seen in Figure 8.9.
- **Level Count.** This option adds 3D levels and its discussion is not be included in this 2D only chapter.
- **Item Count.** This option adds or deletes items (similar to the Ribbon function), or it can just be used to tell you how many items there are in the array, also a useful feature.
- **Fill Angle.** This option (also via the triangle grip) changes the included angle, currently set to 360°. An example is shown in Figure 8.10, where the fill angle has been rolled back to 270°.

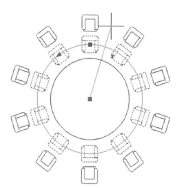

FIGURE 8.8
Stretch Radius in progress.

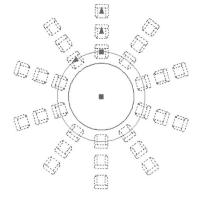

FIGURE 8.9
Row Count in progress.

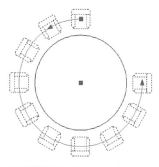

FIGURE 8.10
Fill Angle in progress.

Go over all these features until you are comfortable working with them. Be aware that you also can explode the array, which will make it behave like the "legacy" array, where each piece is separate. A short discussion of the legacy tools follows.

Legacy Polar Array (Pre-AutoCAD 2012)

Because the array command was so heavily redesigned for this release of AutoCAD, it makes sense to briefly mention the legacy polar array tool that will be encountered via a dialog box prior to AutoCAD 2012. Polar array can be accessed via the following legacy methods:

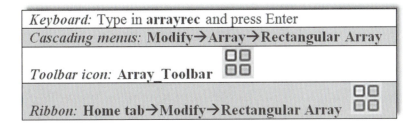

Keyboard: Type in **arrayrec** and press Enter

Cascading menus: **Modify→Array→Rectangular Array**

Toolbar icon: **Array_Toolbar**

Ribbon: **Home tab→Modify→Rectangular Array**

What you then see is the classic Array dialog box (Figure 8.11, taken from AutoCAD 2011). You first need to make sure you have selected the <u>P</u>olar Array radio button at the very top. Then, run through the same steps outlined in the beginning of this chapter and shown here again:

- Select object to array (button at top right).
- Select the center point of the array (button center top right).
- Enter how many objects to create (in the middle).
- Enter how many degrees to fill (just below the previous entry).

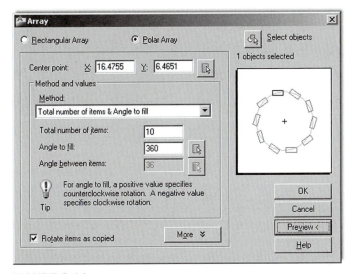

FIGURE 8.11
Legacy Polar Array dialog box.

Preview the array via the Pre<u>v</u>iew < button and press OK to finish off the array.

Note that the entire array is not a block, just a collection of objects, and to redo the whole thing, you need to erase all the copies, leaving just the table and the original chair and repeat everything. That is why the Preview button is so important. Do not be so quick to press OK right away. This array of course has none of the editing options introduced in AutoCAD 2012.

8.2 RECTANGULAR ARRAY

The idea behind a rectangular array is not fundamentally different from a polar one. We are looking to replicate objects, but this time in rows (horizontal) and columns (vertical). In 3D, you can also do levels, which we do not address here. An example, as shown in Figure 8.12, may be columns of a warehouse. If they are spread out evenly, then this is the perfect tool for this. You simply draw one I-beam column, make a block out of it, and array it up and across the appropriate area.

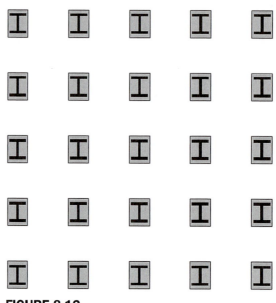

FIGURE 8.12
Rectangular array example.

188

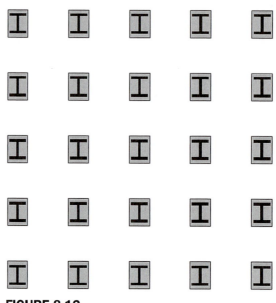

FIGURE 8.13
I-beam for column.

Steps in Creating a Rectangular Array

The first thing to do is to draw a convincing-looking I-beam. Because we need actual distances, you cannot just randomly draw any object; use known sizing this time around. With one of the columns in Figure 8.12 as a model, create a 20" wide by 26" tall rectangle. Inside of it, put an I-beam drawn with basic linework and a solid hatch fill, and shade the background a light gray (Figure 8.13). Finally, make a block out of it. That sets up the column for an array.

Now, what does AutoCAD need to know to create a rectangular array?

- It needs to know *what object(s)* to array (the I-beam column).
- It needs to know *how many rows* and *columns* to create (rows = across, columns = up and down).
- It needs to know the *distance between the rows and the columns* (from centerline to centerline).

Keyboard: Type in **arraypath** and press Enter
Cascading menus: **Modify→Array→Path Array**
Toolbar icon: **Array_Toolbar**
Ribbon: **Home tab→Modify→Path Array**

Step 1. Start up the array command via any of the preceding methods.
  ○ AutoCAD says: `Select objects:`

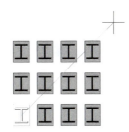

FIGURE 8.14
Creating a rectangular array dynamically.

Step 2. Select the I-beam column block.
- ○ AutoCAD says: `1 found`

Step 3. Press Enter.
- ○ AutoCAD says: `Type = Rectangular Associative = Yes`
  `Specify opposite corner for number of items or [Base point/`
  `Angle/Count]`
  `<Count>:`

Step 4. You have now successfully selected the *object to array*. Now you need to specify the number of rows, columns, and of course the distance between them. You immediately notice that you can do this "dynamically" by moving the mouse around. A diagonal path is most effective, as seen in Figure 8.14.

Step 5. This dynamic method is not really appropriate for setting distances between the columns, which are random values for now, but it does make creating the pattern easy. If you want to specify actual values, you need to enter c for `Count`.
- ○ AutoCAD says: `Enter number of rows or [Expression] <4>:`

Step 6. Enter a value for the rows: 5 in this example.
- ○ AutoCAD says: `Enter number of columns or [Expression] <4>:`

Step 7. Enter a value for the columns: 5 in this example.
- ○ AutoCAD says: `Specify opposite corner to space items or [Spacing]`
  `<Spacing>:`

Step 8. Enter s for `Spacing`. This indicates that you want to manually enter the appropriate distances. Note that, once again, you could do this dynamically.
- ○ AutoCAD says: `Specify the distance between rows or [Expression]`
  `<39.000>:`

Step 9. Enter a value for the row distance, such as 60. Knowing the I-beam sizing pays off here, because a value significantly less than 60 would have had the rows sitting on top of each other.
- ○ AutoCAD says: `Specify the distance between columns or [Expression]`
  `<30.000>:`

Step 10. Enter a value for the column distance, such as 60 again. Knowing the I-beam sizing pays off here as well, because a value significantly less than 60 would have had the same "crowding" effect on the columns.
- ○ AutoCAD says: `Press Enter to accept or [ASsociative/Basepoint/`
  `Rows/Columns/Levels/eXit]<eXit>:`

Step 11. Press Enter to accept the array as is. If you did everything right, the final result is what you would expect and is similar to what is shown earlier in Figure 8.12.

Additional Operations with Rectangular Array

After a lengthy discussion on additional operations with the polar array, you will find these options with the rectangular array to be quite familiar. Click on the new array of I-beam columns, and notice a set of grips, a menu, and an Array tab in the Ribbon, all similar to the earlier polar array (Figure 8.15).

189

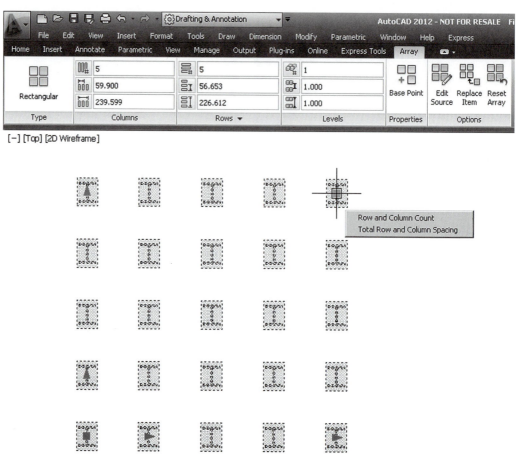

FIGURE 8.15
Rectangular array editing.

FIGURE 8.16
Rectangular Array tab.

The Ribbon has three areas of interest this time: the Columns, Rows, and Options categories, as seen in Figure 8.16. The Columns category allows you to adjust the number of columns, distance between them, and the total distance. The Rows category is the exact same thing but for rows. Finally, in the Options category, we already discussed the functionality of Edit Source, Replace Item, and Reset Array. They are all exactly the same as with the polar array.

Next we have the menu seen when the mouse is held over the top right I-beam column grip. The two choices are Row and Column Count and Total Row and Column Spacing. Their functions should be relatively easy to see by just trying them out. If you activate the first choice and move the mouse, the number of rows and columns increases, but the spacing remains the same. If you activate the second choice, the spacing increases but not the number of rows and columns.

Finally, try out the other grips on the array, specifically the ones in the lower left corner. You find that the square grip moves the entire array and the inner triangles increase the spacing. The outer triangle grips (toward the edges of the array), however, are used to increase the count. Experiment with all the options.

Finally it is worth mentioning that you can create a rectangular array where all the elements are at an angle by selecting a for Angle in Step 3. It is up to you to try this out, but the result (for 45°) is shown in Figure 8.17.

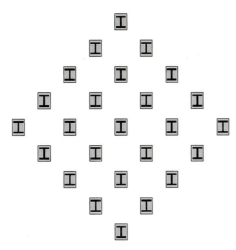

FIGURE 8.17
Rectangular array at a 45° angle.

Legacy Rectangular Array (Pre-AutoCAD 2012)

The legacy rectangular array is accessed via the same methods as the legacy polar array. When the dialog box appears, make sure the Rectangular array radio button is selected and you see what is shown in Figure 8.18 (also taken from AutoCAD 2011). The procedure remains basically the same. You must select the object, then enter the number of rows and columns, followed by the distances between those rows and columns. Do a preview and press OK.

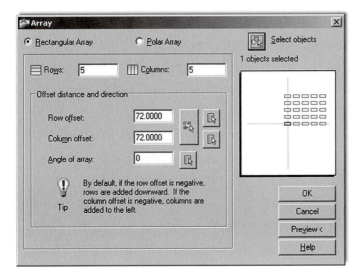

FIGURE 8.18
Legacy Rectangular Array dialog box.

Note again that the entire array is not a block, just a collection of objects, and to redo the whole thing, you need to erase all the copies, leaving just one I-beam column and repeat everything. This array of course has none of the editing options introduced in AutoCAD 2012 either.

In general, polar arrays are used more often than the rectangular ones, but both are important to know.

8.3 PATH ARRAY

The path array is the new kid on the block. Requested by users, this feature was finally added in AutoCAD 2012. The idea is simple, and it is surprising that this feature was not added earlier. A path array creates a pattern along any *predetermined pathway*, of any shape, as opposed to a strictly circular or linear one. This path can be an arc, a pline, or a spline. Because plines and splines have not yet been discussed (a Chapter 11 topic), we focus on arcs as the pathways. A path array can also be created in 3D, but we do not get into that right now. An example of a path array in progress is shown in Figure 8.19.

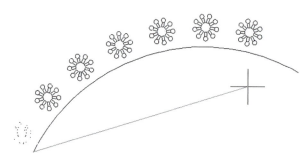

FIGURE 8.19
Path array in progress.

Steps in Creating a Path Array

Boiled down to its essence, the path array just needs an *object to array* and a *path to follow*. Much like the previous two arrays, the objects can be spread out dynamically with mouse movements (as seen being done in Figure 8.19), or with an input of a specific value. All other options flow from this basic approach. Let us give it a try. Create an arc and some random object to array; a rectangle will do, as seen in Figure 8.20.

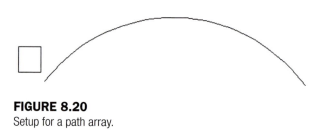

FIGURE 8.20
Setup for a path array.

So, what does AutoCAD need to know to do the array?

- It needs to know *what object(s)* to array (the rectangle).
- It needs to know *the path* (the arc).
- It needs to know *how many* objects to create (let us stick to ten or so).

Keyboard: Type in **array** and press Enter	
Cascading menus: Modify→Array...	
Toolbar icon: **Modify** toolbar	⬜⬜ ⬜⬜
Ribbon: **Home** tab→Modify	⬜⬜ ⬜⬜

Step 1. Start up the array command via any of the preceding methods.
- ○ AutoCAD says: `Select objects:`

Step 2. Select the rectangle.
- ○ AutoCAD says: `1 found`

Step 3. Press Enter.
- ○ AutoCAD says: `Type = Path Associative = Yes`
`Select path curve`

Step 4. You successfully selected the *object to array*. Now, select the path by clicking on the arc.
- ○ AutoCAD says: `Enter number of items along path or [Orientation/ Expression] <Orientation>:`

Step 5. Here, you are asked to specify how many items you wish to fit along the path. You can do this dynamically via the mouse (as in Figure 8.19) or type in a value, let us say 10.
- ○ AutoCAD says: `Specify the distance between items along path or [Divide/Total/Expression] <Divide evenly along path>:`

Step 6. Here, you can specify the distance between objects, but usually you just want to accept the default choice of dividing evenly along the path, so press Enter.
- ○ AutoCAD says: `Press Enter to accept or [ASsociative/Basepoint/ Items/Rows/Levels/Align items/Z direction/eXit]<eXit>`

Step 7. This final menu introduces some options, but we explore them separately in a moment so just press Enter once again. The completed array is seen in Figure 8.21.

FIGURE 8.21
Path array completed.

Additional Operations with Path Array

Just as with the previous two arrays, the path array has its own Ribbon tab, which is revealed by selecting one of the elements in the array, as well as its own menu, revealed as you hover the mouse over the grip as seen in Figure 8.22.

The Items category is of some interest and at this point should be familiar. You can vary the number of elements, distances between them, and total distance. Under the Properties category experiment with redefining the base point—this shifts your pattern around. Also check out Align Items—this makes all your elements face in the same direction versus angling with the curve. Edit Source, Replace Item, and Reset Array all have the same meanings as with the previous polar and rectangular arrays. Spend a few moments understanding the cause and effect of these options.

Finally, let us take a look at the grip menu. It also should be more familiar after your experiences with the previous two arrays. Move moves the array away from its originating path and triggers a warning along the way. It is shown in Figure 8.23, and you may have seen it with earlier arrays as well, if you tried to move them around. Simply press Continue and carry on.

Row Count creates additional rows of the pattern, as seen in Figure 8.24. This option also repeats what is available with the other arrays. Level count is not covered here, as it is a 3D function.

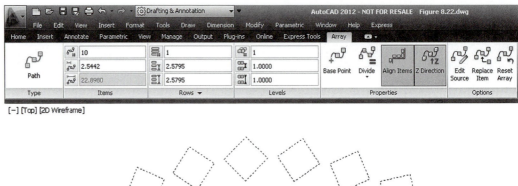

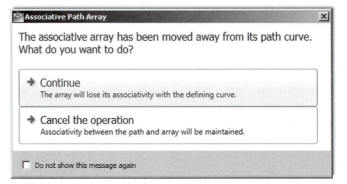

FIGURE 8.22
Path array editing.

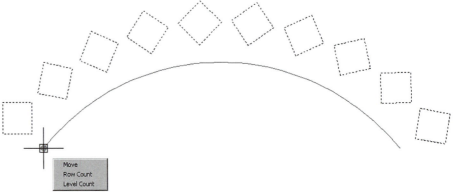

FIGURE 8.23
Associative Path Array warning.

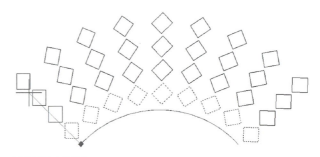

FIGURE 8.24
Row Count in progress.

8.4 IN-CLASS DRAWING PROJECT: MECHANICAL DEVICE

Let us now put what you learned to use by drawing a mechanical gadget (a suspension strut of sorts), created years ago for a class. Although what it is exactly is still debatable, the gadget proved to be an excellent drawing example and features quite a bit of what we covered in this and preceding chapters as well as a few minor nuances of AutoCAD drafting that, once learned, raise your skill levels considerably. The trick is to make your drawing look exactly like the example. If you approximate and cut corners, you miss out on some of the important concepts that give a more professional look to your drawings.

A full-page view of the completed device is found toward the end of the chapter (Figure 8.34). Take a good look at it and decide on a strategy to draw it. Over the course of the next few pages, you will see a step-by-step process to actually do it. Follow the steps carefully, and do not worry about the thick border; that is a polyline, to be covered in Level 2. The title block is otherwise just a simple collection of lines and text.

Step 1. For starters open a new file, give it a name (Save As…), and set up your layers. We need, at a minimum:

M-Part, Color: Green (for most of the parts of the device).
M-Text, Color: Cyan.
M-Dims, Color: Cyan (for the dimensions).
M-Hidden, Color: Yellow, Linetype: Hidden (for all hidden lines).
M-Center, Color: Red, Linetype: Center (for the main centerlines).
M-Hatch, Color: 9 (a gray for cross-section cutaways).

You may also want additional layers for the spring and the title block. You can decide on the exact layers and their colors; the preceding is just a guide. Also, the images in this chapter remain black and white, for clarity on a white page.

Step 2. Set up all the necessary items for smooth drafting, such as the Text Style (Arial, .20), the Dim Style, and Units (either Architectural or Mechanical is fine), and get rid of the UCS icon if you do not like it.

Step 3. To begin the drawing itself, we need to start with the hex bolts. Go ahead and draw one of them according to the information in Figure 8.25. Be *very* careful, all sizes are diameters *not* radiuses; you must tell AutoCAD this by pressing d for Diameter while making a circle. This has tripped up many students in the past. Also do not forget to make layer M-Part current. If the bolt is anything but green in color, you missed this step. The dimensions are only for your reference for now; there is no need to draw them.

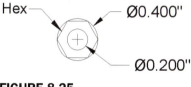

Hex — Ø0.400"
Ø0.200"

FIGURE 8.25
Bolt layout.

Step 4. Now make a block out of the bolt, calling it Bolt—Top View.
Step 5. Complete the rest of the top view by drawing the remaining circles and positioning the bolt at the top, as shown in Figure 8.26. Use accuracy (OSNAPs), although the bolt can be any distance from the top of the circle. Be careful to use diameters as indicated.
Step 6. Array the bolt around the top view four times, as shown in Figure 8.27.
Step 7. You now have to project the main body of the part based on this top view. The best way is by drawing two horizontal guidelines, of any length, starting from the top and bottom quadrants of the big circle; then draw an arbitrary vertical line to mark the beginning of the part on the right, as shown in Figure 8.28.

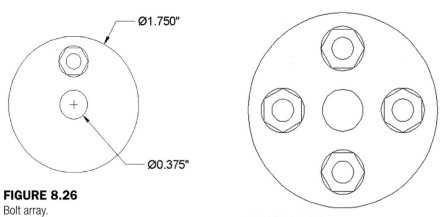

FIGURE 8.26
Bolt array.

FIGURE 8.27
Bolt array completed.

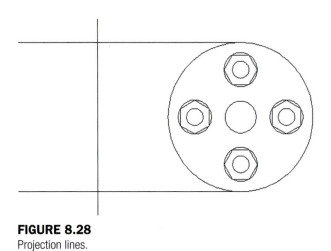

FIGURE 8.28
Projection lines.

Step 8. Trim or fillet the main vertical guideline and continue offsetting, based on the given geometry, to "sketch out" the basic outline of the shape, as seen in Figure 8.29. Do not yet add dimensions; they are for construction only.

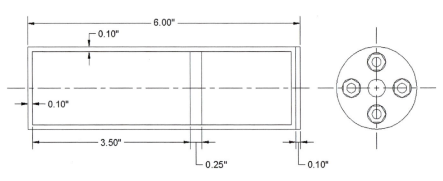

FIGURE 8.29
Body of part.

Step 9. Continue adding the basic geometry to the side view, as well as hidden and centerlines, as shown in Figure 8.30. Set LTSCALE to .4.

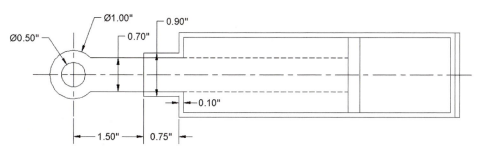

FIGURE 8.30
Creating side view.

Step 10. Add the bolt projections, as seen in Figure 8.31 (the depth into the side view can be approximated). Finally, add hatch patterns, using the Angle option to reverse the hatch directions, which is a common technique in mechanical design to differentiate the parts from each other in a cross section.

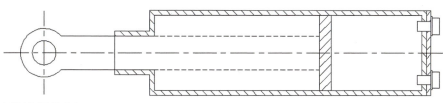

FIGURE 8.31
Hatching side view.

197

Step 11. Now, how do you do the spring? Well, this is where the `Circle/Ttr` comes in. Although we have not practiced this version of the circle command, it is very simple to do. Start it up and select the `Ttr` option. Then, pick the vertical and the horizontal lines where the circle needs to go (seen in Figure 8.32) and specify a radius (0.10"). One way of proceeding is to draw a straight arbitrary line down from the center of that circle, rotate the line 15° about that center point, then offset 0.1" in each direction, erase the original line, extend the two new ones to the bottom (if necessary), and repeat the process, as seen in Figure 8.32. Once a set is done, you can mirror the rest and trim carefully to create the impression of a coiled spring.

Step 12. Finally add dimensions (Figure 8.33) and a title block (Figure 8.34) by use of rectangles. The title block and its contents are typical of a mechanical engineering drawing.

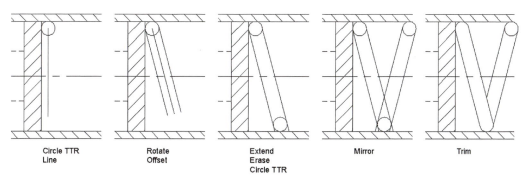

Circle TTR
Line

Rotate
Offset

Extend
Erase
Circle TTR

Mirror

Trim

FIGURE 8.32
Drawing the spring.

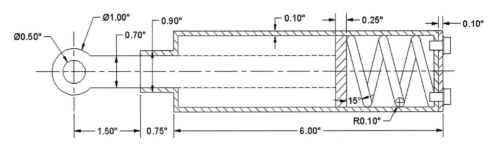

FIGURE 8.33
Completed side view.

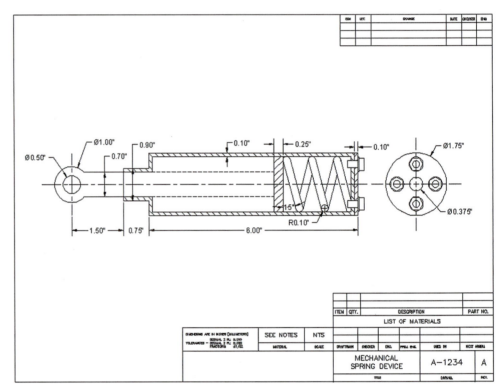

FIGURE 8.34
Mechanical project—final.

SUMMARY

You should understand and know how to use the following concepts and commands before moving on to Chapter 9:

- Polar array
  - ○ Select object
  - ○ Select center point of array
  - ○ How many objects to create
  - ○ Angle to fill
  - ○ Additional operations and options
- Rectangular array
  - ○ Select object
  - ○ How many rows and columns
  - ○ Distance between rows and columns
  - ○ Additional operations and options
- Path array
  - ○ Select object

○   Select path
○   Additional operations and options

REVIEW QUESTIONS

Answer the following based on what you learned in Chapter 8.

1. List the three types of arrays available to you.
2. What steps are needed to create a polar array?
3. To array chairs around a table, where is the center point of the array?
4. What steps are needed to create a rectangular array?
5. What steps are needed to create a path array?

EXERCISES

1. In a new file, draw this slightly altered version of the image found at the opening of this chapter, shown here with all the necessary dimensions. (Difficulty level: Easy; Time to completion: 10 minutes)

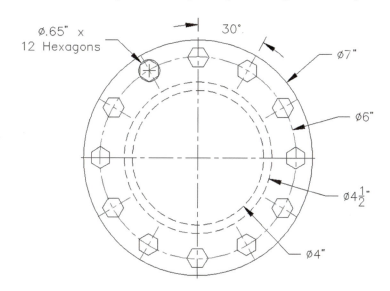

Ø.65" x
12 Hexagons

30°

Ø7"

Ø6"

Ø4½"

Ø4"

2. In a new file, draw the following five row by five column rectangular array. The chairs are 20" by 20" and the offset between them is 50". The array is also at a 45° angle. (Difficulty level: Easy; Time to completion: <10 minutes)

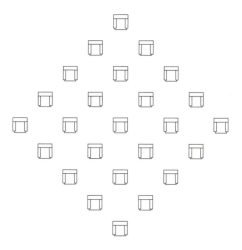

3. In a new file, draw the following set of path arrays. The two paths are simple arcs, and the symbols used are scaled up and shown next to the patterns. (Difficulty level: Easy; Time to completion: <10 minutes)

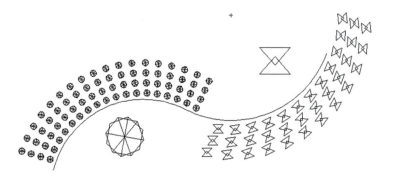

4. In a new file, draw the following gear array. Begin with circles of the indicated diameter. Then, focus on creating the first tooth as seen in the detail. Make sure it is centered and aligned. Finally, array the tooth (24 teeth are used in the example) and clean up the design with trimming. The final result is shown below. (Difficulty level: Moderate; Time to completion: <20 minutes)

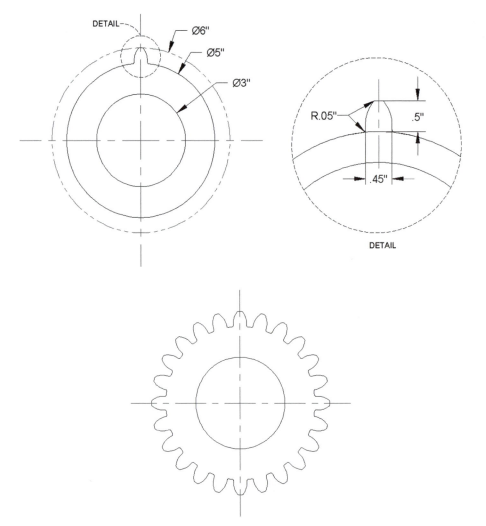

5. The final exercise of this chapter has no arrays but is a basic floor plan to keep your drawing skills sharp. It also features some electrical symbol blocks; get used to them, as they are standard in any electrical layout. The numbers inside the rectangles are the doorway dimensions. All walls are 5½". A few secondary dimensions are not given due to space limitations; you need to make a best estimate. Use proper layers and make a block of all the electrical symbols. (Difficulty level: Moderate; Time to completion: 90–120 minutes)

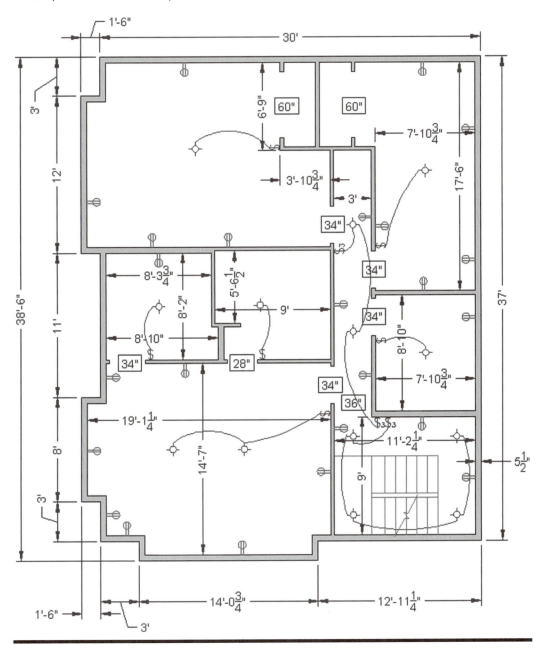

201

CHAPTER 9

Basic Printing and Output

LEARNING OBJECTIVES

In this chapter we introduce the Plot dialog box and the Page Setup Manager and go through all the associated options to produce a print or a plot, including the following:

- What *printer* or *plotter* to use
- What *paper size* to use
- What *area* to plot
- At what *scale* to plot
- What *pen settings* to use
- What *orientation* to use
- What *offset* (if any) to use
- The Page Setup Manager

Upon the completion of this chapter you will be able to recognize and set up all the features needed to print or plot any type of drawing from any AutoCAD workstation.

Estimated time for completion of chapter: 1 hour.

9.1 INTRODUCTION TO PRINTING AND PLOTTING

Designers and their clients usually want something tangible to look at and use. Because of this, most design work in AutoCAD must be created with the ultimate goal of displaying it on a piece of paper, not just on screen. Indeed, the fundamental immediate "product" of most design work is paper output. This output is then used to manufacture, fabricate, build, or assemble the design. It also may be needed for pre- and post-design purposes, such as sales or archiving. These needs influence the evolution of a drawing from the outset, and a skilled AutoCAD designer selects paper size, output scales, and lineweights early in the process so the inevitable output proceeds smoothly.

Professional-looking output in AutoCAD is no accident and comes with its own set of challenges. Although you may be able to walk up to an expertly set up drawing and just select **File→Plot→OK**, such smooth results come from only a thorough understanding of what is involved. More important, if the requirements call for a different output of the same drawing (e.g., for an informal check of a small section of the drawing versus a final full-size submission to the client), you must know what to change and by how much.

AutoCAD allows for a wide array of output, each method with its own tricks and details. Output spans this chapter as well as part of Chapter 10 ("Advanced Output—Paper Space") and Chapter 19 ("Advanced Output and Pen Settings"). Our goal right now is only to familiarize you with the essentials and what AutoCAD needs to know to print or plot a drawing.

Take note, as you go through this chapter, that although some specifics, such as type of printer or plotter, may be unique to your particular office, home, or school, most of what you need to know stays constant from situation to situation and from one release of AutoCAD to another (even if the locations of the buttons and menus change). Therefore, it is important that you memorize and understand the essential underlying concepts, not just "which buttons to push."

9.2 THE ESSENTIALS

You need to give AutoCAD the following information when setting up a drawing for printing or plotting:

- What *printer* or *plotter* to use.
- What *paper size* to use.
- What *area* to plot.
- At what *scale* to plot.
- What *pen settings* to use.
- What *orientation* to use.
- What *offset* (if any) to use.

There are of course some other subtleties, but we focus on these essential ones for now. Note that we do not look at the actual Plot dialog box until later, as you first need to understand what the preceding means.

What Printer or Plotter to Use

This is where you select what output device to use. In most architecture offices, engineering companies, and schools, you have a plotter (color or black and white) for C-, D-, and E-sized drawings (paper sizes are discussed next), and a LaserJet printer (color or black and white) for A- and B-sized drawings. Your typical AutoCAD station usually is networked to access both, as requirements may change daily. You need to select which device is getting the drawing (AutoCAD does not care—it will print a floor plan of a sports stadium on a postcard if you let it). This choice is of course tied in with the scale and the ultimate destination of the printed drawing (is it a check plot or for a client?). We discuss this in more detail later.

What Paper Size to Use

You would think that paper sizing is a cut-and-dried, straightforward matter (somewhere a professional printing company manager is laughing), but this is certainly not the case. Paper comes in a bewildering array of international sizes and standards in both inches and metric (millimeter) sizing. Some common standards include ANSI, ISO, JIS, and Arch. For the purposes of this discussion, we need to simplify matters and present all paper as existing in the following basic sizes as viewed in Landscape mode, meaning the first (larger) value is the horizontal size:

- A size: 11" × 8.5" sheet of paper (letter).
- B size: 17" × 11" sheet of paper (ledger).
- C size: 22" × 17" sheet of paper.
- D size: 36" × 24" sheet of paper.
- E size: 48" × 36" sheet of paper.

Generally speaking, these are the maximum paper dimensions. Some standards list essentially the same sizes but with the border accounted for, so a D-size sheet is 34" × 22" and so on. Note that legal (14" × 8.5") size paper, while certainly can be used, is not common in engineering or architecture.

As a side note, we can now define precisely the difference between printing and plotting.

Printing usually refers to the output of drawings on A- or B-sized paper, which is easily done on most office LaserJet printers. Most home printers can do only A size, but this of course varies with the size (and cost) of the printer.

Plotting usually refers to output of drawings on C-, D-, or E-sized paper, which has to be done on plotters, as a LaserJet generally cannot accept paper of this size.'

Memorize the paper sizes presented here, as it is pretty much the industry standard as far as AutoCAD design is concerned. You will hear these sizes asked for and referred to, come printing time; and if there is any deviation from these sizes, rest assured it will not be anything radical, as printers and plotters can accept only so much flexibility, at least among the models that small companies and schools can afford.

What Area to Plot

What "area to plot" can be taken quite literally, as in what do you want to see on the printed paper? The essential choices really boil down to two. Do you want to see the entire drawing (usually) or parts thereof (sometimes). Your choices in that regard include the following.

EXTENTS

This option prints everything visible in the AutoCAD file. However, this means everything visible to AutoCAD (and not necessarily to you), so if you have a floor plan with some stray lines that you did not notice located some distance away, AutoCAD shrinks the important floor plan to make room for those unimportant stray lines. This is exactly the same as typing in **Zoom→Extents** (from Chapter 1). So it is a very good idea to zoom to extents prior to printing. Remember this acronym: WYSIWYG—what you see is what you get. If you see it on

205

the screen, it will print unless prior arrangements are made (freezing the layer, etc.) or the geometry is on the Defpoints layer. For a carefully drawn layout, Extents is the only choice when you want to see the entire drawing.

WINDOW

As you may have guessed, this is for those situations where you want to see just a piece of the drawing. You simply select that choice, click the Window button that appears to the right of it, and draw a selection window around what you want to see.

DISPLAY AND LIMITS

The remaining two choices are of little use in most situations. Limits plots the entire drawing from 0,0 to the preset limits, something most users do not set anyway. Display plots the view in the current viewport in the Model tab or the current paper space view in a Layout tab, something we have not covered yet. One can generally use Window in those situations, but Display is fine as well if you want only the current viewport printed.

At What Scale to Plot

Of all options in Plot, this one causes the most confusion to a beginner AutoCAD user, for the simple reason that scale is closely tied in with a very important topic: paper space. The essential idea here is that while, yes, you can assign a scale from the Plot dialog box, in reality this is not usually done for most drawings, except simple and small mechanical pieces, such as bolts and brackets. What is done instead is that a title block is set up in paper space and the design (in model space) is scaled by means of viewports (much more on this in Chapter 10), so, in the Plot dialog box, you can select 1:1 (one-to-one) and a properly set up drawing prints to scale.

What if you do not care about scale and just want a print to look over? Here, the pendulum swings in the other direction and things become very easy. Just check off the Fit to Paper box above the scale and the drawing fits on whatever paper you are using, with no regard to scale whatsoever. While this technique is *not* for final prints of to-scale architectural layouts, it is used very often for non-scaled work, such as electrical wiring schematics, network design, conceptual layouts, and of course check plots.

What Pen Settings to Use

This is a topic that has an entire chapter all to itself (Chapter 19), and we just do a brief overview here. Proper pen settings are a major factor in giving your drawings a professional look. The concept is simple, usually addressed in hand drafting, yet often overlooked in the computerized world of AutoCAD. It deals with the fact that linework in a drawing is not all the same size. Primary features, such as walls on a floor plan, should appear darker on a print, while secondary features, furniture, for example, should appear slightly lighter. This was the reason for the existence of 2H, 4H, and other pencil leads, for those students old enough to remember, and the reason today for the existence of pen settings.

With AutoCAD, different colors represent the different pencil leads, and a strict standard needs to be followed as to what drawing features have what colors assigned to them. In this manner, drawings come out with different shades of linework, something noticed by the eye, and as a result, the drawings are much more readable and professional looking. These settings are created, customized, adjusted, and saved under the `ctb` extension, for all to use according to company standards set by management.

For now, select `monochrome.ctb` as your `ctb` file; the entire set of customization tools are presented in Chapter 19.

206

What Orientation to Use

This is simply a question of whether you want the drawing to be oriented Portrait or Landscape; in other words, do you want the longer side of the sheet of paper oriented up/down or across? This should be familiar to just about every computer user who has printed anything from a word processor or other software.

One twist here is that you can check off a box to plot upside down. A useful application for this option appears when you are printing many sets of similar-looking (but different) drawings. If you want to know which drawing is coming out of the plotter without waiting until it plots completely, send it in backwards. That way the title block (with the identifying drawing number) comes out first and that question is instantly resolved.

What Offset to Use

This refers to whether or not you need to shift your drawing around to account for a border and stapling area on the left (or elsewhere). If you check off Center the plot, it will be automatically centered on the paper, and in most cases this is what you would do.

Miscellaneous

There are a number of other features in Plot. For example, you can plot to a file (`*.plt`) and take it with you to another computer. This is useful for large volume drawing duplication at stores that have no AutoCAD, only printers, such Kinko's®. You can also turn on a plot stamp so information such as date and time of the printing can appear in the border. Some plotting features are unique to 3D and solid modeling. We explore some of these topics as well in Chapter 19. Finally there is a Preview window and a Number of copies to print selection, the functions of those being self-explanatory.

207

9.3 THE PLOT DIALOG BOX

Let us now take a close look at the plot dialog box and identify the locations of all the features described earlier. Open up the floor plan from the earlier chapters and bring up Plot via any of the following methods.

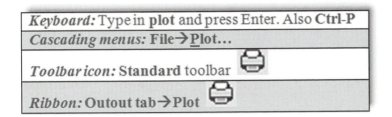

Keyboard: Type in **plot** and press Enter. Also **Ctrl-P**
Cascading menus: File→Plot...
Toolbar icon: **Standard** toolbar
Ribbon: **Outout** tab→Plot

The Plot - Model dialog box appears, as seen in Figure 9.1. If it is not full size, press the small arrow at the bottom right to expand it fully. The features we discuss are in Figure 9.2.

Go through all of the settings, referencing the explanations in the previous discussion, until you completely understand what each one does. Set the final settings as shown next.

- **What printer or plotter to use.** Set whatever printer is available at your location (it likely differs from what you see here). If you have a plotter, ignore it for now and use a printer instead, as we are using A size for this exercise.
- **What paper size to use.** Select from the drop-down list. We want any A size available; Letter (8.5" × 11") or similar is fine.
- **What area to plot.** Select Extents.
- **At what scale to plot.** Check off Fit to paper for this exercise.

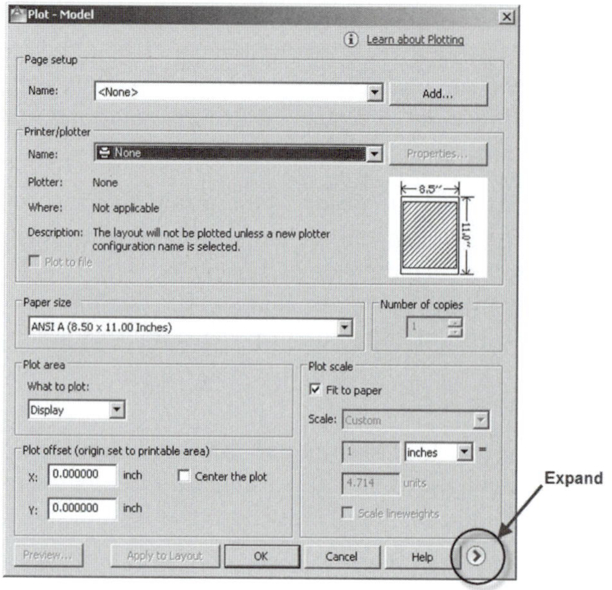

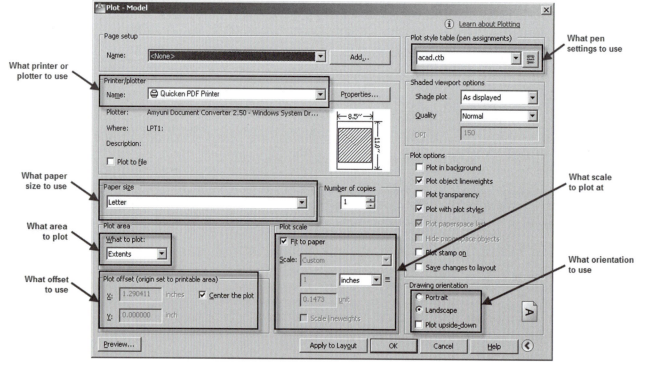

FIGURE 9.1
Plot dialog box.

FIGURE 9.2
Plot features.

- **What pen settings to use.** Select acad.ctb, and click on Yes if prompted to use this for all layouts.
- **What orientation to use.** Select Landscape for this example.
- **What offset (if any) to use.** Check off Center the plot.

Do *not* press OK yet. The Plot box should look like (or close to) Figure 9.2. We need to first discuss an important feature of printing—the preview.

Preview

Much like the situation with creating an array or hatch, a Preview button is available. With printing and plotting drawings, it is crucially important to preview the drawing before sending it to the printer or plotter. Paper and ink cost money, and everyone regardless of skill level makes mistakes in setting up printing. This really should become a habit, as errors can be costly, especially on large print jobs. On an industry wide scale, this simple Preview button actually saves tons of paper and gallons of ink. Not only printing setup errors but even linework drawing errors can be caught during the preview process.

Go ahead and press Preview… (lower left corner) and the Preview screen in Figure 9.3 appears. You can zoom and pan around using standard mouse techniques, and when you are ready to send the drawing to the printer (or return to Plot for adjustments), right-click and select Exit or Plot, respectively.

Go ahead and plot. If everything was set up correctly and your computer is connected to the printer, the floor plan should print pretty much as expected and seen in the preview of Figure 9.3.

If something does not look right anywhere in the process, review the entire set of steps, one by one, returning to the Preview screen to check up on the effects of the changes.

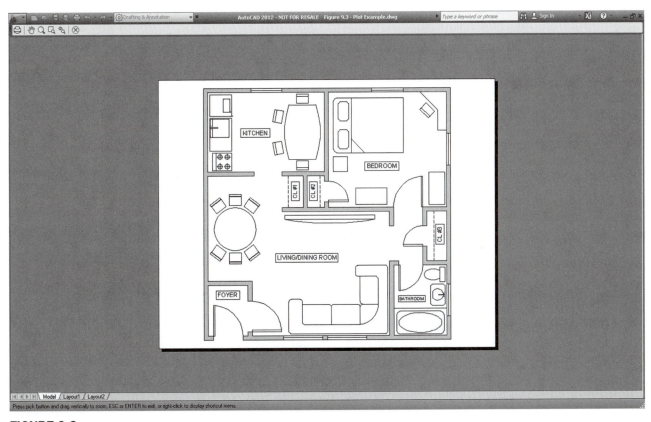

FIGURE 9.3
Plot preview.

9.4 PAGE SETUP MANAGER

There is one last topic to learn in this chapter. The settings on the print job you just sent are not saved in memory. Test this by typing in `plot` again. Unless the settings have been saved somehow, you have to reset them all again. The technique to "save" your settings involves another feature of plot called *Page Setup Manager*.

The idea here is as follows. You need to enter all the data for a print job in another, similar-looking print dialog box and save it under a descriptive name. Then, save the drawing with these settings, and when you do the actual plot, the name you saved it under appears under Page setup (upper left corner). No more entering settings one item after another; just select the Page setup (it may already be cued up), preview, and print.

This has two advantages. The first is obvious: It speeds things up a lot by keeping all the drawing's print settings ready for use. The second may not be as obvious. Because you can save as many setups as you want, you can have one drawing set up for multiple printers or multiple settings.

For example, your office may need the final output sent in color to a D-sized plotter at a certain scale, but in-house checks are done on 17" × 11" sheets, in black and white and no scale. You can easily set up two distinct setups and name them `Client_Layout` and `Check_ Layout`. It is not unusual to see up to half a dozen setups on one drawing, each with its own purpose: presentation, archiving, optimized for color, check plots, and so forth.

A question usually arises when this topic is presented. Does this need to be done every time a new drawing is created from a blank file? The answer is yes, but it is something that needs to be done only once and saves you much time down the road. Obviously, using templates negates the need to do this at all, as all settings carry over.

The Page Setup Manager is found using the cascading drop-down menu: **File→Page Setup Manager**… or the Output tab on the Ribbon and is shown in Figure 9.4.

Press the New… button and the New Page Setup box appears, as shown in Figure 9.5. Enter the name of the new setup and press OK.

210

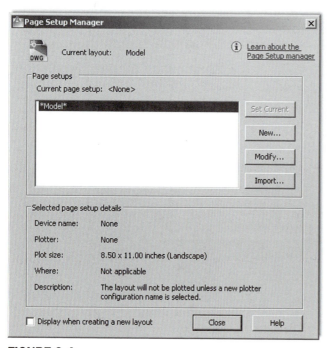

FIGURE 9.4
Page Setup Manager.

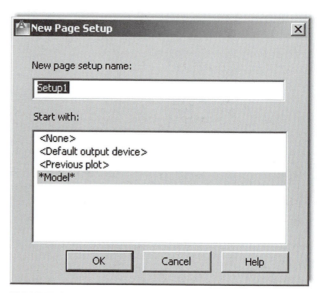

FIGURE 9.5
New Page Setup.

You are then taken to the now familiar-looking Page Setup dialog box. Upon completion of the settings, a preview (yes, still needed), and a press of the OK button, you are redirected to the Page Setup Manager, ready for another setup if needed. If not, press Close.

To see the benefit of this, go to Plot and select the name of the setup from the Page setup in the upper left corner; all your settings appear. Plot the drawing and save it. You never have to enter settings again (unless something changes, like the arrival of a new printer). So finally, after learning all this new material, you can now reduce printing to a pressing of the Preview and the OK buttons—our goal all along.

As stated before, there is still a lot we have not covered, and we revisit this topic again in Chapters 10 and 19. Printing is something you usually receive assistance with at a new job, training center, or consulting gig. However, the most you should expect to be told is what printer is commonly used (up to ten connected to one computer is not unheard of) and what is the standard `ctb` file in use. The rest is up to you, and with the knowledge you gained in this chapter, you should be proficient in all printing matters just short of the level of AutoCAD management and administration.

SUMMARY

You should understand and know how to use the following concepts and commands before moving on to Chapter 10:

- What printer or plotter to use
  - Differences between plotting and printing
- What paper size to use
  - A size
  - B size
  - C size
  - D size
- What area to plot
  - Extents
  - Window
  - Display
  - Limits
- At what scale to plot
  - 1:1
  - Fit to paper
  - Other scale
- What pen settings to use
  - `ctb` files
- What orientation to use
  - Portrait or Landscape
- What offset (if any) to use
  - Center the plot
  - X or Y offset
- Preview
- Page Setup Manager

REVIEW QUESTIONS

Answer the following based on what you learned in Chapter 9.

1. List the seven essential sets of information needed for printing or plotting.
2. List the five paper sizes and their corresponding measurements in inches.

211

3. What is the difference between printing and plotting?
4. List the four plot area selection options.
5. What scale setting ignores scale completely?
6. What is the extension of the pen settings file?
7. Explain the importance of Preview before printing and plotting.
8. Why is the Page Setup Manager useful?

EXERCISES

1. Open the mechanical device drawing from Chapter 8 and practice printing with a variety of scales and other settings. (Difficulty level: Easy; Time to completion: <10 minutes)
   ○ Select printer (specific to your school, home or office).
   ○ Select paper size; stay with A-size.
   ○ Select Extents for the area; observe the results in Preview.
   ○ Select Window for the area; select an area and observe the results in Preview.
   ○ Try a variety of scales such as 1:1, 1:2, and so on; observe the results in Preview.
   ○ Pick the `monochrome.ctb` file.
   ○ Try both Landscape and Portrait orientation.
   ○ Center the plot.

   The following figures show the mechanical device in an Extents Preview (top) and Windows Preview (bottom).

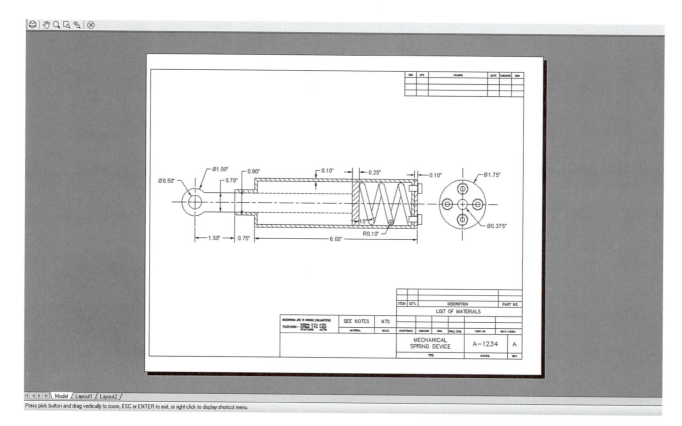

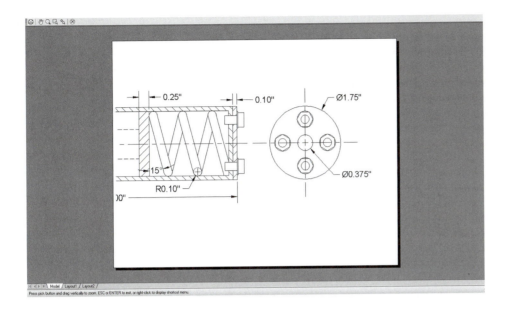

2. Open a new file and set up layers of your choosing, followed by text (Arial_0.18) and dimension styles (Decimal, 0.00 precision, with " suffix, Fit - 1.5). Then, create the following electrical power strip according to the critical dimensions shown. Where no dimensions are given, assume an appropriate value. Add hatching, text, and the dimensions themselves. Plot the exercise to Extents and a section of it via the Window option. (Difficulty level: Easy; Time to completion: 20–30 minutes)

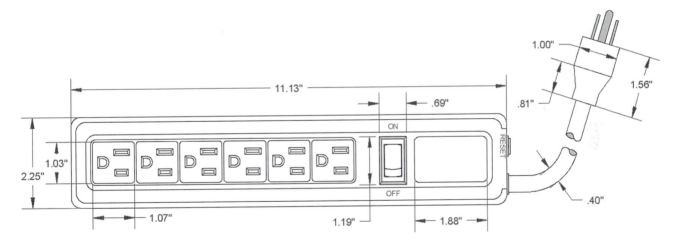

3. Creating architectural details is a common task for any architectural draftsperson. For the final exercise of this chapter, set up your file with the appropriate layer, text, and dimension settings and draft the following detail of roof and support beams. All necessary dimensions are found either in the main detail or the two supporting details (which you do not need to draw separately). Where there are break lines, no dimensions are given as they are not needed. The 6/12 above the roof indicates its slope and may be used to calculate the angle via some basic trigonometry, if needed. (Difficulty level: Moderate; Time to completion: 30–40 minutes)

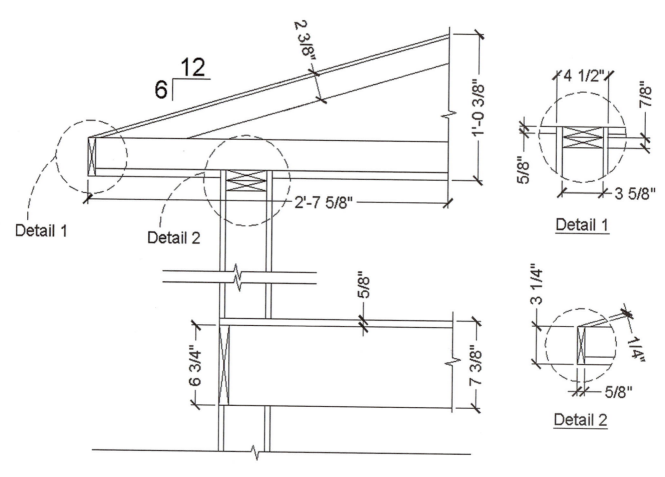

Advanced Output—Paper Space

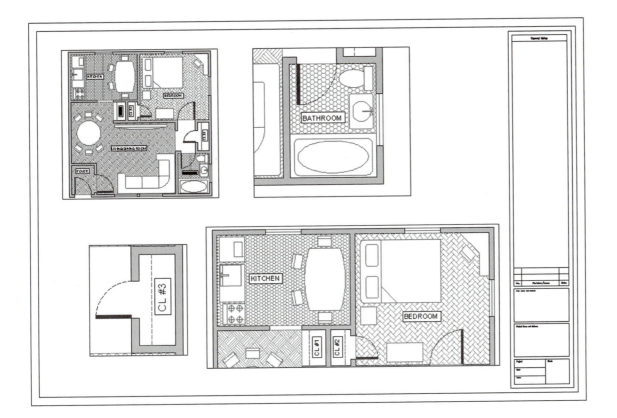

LEARNING OBJECTIVES

In this chapter we introduce and thoroughly cover the concept of Paper Space.
We specifically introduce:

- The primary reasons for using Paper Space
- Entering Paper Space and setting up title blocks (Layouts)
- Viewports and their properties
- Viewport scale assignments
- Controlling what is visible in viewports
- Setting up text and `dim` styles based on the viewport scale
- The new Annotation feature that simplifies setup of text and `dims`

Up and Running with AutoCAD® 2012: 2D and 3D Drawing and Modeling.

By the end of the chapter, you will be well versed in applying this critical concept to your design work.

Estimated time for completion of chapter: 3 hours.

10.1 INTRODUCTION TO PAPER SPACE

Paper Space is an extensive and critically important topic in AutoCAD. Learning it is a rite of passage for all AutoCAD designers, and proper setup and use of Paper Space is a mark of an expert user. The topic is not necessarily difficult but can be somewhat tedious, requiring care and attention from the designer who really has to "dot the *i*s and cross the *t*s" to assure everything is done right. It is also a topic that needs to be understood well in theory, first and foremost. The actual "button pushing" process to set things up is much easier when you understand what Paper Space is and why it is used. The result of learning all this, however, is a professional drawing set that is accurate and easy to read, understand, and print. This is a necessity and a goal well worth striving for. We explore Paper Space from all angles, one step at a time. Read and understand each section before moving on to the next as the knowledge is cumulative.

What Is Paper Space?

When you create a design of a building with pencil and paper, this design needs to be drawn to a certain scale, as opposed to full-size, for the simple reason that it would not fit on a piece of paper if you did not do this. This is the reason why architects carry with them an architectural scale, a portion of which is shown in Figure 10.1.

Then, they can simply pick the appropriate scale for the size of their intended design ($\frac{1}{8}$" = 1'−0" is common), and now, because 8 feet are represented by 1 inch, the building floor plan can fit nicely on a large sheet of paper.

In AutoCAD, however, you do not draw "to scale." Because you now have an infinite-size canvas upon which to create your design, everything is drawn one to one (1:1) or "life-size," meaning that, if you were to jump into your design through the computer monitor, you would find yourself in proportion to (and would be able to freely walk around inside) your building.

This is one of the Golden Rules of AutoCAD, and indeed in all of computer-aided design, and marks a significant departure in philosophy and method from the previous hand-drafting approach. In fact, attempting to draw "to scale" results in severe complications and

FIGURE 10.1
Architectural scale.

216

potential for error when creating, and later changing, the design and is simply not done in practice.

Drawing one to one, however, creates a whole new set of problems. How do you output a design that may be dozens or hundreds of feet across, onto a paper that is generally only 3 or 4 feet wide? Furthermore, this design has to be framed by a title block or title page that itself has to be drawn one to one, so as to not violate the Golden Rule. The solution is elegant, simple in principle, and has stood the test of time.

The idea is as follows. Draw the design in one area of AutoCAD, called *Model Space*, and then draw the title block in another unconnected and unrelated area, called *Paper Space*. Finally, connect the two together via a tool called a *viewport*, which is simply a "window" looking in from Paper Space onto Model Space. This viewport is then assigned an appropriate scale; and if everything is drawn (and printed) one to one from Paper Space, then the design comes out to scale and the problem is resolved.

This in essence is Paper Space: nothing more than a technique to separate the two worlds and reconcile them in a meaningful way to get around the tricky problem of a finite-size piece of paper holding a much larger design.

Be sure you understand the previous few paragraphs completely. Everything outlined from here on is easier to understand once you have a good idea of why we are doing what we are doing.

10.2 PAPER SPACE CONCEPTS

The devil, as they say, is in the details, and the study of Paper Space involves a number of separate concepts, each of which is an important part of the complete picture and needs to be looked at closely:

- **Layouts.** Entering Paper Space and setting up title blocks.
- **Viewports.** Viewports and their properties.
- **Scaling.** Viewport scale assignments.
- **Layers.** Controlling what is visible in viewports.
- **Text and dims.** Setting up text and `dim` styles based on viewport scale.
- **Annotation.** A new feature that simplifies setup of text and `dims`.
- **Summary.** Putting it all together.

These topics are listed roughly in the order a designer would go through them as he or she sets up and does the project. As an example drawing, we continue using the architectural floor plan that you have been working on since the start of Level 1. It happens to fit nicely on a D-size (36" × 24") sheet of paper at ½" scale to illustrate the concepts. If you want to use another design, perhaps one drawn in one of the end-of-chapter exercises or one your instructor assigned to you, by all means use that. If you stick with the apartment floor plan, then thaw all the layers. It may be messy, but we fix that later. Let us begin with the first topic, layouts.

Layouts

The first step involves switching over to Paper Space and familiarizing yourself with the layout tabs. They are found at the bottom of your drawing area and resemble Microsoft Excel tabs, say, Model, followed by Layout 1 and Layout 2, as seen in Figure 10.2.

Go ahead and switch into Paper Space by clicking on Layout 1. You notice a different environment, as shown in Figure 10.3. For starters, the UCS icon is a triangle and the Layout 1 tab is highlighted. Your background color is white, and finally you have a variety of background features (rectangles, shadows, viewports, and others). We modify this

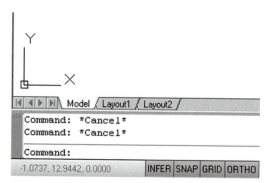

FIGURE 10.2
Layout tabs.

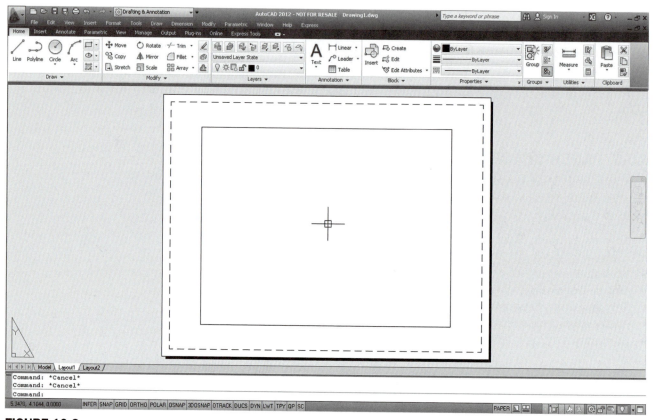

FIGURE 10.3
Paper Space.

environment in a minute, but in the meantime, click on the Model tab and jump back into Model Space. You now know how to enter and exit Paper Space.

There really is no need for Paper Space to look any different from Model Space. Nothing is inherently special about Paper Space. Remember, it is just a technique in AutoCAD to create two separate worlds. You can actually draw in Paper Space as if it is your regular drawing area with no adverse effects, but of course that would defeat the whole purpose of what we are doing.

Based on this, it is highly recommended that the Paper Space Shadow and other features are turned off. Indeed, most experienced users prefer their Paper Space to look as close to Model

Space as possible. It is ultimately up to you if you want to do this, but all screen shots in the remainder of this chapter feature the plain Paper Space environment. To do this you need to right-click to Options…, then select the Display tab and take a look at the Layout elements (lower left). Uncheck all but the very first choice, as seen in Figure 10.4, and set the Sheet/Layout color to black in Colors… (though in the textbook the background remains white to conserve ink and for clarity). It is also a good idea to set your crosshairs to 100 for reasons to be explained later in the chapter. Ask your instructor for assistance if needed. Finally, click on OK.

After this modification, Paper Space should look like what is seen in Figure 10.5 (except of course for the screen color, which should be black, as preferred by most users). Notice that now the only indication that you are in Paper Space is the triangle UCS icon and the highlighted Layout 1 tab. Also the viewport rectangle is erased. Note of course that, if you already had this environment (or a variation of it) set up, you could have skipped the previous few steps.

FIGURE 10.4
Paper Space preference settings.

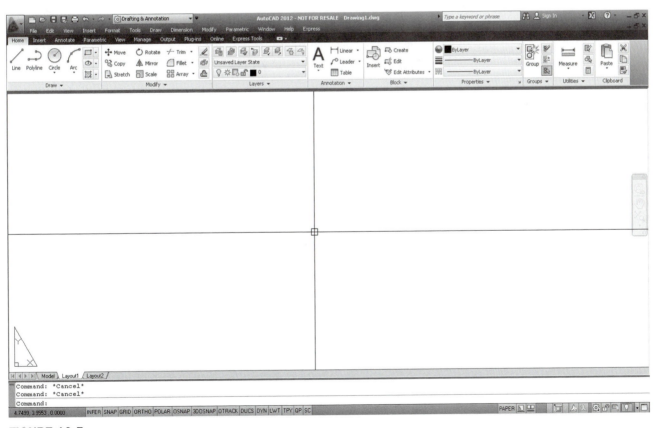

FIGURE 10.5
Paper Space, modified.

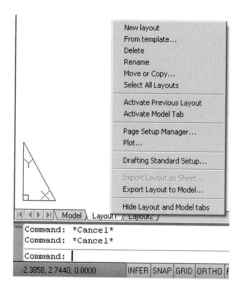

FIGURE 10.6
Layout tab menu.

So, what are these layout tabs in real life? They are actually sheets of paper when the job is completed and ready for printing. The design (done in Model Space) is "presented" in Paper Space with each layout being one sheet of the total drawing set, that is, the Architectural, Mechanical, Electrical, Details, and so on.

If you are thinking that we need to learn how to make title blocks at this point and perhaps create more Layouts, then you are correct. We create a title block from a built-in AutoCAD template. This may or may not be the way you will do it at your job. If you are in an architecture office, there may be a pre-drawn title block that you can just retrieve and paste in, but for our learning purposes we create a new one.

Select the Layout 1 tab (left-click) then *right-click* to bring up the menu shown in Figure 10.6.

We work with this menu for the next few pages. Let us first of all try creating a title block from a template by clicking the **From template**… choice. You get the dialog box seen in Figure 10.7. Pick the `Tutorial-iArch.dwt` template and click Open.

Another small dialog box, Insert Layout(s), appears, as shown in Figure 10.8. It is simply informing you that the layout is a D-size one. Click on OK.

Notice what happened: You have a new tab, called *D-Size Layout*, positioned right after the other three layouts. Click on it and you see a decent title block, much improved in AutoCAD 2010 versus older releases (AutoCAD 2012 continues to use it). It is a reasonable representation, minus the fancy logos, of what a typical architectural office may use. Inside the title block is a blue rectangle through which you may or may not see your building; this is the viewport, and we cover this topic next. For now, go ahead and delete it. We make our own soon. Also, explode the title block as we want to modify it.

What you have now is shown in Figure 10.9 (except that your screen background likely is colored black).

The next step is to modify the layout tabs by renaming them to something more descriptive and removing the unnecessary ones. Right-click on the new D-Size layout tab to get the menu to appear again. Select the **Rename** choice and the tab's name becomes editable; type in `Architectural Layout`. Now get rid of the Layout 1 and Layout 2 tabs, as they are empty and not needed. Begin by clicking on Layout 1, then right-click to get the menu up again.

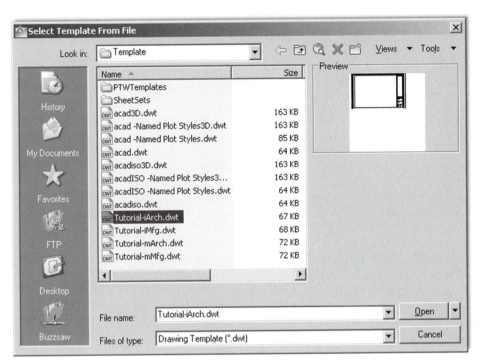

FIGURE 10.7
Select Template From File.

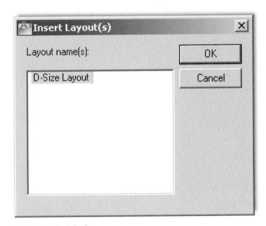

FIGURE 10.8
Insert Layout(s).

Select **Delete** and press OK when the warning box pops up. Repeat the procedure for Layout 2. After these steps, we have only two tabs remaining, Model and Architectural Layout, as seen in Figure 10.10.

Take a moment to look over the title block in the Architectural Layout tab. As a reminder once again, delete the original viewport (the blue rectangle), as we will make our own from scratch. Then, explode the title block if you have not done so yet. Modify anything you may want to look different (for example, delete the USER, REVDATE, and the FNAME text in the lower left corner and added sheet A1 in the lower right corner of the title box), but this is optional.

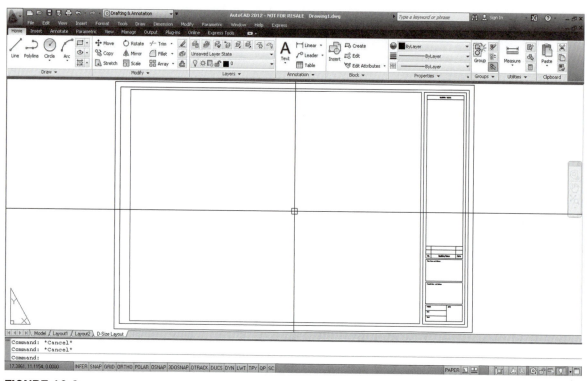

FIGURE 10.9

`Basic Tutorial-iArch.dwt` title block.

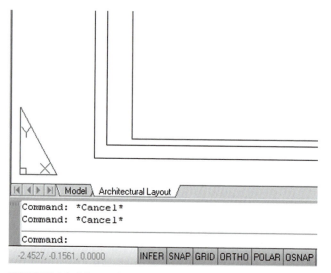

FIGURE 10.10

Model and Architectural Layout tabs.

Now, we need to create more of these title sheets, for the Mechanical and Electrical layouts. Go back to the menu (right-click on the Architectural Layout tab) and select the Move or **Copy…** choice. The dialog box in Figure 10.11 appears.

Be sure to do two things. First of all, check off the Create a copy box at the bottom left and click on the (Move to end) choice, which simply places the new title sheet next in line after the existing Architectural Layout and may be important later when it comes to the printing

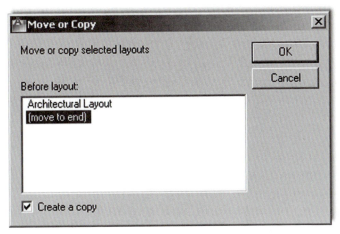

FIGURE 10.11
Move or Copy.

FIGURE 10.12
All tabs.

order. Finally, click on OK and observe what happens with the tabs. A new one appears, saying Architectural Layout (2), and it is a perfect copy of the previous existing tab. Go ahead and rename it Mechanical Layout. Repeat the whole procedure to create a third tab, called *Electrical Layout*, and finally a fourth tab, called *Landscaping Layout*. The final result is in Figure 10.12.

Review this entire section; it is an important first step to learning Paper Space and is a common procedure needed to set up a project that has multiple pages, which is pretty much the case with most projects. It should be clear as to what each sheet (tab) represents. Next, we work on the sheets to set up viewports so the design can be viewed at the proper scale.

Viewports

Viewports, as mentioned in the introduction, are simply windows from the Paper Space world looking into the Model Space world. It is as if we ripped a hole in the fabric of Paper Space and observed what was behind there. We assign scales to these viewports later on and in this manner reconcile the Paper and Model Spaces. For now, we need to first learn how to create and modify the viewports and go over some of their properties. Your new layouts already have a premade viewport (the blue rectangle), but as mentioned earlier, you should have now erased it; we will learn to create a new one from scratch.

Just a reminder: To go through the following steps, make sure you have a design open to look at, which you already should have, as suggested earlier in this chapter. For an example in this text, the apartment plan from the previous chapters continues to be used. It is slightly modified to include both furniture and floor hatching. You are welcome to duplicate the same thing. If all is well, then let us proceed to learning about the viewport or Mview window.

Click on the Architectural Layout tab, entering Paper Space, type in mview, and press Enter. This is the key command for creating new viewports. Alternatively, you can use the cascading menus **View→Viewports→1 Viewport**. *Note here that you can create more than one viewport at a time, but we do only one for now.*

- AutoCAD says:
  ```
  Specify corner of viewport or
  [ON/OFF/Fit/Shadeplot/Lock/Object/Polygonal/Restore/LAyer/2/3/4] <Fit>
  ```

Then simply click and drag the mouse across your screen, effectively drawing a large rectangle from one side of the sheet to the other (avoiding the actual title block in the lower right corner). AutoCAD says: Specify opposite corner: as you are doing this and

223

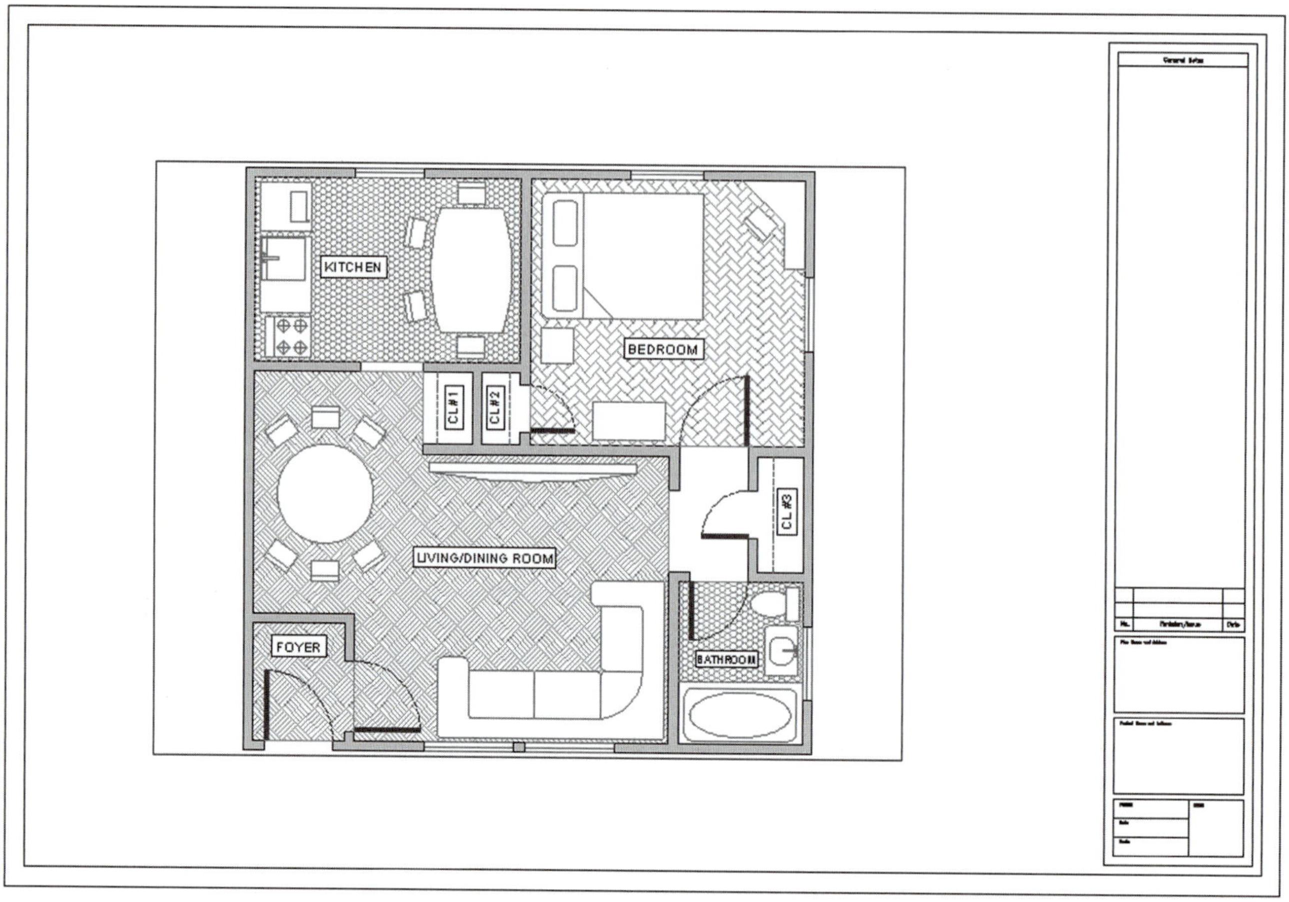

FIGURE 10.13
New viewport.

`Regenerating model` as you complete the viewport. You should now be able to see the design inside the "rectangle" you created, as shown in Figure 10.13.

You can create more viewports in the other layouts if you like for practice, but we focus on this one, in the Architectural Layout.

FLOATING MODEL SPACE

Double-click inside the viewport. Its border becomes thicker and you notice your mouse is "inside" the viewport, as shown in Figure 10.14. The advantage of having full width crosshairs (set at 100) is evident here, as the viewport "clips" the ends of the crosshairs and it becomes clear that you are in this mode. Pan and zoom while inside the viewport and you notice the design shifts around while the Paper Space title sheet stays still. This Floating Model Space mode is sort of a "no man's land" between the Model and Paper Spaces and is critical in scaling the design, as we shall soon see.

For now, double-click anywhere outside of the viewport to get back to Paper Space. Be careful not to double-click on the actual viewport itself. That takes you to the vpmax mode (maximum viewport size), which can be used for editing but is not needed here. If you do accidentally double-click on it, press the button shown in Figure 10.15 to get back (bottom right of the screen).

Practice jumping in and out of the viewport several times; this is very important for the rest of the chapter. Note also the big picture discussed so far. You are in Paper Space and created

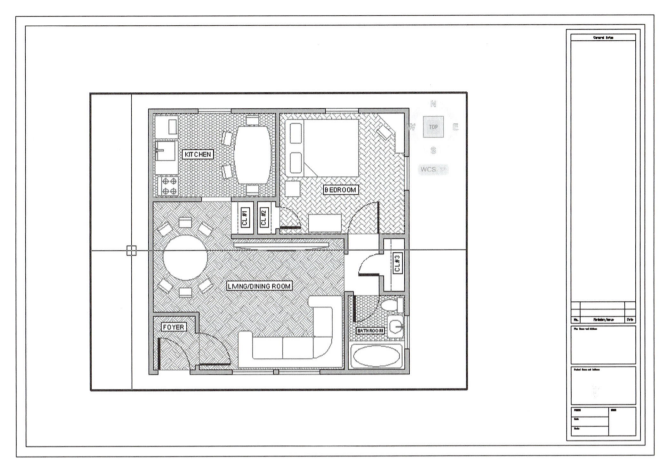

FIGURE 10.14
Floating Model Space.

the viewport to look onto the design in Model Space. When you are inside the viewport, you are technically in Model Space, but still in Paper Space as well. You need this for scaling purposes only. Design work on either the floor plan or the title sheet or block is done in their respective spaces.

Next, we need to learn how to modify the viewport(s). Note that they are *not* static displays but rather dynamic tools you can and should mold and shape to your needs, as outlined in the next section.

FIGURE 10.15
Minimize viewport.

MODIFYING THE VIEWPORT

First, make sure you are in Paper Space, and then click once on the viewport (not inside of it but on the lines of the actual viewport rectangle). The viewport becomes dashed and grips appear. Using them you can resize the viewport to any size desired by clicking on any grip, activating it (it turns red), and adjusting the viewport's size and shape. In general, you can

- Scale the viewport(s) up and down to any size.
- Adjust the viewport(s) to any shape.
- Move the viewport(s) anywhere you like.
- Erase the viewport(s).
- Copy the viewport(s) as many times as needed.
- Make new ones as needed.

Take a few minutes to go through *all* these features. Make your viewport smaller, and copy it three times. Move all four around the screen, and erase one. Finally, create it again with

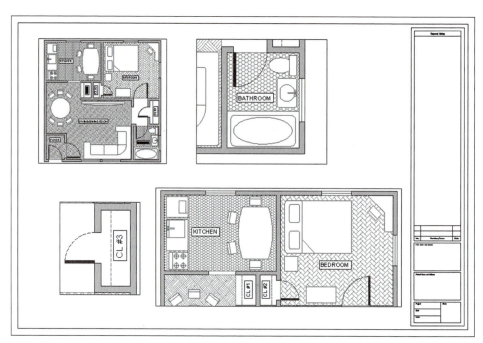

FIGURE 10.16
Four random viewports.

mview. In the end, try to have what is shown in Figure 10.16 and make sure you understand viewport manipulation, the mview command, and getting in and out of Floating Model Space. When you are inside the viewport, do not forget to also scale the drawing up or down and zoom to extents. We focus on a precise scale later.

POLYGONAL VIEWPORT

The shape of the viewport need not be rectangular or square; it can be any shape necessary by using the Polygonal option. To practice this, click over to the Mechanical Layout and create a new viewport using the mview command; however, before clicking to draw the actual viewport, select P for Polygonal in the lengthy menu.

• AutoCAD says: Specify start point: and after the first click Specify next point or [Arc/Length/Undo]:

Click your way around in some random fashion (do not overlap the clicks) and press C for Close to close the viewport. What you get after creating a starburst sort of pattern is shown in Figure 10.17. You can see the design inside the oddly shaped viewport.

Of course, this is not really what this tool is for. Rather it is to make viewports that weave their way around possible revision blocks or notes on the drawing, thereby freeing you from being forced to use a rectangle and wasting space. An example is shown in Figure 10.18, and this is often the case in real-life designs, making polygonal viewports very important for creating an efficient layout.

There are a few other important items to cover with viewports, such as locking them and making them invisible, but we introduce these features a bit later, as the need arises. What we need to look at next is the "core" of Paper Space: scaling the viewports and their contents.

Scaling

Here, we finally put everything together, because really what we have done so far was just the "setup." We have our design in Model Space and our title blocks and sheets in Paper Space,

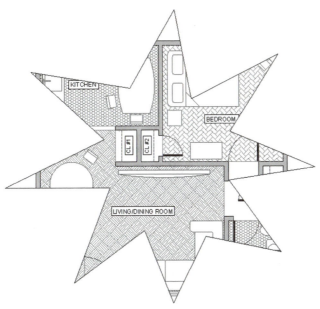

FIGURE 10.17
Polygonal viewport 1.

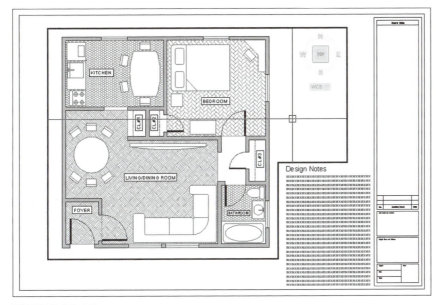

FIGURE 10.18
Polygonal viewport 2.

connected together by one or more viewports. Now, we need to give these viewports a scale, which was our goal all along.

To make things simple, let us just have one viewport in our Architectural Layout, so delete any others you may have, including polygonal ones, and leave one large viewport similar to Figure 10.13.

So, what is scaling? Well, you may have already guessed it has something to do with zooming. After all, as the design is zoomed in and out (while in Floating Model Space), the scale continuously changes and is always some value. So scaling the design properly is just a matter

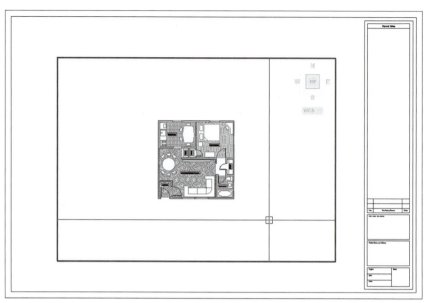

FIGURE 10.19
Design zoomed to ¼" = 1'−0" scale (1/48×p).

of knowing how much to zoom it in the viewport. The big question of course is, How much? Another way to put it is, What scale factor do we apply?

There are two ways to assign scale to any viewport. The first one is a technique used for just about every AutoCAD release up until the last few. It is a bit more involved but really gets you to understand what is going on. Then, we cover the other, easier method.

Method 1. Double-click inside the viewport, type in `zoom`, and press Enter.

- AutoCAD says: `Specify corner of window, enter a scale factor (nX or nXP), or [All/Center/Dynamic/Extents/Previous/Scale/Window/Object] <real time>:`

Here is where you used to pick `E` for `Extents` and see the entire drawing fill up your screen. However, notice the `enter a scale factor (nX or nXP)` choices. This is what we are interested in right now, specifically nXP.

What `nXP` means is that you need to enter a ratio relative to Paper Space (which is the 1:1 "anchor," so to speak) and the design scales to that ratio. So, what is it? Well for ¼" = 1'−0" the *n* is $\frac{1}{48}$ (followed by the xP). So type in `1/48xp` and press Enter. The design zooms to precisely ¼" = 1'−0" scale, as shown in Figure 10.19.

Now this seems to be too small, doesn't it? No problem, just type in `zoom` and press Enter again. Then type in 1/24xp and press Enter again. The design will zoom to ½" = 1'−0", as shown in Figure 10.20, which seems to be a better size for this design and title sheet.

So, where did `1/48xp` and `1/24xp` come from? They are just ratios developed from the fact that there are 12 inches in 1 foot and the scale is 1/4 or 1/2. Therefore, for 1/4, it is 1/48 (⅛ inch scale × 1/12 inches in a foot = 1/48 relative to the 1:1 Paper Space). For 1/2 it is 1/24 (½ inch scale × 1/12 inches in a foot = 1/24 relative to the 1:1 Paper Space) and so on. There are additional scales such as ¹/₁₆" ⅛", ³/₈", ⁵/₈", and others, all of which can be derived in a similar manner. Now that you understand basic viewport scaling, let us apply this method in a somewhat easier way.

Method 2. Double-click back to Paper Space and bring up the Viewports toolbar, shown in Figure 10.21.

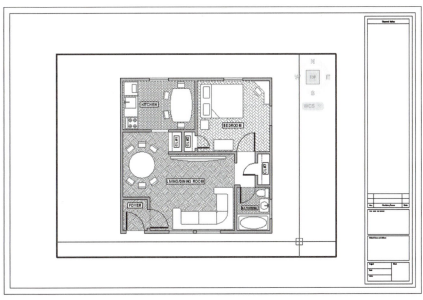

FIGURE 10.20
Design zoomed to ½" = 1'−0" scale (1 / 24 x p).

FIGURE 10.21
Viewports toolbar.

Now click on the viewport once (not inside of it but on the lines of the actual viewport rectangle). When the viewport becomes dashed, the scale associated with that viewport is shown in the white field on the right side of the Viewport toolbar. Click the down arrow and you see a wide variety of choices. Pick the scale you want and the viewport takes on that scale, as shown in Figure 10.22.

Layers

We have done a lot already in setting up the drawing in Paper Space. We now turn our attention to several important details that are either new or change significantly when dealing with Paper Space.

One of those is layers. While the fundamental layer concepts do not change, a new ability is added to the Layers dialog box. It turns out that, when viewports are present, you can freeze and thaw layers in those viewports independent of each other and the main drawing. This is very significant, as you want to be able to display different items as you go from viewport to viewport.

Perhaps the most important example of that is when you jump between the Architectural, Mechanical, and Electrical Layouts. Each tab is a new viewport, yet all views of the design originate from one actual copy of that design. In each layout, you need to see only what is relevant to that layout, and it would be a major problem if you could turn layers on and off only in the main Model Space drawing. Then, each tab would display the exact same thing.

Fortunately, this is not the case, and in each viewport, you can turn layers on and off as needed to get the design points across. Therefore, the Architectural Layout has only architectural layers and so on. It is the same idea with multiple viewports on one layout (such as the case with detail sheets). Each viewport is truly a world unto itself. To see an example of this let us create four viewports in the Architectural Layout, as shown in Figure 10.23. The scales assigned are random.

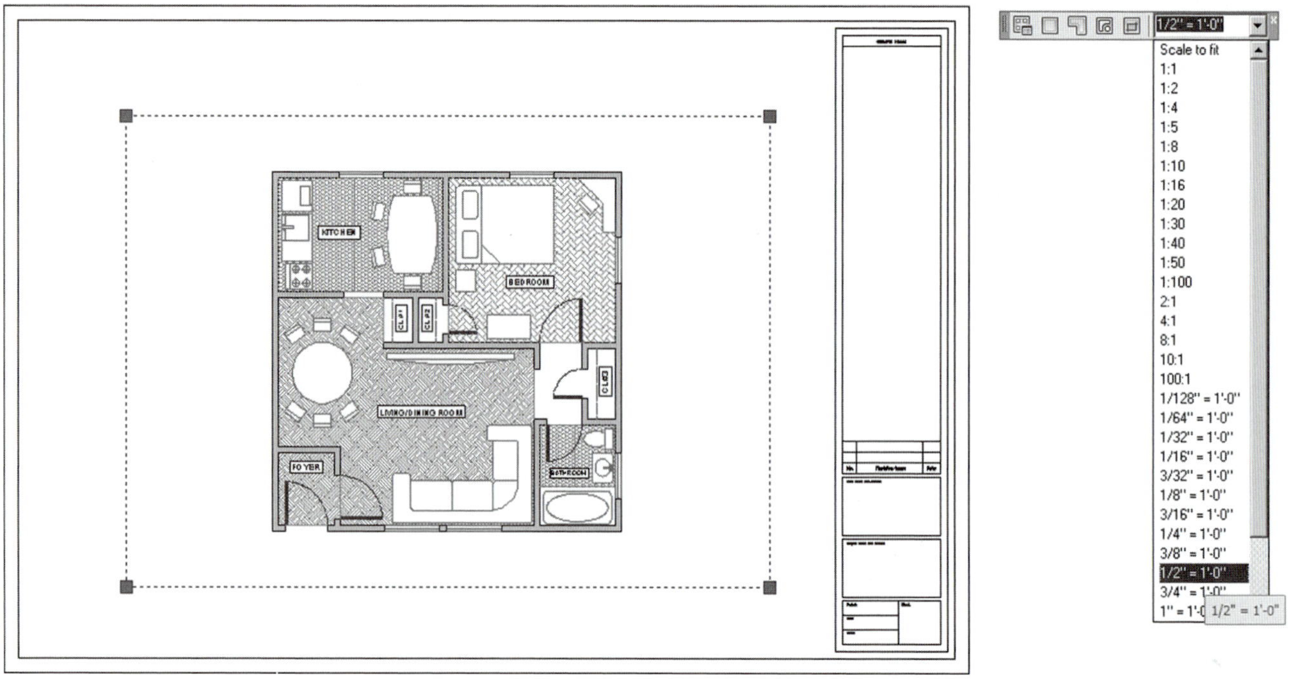

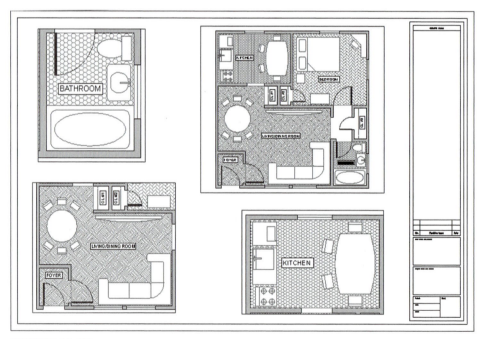

FIGURE 10.22
Viewport scaling via toolbar.

FIGURE 10.23
Four viewports for layers demo.

Double-click into the top left viewport and bring up layers. The Layers dialog box (with some extra columns) appears, as shown in Figure 10.24. All relevant columns have been expanded for maximum visibility of the contents. Look for the VP Freeze column (circled); this is the column in which we are interested right now.

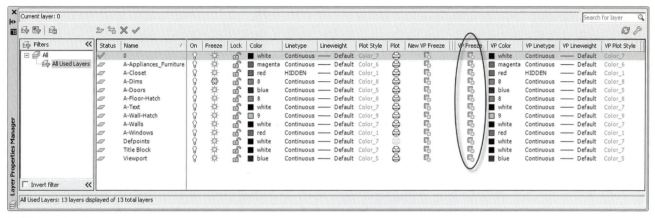

FIGURE 10.24
Layers with VP Freeze column.

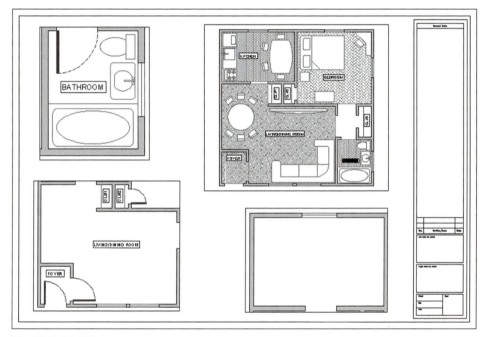

FIGURE 10.25
VP Frozen layers.

In that column, you can freeze all the unneeded layers for that layout without affecting either the main layout or the other viewports. Go ahead and freeze the A-Floor-Hatch layer. The result is shown in Figure 10.25. Note how the neighboring viewports are not affected by what you did in the top left one. Now click over to the top right one and freeze the A-Doors layer only. Click over to the bottom left viewport and freeze the A-Floor-Hatch, and the A-Appliances_Furniture layers. Finally, in the bottom right viewport, freeze everything except the walls and windows. Note the distinctly different layer settings in each viewport; they are completely independent.

LOCK VP

We now cover two other notable features of viewports. The first is the ability to lock the viewport, which simply means you cannot change the viewport's scale and all zooming attempts fail (the entire Paper Space drawing zooms instead). This can be useful to prevent accidental zooming while inside the viewport and should be set after the design is laid out

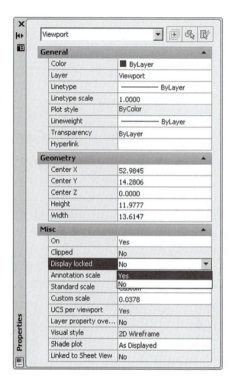

FIGURE 10.26
Locking the viewport.

and set up. To do this, bring up the Properties box. Then click the viewport once, and select (under Misc.) the Display locked option, setting it to Yes, as shown in Figure 10.26.

FREEZE VP

This next feature allows you to either "freeze" the viewport or simply prevent it from printing. The reason for this is that generally you do not want a big fat rectangle on your final plots that will go to the client. This is especially true of the main layouts that feature just one large viewport. The visible viewport rectangle can be mistaken for part of the border and is unnecessary. Even with detail sheets, each of which may display many details, the trend is to just separate them adequately but not have any rectangles visible. If you "did not do it by hand in the old days," then there is no need to introduce it in AutoCAD.

Having said this, however, some designers prefer to see the viewports all the time and have them disappear only upon printing. So you have several options in this regard.

First of all, we need to put the viewports on an appropriate layer if they are not there already. Designers generally create a _VP layer for them. Notice a trick here. The VP layer has an underscore in front of it. This forces the layer (or any layer that has that underscore) to rise to the top of the layer list, above the 0 layer. This makes it easy to find the title block and sheet or the VP layers, as long as you do not overdo it and stick too many up there. Go ahead and change the color of the layer to gray and freeze it. Then finally change all the viewports to that frozen _VP layer. If you did everything correctly, then the viewport boxes disappear, as shown in Figure 10.27.

There are other alternatives of course for those who want to see the viewports but not print them. You can put them on the Defpoints layer. This layer does not print (it is the definition points from dimensions). Another method is to just put a slash through the printer icon in the Layers dialog box corresponding to the viewport layer, which leads to the same result.

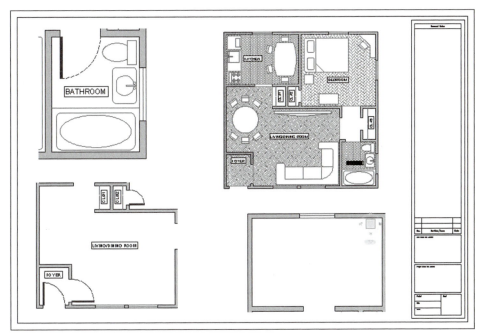

FIGURE 10.27
Viewports frozen.

We are getting to the end of Paper Space, and only two major topics are left: Text and `dims` and the Annotation tools.

Text and dims

This topic is probably the most confusing one in Paper Space and really requires the AutoCAD designer to not only understand everything thoroughly but also work carefully and deliberately. In practice, much of this is done only once and saved as part of a template, but for educational purposes, we assume you are being tasked with the text and dims setup from scratch and must know all the steps.

Our goal here is quite straightforward: We want all the text and dimensions to be the same size when printed, regardless of in what scale viewport they appear. This uniformity is a basic requirement of a professional drawing. Architects and engineers specify something to the effect of: "I want the text height to be ¼" on all drawings, both in layouts and in all details," and it is up to the AutoCAD designer to ensure that this happens. With hand drafting, this was not an issue; the text and dims were drawn all at once using the same size template.

With AutoCAD, due to our use of Model Space and Paper Space, things are not so simple. The basic fact remains that all text and dimensions have a certain fixed measurable size when drawn in Model Space and appear as different sizes when viewed through the various scaled viewports, just as the design itself does. Some designers try to simplify everything by dimensioning the design in Paper Space to achieve uniformity (AutoCAD lets you do this), but this is *not* correct. The dimensions and all text must remain *with* the design in Model Space. The reasons for this are many, not the least of which is that any movement of the design misaligns everything.

So the text and dims need to stay where they are supposed to be, and we need to understand how to properly size text and dimensions based on the anticipated scale of the viewport in which they appear. This all really just boils down to a game of "match-up." The basic theory is as follows. You already know that viewport scale is a function of zooming. Therefore, when you have uniform text in Model Space and view the same set of text through the ¼" = 1'–0"

233

VIEWPORT SCALE	DRAWING TEXT HEIGHT			SCALE FACTOR	LT SCALE
	1/8	3/16	1/4		
Architectural					
1 1/2" = 1'-0"	1	1.5	2	8	3
1" = 1'-0"	1.5	2.25	3	12	4.5
3/4" = 1'-0"	2	3	4	16	6
1/2" = 1'-0"	3	4.5	6	24	9
3/8" = 1'-0"	4	6	8	32	12
1/4" = 1'-0"	6	9	12	48	18
1/8" = 1'-0"	12	18	24	96	36
1/16" = 1'-0"	24	36	48	192	72
1/32" = 1'-0"	48	72	96	384	144
Engineering					
1" = 10'	15	22.5	30	120	45
1" = 20'	30	45	60	240	90
1" = 30'	45	67.5	90	360	135
1" = 40'	60	90	120	480	180
1" = 50'	75	112.5	150	600	225
1" = 60'	90	135	180	720	270

FIGURE 10.28
Text and scale chart.

versus the ½" = 1'−0" scale viewport in Paper Space, the text in the ½" viewport appears two times as large, as the viewport is a larger scale. To counter this effect, you need to make sure that the text that appears in that ½" viewport is drawn to be half as big, so it appears uniform with the ¼" text (just as an example).

The trick is to be consistent and careful, so that the wrong size text does not appear with the wrong viewport scale. So, how is this done in practice? How do you know what size text and `dims` are appropriate for a particular scale? We have tables for this, such as the one shown in Figure 10.28.

To use this chart, you have to know what scale viewports you are using. With the main layouts, such as the Architectural or Mechanical, done earlier, it is very easy. These layouts typically feature only one viewport of a known scale ($\frac{1}{8}$" = 1'−0" is commonly used); therefore, if the architect desires all text output to be $\frac{3}{16}$" in height (also a common size), then by reading the chart from left to right until we meet the $\frac{3}{16}$" column, we arrive at a height of 18 inches, which is to be used for all text in that drawing. For the ¼" = 1'−0" scale, that same text size is now 9 inches, half as big, since the new scale, ¼" = 1'−0", is twice as large.

With details, which may come in all the scales listed, the procedure is no different; you just have to be very careful in knowing ahead of time what scale viewport you are using, to use the appropriate text height. All this comes with experience, and after much practice, you will be able to quickly assess the necessary scale of a given detail to ensure that all aspects of it are visible and clear to whoever is reading the plans. This is the only factor in determining the appropriate text size.

The chart shows not only the text height but also the dimension sizes under the Scale Factor column. Here, you retrieve the appropriate size (24 in the ½" = 1'−0" example) and enter that value in the Use overall scale of: text field under the Fit tab of the DDIM dialog box. Also shown is the appropriate scale for linetypes (9 in the case of ½" = 1'−0"). You need this to ensure you can see a hidden line, for example, under this scale.

Most of the information contained in Figure 10.28 needs to be entered into AutoCAD in the form of text styles and `dim` styles. This is the job of the CAD manager on initially setting up an architectural or engineering office's CAD standards. Yes, it is as tedious as it sounds; the chart

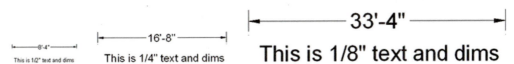

FIGURE 10.29
½", ¼", and ⅛" text and dimensions.

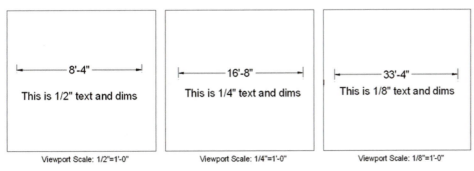

FIGURE 10.30
Paper Space text and dimensions uniformity.

is abbreviated and does not contain every scale used, so the work is substantial. Fortunately, it needs to be done only once, and a template is formed for all users to follow from there on out. Enforcing this usually turns out to be the hardest part of all, as the "system" is only as good as the people who follow it. You must be careful to use the proper style (*Arial_.25*, for example), and be sure to set that style as current while working on the relevant design or detail. The payoff comes later in the process, when you view the design in Paper Space. Uniformity and professionalism is the final goal. Go ahead and practice these concepts.

Step 1. Go to the Style dialog box and create three new text styles, all Arial fonts. They are set up for a final text height of ¼" and are viewed through three viewports of the following scales:
½" = 1'−0" (therefore, size: 6").
¼" = 1'−0" (therefore, size: 12").
⅛" = 1'−0" (therefore, size: 24").

Step 2. Go to the DDIM dialog box and create three new dim styles. All must use the appropriate text from the previous procedure (therefore, the ½" = 1'−0" dimension style uses the Arial_.5 style of 6" height, etc.). Then, set the fit as shown on the chart and reproduced here:
½" = 1'−0" (therefore, Use overall scale of: 24").
¼" = 1'−0" (therefore, Use overall scale of: 48").
⅛" = 1'−0" (therefore, Use overall scale of: 96").

Step 3. One by one, in Model Space, set each text and dimension style as current and create one set of text and one set of random dimensions next to each other, as shown in Figure 10.29.

This may take you a few minutes to set up and execute. Be sure to work carefully, checking and rechecking all your settings. As expected, the three scales go up in size twice for each jump from ½" to ¼" to ⅛". We now want to see them uniform in size in Paper Space.

Go into Paper Space and create three viewports next to each other. Use the tools learned previously to set the first one to the ½" = 1'−0', the second to ¼" = 1'−0", and the third to ⅛" = 1'−0". Pan around until the appropriate text and dims combination is in the respective viewport (you may have to go to Model Space and move them apart slightly). The result is shown in Figure 10.30.

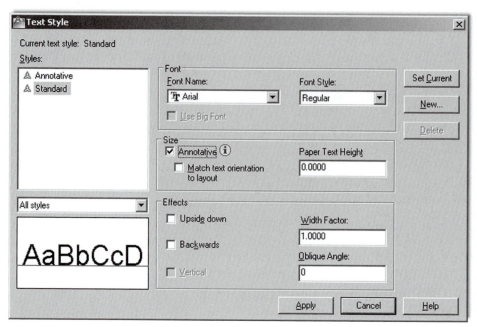

FIGURE 10.31
Annotative style.

Notice the complete uniformity of the text and `dims` in Paper Space, as opposed to how they looked in Model Space in Figure 10.29. This is the essence of what we are doing here. If you can set up these three viewports and all the Text and `dims` correctly, you can do any drawing. Be sure to review and learn everything thus far presented. Ask your instructor for assistance if needed.

Annotation

The final Paper Space feature we discuss is relatively new and first appeared in AutoCAD 2008. Called *Annotation*, its purpose is to present an alternative approach to the methods of the previous section "Text and `dims`." The new approach is different in that it allows you to set multiple scales for the same text, so the proper scales are visible in the proper viewports with no further input from the user. Several items can be annotated: styles, dimensions, blocks, and hatches. We explore text as outlined in the following procedure, but it is similar with dimensions.

Open a brand-new file, type in `style`, and press Enter. The familiar Style dialog box comes up. Create a new style, calling it *Arial_Anno*, font Arial. Check off the Annotative box under Size and leave the Paper Text Height as 0. The new text has a triangular symbol next to it, as seen in Figure 10.31. Be sure to set it as Current.

Now switch to Model Space and create a line of text. The first time you do this, a dialog box, such as shown in Figure 10.32, appears. Set the scale at which you anticipate this text to show up in (such as inside a ¼" = 1'−0" viewport, as done here). Then press OK and finish typing the text.

Now, switch to Paper Space and create a viewport and set it to the ¼" scale. Expand or contract it to fit nicely around the text, as shown in Figure 10.33.

Copy that viewport to the right and change its scale to ⅛". The text briefly shrinks to a smaller size then balloons back up to the same size you set (¼"), as shown in Figure 10.34.

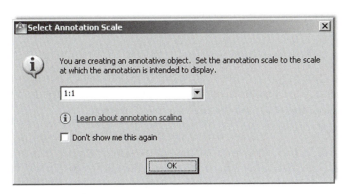

FIGURE 10.32
Select Annotation Scale.

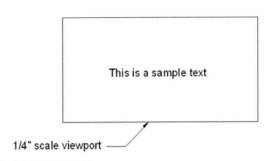

1/4" scale viewport

FIGURE 10.33
Annotation text in sample viewport.

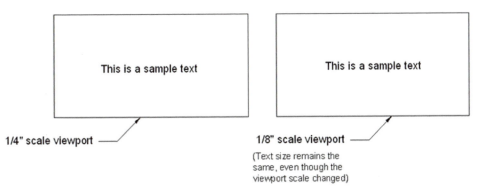

This is a sample text

1/4" scale viewport

This is a sample text

1/8" scale viewport

(Text size remains the same, even though the viewport scale changed)

FIGURE 10.34
Annotation text in sample viewports.

237

Annotation Scale

MODEL

FIGURE 10.35
Annotation scale.

This is a sample text

This is a sample text

FIGURE 10.36
Annotative text copy.

This is the essence of annotation. The Annotation Scale is adjusted at the bottom right of the taskbar, as shown in Figure 10.35, so when new text is being created, you need to set its anticipated viewport scale right away.

You can also set different annotation scales to the same text if you anticipate it being visible in two or more viewports of different scale. Click on the text and a "double" of it pops up (Figure 10.36). This double is not a true copy but a temporary twin. At this point, go ahead and set the new annotation scale from the pop-up list in Figure 10.35 and move the text to a new position (usually right on top of where the old one was).

Everything done here with text can be done with dimensions and the other items listed previously. Experiment with this tool to see exactly what AutoCAD's designers had in mind. It is not necessarily easier than the previous method but may be somewhat simpler—opinions vary. Note the Annotation Visibility option (scale symbol with a light) next to the Annotation Scale list. This shows all objects in all scales; the symbol next to it is for automatic scale addition.

SUMMARY

We go over a lot of information in this Paper Space chapter. Ideally, you have absorbed all of it and are able to use this valuable and important tool. Many use it improperly or not at all, and being an expert in Paper Space sets you apart from the crowd and adds a true level of professionalism to your designs. This summary serves as a capstone to the chapter and presents the big picture once again, now that the details have been mastered.

Do not think of Paper Space as a rigid or fixed system, where buttons are pushed and results pop out. While an element of this is present, the proper way of looking at it is as a flexible and loose collection of tools that allow you to shape and mold the presentation of your design in a way that is efficient, clear, professional, and of course pleasing to the eye. There is no *one* way to do things. Rather there is a constant tug of war between (sometimes conflicting) requirements. For example, you have to consider

- **The size of the design.** This is the first and foremost concern. Your design takes priority over everything else, so focus on that in Model Space, creating your architectural layout or engineering device, as per client need.
- **The size of the available plotter and paper.** After the design is completed, output is considered. What do you have to work with? Most companies have a plotter capable of 36" × 24" plots. Some have the larger 48" × 36" plotters. Know your paper options. The larger plotters take E-sized paper, the smaller only up to D size. You cannot print on what you do not have unless you outsource to a print shop.
- **The scale needed to clearly see all facets of the design.** This is the next important step. Take a look at how your design will look on the D- or E-size sheets. Can you see everything? Are any parts of the design not clearly visible? Generally, a 25' × 25' building (such as the one used as an example in this chapter) fits nicely on a D-size sheet at ½" = 1'−0" scale. If it is much bigger, then you can either switch to an E-size sheet (if possible) or switch to another scale. Another scale, however, makes the small details of the design hard to see, so an option is to stay at the previous scale and "split" the drawing into halves or quarters and use match lines for alignment.

As you can see, many variables come into play in setting up Paper Space properly; and with experience, all this will be in your head at the start of a project and you will know instinctively what paper, scale, and viewports to go with, based on the size and expected overall shape of the design.

As of completion of this chapter, you have graduated from a beginner to intermediate level of AutoCAD knowledge. If you have understood everything up to now and have not skipped any exercises or reading, then great job! If all you need AutoCAD for is basic drafting or just knowing enough to oversee other full-time AutoCAD designers, then you may stop here, you have what you need. If your work requires expert level knowledge, you need to go on to Chapters 11 through 20. See you there!

You should understand and know how to use the following concepts and commands before moving on to Level 2 and Chapter 11:

- Paper Space
- Layouts
- Viewports
- Scaling a design inside of a viewport
- Working with layers in individual viewports
- Scaling text and `dims` in viewports
- The Annotative concept

REVIEW QUESTIONS

Answer the following based on what you learned in Chapter 10.

1. What is Paper Space?
2. What are layouts?
3. How do you create, delete, rename, and copy layouts?
4. What are viewports?
5. How do you create, scale, and shape viewports?
6. How do you scale a design inside a viewport?
7. How do you turn layers on and off in individual viewports?
8. How do you scale text and dims in viewports?
9. Describe the Annotative concept.

EXERCISES

1. Set up a total of five title blocks in Paper Space using the `Tutorial-iMfg.dwt` template file. Name the tabs as seen next, and delete any extra tabs. (Difficulty level: Easy; Time to completion: <10 minutes)
   ○ Front View
   ○ Side View
   ○ Section A-A
   ○ Section B-B
   ○ Section C-C
2. Bring up the floor plan you did in Chapter 4, Exercise 6 (or redraw it). Then, go into Paper Space and create an Architectural Layout tab, followed by several viewports, as in the top figure that follows. Assign each of the viewports a scale as shown in the bottom figure. Freeze the Dims layer in the two detail viewports. Note that the floor plan has been rotated 90° clockwise for this exercise. (Difficulty level: Easy; Time to completion: 10–15 minutes if floor plan is completed, 55–75 if not)

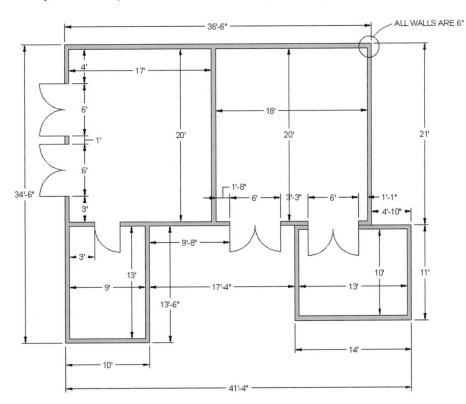

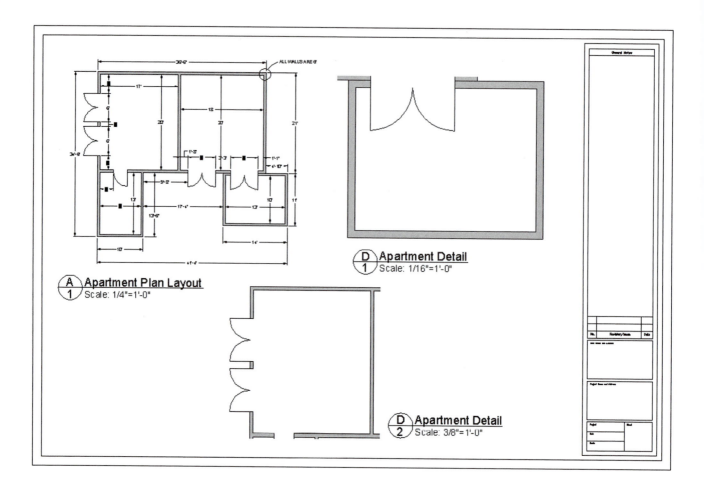

A) Apartment Plan Layout
1) Scale: 1/4"=1'-0"

D) Apartment Detail
1) Scale: 1/16"=1'-0"

D) Apartment Detail
2) Scale: 3/8"=1'-0"

240

3. The final exercise of this chapter again focuses on the basic drafting of an architectural detail. Set up the proper layers and draft what you see based on the given primary dimensions. Where none are shown, you have to do a reasonable estimate. Note how lines and arcs were used to create borders for some of the hatches, then deleted. This is a common technique that we try again with plines in the next chapter to achieve a smoother effect. (Difficulty level: Moderate; Time to completion: 60–90 minutes)

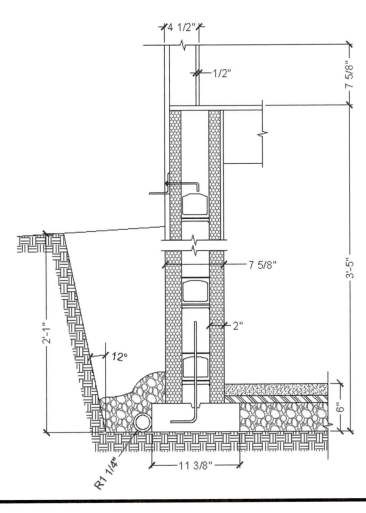

Answers to Review Questions

CHAPTER 1 REVIEW QUESTIONS

1. List the four Create Objects commands covered.
    Line, circle, arc, and rectangle.

2. List the ten Edit/Modify Objects commands covered.
    Erase, move, copy, rotate, scale, trim, extend, offset, mirror, and fillet.

3. List the three View Objects commands covered.
    Zoom, pan, regen.

4. What is a workspace?
    A workspace is a customized screen layout.

5. What four main methods are used to *issue a command* in AutoCAD?
    Command line, toolbars, cascading menus, and Ribbon.

6. What is the difference between a selection Window and Crossing?
    A Window is a solid rectangle produced by moving the mouse to the right of an imaginary line and selects only what is completely inside of it. A Crossing is a dashed rectangle produced by moving the mouse to the left of an imaginary line and selects anything it touches.

7. What is the name of the command that forces lines to be drawn perfectly *straight* horizontally and vertically? Which F key activates it?
    Ortho, F8.

8. What six main OSNAP points are covered? Draw the snap point's *symbol* next to your answer.
    Endpoint, Midpoint, Intersection, Center, Quadrant, and Perpendicular. See the text for the symbols.

9. What command brings up the OSNAP dialog box?
    Osnap.

10. Which F key turns the OSNAP feature on and off?
    F3.

11. How do you terminate a command-in-progress in AutoCAD?
    Esc. key.

12. What sequence of commands quickly clears your screen?
    Erase → All, Enter, Enter.

13. What sequence of commands allows you to view your entire drawing?
    Zoom → E.

14. How do you undo a command?
    Type in u for undo, or use the Undo arrow in the Standard toolbar.

15. How do you repeat the last command you just used?
    Press the space bar or Enter.

244

CHAPTER 2 REVIEW QUESTIONS

1. How do you activate grips on AutoCAD's geometry?
    Click on any of the objects.

2. How do you make grips disappear?
    Press the Esc. key.

3. What are grips used for?
    To modify the shape or position of an object.

4. What is the command to bring up the Units dialog box?
    Units.

5. The snap command is always used with what other command?
    Grid.

6. Correctly draw a *Cartesian coordinate system* grid. Label and number each axis appropriately, with values ranging from −5 to +5.
    See the text, Figure 2.4.

7. On the preceding *Cartesian coordinate system* grid, locate and label the following points:
    (3,2) (4,4) (−2,5) (−5,2) (−3,−1) (−2,−4) (1,−5) (2,−2).
    Student activity.

8. What are the two main types of geometric data entry?
    The manual method and the dynamic method.

9. Describe how Dynamic Input works when creating a line at an angle.
    With DYN activated, begin drawing a line, entering a distance value in the first highlighted box. Then, tab over to the second highlighted box, entering an angle value.

10. List the six Inquiry commands covered.
   Area, distance, list, ID, radius, and angle.

11. What command simplifies a rectangle into four separate lines?
   Explode.

12. What is the difference between an inscribed and a circumscribed circle in the polygon command?
   Inscribed means the polygon appears inside a given circle, with vertices tangent. Circumscribed means the polygon appears outside a given circle, with the sides tangent.

13. What is a *chamfer* as opposed to a *fillet*?
   A chamfer creates an angled edge, whereas a fillet creates a smooth circular edge or none at all.

14. What *template* should you pick if you want a new file to open?
   `acad.dwt`.

15. What is the idea behind limits?
   Limits define the outer boundaries of your work area.

16. Why should you *save* often?
   Because all software and computers are subject to crashes and malfunctions.

17. What is the graphic symbol for save?
   A computer floppy disk.

18. How do you get to the *Help files* in AutoCAD?
   By pressing F1 at any time.

CHAPTER 3 REVIEW QUESTIONS

1. What is a *layer*?
   Layers is a concept that allows you to group drawn geometry in distinct and separate categories according to similar features or a common theme.

2. What is the main reason for *using layers* in AutoCAD?
   To maintain control over your drawn geometry.

3. What typed command brings up the *Layers dialog box*?
   Layer or just `la`.

4. List two ways to *create* a new layer.
   New Layer icon or, once a new layer exists, just press Enter while selecting any layer.

5. How do you *delete* a layer?
   The red X.

6. How do you make a layer *current*?
   Double-click on it or select it and press the green check mark symbol.

7. What three available *Color tabs are available*?
   Index Color, True Color, and Color Books.

8. What does *freezing* a layer do?
   It makes all objects on that layer invisible.

9. What does *locking* a layer do?
   It makes all objects on that layer un-editable.

10. What are some of the *linetypes* mentioned in the text?
Hidden, Center, Dashed.

11. What are three ways to bring up the *Properties toolbar*?
Double-click on any object, press Ctrl + 1, or type in ch.

12. What two other ways to *change properties or layers* are discussed?
Match Properties and Layers toolbar.

CHAPTER 4 REVIEW QUESTIONS

1. What are the two types of *text* in AutoCAD? State some *properties* of each, and describe when you would likely use each type.
Text—single line, limited formatting. Mtext—paragraph form, advanced formatting.

2. What is the easiest method to *edit text* in AutoCAD? What was another typed method?
Double-click on the text. Ddedit.

3. What feature allows you to add extensive *symbols* to your mtext?
Character Map.

4. What are the three essential features to input when using *Style*?
Font name, font, and size.

5. What is the *width factor*?
How wide a font is; 1 is standard width.

6. What is an *oblique angle*?
The "tilt" of the font.

7. How do you initiate *spell check*?
Right-click on the underlined misspelled word, or type in spell.

8. What new OSNAP is introduced in this chapter?
NEArest.

9. How do you get rid of the UCS icon?
Type in ucsicon, then off.

10. What can you do if you cannot see the dashes in a Hidden linetype?
Use ltscale.

CHAPTER 5 REVIEW QUESTIONS

1. List the four steps in creating a basic *hatch pattern*.
Pick hatch pattern, indicate where you want it to go, fine-tune pattern via scale and angle, and preview pattern.

2. What two methods are used to select *where to put* the hatch pattern?
Directly picking an object or picking a point inside that object.

3. What commonly occurring problem can prevent a hatch being created?
A gap.

4. What is the name of the tool recently added to AutoCAD to address this problem?
Gap tolerance.

5. How do you *edit* an existing hatch pattern?
Double-click on it.

6. What is the effect of *exploding* hatch patterns? Is it recommended?
   The hatch breaks up into individual components. No.

7. What especially useful pattern to fill in walls is mentioned in the chapter?
   Solid Fill.

CHAPTER 6 REVIEW QUESTIONS

1. List the 11 types of *dimensions* discussed.
   Linear, Aligned, Diameter, Radius, Angular, Continuous, Baseline, Leader, Arc Length, Ordinate, and Jogged.

2. What command is best for *editing* dimension values?
   Ddedit.

3. What is the difference between *natural* and *forced* dimensions?
   Natural dimensions show the actual value, while forced dimensions are added manually and may be false values.

4. How do you *restore* a natural dimension value when you forgot what it was?
   Type in < >.

5. List the four items that needed to be *customized* in our dimensions.
   Units, font, arrowheads, and fit.

6. What command brings up the *Dimension Style Manager*?
   Dimstyle or ddim.

7. What type of *arrowhead* do architects usually prefer?
   Architectural tick.

247

CHAPTER 7 REVIEW QUESTIONS

1. What are *blocks* and why are they useful?
   A block is a collection of objects grouped together under a common name. They are useful to keep multiple objects together and for reuse via symbol libraries.

2. What is the difference between *blocks* and *wblocks*?
   A block is internal to a file, and a wblock is external.

3. What are the essential steps in making a *block*? A *wblock*?
   For a block, you need to specify a name, select objects, and pick a base point. For a wblock, you also need to specify a path.

4. How do you bring the block or wblock back into the drawing?
   Via the insert command.

5. Explain the purpose of *purge*.
   Purge rids the drawing of unseen elements present in the file, such as blocks, layers, or fonts.

6. Explain the basic idea behind *dynamic blocks*. What steps are involved?
   Dynamic blocks are intelligent blocks that can be modified without being redefined. To create a dynamic block, you add parameters and actions.

7. What are the advantages and disadvantages of using groups?
   Advantages of a group: A group can have objects added and subtracted from it and can have individual objects (and their properties) edited. Disadvantages of a group: A

group cannot be transferred between drawings (for "local" use only), and an update to one group does not update other groups with the same name.

CHAPTER 8 REVIEW QUESTIONS

1. List the three *types* of arrays available to you.
   Polar, rectangular, and path arrays.

2. What steps are needed to create a *polar* array?
   You need to specify what object(s) to array, what the center point of the array is, how many objects to create, and whether you want the pattern to fill out a full circle or part thereof.

3. To array chairs around a table, where is the *center point* of the array?
   The center of the table.

4. What steps are needed to create a *rectangular* array?
   You need to specify what object(s) to array, how many rows and columns, and the distance between those rows and columns.

5. What steps are needed to create a *path* array?
   You need to specify what object(s) to array, the path, and how many objects to create.

CHAPTER 9 REVIEW QUESTIONS

1. List the seven essential sets of *information* needed for printing or plotting.
   What *printer* or *plotter* to use, what *paper size* to use, what *area* to plot, at what *scale* to plot, what *pen settings* to use, what *orientation* to use, and what *offset* (if any) to use.

2. List the five *paper sizes* and their corresponding measurements in inches.
   A size: 11" × 8.5" sheet of paper (letter), B size: 17" × 11" sheet of paper (ledger), C size: 22" × 17" sheet of paper, D size: 36" × 24" sheet of paper, and E size: 48" × 36" sheet of paper.

3. What is the difference between *printing and plotting*?
   Printing usually refers to the output of drawings on A- or B-sized paper and *plotting* usually refers to the output of drawings on C-, D-, or E-sized paper.

4. List the four *plot area* selection options.
   Extents, Window, Display, and Limits.

5. What *scale setting* ignores scale completely?
   Fit to paper.

6. What is the extension of the *pen settings* file?
   `ctb`.

7. Explain the importance of *Preview* before printing and plotting.
   Mistakes can be spotted before a print is sent to the printer or plotter, thereby saving paper.

8. Why is the *Page Setup Manager* useful?
   Because it allows you to set up and save print and plot settings for future use.

CHAPTER 10 REVIEW QUESTIONS

1. What is Paper Space?
   Paper Space is an AutoCAD concept that allows you to draw your design in full size (1:1) units and still print it to scale using viewports.

2. What are layouts?

Layouts are Paper Space pages that hold the title block and viewport.

3. How do you create, delete, rename, and copy layouts?

All these tools are accessed by right-clicking on the Layout tabs.

4. What are viewports?

A viewport is an AutoCAD concept that allows you to link together Model and Paper Space.

5. How do you create, scale, and shape viewports?

You can create a viewport via the mview command. Scaling and shaping it are accomplished via the viewport grips.

6. How do you scale a design inside a viewport?

You can use the icons on the Viewport toolbar or via keyboard entry using scale, then `1/xp`.

7. How do you turn layers on and off in individual viewports?

By clicking inside a viewport, calling up the Layer Manager, and freezing the layers under the VP Freeze column.

8. How do you scale text and dims in viewports?

By using the appropriate charts and matching the text and dim sizes with the correct scale.

9. Describe the Annotative concept.

Annotative gives text, dims, and other AutoCAD features the ability to scale to viewports automatically based on the assigned scaling of those viewports.

Chapters 11–20

Now that you have completed Level 1, beginner to intermediate, you should have the basic skills necessary to work on a drawing of moderate complexity with a certain amount of speed and accuracy. Level 2 is the second part of your journey in becoming proficient in AutoCAD. The goal at the end of this level is for you to learn most of the remaining practical information about AutoCAD 2012 in two dimensions. This will allow you to not only work on very complex drawings but also set up and manage them. Then, with some additional work experience, you will have the skills of a CAD manager or administrator, a necessity in many engineering and architectural firms, even if your job is not just AutoCAD.

Starting with the next chapter, this textbook does not feature any of the basic introductions concerning AutoCAD and what it is. It is assumed you have already read through all the introductory information and are familiar with all the material in Chapters 1–10. There are no further explanations or reviews of earlier material, so if you need to review anything from Level 1, do so before proceeding. With this book, we dive into the new material right away and do not look back, as there is much more to cover.

Level 2 consists of several parts. Chapters 11 through 13 are all about advanced features of basic concepts: lines, layers, and dimensions. We add significant depth to these Level 1 topics. Chapters 14 through 16 cover a broad set of topics that introduce various advanced features, such as interacting with other software, the Design Center, Express tools, the CUI, and many shorter topics. This then clears the way for some remaining "big ideas" in AutoCAD: External References, Attributes, and Advanced Output and Pen Settings, which make up Chapters 17 through 19. These are very important topics, especially external references, as they are critical to many architectural designs. We complete Level 2 with isometric design, which is a very useful tool in its own right and a great lead-in to 3D if you go on to learn that aspect of AutoCAD.

The pace of the course picks up somewhat as well, and more responsibility is placed on you to follow up and do your own research into additional AutoCAD functions. We also introduce a drawing project that spans the entire level, from Chapter 11 to Chapter 20. It is a more in-depth floor plan that, while still greatly simplified, attempts to replicate a real-life architectural layout. There also are fewer generic drawings to do, in favor of more topic-specific exercises at the end of each chapter. One constant remains: You still need to practice to ensure success.

CHAPTER 11

Advanced Linework

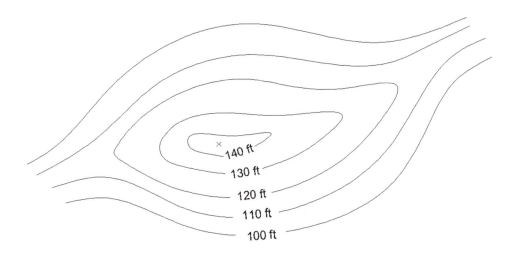

LEARNING OBJECTIVES

In this chapter, we revisit the line concept and expand on it by discussing the following:

- Polyline and pedit
- Xline
- Ray
- Spline
- Mline, mlstyle, and mledit
- Sketch

By the end of the chapter, you will learn how to use and apply these new types of lines and be able to incorporate them into future design work.

Estimated time for completion of chapter: 2–3 hours.

11.1 INTRODUCTION TO ADVANCED LINEWORK

We begin Level 2 in the same way we started Level 1, with the simple line command. At the time, this was enough to start with, and no mention was made of the many other types of lines

AutoCAD has. Classroom experience had shown that introducing all the other types of lines at this stage was a sure way to scare off brand-new students, and the other lines were not needed anyway. Now that you are more advanced, we can come back and look at lines in depth.

As it turns out, while the line command is useful for most drawing situations, it is inadequate or inconvenient for a small but important number of other applications. Here, we look at other variants of the line including the *pline* (polyline), the *xline* (construction line), its close relative the *ray*, a special case of the polyline called a *spline*, an *mline* (multiple line), and the most unusual one of all, the *sketch* command.

11.2 PLINE (POLYLINE)

A pline is a regular line with some unique properties:

- Property 1. A pline's segments are joined together.
- Property 2. A pline can have thickness.

These two properties add tremendous importance and versatility to the old line command, which creates only individual segments that maintain a constant zero thickness. Figure 11.1 provides an illustration of the first property. The top "W" is drawn with four line segments. As you can see, they are all separate, with only the left segment clicked on, and the grips reflect that. The "W" below it is drawn using the pline command and is all one unit, as seen by clicking it once. Notice the slightly compressed center grips.

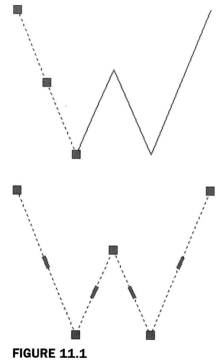

FIGURE 11.1
Line versus pline command (Property 1).

One example of applying this feature may include drawing lengthy runs of cabling or piping. Then, if the run needs to be deleted, only one click is necessary to remove all the segments. Let us try to draw a pline.

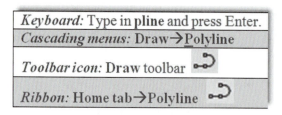

Keyboard: Type in **pline** and press Enter.
Cascading menus: Draw→Polyline
Toolbar icon: Draw toolbar
Ribbon: Home tab→Polyline

Step 1. Start the pline command via any of the preceding methods.
- ○ AutoCAD says: `Specify start point:`

Step 2. Left-click anywhere on the screen.
- ○ AutoCAD says: `Current line-width is 0.0000`
  `Specify next point or [Arc/Halfwidth/Length/Undo/Width]:`

Step 3. Ignoring the lengthy menus for now, click around until you get the same W shape seen earlier.
- ○ AutoCAD says: `Specify next point or [Arc/Close/Halfwidth/Length/`
  `Undo/Width]:`

When done, press Enter or Esc. Then click on your new pline to see that it is all one unit, not separate lines.

Pedit

What if you wanted to apply thickness to the pline (Property 2) or create a pline out of a bunch of lines? For this, you need pline's companion command called *pedit* or polyline edit. Let us try this out on the pline you created and give it thickness. You need to add another toolbar to your collection, called *Modify II* (Figure 11.2).

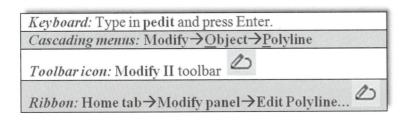

Keyboard: Type in **pedit** and press Enter.

Cascading menus: Modify→Object→Polyline

Toolbar icon: Modify II toolbar

Ribbon: Home tab→Modify panel→Edit Polyline...

FIGURE 11.2
Modify II toolbar.

Step 1. Start the pedit command via any of the preceding methods.
- ○ AutoCAD says: `Select polyline or [Multiple]:`

Step 2. Click on any of the W polyline segments.
- ○ AutoCAD says: `Enter an option [Close/Join/Width/Edit vertex/Fit/`
  `Spline/ Decurve/Ltype]`

Step 3. Press w for `Width`, and press Enter.
- ○ AutoCAD says: `Specify new width for all segments:`

Type in a small value, say, `0.25`, and press Enter. The pline adopts the new width for all segments, as shown in Figure 11.3. You can then press Esc to completely exit out of the pedit command.

FIGURE 11.3
Pline with thickness (Property 2).

Now, try making plines out of lines. Draw the W once again out of ordinary lines, then follow the procedure shown next:

Step 1. Start the pedit command via any of the previous methods.
   ○ AutoCAD says: `Select polyline or [Multiple]:`
Step 2. Click on any of the W line segments.
   ○ AutoCAD says: `Do you want to turn it into one? <Y>`
Step 3. Press Enter to accept "yes."
   ○ AutoCAD says: `Enter an option [Close/Join/Width/Edit vertex/Fit/ Spline/ Decurve/Ltype gen/Undo]:`
Step 4. Select j for `Join`, and select all the lines of the W, pressing Enter when done. The collection of separate lines is now a polyline and you may press w for `Width` as before if you wish. Press Esc to completely exit out of the pedit command.

Another notable option in the pedit command is `Spline`, and if you press s while in the pedit command, your W will look like Figure 11.4.

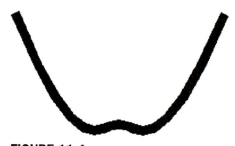

FIGURE 11.4
Pline with the Spline option.

To undo this you can press d for `Decurve`, yet another pedit option. Spline is an important case of a polyline. We look at spline as a separate command later in the chapter.

Exploding a Pline

Plines can be easily exploded using the explode command. The effect is twofold:

- The pline loses its thickness.
- The pline is broken into individual lines.

To put it back together requires the pedit command, so explode and pedit are opposites of each other. Be careful though, and do not explode plines unless absolutely necessary, as there may be many individual lines to account for once exploded.

Additional Pline Options

A few more options are worth mentioning with the pline command. First of all, you can start out drawing a pline with a preset width (or half width). Here is the command sequence for that, presented in a shortened "actual" format, as seen from the command line by the user.

```
Command: pline (via any of the previously shown methods)
Specify start point:
Current line-width is 0.0000
Specify next point or [Arc/Halfwidth/Length/Undo/Width]:w
Specify starting width <0.0000>:.25
Specify ending width <0.2500>:.25
Specify next point or [Arc/Halfwidth/Length/Undo/Width]:
Specify next point or [Arc/Close/Halfwidth/Length/Undo/Width]:
```

As you can see, the W option was chosen and a width of 0.25 was entered for both start and end. If those lengths are not similar, you get something akin to Figure 11.5. In that example, a zero start width was followed by a 0.75 end width to create the triangular first segment of the pline. You can change these widths "on the fly" to create some interesting shapes.

Another option is 1 for Length. This is similar to Direct Distance Entry from Level 1. Finally, there is a for the Arc option. This creates arcs on the fly and is very useful for drawing organic-looking shapes for everything from landscape design to furniture to cabling and wiring, as shown in Figure 11.6.

FIGURE 11.5
Pline with varying width.

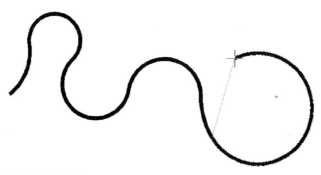

FIGURE 11.6
Pline of 0.2 with Arc option.

Finally, several releases ago, AutoCAD introduced some additional tools to work with plines (including unexploded rectangles, which are plines by definition). These tools come in the form of a new menu, which appears when you hover the mouse over a grip belonging to a pline or a rectangle (Figure 11.7).

257

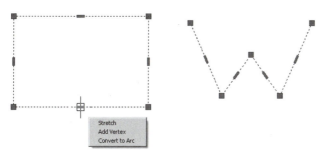

FIGURE 11.7
Grip menu.

As you can see, you can stretch, add a vertex, or even convert the line to an arc, allowing some new design possibilities and even saving some time.

STRETCH

This option, also covered as a separate command in Chapter 15, allows for stretching that particular line and anything attached to it in one of several ways, with one possibility seen in Figure 11.8.

ADD VERTEX

This option allows you to "break" the line with the addition of another control point (grip), as seen in Figure 11.9. You can then create two angled lines out of one straight line.

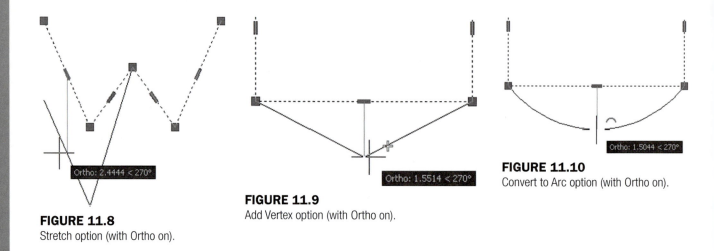

FIGURE 11.8
Stretch option (with Ortho on).

FIGURE 11.9
Add Vertex option (with Ortho on).

FIGURE 11.10
Convert to Arc option (with Ortho on).

CONVERT TO ARC

This somewhat similar option also adds a vertex but one that allows an arc to be formed (Figure 11.10).

11.3 XLINE (CONSTRUCTION LINE)

From the relatively involved pline, we swing over to the very simple xline. This is a continuous line that has no beginning or end. The zoom command has no effect on it. The *construction* part of the name refers back to the days of hand drafting, where you would draw, in light pencil, a guideline that stretched from one end of the paper to the other and was used as a reference starting point for a string of elevations (or anything else) to be drawn one after another.

To use it, make sure Ortho and OSNAP are not activated, and do the following:

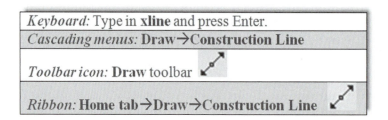

Keyboard: Type in **xline** and press Enter.
Cascading menus: **Draw→Construction Line**
Toolbar icon: **Draw** toolbar
Ribbon: **Home tab→Draw→Construction Line**

Step 1. Start the xline command via any of the previous methods.
   ○ AutoCAD says: `Specify a point or [Hor/Ver/Ang/Bisect/Offset]:`
Step 2. Click anywhere and an xline appears.
   ○ AutoCAD says: `Specify through point:`
Step 3. Click randomly anywhere along the screen and multiple xlines appear. Figure 11.11 shows one result of clicking in a circular manner.

Figure 11.11 is perhaps not the most useful of creations, but landscape designers may find some use for the result of placing a circle at the intersection of those lines and trimming. Then, via some creative scaling, rotating, and copying, you get the reasonable representation of trees or shrubs in Figure 11.12.

As seen in the menu that follows the xlines, you can also create perfectly horizontal, vertical, or angular xlines. The same can be accomplished by having Ortho on, although the options allow you to create multiple xlines more easily. Clear your screen, start up xline again, and

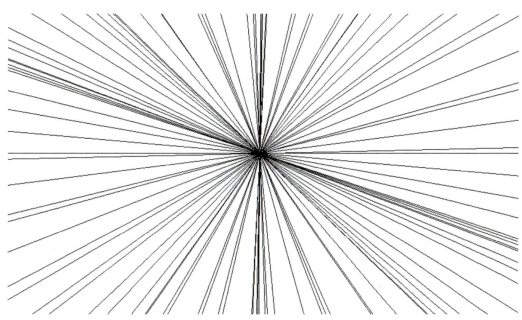

FIGURE 11.11
Random xlines.

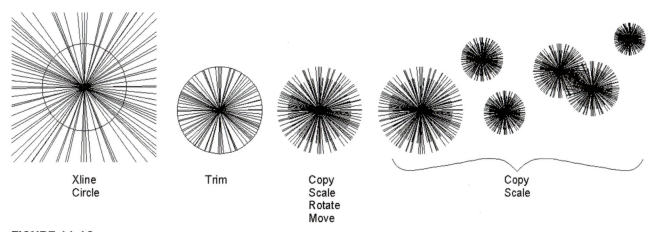

Xline
Circle

Trim

Copy
Scale
Rotate
Move

Copy
Scale

FIGURE 11.12
An application of xline.

when you see the following menu, try out the various options (results are seen in Figure 11.13): `Specify a point or [Hor/Ver/Ang/Bisect/Offset]:`

11.4 RAY

A ray is simply an xline, but infinite in only one direction. It is easy to use and quite similar to xline, except there are no options for horizontal, vertical, or angular; to get a perfectly straight ray, you just turn on Ortho.

Keyboard: Type in **ray** and press Enter.	
Cascading menus: Draw→Ray	
Toolbar icon: none	
Ribbon: Home tab→Draw→Ray	↗

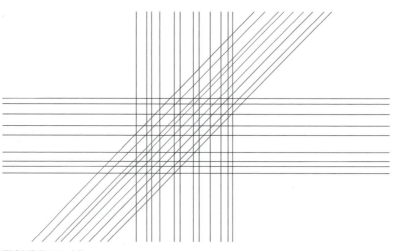

FIGURE 11.13
Hor, Ver, and Ang (set to 45°) options in xline.

Step 1. Start the ray command via any of the preceding methods.
○ AutoCAD says: `Specify start point:`
Step 2. Click anywhere and a ray is created (Figure 11.14).

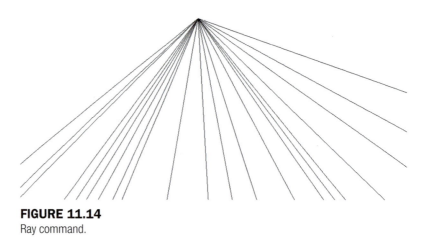

FIGURE 11.14
Ray command.

11.5 SPLINE

Spline is a unique and valuable command. Its origins are in the mathematical field of numerical analysis, and it is called a *piecewise polynomial parametric curve* (or a NURBS curve, non-uniform rational B spline). While knowing the details is not necessary for using this command, what is important is that a spline is ideal for approximating complex shapes through curve fitting. What you do is create a path made out of point clicks, and AutoCAD interpolates a "best fit" curve through those points.

The result is a smooth, organic shape useful for a great number of applications, from contour gradient lines in civil engineering to landscape design to electrical wiring and many other shapes and objects. The term *spline*, by the way, comes from the flexible spline devices used by shipbuilders and draftspersons to draw smooth shapes in pre-computer days. To draw a spline, do the following.

| Keyboard: Type in **spline** and press Enter. |
| Cascading menus: Draw→Spline |
| Toolbar icon: Draw toolbar ∿ |
| Ribbon: Home tab→Draw→Spline ∿ |

Step 1. Start up the spline command via any of the preceding methods.
- AutoCAD says: `Current settings: Method = Fit Knots = Chord`
  `Specify first point or [Method/Knots/Object]:`

Step 2. Click anywhere and a spline appears.
- AutoCAD says: `Enter next point or [start Tangency/toLerance]:`

Step 3. Click on another point somewhere on the screen, in an up and down wave pattern, as seen in Figure 11.15.
- AutoCAD says: `Enter next point or [end Tangency/toLerance/Undo/`
  `Close]:`

  As early as this step, you can select c for `Close` to close the spline tool. In this case, however, continue with the spline curve to add more points.

Step 4. Click on a few more points in a similar wave pattern. AutoCAD keeps saying the same thing as in Step 3. Continue clicking several more times (refer again to Figure 11.15). When done, simply press Enter.

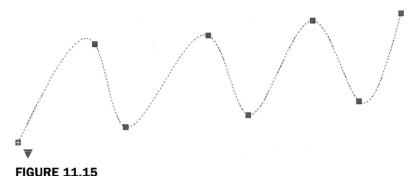

FIGURE 11.15
Spline command, grips enabled.

261

This is your basic "out of the box" spline. Notice how, as you click on points, AutoCAD fits a smooth curve that best approximates the path. Each grip in Figure 11.15 represents a click of the mouse, a control point. You can manipulate the positions of these grip points in the standard manner, with the spline following suit. This makes the spline a very flexible tool, adjustable to the exact demands of the designer. Try activating the grips and moving them around.

The command was redesigned somewhat back in AutoCAD 2011, and many of its advanced features are hidden away in submenus, which is a good thing, as most users need only the basic spline and not extensive options for tangencies and other details.

That is not to say they are not there if needed. Click on the down arrow on the bottom left of the spline (grips enabled), and select Show Control Vertices, which is another way to design the spline, useful for 3D work. Go over the various other options on your own with the Help files; there are too many to go into here, but learning them helps you understand this very intriguing command.

Through the examination of spline, you should now have a good idea of how to create and control it. The points of the spline do not have to be random. You can set up a framework for the shape and use OSNAPs to precisely place where you want the spline to go, as shown in Figure 11.16. What you see here are four lines, representing a framework. Then, using OSNAP points, a spline is added in click by click, from an endpoint to a midpoint to another midpoint, then to the final endpoint, at the bottom left.

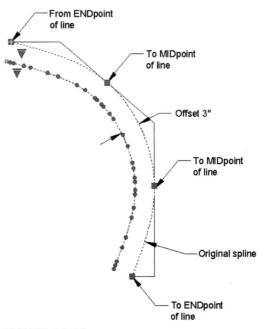

FIGURE 11.16
Spline with framework, offset, and grips.

A spline can then be offset using the standard offset command, although be careful in areas where the spline takes a sharp turn. If the offset distance is too great and the spline cannot make a smooth turn after the offset, it breaks down into separate lines. The offset spline also has many more grip and control points than the original spline. A 3″ spline offset, with all grips enabled, can be seen in Figure 11.16.

Examples of spline use are numerous. Already mentioned were contour lines (the Chapter 11 cover image), electrical wiring, piping, network cables, landscape design (golf courses, curved driveways, and walkways), and interior design (Jacuzzis, swimming pools, furniture), and much more can be constructed with this tool.

11.6 MLINE (MULTILINE)

A multiline is a very useful tool that does pretty much what you would expect from the name: It is used to draw multiple lines all at once. This, in theory, should save you time when drawing everything from architectural walls to multilane highways. As we see soon, the basic mline is easy to draw but not too useful. The trick is to set it up for useful applications using additional tools as outlined in the next sections. Try out the basic mline to get a feel for it, as shown next.

Keyboard: Type in **mline** and press Enter.	
Cascading menus: Draw→Multiline	
Toolbar icon: none	
Ribbon: none	

Step 1. Start up the mline command via any of the preceding methods (no toolbar or Ribbon is available).

　○ AutoCAD says: `Current settings: Justification = Top, Scale = 1.00, Style = STANDARD`
　`Specify start point or [Justification/Scale/STyle]:`

Step 2. A generic mline appears. Click again.

　○ AutoCAD says: `Specify next point or [Undo]:`

Step 3. Click somewhere else again.

　○ AutoCAD says: `Specify next point or [Close/Undo]:`

You can continue in this manner until finished. The default distance between the lines is 1.0, and of course if the mline is too small or too large, you can zoom in or out. Figure 11.17 shows what you should see after properly executing the previous steps.

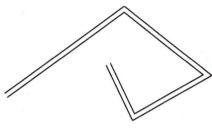

FIGURE 11.17
Basic mline.

Modifying the Mline

While simple, the preceding was not too useful. Specifically, you may want to change one or more of the following properties:

- Number of lines making up the mline.
- Distances between those lines.
- Linetypes making up the lines.
- Colors of the lines.
- Fills, end caps, and other details.

All these and more can be input and modified, thereby making the mline more useful and sophisticated. The key command for setting up mlines is *mlstyle*.

Mlstyle (Multiline Style)

The multiline style command opens a dialog box that allows you to set all the preceding parameters. Type in `mlstyle` and press Enter or select **Format→Multiline...** from the cascading menu (there is no toolbar or Ribbon alternative). The dialog box shown in Figure 11.18 appears.

Let us create a new style. Press New... and you will see a smaller Create New Multiline Style box. Type in a name for your mline (no spaces allowed in the name; use underscores if needed), and press Continue. You then see the main dialog box shown in Figure 11.19. The name created for this example was a generic MY_NEW_MLINE.

We can now go through the essential changes to the mline. Our goal is to create a typical wall used in construction (see Figure 11.20), shown often in cross section in architectural elevations. For simplicity, it is 6" wide and features a 1" drywall panel on either side, with insulation in between, and room for a wooden stud (not shown here).

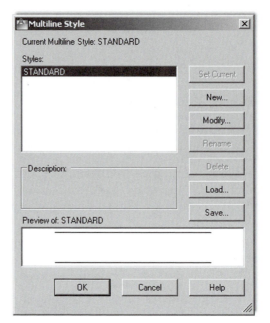

FIGURE 11.18
Multiline Style dialog box.

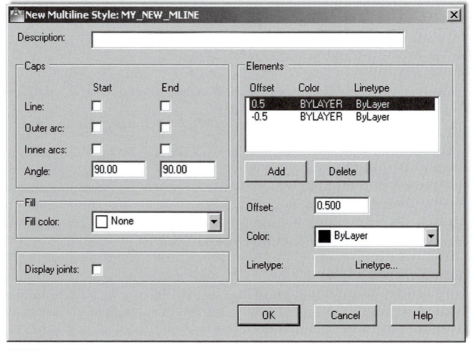

FIGURE 11.19
New Multiline Style dialog box.

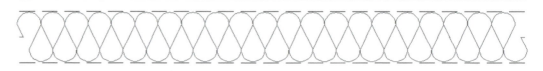

FIGURE 11.20
Architectural wall with insulation.

This mline, once set up, can be used to draw building walls faster and easier than offsetting and changing properties one line at a time. The process to do this follows.

Step 1. You first need to determine the *number of lines* making up the mline, and the *distances* between those lines.

Both of these can be set inside the Elements area (right-hand side) of the New Multiline Style dialog box using the Add button and the Offset field. The idea is to add more line elements (via the Add button) and enter an offset distance for each of these line elements (via the Offset field). This offset distance spaces the lines at the desired interval to create the wall.

Before you do all this, however, you need to decide what the origin of the offset is. This is your starting point for the offsets. The two choices are positioning zero all the way on the leftmost line element (then all values are positive) or in the middle element (then half the values are negative). With the first choice, the outer walls are at 0 and 6, and with the second choice, they are at −3 and +3, as shown graphically in Figure 11.21.

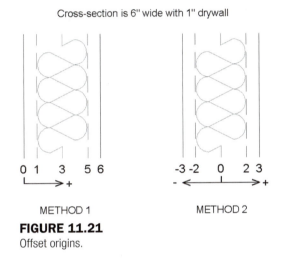

Cross-section is 6" wide with 1" drywall

METHOD 1   METHOD 2

FIGURE 11.21
Offset origins.

For this example, choose Method 1, and after adding additional lines, for a total of six, set the walls to 0, 1, 3, 5, and 6 units (there are no feet or inches in these settings; you use decimal units, representing inches, throughout).

Step 2. Next you need to determine the *linetypes* and *colors* making up the lines of the overall mline.

Both of these can be set in the Elements area using the Color: drop-down menu and the Linetype... button, as shown in Figure 11.22. Simply highlight the line element you wish to change and select a new color or linetype. Remember that you may not have the linetypes loaded yet, so you may need to do this (refer back to Chapter 3 for a reminder, if needed).

In Figures 11.20 and 11.21, the outer wall color was selected to be blue and the inner walls are red with a hidden linetype. The centerline is a special linetype, often used for insulation, called BATTING. Feel free to choose your own colors and linetypes; those shown are just for illustration, but they are a good starting point. Figure 11.23 shows the Elements area as you go through Steps 1 and 2.

Step 3. As the final step you can add in *fills* and *start/end caps*.

End caps, as the name implies, are various ways to finish the start and end of your mlines. You can add regular lines or two types of arcs. Often mlines are joined to each other, so these may not be necessary. You can also add color fills to the mlines, although this is rarely done

265

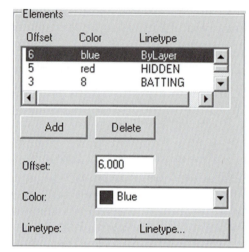

FIGURE 11.22
Color and Linetype settings.

FIGURE 11.23
Elements settings.

266

in practice. All of these can be done via the check-off boxes and drop-down menu on the left-hand side of the New Multiline Style dialog box, as seen in Figure 11.24. You can also display the joints of an mline via a checked box, but that is also rarely called for. This example has caps or fills, but feel free to experiment on your own.

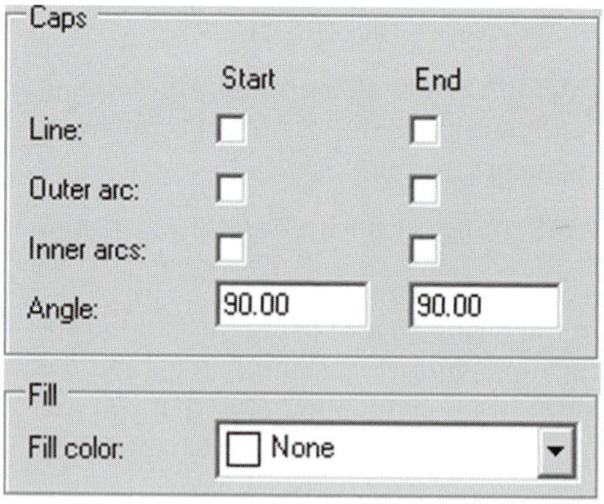

FIGURE 11.24
Caps and Fill.

After your new mline has been created, go ahead and click on OK; you are taken back to the dialog box of Figure 11.18, with your new mline previewing in the window at the bottom. The other choice should be Standard, the default mline you created by just typing in `mline` and saw in Figure 11.17. Set your new mline as the current one (Set Current button), and click on OK. Now go ahead and type in `mline` and draw part of a rectangle, as seen in Figure 11.25.

FIGURE 11.25
The new mline.

The linetypes may not look right in this example. This is due to a low linetype scale being set. If you recall from Chapter 3, the way to fix this is to type in `ltscale` and enter a new, more appropriate value. The value used is 4.8. The result is shown in Figure 11.26.

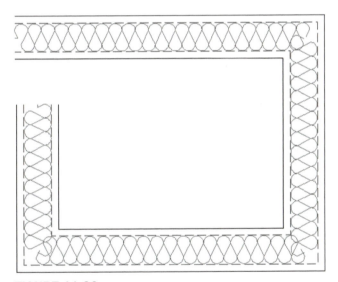

267

FIGURE 11.26
The new mline with an ltscale of 4.8.

Mledit (Multiline Edit)

Multiline Edit is a dialog box that deals with the practical applications of mlines, such as how to trim intersecting mlines and how to deal with joints, cuts, and vertexes. To try out some of the features, draw two intersecting mlines, as shown in Figure 11.27.

Then, type in `mledit`, press Enter (or select **Modify→Object→Multiline** from the cascading menus), and the dialog box shown in Figure 11.28 appears.

Select Closed Cross in the upper left. The dialog box disappears and you are asked to select the first mline, click on it, then the second mline, and click on that also. After the second click, Figure 11.29 appears. As you can see, the mline's intersection has been modified.

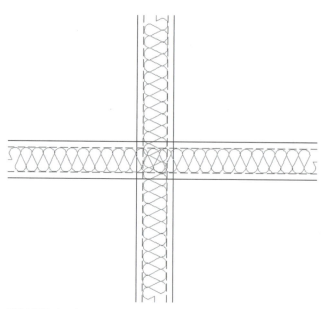

FIGURE 11.27
Intersecting mlines.

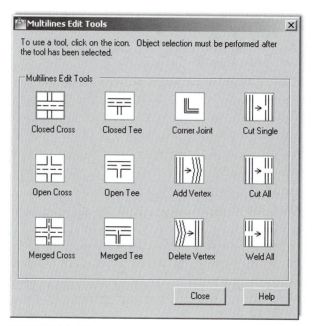

FIGURE 11.28
Multiline Edit Tools.

In the same manner, try out all the features of mledit at your own pace; they are all relatively straightforward and self-explanatory. A cut cuts the line; a vertex adds a "kink" in the mline. Close, Open, and Merge deal with intersections and so on.

Other Mline Properties

- **Grips.** All mlines have grip points at either the top or bottom of the overall line (this is set using Justification when first creating the mline). Feel free to use these grips to stretch the line around as needed.

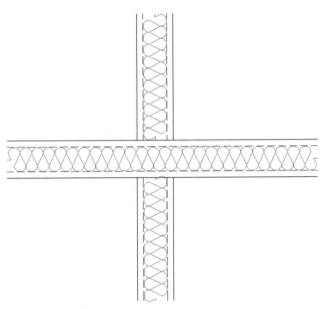

FIGURE 11.29
Multilines edited.

- **Exploding the mline.** Mlines can be exploded, which separates them into individual component lines. This may not necessarily be a bad thing. An architectural floor plan may be quite complex and you may not be successful in getting an mline to do exactly what you want all the time. If the mline is exploded, then you can use the fillet, trim, and extend commands to coax it into the shape you want. Much time has already been saved by using an mline in the first place, so exploding it may be the right solution to finish off the floor plan linework.
- **Trim and extend.** Mlines can be trimmed and extended without exploding them (filleting, however, is not possible). However, an option that pops up is: `Enter mline junction option [Closed/Open/Merged] <Open>:`. The `Closed`, `Open`, and `Merged` options have to do with the intersections, so you need to pick one option to continue the trim or extend. Go through them all as practice.
- **Modification of mlines.** Once created, a new mline style cannot be modified once you start using it (but can be beforehand). If you wish to change an mline after you have a few drawn, you need to create a new style that incorporates your desired changes.

11.7 SKETCH

This is perhaps one of the most unusual tools in AutoCAD. It allows you to draw objects in freehand. This may seem counterproductive and unnecessary. After all, you spent so much time learning precision and accuracy; but sketch actually finds a wide array of applications, as we soon see.

To truly sketch in freehand on a computer you need bitmap graphics (such as what Photoshop® does). Then, the maximum resolution is determined by the pixels. In almost all cases, the pixels are so small that it truly seems like you are drawing smooth and continuous freehand lines and shapes. You would have to zoom in pretty close to see what is called *pixilation* or the breakdown of smooth lines into the little squares (pixels) that make up the drawn geometry.

This is all fine and would be useful for AutoCAD if not for one problem: AutoCAD is a "vector program," meaning it does not use pixels. Instead, all the geometry is represented by mathematical formulas, which completely describe the shape, size, and location of everything seen on the screen. This approach is necessary for many reasons, not the least of which is to ensure the integrity of the geometry when zoomed in close. So then how would you sketch in freehand? The solution is simple; instead of pixels, use very short lines, which look continuous if they are small enough and you are not too close to them. This is exactly how sketch works. The command uses small lines, with their size determined by something called a *record increment*, which we discuss in a moment.

The sketch command has another twist to it. A few releases ago AutoCAD allowed you to turn sketched linework into a spline or a polyline. This can occasionally be useful if you are drawing simple freehand shapes and want maximum smoothness, although the pline option is not recommended, as it adds a tremendous amount of control points to the linework. We discus mostly the "classic" sketch command here.

First, we need to set the record increment to a very small value (use 0.01). Let us try this out.

Step 1. Type in `sketch` (there is no toolbar icon, cascading menu, or Ribbon for the classic sketch command) and press Enter.
- ○ AutoCAD says: `Type = Lines Increment = 0.1000 Tolerance = 0.5000 Specify sketch or [Type/Increment/toLerance]:`

Step 2. First, make sure `Type = Lines`, as that is what we want to practice first. If not, press `t` for `Type` and select `l` for `Lines`. Then, press `i` for `Increment`, and press Enter.
- ○ AutoCAD says: `Specify sketch increment <0.1000>:`

Step 3. Type in `0.01` and press Enter.
- ○ AutoCAD says: `Specify sketch or [Type/Increment/toLerance]:`

Step 4. Nothing further needs to be set, so press Enter.
- ○ AutoCAD says: `Specify sketch:`

Step 5. Move the mouse around to begin drawing.
- ○ AutoCAD says (and continues to say): `Specify sketch:`

You will see what is essentially a smooth line, green in color, and completely freehand generated. To "lift the pen," you need to left-click once; then click again to "put it down." You can also click and hold the left mouse button down when you want to draw, which is more realistic for some. To "set" the drawing, press Enter; the lines turn white (or some other color depending on the current layer). Be careful not to press Esc until you set the lines, as everything disappears. Figure 11.30 shows what you can get after a few minutes of doodling.

FIGURE 11.30
Basic sketch.

As mentioned before, you can also turn the sketched lines into a spline by selecting the Spline option. You will still sketch freehand at first, but upon pressing Enter, the lines smooth out into the familiar spline shapes. The process is the same with the Pline option. You should spend a few minutes experimenting with both. You can also access the Spline option directly via the Ribbon's Surface tab (look for the spiral on the Curves panel), but for that you need to change the workspace to 3D Modeling.

Applications of Sketch

I hope whatever you tried to draw was better than the odd-looking little man with a briefcase. It is not that easy to draw something well with a mouse; occasional erasing and fixing up is needed to get a good sketch. Sometimes a graphic design tablet and a pen-like stylus may be used. Zoom in closely on whatever you created and you notice the figure is made up of the small lines we talked about. In the closeup of the hand and briefcase handle in Figure 11.31, you can clearly see the individual line segments. Be careful making a dense drawing. All these lines increase the size of the file very quickly.

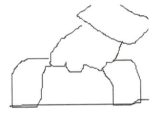

FIGURE 11.31
Line increments visible in the sketch.

Sketch has many applications, which are further expanded via the spline variation of the command. Some examples include

- Architectural elevations featuring sketched people for comparative sizing.
- Sketched people for traffic sign design and civil engineering studies.
- Trees, shrubs, and other landscape design objects.
- Realistic drawings of small cables and wiring (via spline).
- Anytime an organic, non-rigid shape is needed.

The website www.cben.net is a great source of free sketches, both in plan and in elevation. Figure 11.32 shows a few sketch examples from that website.

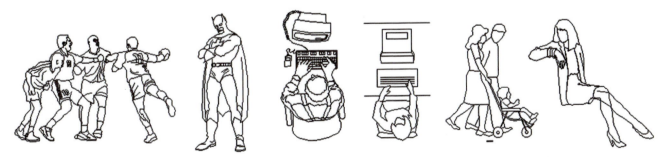

FIGURE 11.32
Creative use of the sketch command.

LEVEL 2 DRAWING PROJECT (1 OF 10): ARCHITECTURAL FLOOR PLAN

Level 2 features a drawing project that is designed chapter by chapter, beginning with this one and completing in Chapter 20. It is an architectural layout, similar to what you did in Level 1, although a bit more involved. The project enables you to put to use all of the cumulative knowledge you acquired up until now and includes features you learn from Level 2 chapters.

The project features many of the basic elements of a "real-world" design and definitely is a step closer to the real thing. You will see a moderately complex floor plan and electrical, HVAC/mechanical, and landscaping plans. You then set up Paper Space and printing with lineweights and other details. A real house design also is likely to feature numerous details, riser diagrams, and sections of doors, windows, and walls. We include some of these elements here and there to increase realism.

Follow all the steps closely, but feel free to embellish the design and get creative. This is excellent practice for the real thing, and if you get through the project smoothly, you will feel right at home with a real design.

Step 1. Open a new file and save it as `House_Plan.dwg`.

Step 2. Set up units as Architectural, and create the following initial layers. The drawing is monochrome in this textbook for clarity, but you need to follow CAD standards as required in the industry (color suggestions shown). For now, you need
- A-Deck (Blue)
- A-Deck-Hatch (Gray_9)
- A-Dims (Cyan)
- A-Walls-Exterior (Green)

Step 3. Draw the floor plan shown in Figure 11.33. The dimensions are not just for your reference anymore; you also need to draw them in. They serve as checks on your design. Use the proper layers and consistent colors. When done and you are sure everything is correct, freeze the dimensions. It should take you about 30 minutes to set up and complete this design.

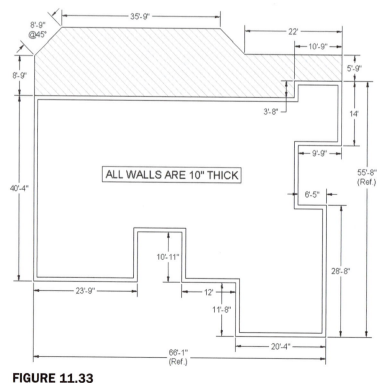

FIGURE 11.33
Level 2 project—overall layout.

SUMMARY

You should understand and know how to use the following concepts and commands before moving on to Chapter 12:

- Pline (polyline)
  - Creating a pline
  - Pedit
  - Setting constant width (thickness)
  - Setting variable width (thickness)
  - Changing lines into plines
  - Adding spline curves and decurving
  - Other pline options (Length, Arc, Offset)
  - Exploding a pline
- Xline (construction line)
  - Creating a random xline
  - Creating a horizontal, vertical, or angular xline
- Ray
  - Creating a random ray
  - Creating a horizontal or vertical ray (with Ortho)
- Spline
  - Creating a random spline
  - Offsetting the spline
  - Grips and modifying the spline
- Mline (multiline)
  - Creating a generic mline
  - Mlstyle
  - Mledit
  - Other mline properties
- Sketch
  - How sketch works
  - Setting a record increment
  - Applications of sketch

REVIEW QUESTIONS

Answer the following based on what you learned in Chapter 11.

1. What are the two major properties a pline has and a line does not?
2. What is the key command to add thickness to a pline?
3. What is the key command to create a pline out of regular lines?
4. What option gives the pline curvature? What option undoes the curvature?
5. What two major things happen when you explode a pline?
6. What other options for pline were mentioned?
7. Define an xline and specify a unique property it has.
8. How is a ray the same and how is it different from an xline?
9. Describe a spline and list its advantages.
10. List some uses for a spline in a drawing situation.
11. What is an mline? List some uses.
12. What are some of the items likely to be changed using mlstyle?
13. What are some of the uses of mledit?
14. Describe the sketch command. What is a record increment? What size is best?

EXERCISES

1. Draw a pline with ten segments, each 2 units long and at right angles to each other. Give the pline a thickness of .2 units. Then, spline the pline. The progression is shown in the following figure. Upon completion, decurve the pline and explode it. (Difficulty level: Easy; Time to completion: <5 minutes)

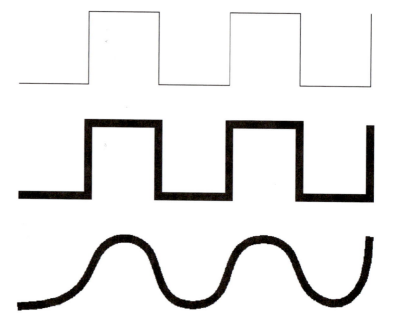

2. Using xline, circle, trim, copy, and rotate, draw a tree symbol as seen in Figure 11.12, reproduced here. Repeat the same procedure using the ray command. (Difficulty level: Easy; Time to completion: <5 minutes)

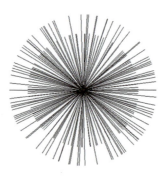

3. Using the spline and text commands, draw contour lines as seen in the chapter cover image and reproduced here. If you wish to create the "gaps" in the splines where the height values are positioned, draw two long vertical lines and trim between them. We cover another method to do this in Chapter 15. (Difficulty level: Easy; Time to completion: 5–7 minutes)

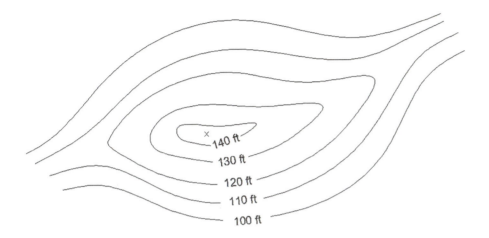

4. Using the mline, line, and hatch commands, draw a representation of a divided highway, as shown. You have to convert feet and inches to just inches, such as 4'–10" = 58", to set up the mline offsets. (Difficulty level: Intermediate; Time to completion: 15–20 minutes)

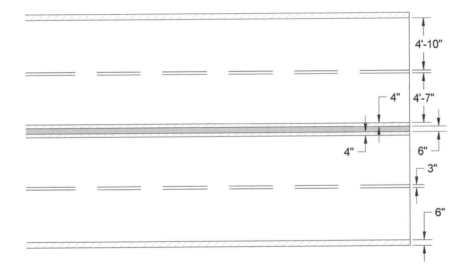

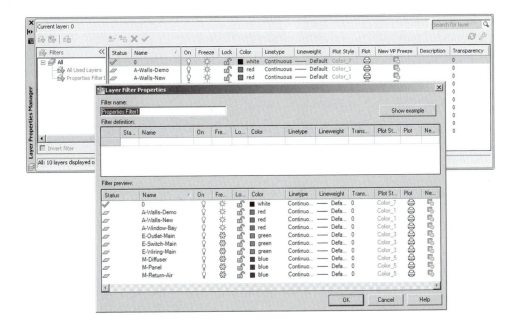

CHAPTER 12

Advanced Layers

LEARNING OBJECTIVES

In this chapter, we introduce advanced layers and discuss the following:

- The need for automation
- Script files
- Layer State Manager
- Layer filters

By the end of the chapter, you will learn how to use and apply these new types of layer tools and be able to incorporate them into future design work where a large number of layers are present in the file.

Estimated time for completion of chapter: 1–2 hours.

12.1 INTRODUCTION TO ADVANCED LAYERS

After the basic commands in Level 1, we introduced the layer concept. This was just a basic primer, showing you how to create them and set a variety of items, such as color and linetype. Also, several methods of control were covered, such as freezing and locking. Here, we look at additional powerful features that allow for unprecedented control over your layers. They are called Script Files, the Layer State Manager, and Layer Filtering.

So, why do you need advanced layer control tools? Well, imagine the following scenario. You work at a large architecture or engineering consulting firm that specializes in complete commercial building and site design. Among your staff there are a number of

- **Architects** and **interior designers** to lay out the overall exterior and interior of the building.
- **Civil engineers** to assist in structural design and site plans.
- **Mechanical engineers** to design the HVAC and mechanical systems.
- **Fire protection** and **security engineers** to design the electrical, sprinkler, and security systems.
- **Landscape designers** to integrate landscape design into the site plan.
- **Additional design professionals** as needed for specialized subsystems.

All these people contribute to the complete design in their own way. Generally, all their work is based on (or ends up in) one main building design file. This common file contains not only the architectural layout of the building but also the data from all these other disciplines. Each discipline has its own set of layers, usually following the AIA convention or another in-house layer naming method. Therefore, architects have their own set of layers starting with an A- (e.g., A-Demo-Wall, A-Bay-Window). Civil engineers have their own (C-Struct-Column, etc.), electrical engineers their own (E-Light-Switch), and so on.

You may already see where this is going as far as layers are concerned. As the project progresses, the layers grow in number. Toward the end, when everyone has added his or her respective design, the file may contain as many as 500 layers (not entirely unusual!). There needs to be a way to impose some order and control over them, so only the needed layers are visible to each discipline. This can certainly be done by manually freezing and thawing layers but is very time consuming. If you guessed that some sort of simple automation is available to you for this purpose, then you are correct.

This in essence is advanced layer management, a process where the layers, once set, are saved under a descriptive name such as Arch_Layout or Elect_Layout, and recalled when necessary by simply clicking on the name. AutoCAD then freezes and thaws the layers automatically, exactly the way you initially told it to. This is what script files, the Layer State Manager, and Layer Filtering do. These tools allow sorting, classifying, and filtering of layers according to a variety of properties and are indispensable when working on a large project.

12.2 SCRIPT FILES

A script, in most general terms, is simply a text file with one command on each line. Its purpose is to automate a sequence of commands when executed. It is perhaps the simplest type of programming you can do. All you need is a text editor (such as Notepad). You next list the commands you want the script to go through, one per line. Then you save it under a name and an `.scr` extension (e.g., `MyScript.scr`), close the file, and run it from AutoCAD's command line by typing in `script` and pressing Enter (alternatively via the Ribbon's Manage tab→Run Script). The Script dialog box appears, the script is double-clicked (executed) on, it does its thing, and if you got the command sequence correct, a short task is automatically performed.

Scripts are generally not used as much now for purposes of layer control as they once were, but we briefly cover them because knowing how they perform gives you insight into the other remaining tools. It is also a good skill to have overall. You can automate a variety of AutoCAD routines. It may be worth your time to read up on further script file details in the Help files, as we do not go too in depth here.

We present a very simple example of a script file. It freezes all layers after first thawing them and setting 0 as the current layer. As scripts do not work with dialog boxes, everything has to be command line driven, such as this version of the layer command (generally any commands that are preceded by a hyphen(-) go to the command line, bypassing any dialog boxes). After the `-layer` is a set of answers to prompts: t means `thaw`, * means `all`, s means `set`, and f means `freeze`. This exactly answers everything AutoCAD asks for if you were to do this by hand at the command line. The last two keystrokes (→) are Enter and Enter.

```
-layer
t
*
s
0
f
*
→
→
```

This reads as Layer, Thaw, All, Set, 0, Freeze, All, Enter, Enter. Try this out with an AutoCAD file that has a bunch of layers. In a real application, you of course have to manually type in the layers deemed not relevant to the particular state (which therefore need to be frozen) in place of the last * and before the two Enters. The steps prior to that (t, *, s, 0) are necessary in case a layer that needs to be frozen is set to current (you cannot freeze the current layer), so the 0 layer is set as current.

This was a lengthy process and is the main reason why script files gradually fell out of favor, replaced by the Layer State Manager for layer control, as shown next. Keep in mind, however, that the Layer State Manager (LSM) works in pretty much the same way, except you select layers and do everything graphically through the Layers dialog box. Now that you have read through the underlying theory, we can rapidly go through the functions of the LSM.

12.3 LAYER STATE MANAGER

As described previously, the LSM is essentially a graphical version of script files. Since you should now have a basic understanding of what you need to do, let us just set up a collection of layers and go through the mechanics of how to set up several layer states. While our example is simple—we set up only three layer states, an Arch_Layout, Mech_Layout, and Elect_Layout—the theory can be easily extended to a large number of layers and states.

Open up a new AutoCAD file and set up the following layers:

A-Walls-Demo
A-Walls-New
A-Window-Bay
E-Outlet-Main
E-Switch-Main
E-Wiring-Main
M-Diffuser
M-Panel
M-Return_Air

Assign color to these layers as shown in Figure 12.1. What we want to do now is create a layer state called *Arch_Layout*. This means only the A- layers are needed, with the electrical and mechanical layers frozen. So go ahead and freeze the six layers beginning with E- and M-, as shown in Figure 12.1. For those of you who notice every detail, the Transparency category has been moved out of the way to the end of the layer palette.

Now that the layers have been frozen, we need to save this "state" under the appropriate name. Click on the Layer State Manager button in the upper left of the Layers dialog box, as seen in Figure 12.2. The LSM then appears (Figure 12.3). It can be expanded (if it is not already) by pressing the arrow in the lower right corner.

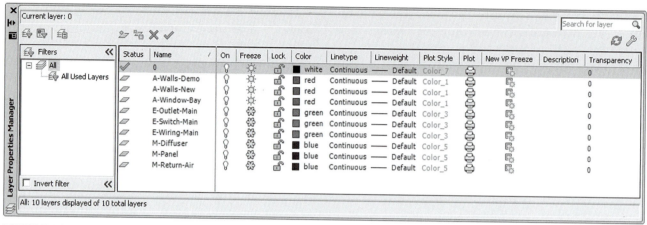

FIGURE 12.1
Layer setup for Layer State Manager (LSM) exercise.

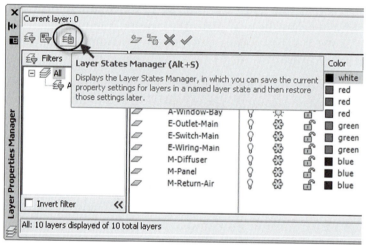

FIGURE 12.2
Layer State Manager icon.

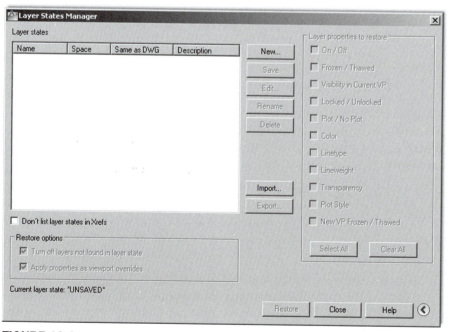

FIGURE 12.3
LSM.

We now create a new layer state by clicking on New… and filling in the name and description (if desired), as shown in Figure 12.4, and clicking on OK. Once the layer state is entered, it appears in the LSM, as shown in Figure 12.5, and you can press Close to exit out of the LSM dialog box.

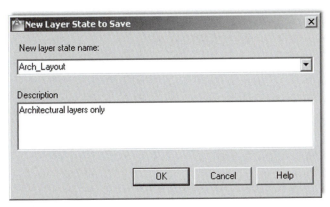

FIGURE 12.4
New Layer State to Save.

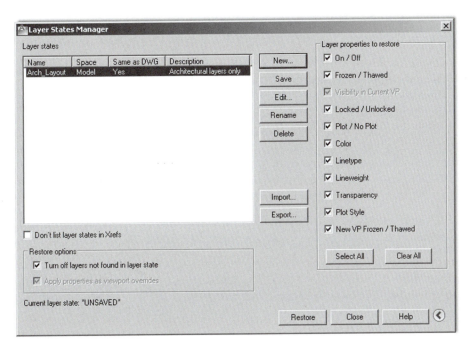

FIGURE 12.5
LSM with new layer state.

In the same manner, go back to the Layers dialog box and create the other two states: Elect_ Layout (only E- layers visible) by turning off the unneeded layers and activating the LSM and Mech_Layout (M- layers visible). The result is shown in Figure 12.6.

To activate these states when needed, you simply go to the Layers dialog box, click on the LSM icon, and when it appears, double-click on the state name or highlight it with one click and press Restore. Instantly, the correct layers are frozen or thawed (you can see this happen in real time in the Layers dialog box as you click on one state or another).

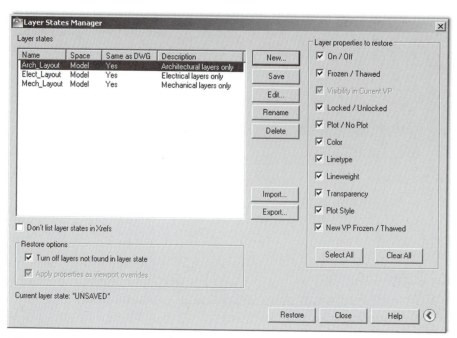

FIGURE 12.6
LSM with remaining layer states.

The purposes of the buttons found to the right of the main LSM window are as follows:

- **New.** Displays the New Layer State to Save dialog box, which we already used.
- **Save.** Saves the selected named layer state.
- **Edit.** Displays the Edit Layer State dialog box, where you can modify the layer state.
- **Rename.** Allows editing of the layer state name.
- **Delete.** Removes the selected layer state.
- **Import.** Layer states (*.las extension) can be imported from another file by means of this option, although this is not done often in practice.
- **Export.** Layer states (*.las extension) can be exported to another file by means of this option, although this is also rarely done in practice.

Some additional options are available in the LSM dialog box, if you click on the small arrow in the extreme bottom right, as previously mentioned (Figure 12.7).

Layer properties to restore is simply a listing of every property a layer typically has, and you can check off different properties if you want them restored. If all are checked, as is more typical in practice, all the layer's properties are affected.

12.4 LAYER FILTERING

Layer Filtering is a tool that allows you to filter out (or selectively choose) layers that fall under a certain description. You can create filters of your own choosing and assign them names. For example, if for some reason you want only the red layers to show or ones that are assigned a hidden linetype, this can be done. Layer names can also be filtered. If you want only those layers that start with an A- to show, this can be arranged, although in this case you are essentially doing what was described previously with the LSM.

The key to layer filtering is to set up filter definitions. These can include full names or partial names (using the wildcard * character) and property definitions (color, linetype, etc.). The entire left side of the Layers dialog box, since AutoCAD 2005, is dedicated to filters. When used properly, these are very effective tools. We highlight some of the more important features

FIGURE 12.7
Additional LSM options.

essential to using them in this chapter. You are encouraged to explore them further on your own as well.

Two types of filters are available as icons on the top left of the Layers dialog box, as shown in Figure 12.8.

283

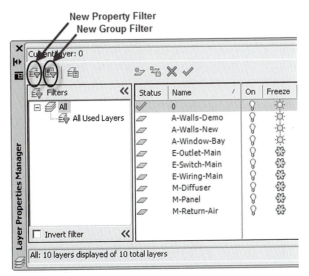

FIGURE 12.8
Layer filters.

The New Property Filter is what you use to set up the definitions. If you press that icon, the box in Figure 12.9 appears, with the previously entered layers visible.

The Layer Group Filter is a general filter that includes all the layers put into the property filter when you define it, regardless of their names or properties. Selected layers can then be added from the layer list by dragging them into the filter. This type of filtering is generally not used

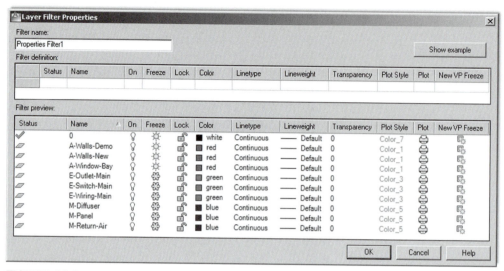

FIGURE 12.9
Layer Filter Properties.

but can be employed to make a new filter based on the layers of another filter. To begin, click on the icon, create a name, and click and drag layer names from the right side of the Layers box into the group filter.

Remember an important point with filters: By themselves, they do little except sort the layers. It is up to you to then do something useful with this, such as freeze or lock them as needed. Therefore, think of filtering as simply a way to get the layers in some sort of order for further action on them.

Here, you first enter the name of your filter in the text field in the upper left. Then, it is just a matter of careful selection of the properties or names (or both) of the layers you want to see. You may use the wild character * in conjunction with typed parts of the layer name to indicate to the filter to include anything before or after the specific letters. Any property can also be used as a filter; color and linetype are common ones.

Go through the filter at your own pace and experiment with the settings. Shown in Figure 12.10, as a simple example, are the layers that are green and have the letter *W* somewhere in the name.

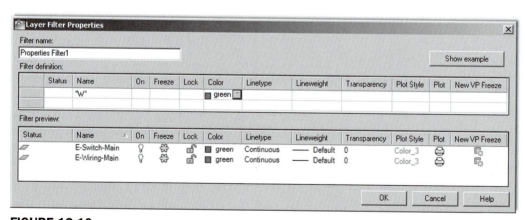

FIGURE 12.10
Example of layer filtering.

LEVEL 2 DRAWING PROJECT (2 OF 10): ARCHITECTURAL FLOOR PLAN

For Part 2 of the drawing project you will continue to add to the walls of the floor plan, with a lot of new work on the interior, as well as adding doors, windows dimensions, and text. We can now also hatch the exterior and interior walls.

Step 1. As always, before starting to draft something new, create the appropriate layers, as shown next.
- ○ A-Doors (Yellow).
- ○ A-Text (Cyan).
- ○ A-Walls-Hatch (Gray_9).
- ○ A-Walls-Interior (Green).
- ○ A-Windows (Red).

Step 2. Add in the exterior windows and all doors. Be sure to follow the layering guidelines carefully. Make all windows and doors blocks for convenience. The exterior dimensions are shown in Figure 12.11. Note that the deck and deck hatching is frozen, as it is not relevant to the new material you are adding.

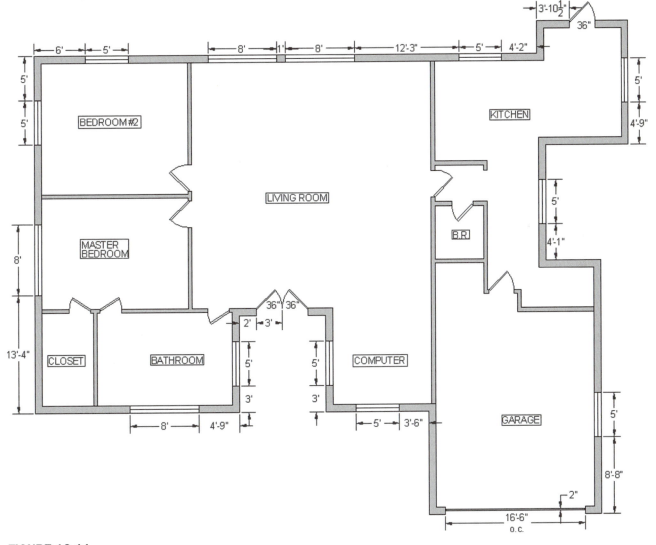

FIGURE 12.11
Exterior windows and doors.

Step 3. Add in the interior walls and doors as shown in Figure 12.12. All interior walls are 5" thick. Also add in the actual dimensions themselves. As mentioned before, they are very useful as a check of your drafting accuracy.

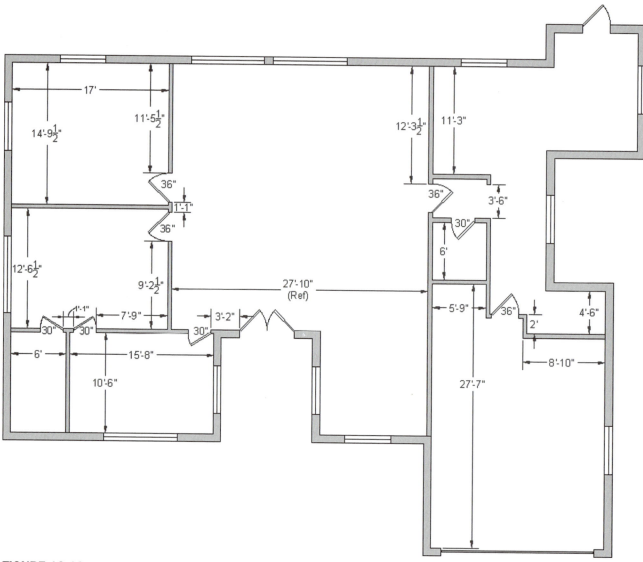

FIGURE 12.12
Interior walls and doors.

Step 4. Finally, add in the hatch (using the light gray color 9). This should be the very last step, because you need to erase the hatch if you find a mistake in the layout.

SUMMARY

You should understand and know how to use the following concepts and commands before moving on to Chapter 13:

- Script files
  - Basic concept
  - Writing a simple script file

- Layer State Manager (LSM)
  - New
  - Save
  - Edit
  - Rename
  - Delete
  - Import
  - Export
  - Restore
  - Layer properties
- Layer filters
  - New Property filter
  - Layer Group filter

REVIEW QUESTIONS

Answer the following based on what you learned in Chapter 12.

1. Describe what is meant by layer management. Why is it needed?
2. Describe the basic premise of a script.
3. What can a script file do for layer management?
4. Describe what the Layer State Manager does.
5. What are the fundamental steps in using the Layer State Manager?
6. What is layer filtering? What are the two types?

EXERCISES

1. Create six layers (and colors) as shown next. Then, write a script file to thaw all the layers; set layer 0 as current. Finally, freeze all and thaw just the even numbered layers. (Difficulty level: Easy; Time to completion: 10 minutes)
   - A-Walls_1 (Green)
   - A-Walls_2 (Red)
   - A-Walls_3 (Green)
   - A-Walls_4 (Red)
   - A-Walls_5 (Green)
   - A-Walls_6 (Red)
2. Using the same six layers, set up layer states Odd_Layers and Even_Layers and run them. (Difficulty level: Easy; Time to completion: <5 minutes)
3. Using the same six layers, create filters to sort and capture only the green layers. (Difficulty level: Easy; Time to completion: <5 minutes)

Advanced Dimensions

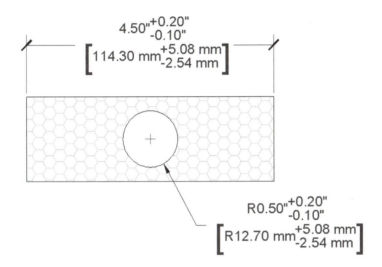

By the end of the chapter, you will learn additional, advanced features of the Dimension Style Manager and be capable of setting up complex dimensions. We also discuss two advanced dimensioning features called *geometric* and *dimensional constraints*.

Estimated time for completion of chapter: 2–3 hours.

Up and Running with AutoCAD® 2012: 2D and 3D Drawing and Modeling.
© 2012 Elsevier Inc. All rights reserved.

13.1 INTRODUCTION TO ADVANCED DIMENSIONS

This chapter is meant to complete your knowledge of AutoCAD's dimensioning features. Specifically, we look at the New Dimension Style (NDS) dialog box and Parametrics. In Level 1, we focused primarily on defining the types of available dimensions and how to properly dimension geometry. Little was mentioned of this dialog box except for the four essential features deemed most important: Arrowheads, Units, Fit, and Text Style (not needed as much anymore, as the better-looking Arial is the default font). In this chapter, we go through the entire NDS dialog box and discuss other options and features, some more than others, according to their usefulness.

The reason you need this knowledge is because AutoCAD allows for a tremendous amount of power, flexibility, and variation with its dimensioning. As a beginner user, you may not come to appreciate what is available, because much of it may already be "set up" for you. As a CAD manager and advanced user (for whom Level 2 is geared), you may be charged with doing this "setting up" and need to know what to do. We conclude the chapter with a new feature introduced back in AutoCAD 2010 called Parametric Dimensions and Constraints. These are routinely found in advanced 3D software, but just recently appeared in AutoCAD.

13.2 DIMENSION STYLE MANAGER

To start off, let us take a closer look at the Dimension Style Manager. You saw it in Chapter 6, so as a reminder here is how we got it up on our screen.

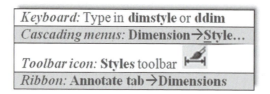

Keyboard: Type in **dimstyle** or **ddim**
Cascading menus: **Dimension→Style…**
Toolbar icon: **Styles** toolbar
Ribbon: **Annotate tab→Dimensions**

Create a new style, by pressing the <u>N</u>ew… button, and name it Sample_Style (Figure 13.1). Click on Continue when done. You then see the New Dimension Style dialog box. We now take a much closer look at the seven tabs, titled Lines, Symbols and Arrows, Text, Fit, Primary Units, Alternate Units, and Tolerances.

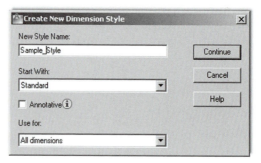

FIGURE 13.1
Create New Dimension Style.

Keep in mind as we go through these that our goal is *not* to cover every possible button and menu in extreme detail. Many of the settings are rarely used and to describe them all completely would take dozens of pages and bewilder even the most dedicated AutoCAD student.

A much better approach is to give a basic overview of each tab and focus on only the features that are of utmost importance and widely used. You will see those features identified and described in more detail, with emphasis on how they fit in with the design process and how they benefit you as an AutoCAD user. If you wish to learn more, use this information as a starting point and consult the Help files.

Let us take a look at the first tab all the way on the left of the New Dimension Style dialog box, called Lines (Figure 13.2). If it is not already visible, click on it to bring it up. Throughout the process, carefully observe the Preview window at the upper right to see the results of changing or adjusting whatever we discuss.

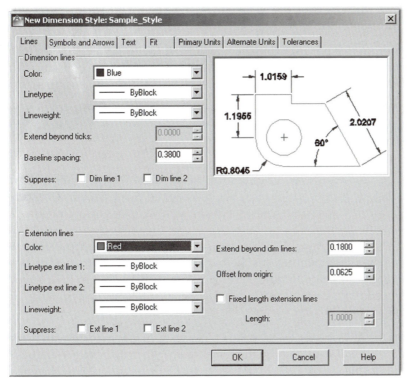

FIGURE 13.2
Lines tab.

291

Lines Tab

This first tab is one that most users leave "as is" and generally do not modify. Here, we can make mostly cosmetic changes to two items specifically: dimension lines (upper left) and extension lines (bottom). It is of course instructive to know what they are. Dimension lines extend from the numerical value and contain arrowheads or another symbol. The color blue is selected for them in Figure 13.2. Extension Lines extend from the part you are dimensioning and "frame" the rest of the dimension. The color red is selected for them in Figure 13.2.

The Lines tab is all about working with these two features. For both the Dimension and Extension Lines, you can change their color, linetype, and lineweight. You simply click the down arrow and change the property. You can also suppress either or both of the Extension or Dimension Lines. This makes them disappear from view. This has some value if, for example, a dimension is sitting between two walls and the extension lines are not visible anyway. You can then suppress them. This however would be a "local" thing, applied to just that one dimension via the Properties dialog box. There is little use in suppressing either Extension or Dimension Lines for the entire drawing. Next, a few values are presented:

- Extend beyond ticks: (grayed out).
- Baseline spacing.
- Extend beyond dim lines.
- Offset from origin.

All these values refer to the various spaces between dimension elements or length of elements. They can all be fine-tuned one at a time. However, it is very important to understand that you do *not* want to do that here in this tab. If you wish to change the size of a dimension, you need to do this by increasing the *overall* scale, not one element or space at a time. We discuss this in great detail later, when we cover the Fit tab. If you try to adjust the values one at a time, you will end up with a lopsided dimension, so leave them all as default.

Finally, we have a check box for fixing the length of an extension line (and assigning a permanent value). This has the effect of placing the dimension a fixed distance from the design but is also rarely used, as it takes away some flexibility in dimension placement.

Just for practice, go ahead and change the colors of the Dimension lines to blue and the Extension lines to red as seen in Figure 13.2.

Symbols and Arrows Tab

The most important feature found under this tab is the ability to change the arrowheads of your dimensions (upper left of tab). You already did this briefly in Chapter 6 and changed your arrowheads to tick marks. Go ahead and do this again, making a note of the other available choices. You can of course make the second arrowhead different from the first, but that is rarely done. The leader is generally left as an arrow since its main purpose is to point at something. Just below the leader is a field to change arrow sizing. This also falls under the "don't do it here" category, as discussed in the previous tab. Arrowheads need to be sized according to the *overall* scale, not individually. We discuss this under the Fit tab.

Moving below the Arrowheads, we have the Center marks. This has to do with measurements of a circle and what you want to see in the center of one when a diameter or radius dimension is added. Mark or Line is typically selected; let us go with Line. Notice how the Preview window reflects both this and the arrowhead changes.

Below Center marks and to the right, we have some relatively minor options. The Arc length symbol concerns the placement of the symbol, and the Radius jog dimension has to do with the angle of the jog "wiggle." Leave them, as well as the Linear jog dimension, as default all around. Do the same thing for the Dimension break field. It sizes up or down according to overall scale settings, so leave it as is for now. Your Dimension Style dialog box should look like Figure 13.3.

Text Tab

This tab is all about what dimension text looks like (appearance) and where it goes (placement and alignment). Take a look at the Text appearance category at the upper left. You already practiced changing the text under Text style. The goal was to make the dimension text the same as the rest of the text in the drawing—same font and usually the same size. In Chapter 6, when this was done, the text you selected on the floor plan was Arial 6".

In this case, create another font from scratch by pressing the button to the right of the Text style: field (button with the three dots) and setting a font style called RomanS .5. You see the Style dialog box, as first outlined in Chapter 4, and need to set the name, font, and size. Review that chapter if you have any problems doing this.

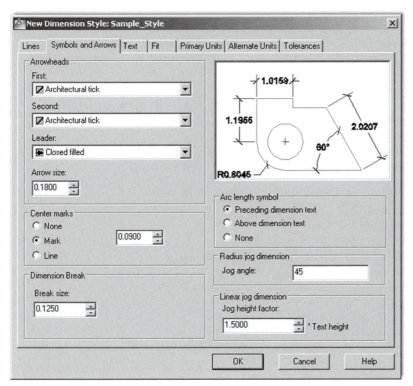

FIGURE 13.3
Symbols and Arrows tab.

Moving down the Text appearance category, we find

- **Text color:** Usually not set separately and remains "By Layer." Here, for practice, we set it to the color Magenta as seen in Figure 13.4.
- **Fill color:** Creates a color highlight behind the text. Try it out for practice, but in this example, no fill is set permanently. To add a frame around the text, there is a check box for this further on down, which we check off. Often a fill and a box are used together.
- **Text height:** (Grayed out.) This is *not* adjusted here. Use the previously mentioned Style dialog box.
- **Fraction height scale:** (Grayed out.) Applies to the size of fractions. This field can be set only when architectural units are being used. We look at this setting in detail soon.

The previous settings take care of pretty much every text appearance issue. Next, we look at how the text is placed in the drawing.

The Text placement category (bottom left) allows you to position the dimension value in every conceivable way near the rest of the dimension. You can place the values above, below, or centered on the dimension line. You can also move the values around the extension line, lining them up. Experiment with all the possible settings, observing the results in the preview window to the upper right. The default setting for both Vertical and Horizontal is Centered, as seen in Figure 13.4. View Direction: flips the values around and Offset from dim line: is best left alone and is adjusted as part of the overall scale settings.

With Text alignment (bottom right), the common default is Horizontal, because people generally do not want to tilt their heads to the side every time they check a dimension on a printout. Do try the other two options (aligned and ISO) to see what they do, though. Your Dimension Style dialog box should look like Figure 13.4.

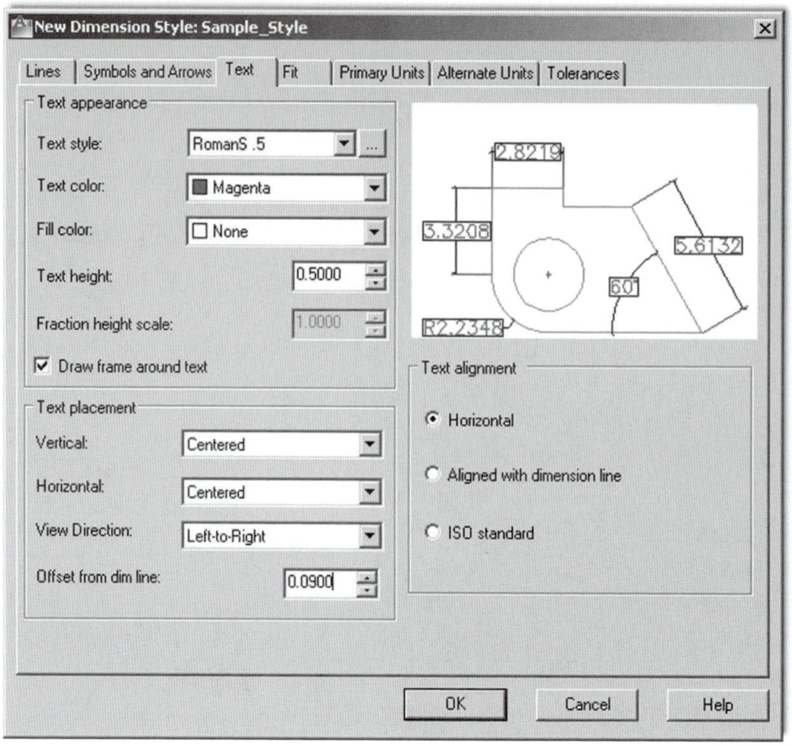

FIGURE 13.4
Text tab.

Fit Tab

The contents of this tab can be a bit confusing at first; after all, didn't we just focus on text placement under the Text tab? Yet, here is another set of categories to fine-tune text placement. You can think of the Fit tab as a continuation of the Text tab with additional Fit options, Text placement, and Fine tuning categories.

Sometimes when you dimension small areas, the extension lines are so close together that the dimension value just does not fit in between them. The Fit options category, at the upper left, simply asks: "If the dimension doesn't fit, what do you want to move?" The best answer here is to select the first option, Either text or arrows (best fit), and let AutoCAD decide. You can of course manually force other choices as seen in the listing (arrows, text, etc.) but then this choice is applied to all cases uniformly, which may not always be correct.

You can even suppress arrows if they do not fit between the extension lines, but that is rarely a good idea and makes for some odd-looking dimensions.

The Text placement category (bottom left) is really just a minor setting and should be left in the default first choice. At the very bottom right, we have the Fine tuning category. These two boxes are not all that important, as you rarely want to place text completely manually (adds extra workload) and asking AutoCAD to always draw a dim line between extension lines makes no sense, as the numerical value is then "crossed out."

The most important category to discuss here is the Scale for dimension features just above the Fine tuning category. You already saw the all-important Use overall scale of: field in Chapter 6 (you entered the value 15). Then, in Chapter 10 (Paper Space), there was a full discussion of what values go in that field and why. Recall that this value boosts (or lowers) the overall sizing of the dimension, and there is no need to adjust all the individual components that we mentioned in the previous tabs. For now, just enter the value 5.0000.

Your Dimension Style dialog box should look like Figure 13.5.

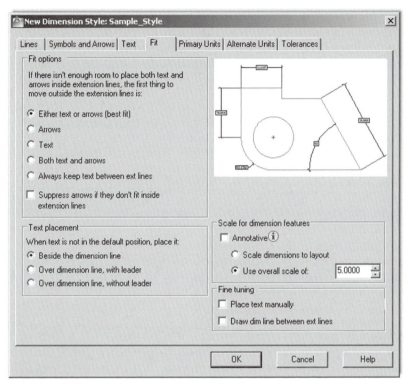

FIGURE 13.5
Fit tab.

Primary Units Tab

This tab features quite a lot of useful settings. In the Linear dimensions category (upper left), you already practiced changing the Unit format: from Decimal to Architectural in Chapter 6. Generally, the dimension units are the same as the drawing units, so if you draw in architectural feet and inches, you would want to dimension in similar units. For this exercise, we leave the units in Decimal mode. In both cases, the Precision setting just below the units can be set to whatever value is appropriate. The range is 0 to 0.00000000 for Decimal and 0'-0" to 0'-1/256" for Architectural. Make a note also of the other units that can be selected, although they are rarely used.

Just below Precision you find Fraction format:, Decimal separator:, and Round off: fields. The first two should be relatively clear in their function. Fraction format gives you a few choices (Horizontal, Diagonal, and Not Stacked) in how you want the fractions to look if using architectural units. This field is grayed out if using decimal dimensions.

The Decimal separator gives you a few choices (Period, Comma, and Space) in how you want the decimal values to look (European engineers do not use a comma when writing 1,000, for example). This field of course is grayed out if using architectural dimensions. You do not want to use Round off: at all. This field of course is grayed out if using architectural dimensions. Some other features further down the column are outlined next.

- **Prefix/Suffix:** Very useful features allowing you to add text, symbols, or values before and after the dimension values. We add the inch symbol in the suffix as an example.
- **Measurement scale:** Scales dimension values; not necessary in common usage.
- **Zero suppression:** Another very useful feature allowing you to suppress zero values either before or after the main significant digits when using decimal units. For example,

a 0.625 value becomes just .625 and a 1.250 becomes 1.25. This is used in conjunction with the Precision setting to "clean up" dimension values, removing extra zeros. With the Architectural units setting, the idea is the same but now 0 feet and 0 inches are suppressed.

The final category is Angular dimensions (bottom right). Here you can select other units for angles such as Gradians, Radians, and Degrees/Minutes/Seconds (useful for surveyors). For most uses however, Decimal Degrees are sufficient and the Precision: field below that allows for setting 0 to 0.00000000. Angular dimensions can also have zero suppression, leading and trailing.

Run through all the options available and, when done, set your Dimension Style dialog box to what is shown in Figure 13.6.

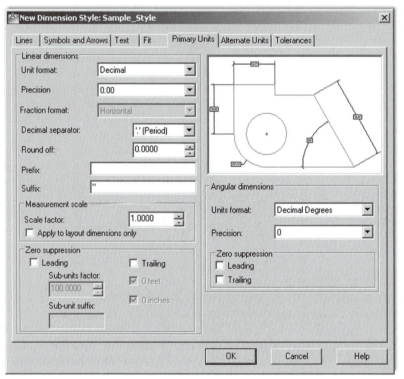

FIGURE 13.6
Primary Units tab.

Alternate Units Tab

This is a tab we did not explore at all in Level 1. Alternate units are a secondary set of dimensions that reside inside parentheses attached to the main set of units. Their purpose is to present the main units in an alternate form such as metric versus English (just one example).

The first thing you must do, as you click over to the Alternate Units tab, is to turn on the alternate units if they are all grayed out. Check off Display alternate units at the upper left and the entire tab activates and the fields become editable. Most of our discussion centers on the Alternate units category (upper left). The Zero suppression and Placement categories should be familiar by now.

Under this category, we have

- Unit format:
- Precision:

- Multiplier for alt units:
- Round distances to:
- Prefix:
- Suffix:

Most of these should also be familiar to you, such as Unit format and Precision. These values generally match the main set of units. Skipping Multiplier, we have Round (never a good idea to use) and finally the previously explained Prefix and Suffix fields.

Understanding alternate units then boils down to fully grasping the Multiplier for alt units: concept. It is a powerful and simple idea. Instead of presenting a collection of "canned" multipliers, AutoCAD allows you to set your own. So, if the main unit is 1 mile, to express this in feet (as an alternate unit), type in 5280 as a multiplier. You are in no way limited as to what sort of value you can enter (inches to miles or millimeters to yards is completely acceptable—if not necessarily practical), but of course you have to know exactly what the multiplier is or else, as they say, "garbage in, garbage out." Therefore, an incorrect multiplier gets you false alternate unit readings.

Explore all the features of Alternate units and try entering several multipliers. For example, the number of yards equal to 1 mile (1760). Note also the default multiplier: 25.4. That is how many millimeters there are in an inch; this is useful for the most common application of alternate units, which is presenting metric units next to English ones.

In our case here, let us go with the amount of centimeters in an inch (2.54) and enter that as the alternate unit. Also add cm in the suffix. Your Dimension Style dialog box should look like Figure 13.7.

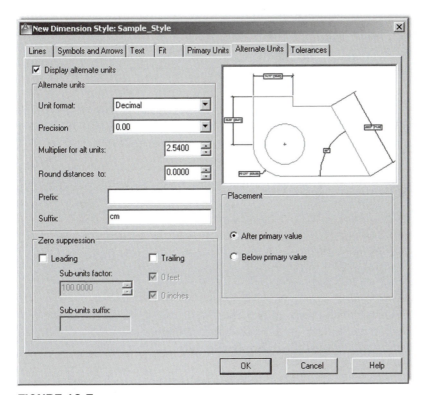

FIGURE 13.7
Alternate Units tab.

Tolerances Tab

This final tab also has not been examined yet. Tolerances are generally defined as deviations from an ideal stated value, as used in manufacturing specifications of engineering devices. Tolerances are relevant mostly to engineering and manufacturing and less so to architecture, but knowledge of this tab is useful for all AutoCAD students.

The underlying need for specifying tolerances hinges on the fact that, in manufacturing, closer tolerances may be costlier to achieve, while larger tolerances, although cheaper, may adversely affect performance of a part. Tolerances are a way of expressing just how much a part may be off the mark and still be acceptable, based on engineering need and cost analysis. Aerospace, automotive, and biomedical engineering tolerances are closer than most consumer products, for example.

Tolerances are a familiar topic to mechanical engineers and machinists. To them no further explanation is needed, and AutoCAD's tolerance choices are familiar: Symmetric, Deviation, Limits, and Basic. Other options under this tab, such as Zero suppression and Placement, should look familiar to all. A short description of each type of tolerance is presented next.

- **Symmetric tolerances:** These indicate the overall deviation, which is symmetric for upper and lower values; therefore you set only one (the upper value).
- **Deviation tolerances:** These indicate nonsymmetrical deviations in either direction with a stacked $\pm$.
- **Limits tolerances:** These are similar in theory and indicate the acceptable limit in one or both sets of values, but no $\pm$ is shown.
- **Basic tolerances:** These are the same as none, the first choice.

Tolerances are an interesting topic to explore further, but because a large percentage of this textbook's readers will not need them in their current or future work (as related to architecture), we do not go any further in depth. If you wish to learn more, there is a tremendous amount of information online. For now, feel free to experiment and set your tolerance to Deviation, as seen in Figure 13.8.

So what did we end up with after all these settings? Well, in the interest of trying out as much as possible, we did not really create a very realistic looking dimension. Let us take a look at it. Click on OK in the Dimension Style box; it will disappear and take you back to the Dimension Style Manager. Then, press Set Current in the next box, followed by Close. Now draw a 10" by 10" rectangle and dimension it using the basic Horizontal dimension. It should look similar or close to Figure 13.9.

Let us try another sample dimension on a hypothetical section of roadway. Draw something similar to Figure 13.10, with the two vertical lines 3 units apart. Run through all the tabs again, using the following data:

- Architectural ticks.
- Red-colored extension lines.
- Blue-colored dimension lines.
- Arial font, .25", color: Red.
- Fit: 1.000.
- Primary units: Decimal. Precision = 0.
- Suffix added on main units→mile(s) and alternate units→feet.
- Alternate units, use 5280 as the multiplier.
- No tolerances.

You should get a result similar to Figure 13.10. Note how simple it is to use alternate units if you know the multiplier.

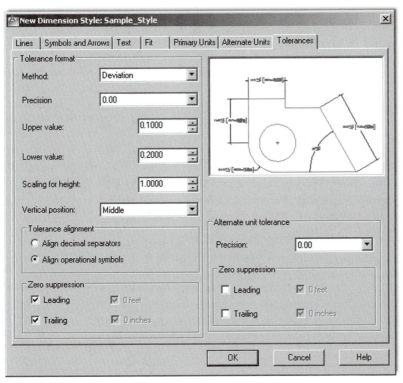

FIGURE 13.8
Tolerances tab.

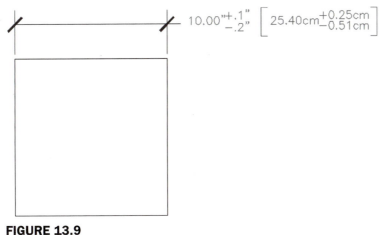

FIGURE 13.9
The new dimension style.

13.3 INTRODUCTION TO CONSTRAINTS

AutoCAD was always a top-notch 2D drafting software application but one with a bit of an inferiority complex. Not in regards to any other 2D competitors but rather toward 3D solid modeling and design software. These sophisticated products were developed in their own world, apart from AutoCAD, via development firms that never competed against or had anything to do with Autodesk or architecture. These programs included CATIA, NX, Pro-Engineer, and SolidWorks among others. Their developers invented quite a few revolutionary concepts and methods that transformed the way engineering design is done. Autodesk eventually jumped on board with Inventor and later Revit, both of which are excellent 3D products (one for engineers and the other for architects).

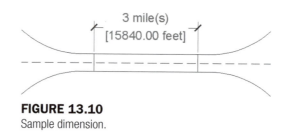

FIGURE 13.10
Sample dimension.

That left the AutoCAD development team looking at how their software could be enhanced by incorporating some of these sophisticated 3D tools without fundamentally altering AutoCAD's identity or formula for success. If you go on to study AutoCAD 3D, you will notice how many "borrowed" concepts found their way into the software as AutoCAD reached ever further into the high-end 3D world.

Some of these concepts also found their way into 2D, such as the ones we are about to discuss, namely Parametrics, which is really two topics: geometric constraints and dimensional constraints; the latter is also referred to as *dimension driven design*. If you are an engineer or an engineering school student who used the previously mentioned 3D solid modeling software, you quickly recognize these concepts. AutoCAD literally took a page out of their playbook. Let us discuss each concept separately.

13.4 GEOMETRIC CONSTRAINTS

The fundamental idea behind geometric constraints, regardless of what software we deal with, is to restrict (constrain) the possible movement of drawn geometry and force that geometry into certain positions. Ortho is a horizontal and vertical constraint on lines. So, if you want two parallel or perpendicular lines, a constraint can be placed on them that forces them to always stay parallel or perpendicular as well as to be that way in the first place. Expanding to more general terms, geometric constraints set "relationships" between geometries such as parallel, perpendicular, tangent, and coincident.

This overall concept is quite important in setting what is referred to as *design intent*. If a drilled hole in a bracket absolutely has to be concentric to the bracket's corner fillet to fulfill the design intent, then that is the main driving factor in its design (*concentric* means the circles and arcs share the same center point). Other aspects of the design, such as perhaps notches and other geometry on the bracket, however, may be less critical. The constraints then "hold" the critical geometry, while you design the rest around it.

This is a very simple example; with a bracket, you can probably keep track of the critical relationships without help. Geometric constraints really earn their keep with more complex designs, where many constraints are needed to preserve design intent. Without them, it would be easy to forget one or two important relationships and discover later down the road that the part you are working on does not function properly because a tangency somewhere was not preserved.

Almost since the beginning of high-end 3D software, geometric constraints were an important part of the design approach, but they were not included in AutoCAD until recently (Release 2010); and AutoCAD designers needed to keep track, and check over, their geometric relationships manually. With these new tools, AutoCAD's engineering and architecture design capabilities have been greatly enhanced.

Types of Geometric Constraints

Listed next are most of the constraints available to you. We do not go through every single one but cover only the more commonly used ones, leaving the rest for you to explore. The best way to set constraints is via the Geometric Constraints toolbar (see Figure 13.11) or the Ribbon's Parametric tab.

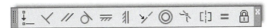

FIGURE 13.11
Geometric Constraints toolbar.

The critical constraints we discuss are

- **Perpendicular.** Constrains two lines or polylines to perpendicular 90° angles to each other.
- **Parallel.** Constrains two lines to the same angle.
- **Horizontal.** Constrains two lines to lie parallel to the X axis.
- **Vertical.** Constrains two lines to lie parallel to the Y axis.
- **Concentric.** Constrains circles, arcs, or ellipses to maintaining the same center point.
- **Tangent.** Constrains a tangency between curves and lines.
- **Coincident.** Constrains two points together (to be discussed along with dimensional constraints).

Adding Geometric Constraints

Adding geometric constraints involves simply picking the constraint you wish to add and selecting the first object (which serves as the reference point) followed by the second object. The constraint then appears next to the objects, reminding you it is set. The constraint tool also forces the objects into that constraint, so if you draw two lines that are not quite parallel, once you set the constraint, they become parallel right away.

Having read this, you may be tempted to completely do away with Ortho and even OSNAP to create accurate geometry, allowing constraints to "fix" drawn shapes. This of course is not generally recommended, as sloppy drafting catches up to you sooner or later (though we do try this in the next example). While geometric constraints are very useful new tools, use them when appropriate, without forgetting or throwing away basic accuracy habits.

Let us try this out with the first constraint on the previous list: Perpendicular.

Keyboard: none

Cascading menus: Parametric→Geometric Constraints→Perpendicular

Toolbar icon: Geometric Constraint toolbar

Ribbon: Parametric tab→Perpendicular

Step 1. Draw two lines that are roughly perpendicular, as seen in Figure 13.12 (left).
Step 2. Select the perpendicular geometric constraint via any of the preceding methods.
- AutoCAD says:
```
Enter constraint type
[Horizontal/Vertical/Perpendicular/PArallel/Tangent/SMooth/
Coincident/CONcentric/COLlinear/Symmetric/Equal/Fix]
<CONcentric>:_Perpendicular
Select first object:
```
Step 3. Select the first line (the vertical one).
- AutoCAD says: `Select second object:`
Step 4. Select the second line.

301

FIGURE 13.12
Perpendicular geometric constraints.

Immediately after the selection of the second line, you see it snap to perpendicular and two sets of geometrical constraints are added next to each line, as seen in Figure 13.12 (right). Note that, if you were to select the bottom line first, then the vertical one would snap to it, not the other way around. Selection order matters.

Let us try one more, the concentric one. We leave the rest as an in-class exercise.

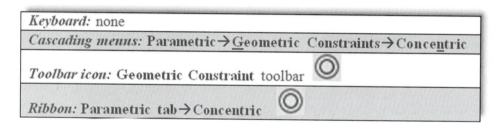

Keyboard: none	
Cascading menus: **Parametric→Geometric Constraints→Concentric**	
Toolbar icon: **Geometric Constraint** toolbar	◎
Ribbon: **Parametric tab→Concentric**	◎

Step 1. Draw a circle, then an arc above it, as seen in Figure 13.13 (left).
Step 2. Select the concentric geometric constraint via any of the preceding methods.
- ○ AutoCAD says:
  ```
  Enter constraint type
  [Horizontal/Vertical/Perpendicular/PArallel/Tangent/SMooth/
  Coincident/CONcentric/COLlinear/Symmetric/Equal/Fix]
  <CONcentric>:_Concentric
  Select first object:
  ```
Step 3. Select the circle.
- ○ AutoCAD says: `Select second object:`
Step 4. Select the arc.

Immediately after the selection of the arc, you see one or both snap to a concentric setting (their centers are the same) as seen in Figure 13.13 (right). Note that, if you were to select

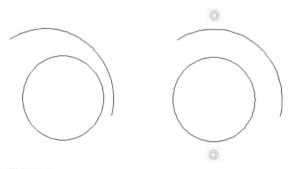

FIGURE 13.13
Concentric geometric constraints.

the arc first, the circle would adjust to it, not the other way around. Once again, selection order matters.

Try out the remaining geometric constraints on your own as an in-class exercise.

Hiding, Showing, and Deleting Geometric Constraints

There are two ways to access these additional options to work with constrains. You can either right-click on one of the constraints to reveal a small menu, as seen in Figure 13.14, or find these same options on the Ribbon, as seen in Figure 13.15, along with the rest of the constraints.

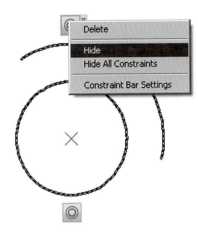

FIGURE 13.14
Geometric Constraints options, right-click menu.

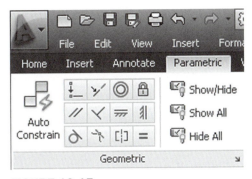

FIGURE 13.15
Geometric Constraints options, Ribbon.

The Constraint Settings dialog box, which can be accessed via the small drop-down arrow at the bottom right of the Geometric tab, is actually just for global settings, for which geometric constraints to show. Leave all the choices selected and the transparency at 50%, as seen in Figure 13.16. Finally, Figure 13.17 shows all seven of the constraints mentioned, as you should see them on your screen after completing the in-class exercise.

13.5 DIMENSIONAL CONSTRAINTS

Since we already discussed the basic idea of constraints, this introductory section is shorter. The idea here is to move from constraining pieces of geometry (and their positions relative to each other) to constraining the actual dimensions of the drawn design. So, if two circles,

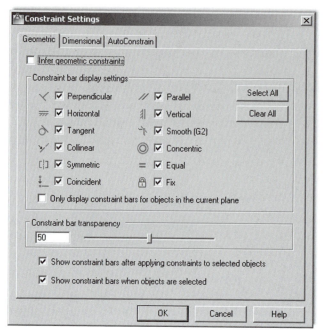

FIGURE 13.16
Constraint Settings.

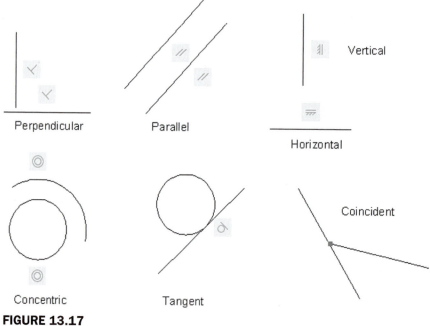

FIGURE 13.17
Geometric constraints.

representing drilled holes, absolutely need to be 3" apart, then constraining that dimension ensures it does not change regardless of what other design work goes on around them.

Dimensional constraints are also a major part of 3D parametric design and can be easily transferred to 2D work, as was done here in AutoCAD. Also, because the geometry has to obey the constrained dimension, you can change the geometry by simply changing the dimension value, which is quite a departure from the normal order of business with AutoCAD. This ability to change geometry by updating dimensions is called *dimension driven design.* With high-end 3D solid modeling software this is really the only way to change the size of an object (usually called a *feature*). AutoCAD lacks this ability in 3D but, as just mentioned, can do this with 2D shapes.

At this point, you may start to see why all the preceding is referred to as *parametric design.* We are designing by changing or assigning parameters to our geometry. The geometry itself is almost secondary to the relationships and data that drive its existence and creation. This type of philosophy and approach is one of several steps critical to assuring that what we have is a valid engineering design, not just a pretty picture. The data behind the design has to make sense.

Working with Dimensional Constraints

Learning the basics of dimensional constraints is straightforward. You can either add them in right away to your design or convert existing regular dimensions to constraints. Dimensional constraints have their own toolbar (Figure 13.18), or they can be accessed through the Ribbon (Figure 13.19).

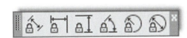

FIGURE 13.18
Dimensional Constraint toolbar.

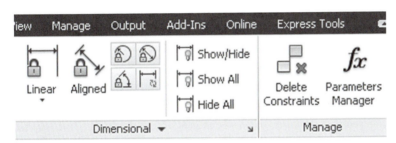

FIGURE 13.19
Dimensional Constraints using the Ribbon.

The available dimensional constraints mirror the regular dimensions closely:

- Horizontal
- Vertical
- Aligned
- Radius
- Diameter
- Angle

To practice them, draw the bracket seen in Figure 13.20, but do not dimension it; use the shown values only to create the geometry.

Now add a horizontal dimensional constraint.

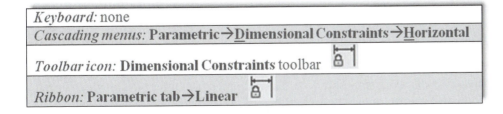

Keyboard: none	
Cascading menus: **Parametric→Dimensional Constraints→Horizontal**	
Toolbar icon: **Dimensional Constraints** toolbar	
Ribbon: **Parametric tab→Linear**	

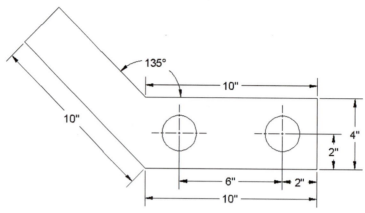

FIGURE 13.20
Bracket.

Step 1. Start up the dimensional constraint via any of the preceding methods.
   ○ AutoCAD says: Specify first constraint point or [Object] <Object>:
Step 2. Pick one end of the bottom part of the bracket (a red circle with an X appears).
   ○ AutoCAD says: Specify second constraint point:
Step 3. Pick the other end of the bottom part of the bracket (a red circle with an X appears again).
   ○ AutoCAD says: Specify dimension line location:
Step 4. Locate the dimensional constraint some distance away from the part, similar to a regular dimension.

AutoCAD displays the value (d = 10.0000) with a graphic of a lock, as seen in Figure 13.21. Press Enter.

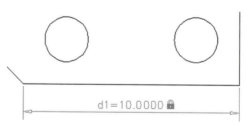

FIGURE 13.21
Dimensional constraint.

In a similar manner, you can add additional dimensional constraints to the rest of the bracket or convert existing dimensions (if you had some) via the Convert button in the Ribbon's Parametric tab. Note the central idea behind this tool. The dimensional values are "locked in," as represented by the lock graphic, and do not budge regardless of how other surrounding geometry may change.

13.6 DIMENSION DRIVEN DESIGN

Let us put the preceding together into a concise approach to design, as well as add in some comments about how to work smoothly with dimensional constraints. First of all, you can change the value of the dimensional constraint by just double-clicking on it and typing in something else (Figure 13.22).

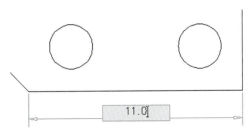

FIGURE 13.22
Altering linear dimensional constraint.

The value forces the line to move to the right, leaving a gap, which is not quite what you intended. Try the same with a circle and diameter constraint (Figure 13.23).

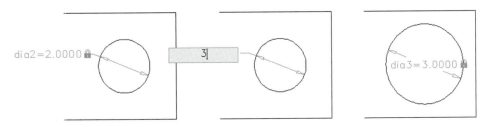

FIGURE 13.23
Altering diameter dimensional constraint.

This leads to much better results; the circle has changed its size. There is a very important point to understand here in order to take full advantage of these tools. With linework and angles, dimensional constraints need to work together with geometric constraints. Used just by themselves, dimensional constraints change only the dimension value and drag the lines along for a ride; but if you add a coincident geometric constraint, the lines are "glued" together so you can make entire sections of the design properly constrained. A coincident constraint asks you to select one of the lines, then the second, and create a point where they intersect. Try this out on our bracket's bottom pieces, and try changing the bottom line's size again, leading to quite a different effect.

A lot more can be written about constraints in general; it is a broad and important topic. You are encouraged to explore further and practice setting up dimensional constraints on critical parts of a design. It will take some getting used to, and applying dimensional constraints intelligently can be awkward at first. The key is to apply these constraints only to the critical geometry and not constrain every single line and circle on the screen. Fortunately, AutoCAD prevents you from doing this and issues a warning that you are over-constraining or there are conflicts, similar to high-end solid modeling applications.

LEVEL 2 DRAWING PROJECT (3 OF 10): ARCHITECTURAL FLOOR PLAN

For Part 3 of the drawing project, you complete the architectural/interior design by adding furniture and appliances to the house.

Step 1. You again need new layers:
  ○ A-Appliances (Magenta).
  ○ A-Furniture (Red).
Step 2. Dimensions for the furniture and appliances are not specifically given, but draw in reasonable-size designs, as seen in the layout. Feel free to add more than is shown. The full plan is in Figure 13.24. Your drawing features the colors listed in Step 1.

307

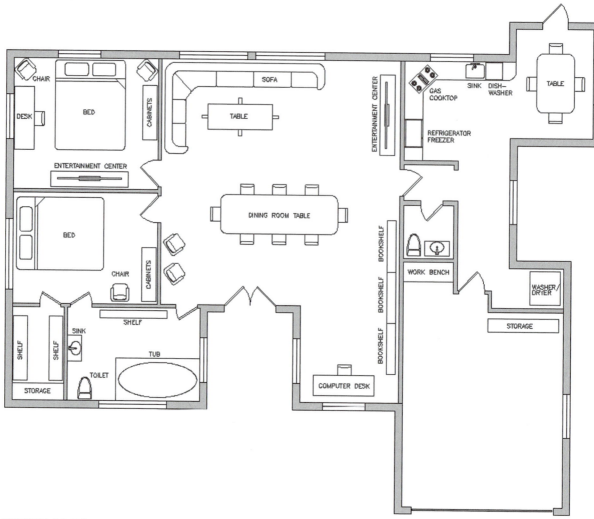

FIGURE 13.24
Furniture and appliances in floor plan.

Step 3. Add in text describing the furniture and appliances. Put it on the text layer. It should take you about an hour to an hour and a half to add in the new features. Use copy often!

SUMMARY

You should understand and know how to use the following concepts and commands before moving on to Chapter 14:

- Creating a new dim style (`ddim`)
- Lines tab
  - No permanent settings, but know functions of all
- Symbols and Arrows tab
  - Architectural tick setting; know functions of all others
- Text tab
  - Text font and size settings; know functions of all others
- Fit tab
  - Fit setting; know functions of all others

- Primary Units tab
  - Units setting
  - Precision setting
  - Prefix setting
  - Suffix setting
  - Zero suppression settings; know functions of all others
- Alternate Units tab
  - Multiplier settings; know functions of all others
- Tolerances tab
  - Tolerance settings (if needed)
- Geometric constraints
  - Perpendicular
  - Parallel
  - Horizontal
  - Vertical
  - Concentric
  - Tangent
  - Coincident
- Dimensional constraints
  - Horizontal
  - Vertical
  - Aligned
  - Radius
  - Diameter
  - Angle

REVIEW QUESTIONS

Answer the following based on what you learned in Chapter 13:

1. Describe the important features found under the Lines tab.
2. Describe the important features found under the Symbols and Arrows tab.
3. Describe the important features found under the Text tab.
4. Describe the important features found under the Fit tab.
5. Describe the important features found under the Primary Units tab.
6. Describe the important features found under the Alternate Units tab.
7. Describe the important features found under the Tolerances tab.
8. Describe the purpose of geometric constraints.
9. What seven geometric constraints are discussed?
10. Describe the purpose of dimensional constraints.
11. What six dimensional constraints are discussed?

EXERCISES

1. Draw the following shape and label it exactly as shown. Be sure to note all the modifications to the dimensions, including arrowheads and the presence of alternate units and tolerances. You may have to get creative with the leader. (Difficulty level: Intermediate; Time to completion: 10–15 minutes)

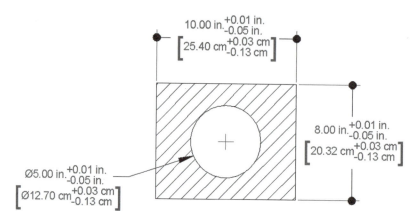

2. Draw a rough parallelogram shape, with one side 20", as shown on the left. Parallel constrain all four sides, filleting if necessary. Then, carefully, coincident constrain all four corners. Finally, dimension constrain one side, using a value of 30 as seen on the right. If everything is done right, the rectangle should increase in size uniformly. (Difficulty level: Easy; Time to completion: 5 minutes)

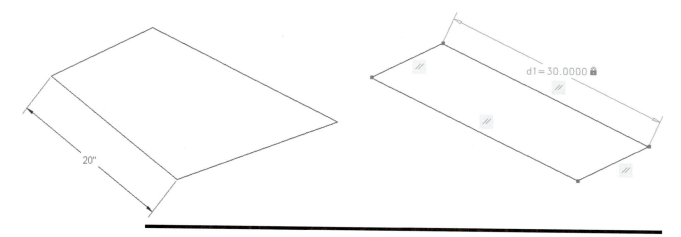

Spotlight On:
Interior Design

Interior design is a major branch of architecture that concerns itself with interior living spaces. The goal is to create a living (or office) space that is both functional and aesthetically attractive by the creative use of colors, lighting, and materials. Interior design is not to be confused with interior decorating. Interior designers are often well versed in the principles of architecture, engineering, and materials and must generally hold at least a four-year degree. They can be called upon to create challenging spaces that are tailored to certain ecological or handicapped-accessible needs. It is a creative profession, but one that still requires analytical skills.

FIGURE 1

Education for interior designers (in the United States) involves a four-year baccalaureate degree in interior design. Masters degrees (MS, MA, MFA, and recently the MID) in interior design are also available, although this advanced degree is less common than the baccalaureate degree. Many professionals pursue advanced degrees in related subjects, such as industrial design, fine art, or education. Doctoral programs in interior design are increasing in number at various institutions of higher education. It is worth noting that some of the more notable interior designers held no formal education, though this is the exception not the rule.

Following formal training, graduates usually enter a one- to three-year apprenticeship to gain experience before taking a national licensing exam or joining a professional association. The National Council for Interior Design Qualification (NCIDQ) administers the licensing exam. To be eligible to take the exam, applicants must have at least six years of combined education and experience in interior design, of which at least two years constitute postsecondary education in design. Once candidates have passed the qualifying exam, they are granted the title of Certified, Registered, or Licensed Interior Designer, depending on the state. Some states require continuing education units to maintain the license.

Interior design earnings vary based on employer, number of years with experience, and the reputation of the individual. For residential projects, self-employed interior designers usually earn a per-hour fee plus a percentage of the total cost of furniture, lighting, artwork, and other design elements. For commercial projects, they may charge per-hour fees or a flat fee for the whole project. The median annual earnings for hourly and salaried interior designers, in 2009, were approximately $61,000, with the top 10% earning over $78,760.

So how do interior designers use AutoCAD and what can you expect? Very often, as far as AutoCAD goes, interior design is tied in with the general architecture design of a building. Therefore, interior designers work side by side with architects on their CAD files. The layering convention often follows AIA standards. You find layers such as A-Furniture, A-Carpeting, and the like. Designers also work with blocks and attributes, and make extensive use of custom hatch patterns and a wide array of colors (think Pantones).

Interior designers often work in 3D, as the whole idea is often to present and sell the design concept. In these cases, they have to master 3D AutoCAD as well as additional rendering software. With 3D work, AutoCAD is not the only choice, and one can find Rhino, Form Z, and other applications running side by side with AutoCAD or even bypassing CAD completely.

FIGURE 2

Shown in Figure 3 is a 2D plan view of a residential interior. It makes use of solid shading and custom hatch patterns to give the client a good idea of the final outcome of the design effort.

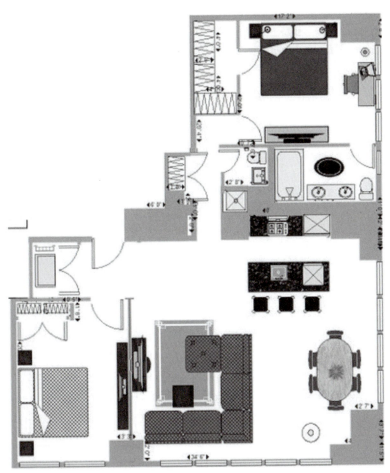

FIGURE 3
2D plan view of a residential interior.

Shown in Figure 4 is a 3D photorealistic rendering of another design. Although AutoCAD can be used to easily create the underlying architecture, additional rendering tools are usually used for this level of sophistication and realism.

FIGURE 4
3D photorealistic rendering of another design.

CHAPTER 14

Options, Shortcuts, CUI, Design Center, and Express Tools

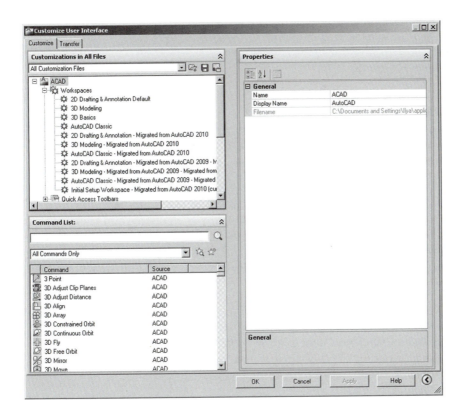

LEARNING OBJECTIVES

In this chapter, we introduce a variety of advanced tools and discuss the following:

- Options dialog box
  - ○ Files tab
  - ○ Display tab
  - ○ Open and Save tab
  - ○ Plot and Publish tab
  - ○ System tab

- ○ User Preferences tab
- ○ Drafting tab
- ○ 3D Modeling tab
- ○ Selection tab
- ○ Profiles tab
- Shortcuts and the `acad.pgp` file
- Customize User Interface (CUI)
- Design Center
- Express Tools

By the end of the chapter, you will learn how to significantly customize your environment and speed up your work via importing pre-drawn blocks and using shortcuts to accelerate command input.

Estimated time for completion of chapter: 2–3 hours.

14.1 OPTIONS

Our topics for this chapter are the Options dialog box, followed by the concept of Shortcuts (the `acad.pgp` file), and the Customize User Interface (CUI). We then look at the Design Center and the Express Tools. These topics introduce you to basic customization, greatly enhance your efficiency, and I hope will dispel any possible remaining frustration with making AutoCAD look and behave how you want it to.

The Options dialog box is your access point for basic AutoCAD customization menus, and through it, you can change many essential settings and tweak AutoCAD to your personal taste and style. Almost every user finds a few items worthy of changing, and knowing this dialog box is absolutely essential. Often, while covering this topic, students remark how they have been annoyed by some feature (or lack of it) and now know how to turn it off or on. This is exactly why we introduce this dialog box, or cover it in more detail if you have already seen it.

Note that our goal is not to go over every single item in Options; it is far too big for that, and it is unnecessary anyway. Instead, we outline the most important features and explain why they are necessary. There are instances where you will be advised against changing the default values of one item or another. If you want to know more, press F1 (Help files) while in each of the tabs for a detailed outline of every single feature. Note, however, that you should first have a clear understanding of the selected items this chapter focuses on before going into more detail.

To access the Options dialog box, right-click anywhere in the drawing area or command line and a menu appears, as shown in Figure 14.1. Click on the last choice: Options…

A sizable dialog box appears, as seen in Figure 14.2, with the following tabs: Files, Display, Open and Save, Plot and Publish, System, User Preferences, Drafting, 3D Modeling, Selection, and Profiles. With the exception of 3D Modeling, all the significant features of each tab are introduced, although with limited discussion in the case of several tabs. Click the very first tab (Files) and let us begin.

Files Tab

This first tab is the easiest to go over for the simple reason that you rarely have to change anything here. What are all these listings then? When AutoCAD, or any software for that matter, installs in your computer, it spreads out all over a certain area of the disk drive (where you indicated it should go on installation) and drops off necessary files as it sees fit.

FIGURE 14.1
Accessing Options....

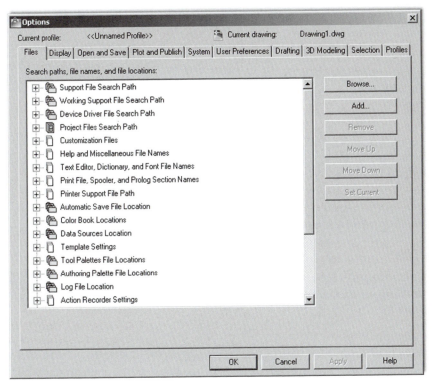

FIGURE 14.2
Files tab.

What you are looking at in Figure 14.2 is a collection of paths to some of those files as they appear after an installation. They are deemed to be appropriate for a user to access. You can view each path and modify it by clicking on the plus signs. There are quite a few of them, indicating locations of a variety of important AutoCAD functions: Help files, automatic saves, and the like. Some of these you may recognize, some you may not.

Be careful though; default settings are a good thing in this case. If you change the end location of some file, often you (or sometimes AutoCAD) may not find it, and difficulties result, such as when you try to move the device driver files. It is strongly advised to leave

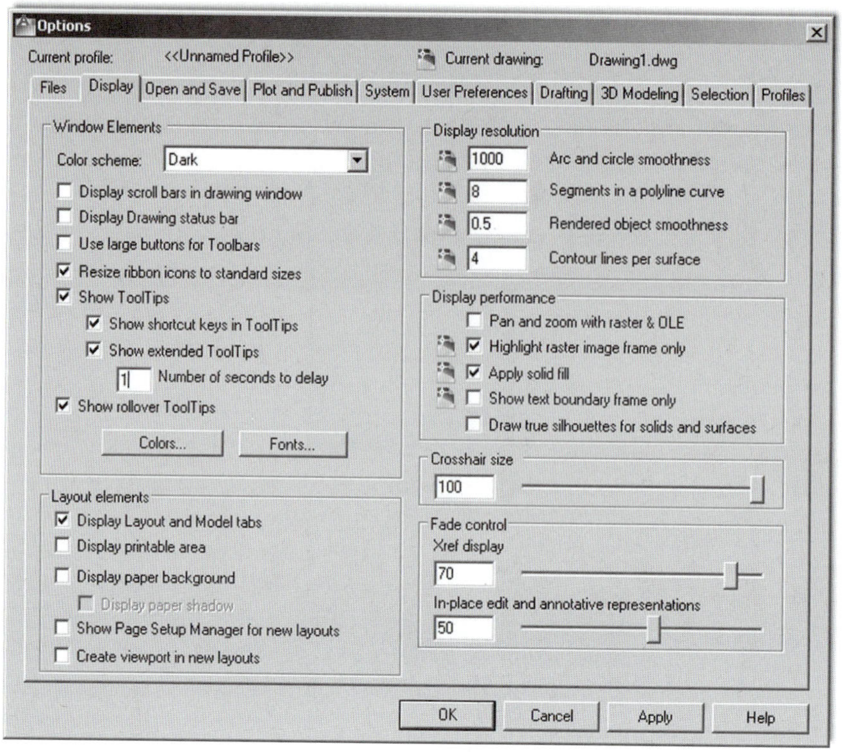

FIGURE 14.3
Display tab.

all these as they are, as there really is no compelling reason to change the preset locations. I have needed to change only a few of these paths in all my years of AutoCAD use and management, and those had to do with log file locations and AutoSave, so the average user probably has little use for this advanced tab. If you want a detailed description of what each path means, press F1 while you have the Files tab open.

Display Tab

This second tab (Figure 14.3) has a few items of interest, specifically under the Window Elements and **Layout elements categories** (on the left side of the box, top to bottom), and **Crosshair size** (on the right).

Under the category **Window Elements**, you do not really need the first three boxes checked unless you really like the scroll bars or need to see larger toolbar buttons (very useful for the visually impaired). Below that is the Ribbon resizing, just a minor adjustment which you can leave checked. The next few boxes feature various ToolTip settings, and these are actually good to leave checked to explain the meaning of new toolbars and other items. Included in there is the delay time, with 1 or 2 seconds adequate.

The **Colors**… button just below ToolTips refers to the color of your screen background. Generally this is colored black, as preferred by most (but not all) users. Black is a good choice for the simple reason that the dark background throws less light at your eyes from the screen, and lessens fatigue. The only catch is that printed paper output is always white, and some people prefer seeing the design as it would appear in real life, on white paper, hence the white background. Obviously it is easy to get used to the black background and pretty much disregard the "paper" aspects of the output, and this is what most users do. In this and most textbooks, however, the screen background color is always white to save ink and blend images smoothly with the white paper.

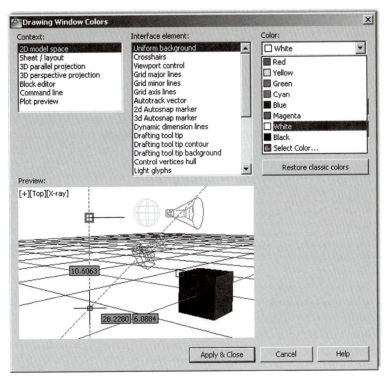

FIGURE 14.4
Drawing Window Colors with White selected.

Other colors are available of course (actually all 16 million colors are available as a background), but as we joke in class, if you see someone using a bright neon green background, stay away—this is probably not a stable individual. To change the color scheme for practice, press the Colors… button and the dialog box in Figure 14.4 appears.

Here you can select from the general Context: menu on the left (only the first choice, 2D model space, for now), and select some element in the Interface element: menu on the right (e.g., the Uniform background). Then pick the color you want this element to have. Figure 14.4 shows white being chosen. Note the sizable number of choices in both the left and right columns. AutoCAD allows for the color of just about anything to be changed; browse through all the choices to get familiar. If you change so much that you forget what you did, you can always press the Restore classic colors button on the right of the dialog box. When done, press Apply & Close.

The **Fonts**… button to the right of Colors… in the Display tab refers to the font present on the command line text, not the font in the actual drawing, so unless you have an objection to the Courier New font (the default), do not change anything.

In the **Layout Elements** category at the bottom left, you can uncheck all the boxes except the top one: Display Layout and Model tabs. Generally, all those options are distractions and unnecessary in Paper Space, as discussed in more detail in Chapter 10.

Moving to the upper right of the Display tab, we have the **Display resolution** category (leave everything as default) and the **Display performance** category (also leave as default). The only thing worth mentioning here is the Arc and circle smoothness value, set at 1000. That was set low for historical reasons that have to do with accelerating regenerations on slower old computers by having arcs and circles be made up of small lines. You can set that to 20,000 maximum on today's fast machines.

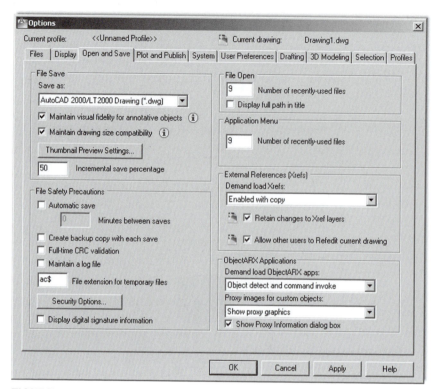

FIGURE 14.5
Open and Save tab.

Moving down we have the **Crosshair size** setting. Crosshair size is generally a personal preference. Many leave it at a "flyspeck" setting, which is 5, but increasing the size to a full 100 is recommended, as this makes life easier when you are in a Paper Space viewport, as detailed in Chapter 10, or checking if a drawn line is straight by comparing it to the straight-edge crosshairs. In the end, it is up to you.

The last category at the bottom right is **Fade control**. This has to do with how dark the xref is when you are editing it in place, a Chapter 17 topic. For now, the default settings are fine.

If you want a more detailed description of what each selection means, press F1 while you have the Display tab open. Otherwise, check to see that your settings reflect what is in Figure 14.3 and let us move on to the next tab.

Open and Save Tab

The Open and Save tab (Figure 14.5) contains several important features worth discussing, all of them contained inside the two categories on the left-hand side: File Save and File Safety Precautions.

In the **File Save** category, let us take a look at the top option, labeled Save as:. This drop-down menu is very important, as it allows you to permanently save your files down to an older version of AutoCAD. The significance of this cannot be overstated because you simply cannot open a file created on a more recent version using an older release. AutoCAD is an often updated software, with new releases every year (since 2004), and you can rest assured that not everyone is running the same version. I have seen AutoCAD 14 (from the mid-1990s) running on a company PC as late as 2006. While that may have been an isolated case, AutoCAD 2004 and 2006 are still in widespread use, and nothing is more frustrating to someone than getting a copy of your files for collaboration and not being able to work with them.

Autodesk of course realizes this and allows you to save files down on a case-by-case basis during the save procedure. AutoCAD even automatically saves one release down anyway. But to be safe (if it is not already set this way), select the AutoCAD 2000/LT 2000 Drawing (*.dwg) choice, as seen in Figure 14.5; that way you are assured of the file being accessible by just about everyone else. We can skip over the rest of the options under the File Save category (leave everything as default) and move on down to the File Safety Precautions category.

The **Automatic save** check box is about as close to controversial as something in AutoCAD gets. The suggestion is this: Do not use it. Others may say, why not, it is there. The argument against it is simple: As a computer user, you need to save your work often, no matter what application you use. To rely on automatic saves is to invite trouble when you use a program that does not have an automatic save (or it is not turned on). As an AutoCAD user, you should be saving your work every 5 to 10 minutes, something students often do not do in class.

There are also some practical difficulties with Automatic save. For one thing, you need to exit a command to save a drawing. AutoSave does this for you, so even if you are in the middle of something, you are kicked out for AutoSave to run (except while in block editing mode). Other issues are where AutoSave files go and what extension they have. You can set where they go using one of the paths under the File tab, but then all AutoSave files go there, regardless of what actual drawing is open, and overwrite existing files. The extension of AutoSave files is .sv$.

In summary, while AutoSave can be set up, it is not advisable to rely on it; often you see veteran programmers and computer users ignore this feature in all computer use and save instinctively every few minutes. If you do wish to use AutoSave, simply turn it on by checking the box and setting a time threshold (Minutes between saves) just below that.

The **Create backup file with each save** check box can also be mildly controversial, as it seems like a good idea initially, yet does not stand up under scrutiny. Backup files (if the option is checked) are created every time you save a file. A backup file (.bak) is essentially a regular drawing file (.dwg) that is cloaked with an extension Windows does not recognize, but you do (see Appendix C for more info on all extensions). If the .dwg file is lost, then the .bak file can be renamed and, voila, the drawing is there again. So, what is wrong with this? Well, in theory nothing, but in real-life use there are a few issues.

The main problem is that backup files actually double the size of your project folders; it is like having a copy of every drawing file next to the original, and they just are not necessary, due to the combination of cheap storage media and a heightened sense of risk in recent years that has prompted just about every company to do daily and weekly backups. Also, AutoCAD's drawings rarely "go bad" to the point of being unrecoverable by the Audit and Recover commands. Finally, backup files reside right next to the drawing files themselves, and if you lose the entire folder, they all go. Rarely do you lose just the drawing, and even then, going to the previous night's tape backups is enough to get them back. So, while some would argue that no harm is done by keeping the *.bak files, one could also say: Why add the extra junk to your company's server? Ultimately, the choice is up to you, although sometimes company policy may dictate the use of backup files, and this is beyond your control.

Most users do not need **Full-time CRC validation**. CRC stands for cyclic redundancy check.

Maintain a log file makes AutoCAD write the contents of the text window to a log file, also an unnecessary feature for most users.

Security Options… is a relatively new feature in AutoCAD that addresses the need to secure or authenticate the drawings. This is not often used, as most AutoCAD drawings are not secret in nature, but the idea is intriguing and worth a look. When you click on the button, the box in Figure 14.6 appears.

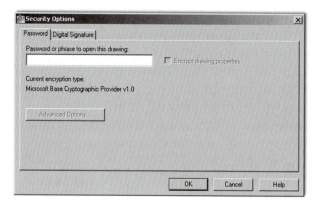

FIGURE 14.6
Security Options.

The first tab allows you to add a password to the drawing. Then the drawing can be opened only by someone who knows the password; although be careful—if the password is forgotten, the drawing is lost. The second tab has to do with a digital signature. This is in itself a lengthy topic and is sometimes used when collaborating with others. In short, a user installs a certificate from a digital security vendor (VeriSign® in this case), which allows for a "digital signature," which in turn prevents changes to the drawing or keeps track of them. This is of course a very simplified version of the concept, but if your company already uses this, then no further explanation is necessary; if not, you will likely not care about this feature anyway.

We do not go over anything in detail in the right-hand column of the Open and Save tab. The File Open category is self-explanatory, and the other settings under the Application Menu, External References (Xrefs), and ObjectARX Applications categories are best left as default. If you want a detailed description of what each selection means, press F1 while you have the Open and Save tab open. Otherwise check to see that your settings reflect what is in Figure 14.5 and let us move on.

Plot and Publish Tab

While there is nothing critically important under the Plot and Publish tab (Figure 14.7), a few features are worth a look. We run through everything with just general comments, while expanding on what proved to be useful over the years and why. Starting from the top left, moving down, we have the following options.

The **Default plot settings for new drawings** category selects the default plotter or adds a new one. However, this should not be done from here but rather from the Windows control panel. When new plotters or printers are added, AutoCAD notices them, as well as what the chosen default is. In the end however, this is of little significance. Recall from Chapter 9 that the printer settings for each drawing are set individually. Therefore, leave everything here as default.

In the **Plot to file category** a drawing can be plotted to a plt (plot) file and saved in this location. While this is generally an outdated legacy command from the old days of plt files being sent (or physically carried on a floppy disk) to plotters, it has found some use today when taking drawings to a printing company (such as Kinko's). The company may not have AutoCAD, or anyone qualified to operate it, so instead they accept the plt files, which are nothing more than ready-to-print code that can be interpreted by any plotter.

The Background processing options category has to do with being able to plot in the background while working on a drawing. With the Plotting option unchecked, you must wait until the plotting is complete before continuing.

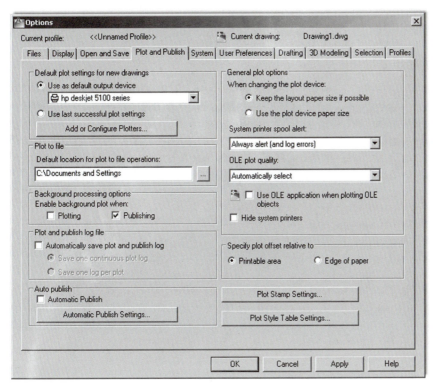

FIGURE 14.7
Plot and Publish tab.

The **Plot and publish log file** category is an option that creates (or turns off) the log feature, which is nothing more than a record of the Job name, Date and time started and completed, Full file path, Selected layout name, Page setup name, Device name, Paper size name, and a few others. It is recommended to disable (uncheck) this feature unless there is a specific reason to create a record of all this information.

The Auto publish category is an option that allows you to automatically create DWF files. DWF stands for design web format and is a topic all unto itself. The basic idea is to generate vector files that can be viewed by others who have the Autodesk Viewer or, if not, then through a browser.

Moving to the upper right of the Plot and Publish tab, we have the **General plot options** category. This has to do with the output quality of OLE (object linked embedded) items, a topic to be discussed in Chapter 16. The OLE plot quality drop-down choices can be adjusted depending on what is embedded. Leave them as default in this case, and all the boxes unchecked.

Leave the **Specify plot offset relative to** category as default.

The Plot Stamp Settings… button at the bottom right is an item of interest. A plot stamp is simply a text string that appears on all output when activated in the Plot dialog box. This text string has information on a variety of parameters, but the ones of most interest include the drawing name and the date and time. This can be useful for internal check plots (obviously not for final output), to show where the drawing is on the company server or to indicate which paper output is the latest based on the date and time. Use the Advanced button in the lower left (Figure 14.8) to further refine the plot stamp.

Leave everything under the **Plot Style Table Settings**… button as default. These settings are addressed elsewhere. If you want additional detailed descriptions of each of the previously discussed options under this tab, press F1 while you have the Plot and Publish tab open. Otherwise, check to see that your settings reflect what is in Figure 14.7 and let us move on.

System Tab

There is not much to discuss under the System tab (Figure 14.9). All settings basically remain as default. A few items are given a brief overview as seen next. We start on the left-hand side again and proceed top to bottom.

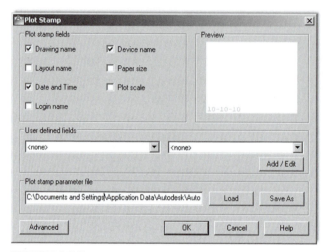

FIGURE 14.8
Plot Stamp.

324

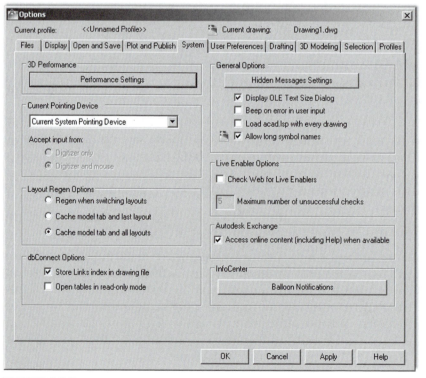

FIGURE 14.9
System tab.

There is nothing to discuss under the **3D Performance** category, as we are working only in the 2D world here.

The Current Pointing Device category allows for setting up digitizer operations, a topic not relevant to most users today.

The **Layout Regen Options** category should be left as default; it refers to regeneration of layouts.

The **dbConnect Options** category should be left as default. The *db* refers to database connectivity, which is also not relevant to most users but may be important in some high-end custom applications, where AutoCAD is linked to a database.

Leave everything in the **General Options** category as default. Here resides perhaps one of the most annoying features of AutoCAD: The "Beep on error in user input" option. This may push a student just starting out and making lots of errors over the edge—not recommended!

The **Live Enabler Options** category tells AutoCAD to check for object enablers. They can be used to display and use custom objects in drawings when the ObjectARX application that created them is unavailable. This is not relevant to most users and may sometimes cause a program stall if there is no internet service to the PC.

Autodesk Exchange is a new feature added to AutoCAD 2012. As first seen in Chapter 1, it is an online repository of useful data, updates, applications, and other info. If your computer is regularly connected to the internet, then leave this option checked.

Finally, we have the **InfoCenter** and the **Balloon Notifications.** These are various information balloons that appear with helpful updates or tidbits of information. Although an advanced user may find them annoying, as a student you should keep them around. Briefly explore the notification button to see what is there.

If you want additional detailed descriptions of each of the previously discussed options under this tab, press F1 while you have the System tab open. Otherwise, check to see that your settings reflect what is in Figure 14.9 and let us move on.

User Preferences Tab

This tab (Figure 14.10) has one very important setting under the **Windows Standard Behavior** category in the upper left. The rest of the categories are not relevant to most users, although we briefly mention them.

The Windows Standard Behavior category concerns speed, which is a topic near and dear to the hearts of most AutoCAD users. Since time equals money, whatever technique accelerates drafting speed is always welcome. One of the more important ones (shortcuts) is discussed later in this chapter. Another important "trick" is to set your mouse buttons to the most likely and needed settings for fast drafting. As such, **Right-click Customization**… is very important. Click on that button and the dialog box in Figure 14.11 appears.

Your mouse settings should be as shown in Figure 14.11, with the last one, Command Mode, set to ENTER. This subtle but powerful setting speeds up your basic drafting by allowing you to instantly finish commands by using the right mouse button instead of having to press Enter on the keyboard. Also set Edit Mode to Repeat Last Command and Default Mode to Shortcut Menu. Nothing else needs to be adjusted. Press Apply & Close.

The **Insertion scale** category controls the scale for inserting blocks; leave the default values. The **Hyperlink** category controls settings of hyperlinks; leave this box checked. We cover hyperlinks in the next chapter. The **Fields** category sets field preferences, which are not relevant to most users; leave as default. Moving to the top right, the **Priority for Coordinate Data Entry** category controls how AutoCAD responds to coordinate data input, also not necessary to change for most users.

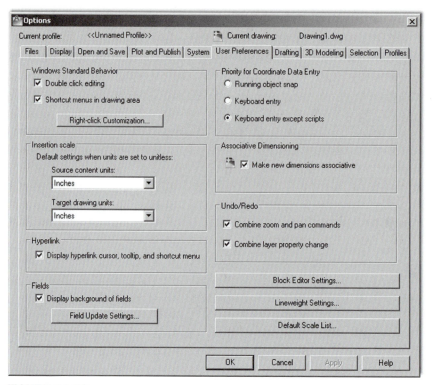

FIGURE 14.10
User Preferences tab.

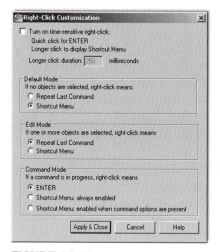

FIGURE 14.11
Right-Click Customization.

Leave the check box under the **Associative Dimensioning** category checked, since all dimensions should be associative (meaning, if the geometry changes, so do the dimensions associated with it). The **Undo/Redo** category controls Undo and Redo for zoom and pan; leave both boxes checked off.

Below these settings are four rectangular buttons that you can press to set certain preferences. They are described next, as a few useful item are there.

The **Block Editor Settings**… button changes the environment of the block editor. You worked with that editor when you first learned dynamic blocks. The **Lineweight Settings**…

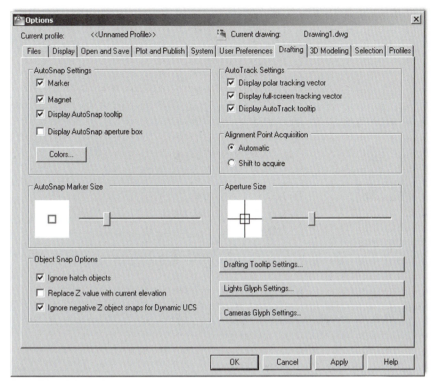

FIGURE 14.12
Drafting tab.

button displays the Lineweight Settings dialog box. These settings can be adjusted when setting up the .ctb files, as described in Chapter 19; they are not needed here. The **Default Scale List**… button displays the Edit Scale List dialog box. Additional scales are rarely needed but can be added here, if necessary.

If you want additional detailed descriptions of each of the previously discussed options under this tab, press F1 while you have the User Preferences tab open. Otherwise, check to see that your settings reflect what is in Figure 14.10 and let us move on.

Drafting Tab

The entire left side of this tab (Figure 14.12) is devoted to snap point markers, those colored geometric objects that represent endpoints, midpoints, and the like. Leave everything in the **AutoSnap Settings** category as default, including Colors, unless you prefer something different. Leave the **AutoSnap Marker Size** as default, unless you have trouble seeing the symbols or find them too big. Also leave everything under the **Object Snap Options** category as default.

Moving to the upper right, we have the **AutoTrack Settings** category. It has to do with AutoCAD's tracking tool OTRACK, which is briefly looked at in Chapter 15 along with other miscellaneous topics. Leave these settings as default for now. Leave the setting under the **Alignment Point Acquisition** category as default.

Aperture Size is perhaps the one item that may occasionally be adjusted to suit the user's taste, but generally keep it the same size as the AutoSnap marker. If the aperture is too big, it may be distracting or may actually be confused with small rectangles on the screen. At the very bottom right, we have three rectangular buttons.

The **Drafting Tooltip Settings**… button features minor nuanced adjustments to drafting tooltips (such as distance or angle displays) and their visibility to the users (the transparency).

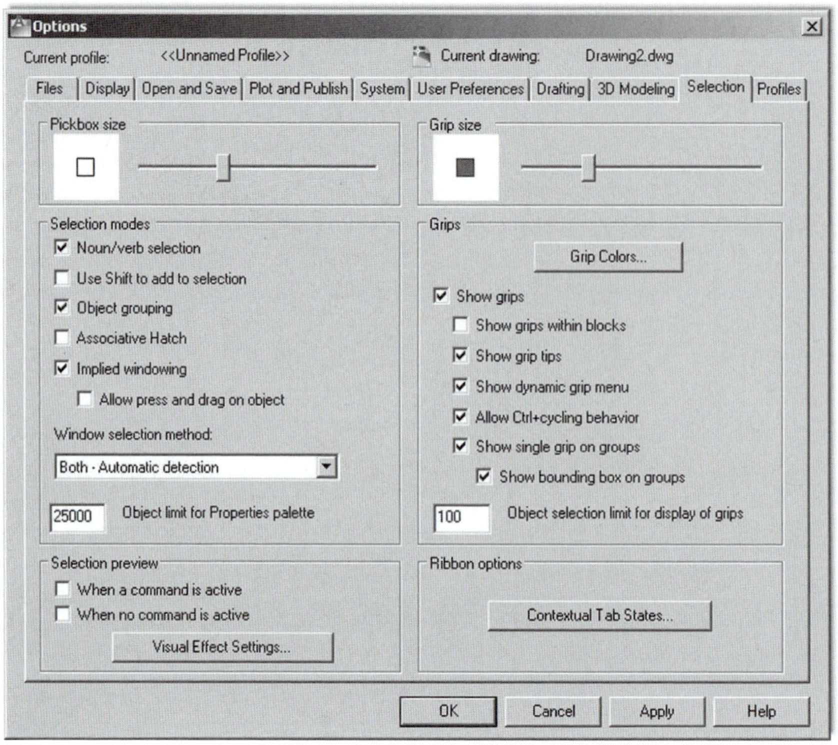

FIGURE 14.13
Selection tab.

Leave all them as default. The **Lights Glyph Settings**… and **Cameras Glyph Settings**… buttons are advanced settings that are not relevant at this point; leave them as default.

If you want additional detailed descriptions of each of the previously discussed options under this tab, press F1 while you have the Drafting tab open. Otherwise, check to see that your settings reflect what is in Figure 14.12 and let us move on.

3D Modeling Tab

Settings and adjustments under this tab are not discussed in this 2D-only level of the book.

Selection Tab

This entire tab (Figure 14.13) is devoted to minor adjustments to your drafting environment, which may or may not be important depending on your personal tastes and style of drafting.

Starting at the top left, the **Pickbox Size** category is simply what the cursor becomes when you are in the process of picking something. Set it too small and you have trouble selecting the geometry. Set it too big and you select more than you wanted. Generally, leave it at the default size shown in Figure 14.13.

The **Selection modes** category is just more adjustments to object selection. Leave them all as default, as seen in Figure 14.13, including the Window selection method choice. Object limit for Properties palette can be left as the ridiculously high default of 25,000. Rarely will you have that many objects selected or even present in your drawing.

The **Selection preview** category is a relatively new feature that causes objects to "light up" when your mouse hovers over them. What exactly constitutes "light up" and a few other items is addressed with the Visual Effect Settings… button. When you press it, you see what is shown in Figure 14.14.

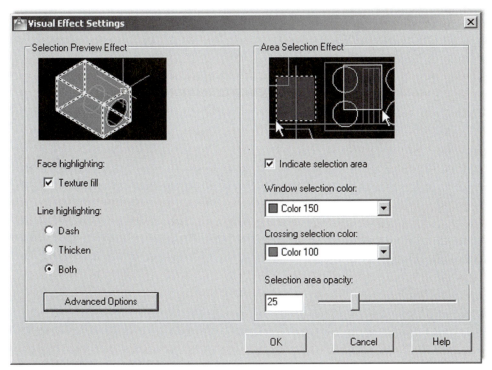

FIGURE 14.14
Visual Effects Settings.

Here you can set whether you want the lighting-up effect to be dashed linework, thicker lines (as if a highlighter magic marker went over them), or both. On the right is an interesting setting that has to do with the internal colors of the familiar Window/Crossing. You can change those colors as well as their intensities. You can also eliminate them, as the colors are technically unnecessary; what is important is whether the selection rectangle is dashed or solid.

Moving to the upper right, the **Grip size** category with the slider adjustment should be self-explanatory. Leave the grips as default size unless you prefer bigger or smaller ones.

The **Grips** category is yet more grip settings for colors and other adjustments. There is really no reason to change grip color. Blue means cold (unselected), red means hot (selected), and hover can be any color (default Color 11 is just fine). Leave the other settings as default; you want the grips enabled but not inside blocks. Leave the maximum amount of grips that can be shown at 100 so as to not overwhelm the drawing if you select more.

Finally, the **Ribbon options** category provides some options regarding contextual tab states. Leave them as default.

Really most of these settings can be left alone. They are included for the benefit of the handful of people who may want to change something and feel that, with a $4000 software program, they ought to be able to alter *anything*.

If you want additional detailed descriptions of each of the previously discussed options under this tab, press F1 while you have the Selection tab open. Otherwise, check to see that your settings reflect what is in Figure 14.13 and let us move on.

Profiles Tab

This tab (Figure 14.15) is simply there to save your personal settings, so if someone else using your computer and AutoCAD prefers other settings (maybe that neon green

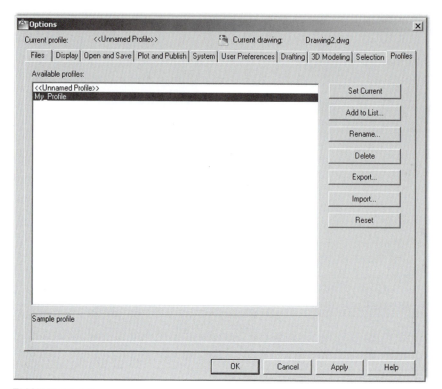

FIGURE 14.15
Profiles tab.

FIGURE 14.16
Add Profile.

background color), that person can set his or her own and save them as well. Simply press the Add to List... button and enter a name for the profile, as seen in Figure 14.16, and finish by press Apply & Close.

The profile is added to the list (as many as you want can be added) and can be restored by selecting it and pressing Set Current. Profiles can be exported (as ARG files), imported, renamed, and deleted as needed. Profiles were more relevant in the days when PCs were more expensive and each one in a company may have had a few users, but some people still use them to avoid office arguments about exactly how AutoCAD "should look" if one were to "borrow" a PC for the day, change AutoCAD to suit his or her taste, and forget to reset it to how the original owner or user likes it.

14.2 SHORTCUTS

Recall that it was mentioned in the previous pages how important speed is to an AutoCAD designer. One of the tools to help you work faster was already mentioned, the mouse Right-click settings under the User Preferences tab. Shortcuts, however, are another important way to accelerate your input. They are useful only if you type. If you are a die-hard Ribbon or toolbar user and abhor the keyboard, then you have no need for them. Many students, however, end up typing quite a lot, even if they also integrate some toolbar and Ribbon use. The reasons are simple; typing is a fast and very direct method to interact with AutoCAD and remains quite popular despite other methods. However, to take advantage of this, you need shortcuts, as typing out the full command defeats the speed advantage.

You may have already been using shortcuts all along. Instead of typing in `arc`, you may have just pressed the letter `a`, instead of `move`, the letter `m`. This is exactly what is meant by shortcuts; it is the process of shortening the typed commands so they can be entered faster. What we want to do here is put all this on solid footing and outline exactly how to set your own shortcuts to make best use of them.

pgp File

Here is a basic primer on shortcuts. They have been a part of AutoCAD since the early days, when typing commands was the best (and in the very beginning the only) way to interact with the software. Abbreviating the commands was the logical next step to speed up typing, and that capability was integrated into AutoCAD as well.

The abbreviations are stored in a file called `acad.pgp`, which resides in the Support folder of your AutoCAD installation. AutoCAD comes with a wide range of preinstalled shortcuts, but the file is easily found and meant to be modified to suit the user's taste and habits. The key here is that you need to have your shortcuts memorized (or it makes no sense to use them in the first place), and of course you need to pick the most efficient abbreviations of the commands. To that end, AutoCAD does not care what you use—it can be numbers or names of your relatives—but obviously you want to use something that makes sense to you and shortens the typing effort. Ultimately, the first letter, first and second letters, or first and last letters are used for all but a few of the commands.

Here are some suggestions for the Modify commands:

Erase: `e`
Move: `m`
Copy: `c`
Rotate: `ro`
Scale: `s`
Trim: `t`
Extend: `ed`
Fillet: `f`
Offset: `o`
Mirror: `mi`

The idea here is to abbreviate the most often used commands in the shortest way possible. If, for example, there is a conflict between two commands that start with the same letter (e.g.: circle and copy), then you determine which command gets used more often and give *that* command fewer letters. Copy is generally used more often, so it would be `c` and circle would be `ci`.

Sometimes, doubling up of letters is done, such as with Array (`aa`), because `a` is taken by arc, and so on. Listed in Figure 14.17 is a sample `acad.pgp` file. It is presented here only for illustration purposes and may or may not match what you may want in yours, although it can be a good starting point to creating your own.

A,	*ARC
AA,	*ARRAY
B,	*BHATCH
BL,	*BLOCK
C,	*COPY
CF,	*CHAMFER
CH,	*CHANGE
CI,	*CIRCLE
D,	*DTEXT
DD,	*DDEDIT
DI,	*DIST
DDC,	*PROPERTIES
DDO,	*DDOSNAP
DDS,	*DDSTYLE
DDU,	*DDUNITS
E,	*ERASE
ED,	*EXTEND
EL	*ELLIPSE
EX	*EXPLODE
F,	*FILLET
I,	*INSERT
L,	*LINE
LA,	*LAYER
LF	*LAYFRZ
LI,	*LIST
LW,	*LAYWALK
M,	*MOVE
MI	*MIRROR
ML,	*MULTILINE
MM	*MATCHPROP
O,	*OFFSET
P,	*PLOT
PE,	*PEDIT
PL,	*PLINE
PP,	*PURGE
Q,	*QSAVE
R,	*REGEN
RO,	*ROTATE
RE,	*RECTANG
S,	*SCALE
SP,	*SPLINE
T,	*TRIM
U,	*UNDO
UU,	*UCSICON
W,	*WBLOCK
XL,	*XLINE
Z,	*ZOOM

FIGURE 14.17
Sample `acad.pgp` file.

Altering the pgp File

So, how do you find and alter the pgp file to suit your tastes? Prior to AutoCAD 2004, you had to look for it in the C:\Program Files\AutoCAD 2002\Support folder (for ACAD 2002 in this example). You no longer have to do this and can access the file automatically through the drop-down menus as follows: **Tools→Customize→Edit Program Parameter (acad. pgp)**. The pgp file then appears on your screen (opened as a Notepad file, which is just a simple ASCII text editor, if you never used it before). Scroll down the file until you see the rather lengthy built-in listing of shortcuts. At this point, you may do one of three things:

1. Leave the entire file as it is, and use it without modification.

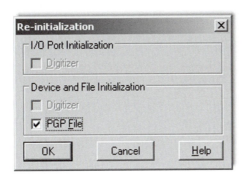

FIGURE 14.18
Reinitialization of the pgp file.

2. Erase or change the shortcuts for commands that you either do not use or would like to modify.
3. Erase the entire list and start fresh, entering the shortcuts as you would like them to be, one by one. If this is the case, be sure to follow the format given where you enter the shortcut, then a comma, then a few spaces, and finally the command preceded by a *, as seen in Figure 14.17.

When done, save the acad.pgp file and close Notepad. At this point, for the changes to take effect, you can either restart AutoCAD or (if you are in the middle of a drawing and do not want to do that) type in reinit and press Enter. The dialog box in Figure 14.18 appears.

Check off the PGP File box and press OK. The pgp file is ready to go, and you can use it right away. To add or subtract commands, just repeat the steps at the beginning of this section, save the pgp file, and reinitialize.

It is strongly recommended that you at least try to modify and use the pgp file. It is almost always faster to type the shortcuts than to click on icons. You can eventually learn to type quickly and without even glancing at the keyboard (similar to a good secretary or court reporter), assuming you do not already type fast, as many students do. Another advantage is that, by typing commands, you free up space on your screen where toolbars normally go.

14.3 CUSTOMIZE USER INTERFACE

The Customize User Interface dialog box (referred to from here on as just the CUI) is the next thing we examine this chapter. The CUI has to do with changing the look and functionality of menus, toolbars, and other items and, as such, can be of importance to a small set of users who may need to do this for aesthetic or functional reasons.

The CUI is an advanced and somewhat mysterious concept and one not taken advantage of by most users, as either they do not know much about it or do not wish to change their toolbars or menus. A few though find this customization ability of great interest and spend time customizing the look of AutoCAD to the point of almost having AutoCAD look like another program entirely.

We take the middle ground here and present some CUI basics, demonstrating the creation of a new custom toolbar and leaving the student with some CUI ideas to pursue further if he or she is interested in doing so.

Let us introduce the basic CUI as it opens in default mode. Type in cui, press Enter (or alternatively use the Ribbon's **Manage tab→User Interface**), and after a few seconds, the

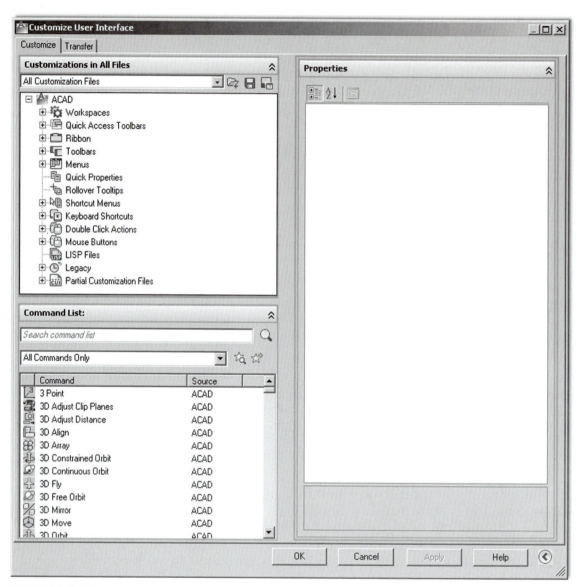

FIGURE 14.19
The Customize User Interface (CUI) dialog box.

dialog box in Figure 14.19 opens up. Press the expansion button (small arrow, bottom right) to make it full size if it does not do this automatically.

In general, the CUI is your window into customizing virtually anything in AutoCAD that you can push, select, or type in. The full list of candidates is in the upper left of the CUI box: Workspaces, Toolbars, Menus, Dashboard Panels, and all the other choices. You may need to minimize the Workspace menu to see all the rest of them. Just to get your feet wet, let us create a custom toolbar that has only the buttons you want and none of the ones you never use. This is good idea for 3D, where toolbar use is more of a necessity than in 2D.

Step 1. In the upper left-hand corner, find the Toolbars expandable menu and right-click, selecting New Toolbar. Give your new toolbar a name in the name field (My_New_Toolbar, in this example). Once it is done, you can click it and a Toolbar Preview Properties appears on the right of the CUI window, along with an expanded Properties palette.

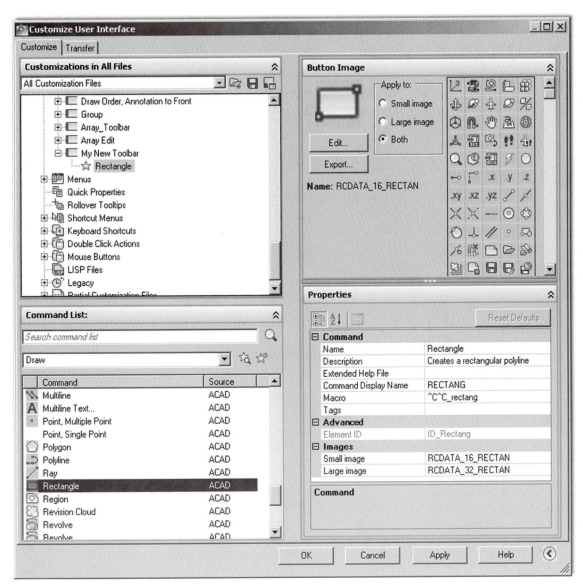

FIGURE 14.20
MyToolbar, Add Rectangle.

Step 2. On the bottom left begin selecting the commands you want your new toolbar to have (you can filter the commands to find them easier), and drag them into the new toolbar name on the upper left. As soon as you begin doing this, a toolbar preview and Button Image category appear on the right-hand side, as seen in Figure 14.20, when "rectangle" was added as the first command.

Interestingly, you can even modify the image itself by pressing Edit… in the Button Image category at the upper right, which takes you into the dialog box shown in Figure 14.21. Here, you can use pixel by pixel editing tools to edit the image. You have some additional tools such as a grid and a Preview window to see what you have created. Finally, you can save your work and return to the CUI. You can draw some rather creative-looking buttons if you so desire (a few practical jokes from years gone by do come to mind here). Admittedly, this is not done very often, as most users have better things to do with AutoCAD.

Finally, after adding a few more commands (Revcloud, Circle, and Spline) and pressing Apply and OK, the new toolbar is born, as shown in Figure 14.22; this is certainly a good trick if you like toolbars and want to see only what you use on yours.

FIGURE 14.21
Button Editor.

FIGURE 14.22
The new MyToolbar.

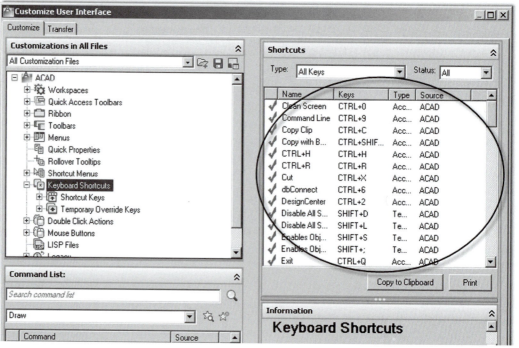

FIGURE 14.23
Accelerator keys.

This is of course just the tip of the iceberg. You can modify all the rest of the choices. Shortcut Keys under the Keyboard Shortcuts are quite useful. This refers to shortcuts such as Ctrl + C for Copy Clip (also known as *Accelerator keys*). You can set up your keyboard to accelerate many commands. Just some of the preset ones are shown in Figure 14.23 at the upper right of the CUI dialog box.

It may be worth your time to look through the Help files and investigate the CUI further; it really is a bottomless pit of customization, although be careful not to change something just for the sake of change but only if you feel it may help your productivity.

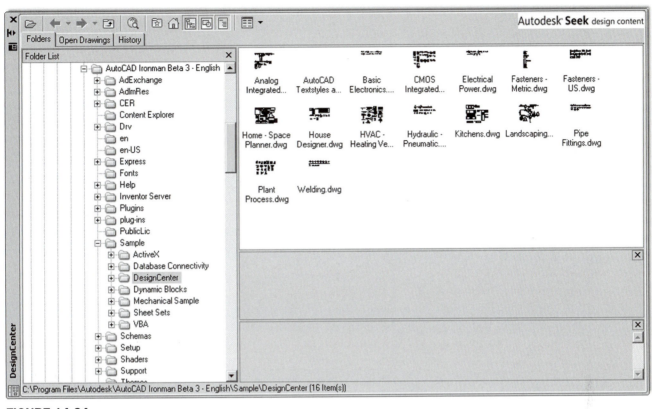

FIGURE 14.24
AutoCAD 2012 Design Center.

14.4 DESIGN CENTER

You can probably guess at one of the purposes of the Design Center (DC) as soon as you open it via the cascading menus **Tools→Palettes→DesignCenter** (also via typing in dc and pressing Enter, or Ctrl+2). Be sure to dock it to the left side of the screen (if it is floating) and expand all the sections for good visibility, as shown in Figure 14.24.

What you see in the far left window is basically the Windows Explorer used regularly to view files (such as when you click on My Computer). This is a big clue to the first of two functions of the DC, which are to view the contents of one drawing on the left and, as we soon see, transfer parts of that drawing into the active drawing on the right. Let us give this a try using the process outlined next.

By default the DC opens the //Sample/DesignCenter file folder on the left, revealing a collection of sample drawings. You can of course browse for and choose an actual work drawing you created (that is the whole point, after all), but for our purposes right now, one of the sample drawings is just fine. Go ahead and pick the HVAC one. Clicking on the plus sign next to the file name and expanding the folder reveals a collection of useful data, such as Blocks, Dimstyles, Layers, and Layouts. Now click on a category, such as Blocks. A collection of blocks present in this drawing are revealed in the middle window, next to the file tree. Click and drag any block from the choices given on the left into the drawing space on the right, as seen in Figure 14.25.

In the same manner, you can transfer layers and anything else lifted from one drawing into another. This is very useful when you need only a few items as opposed to everything and is a great visual alternative to the usual Copy/Paste (Ctrl + C and Ctrl + P).

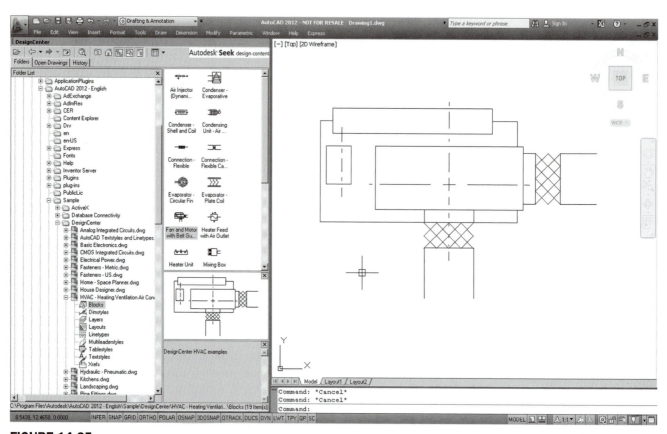

FIGURE 14.25
Design Center transfer of blocks (Ribbon and toolbars are turned off).

Autodesk **Seek** design content

FIGURE 14.26
Autodesk Seek.

The next two tabs of the DC are not all that significant. Open Drawings simply looks at the content of the current active drawing, and History is exactly that, the history of all the open drawings thus far in the drawing session. There is however another function to the Design Center that we have yet to explore.

This function is the ability of the Design Center to present to you an extensive array of drawing blocks for you to use as generic items in your floor plan. As of AutoCAD 2011, you can access all this online, as opposed to clicking another tab that leads you to blocks built into AutoCAD and installed on your computer. This of course greatly expands the available library and allows Autodesk to update it in real time. This all continues with AutoCAD 2012.

Assuming you have internet access on your computer, you can access Seek by clicking on the Autodesk Seek design content logo in the upper right of the Design Center (Figure 14.26).

After a moment or two, this leads you to Autodesk's online database of product specs and design files. If you want to access the site without using the Design Center, here is the web address: http://seek.autodesk.com/. You then see the screen in Figure 14.27.

As you can see from inspecting the website, you have quite a lot of categories from which to choose. Pick one (for example, refrigerators) to see what is available. You can then choose by

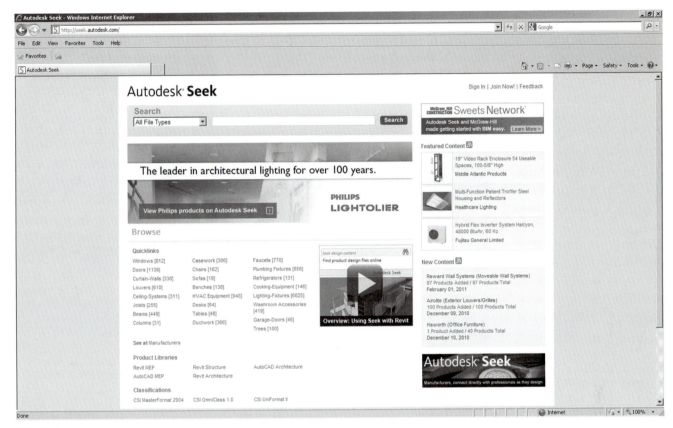

FIGURE 14.27
Autodesk Seek web page.

generic or manufacturer. If you go to Browse, then Quicklinks, then pick refrigerators, you get the screen shown in Figure 14.28.

You can then choose a file type on the right (many are available, not just AutoCAD drawing) and actually download the file. Explore this extensive website further; it is truly a great resource.

14.5 EXPRESS TOOLS

Express Tools are a set of commands grouped together under the Express drop-down menus. These commands add extra functionality to AutoCAD, and while some of them are redundant or not that useful, others became so indispensable that they were later incorporated into regular AutoCAD menus and command sets. It really is a mixed bag, and our goal here is to highlight only the most useful ones, leaving the rest for you to explore if you so choose.

Express Tools were originally called *Bonus Tools* and got their start around Release 14 in the mid-1990s as a collection of useful routines and small programs. Where they originally came from is a matter of debate, but Autodesk likely acquired them from other third-party developers and possibly regular programming-savvy users who felt that AutoCAD just did not have that perfect command and developed their own as a "plug-in."

However it happened, these tools found their way to Autodesk and the company began to offer them as an "extra" to AutoCAD (not technically part of the core software). Sometimes, these tools were offered for sale; sometimes, they were free to VIP subscribers, and more recently, they became free to all customers, although they often still needed to be installed in a separate and deliberate step while installing AutoCAD itself.

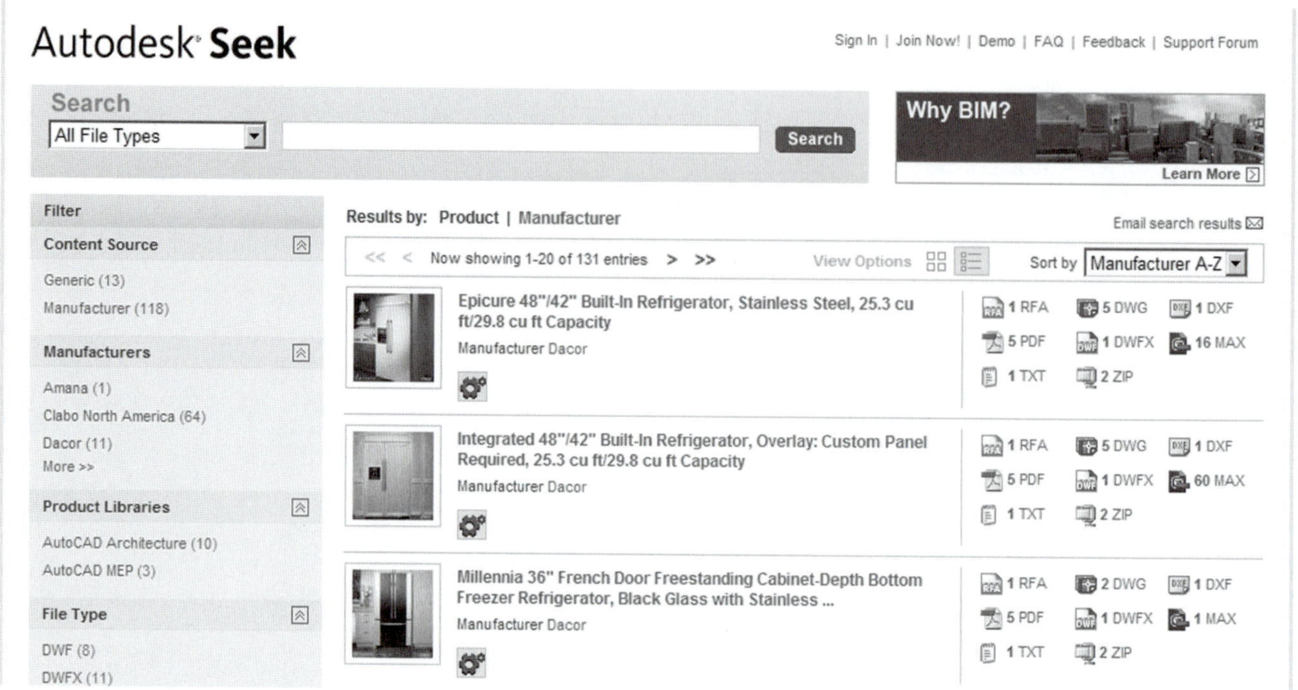

FIGURE 14.28
First few choices in the refrigerator category.

340

FIGURE 14.29
Express Tools cascading menu.

The first step in learning Express Tools is to check that you have them (you should with AutoCAD 2012), and we proceed from here on with the assumption that all is well and you are ready to go over them. Figure 14.29 shows what the topmost listing looks like when dropped down from the top cascading menu. This is the easiest way to access these tools, although there are a few toolbars as well.

Next is a rundown of the most useful Express Tools. If you would like info on those commands that are not specifically covered, refer to the Express Tools Help files for detailed description of each one. The Help files for the Express Tools are separate from the regular AutoCAD Help files. They are accessed via Help at the bottom of the Express Tools cascading

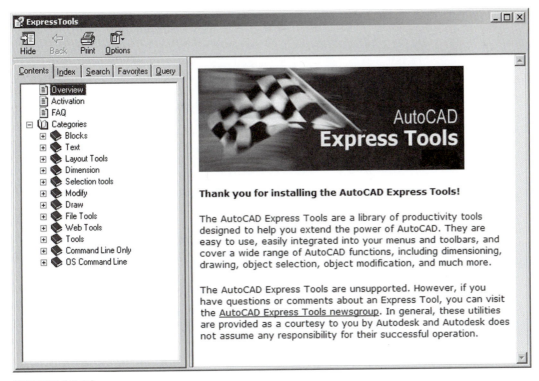

FIGURE 14.30
Express Tools Help files.

menu and shown in Figure 14.30. Sometimes, you may find the functionality of an express tool already a part of AutoCAD in some other way, shape, or form. Occasionally, there are also typed commands for these tools as well. These will be discussed as they come up.

Blocks

The first category from the top is Blocks (Figure 14.31).

Here, you find ten menu choices, all having to do with blocks, nested objects, shapes, and attributes. Do explore all of them, but pay special attention to the ones described in more detail next.

- **Copy Nested Objects.** This command is useful for plucking out nested items, such as those embedded in xrefs (Chapter 17 topic) or blocks. These objects are otherwise stuck inside and not easily accessible for copying. The command line equivalent for this is ncopy and does the same thing. Additional commands below this one allow you to trim and extend to these nested objects as well.
- **Explode Attributes to Text.** When you explode attributes (Chapter 19 topic), they do not turn into regular text; instead they take the shape of attribute definitions. This useful tool forces them to revert to ordinary text so you need not retype anything.

Text

The second category from the top is Text (Figure 14.32).

Here, you find 12 menu choices, all having to do with text and mtext. Do explore all of them, but pay special attention to the ones described in more detail next.

- **Convert Text to Mtext.** This command pretty much does what it advertises; it converts lines of text to a paragraph of mtext. It is faster and more efficient than copying the text

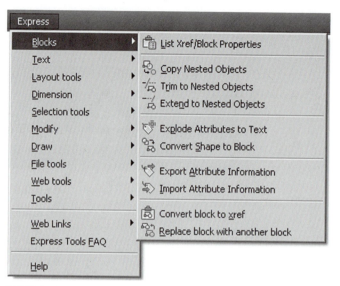

FIGURE 14.31
Express Tools (Blocks).

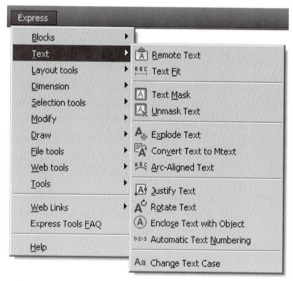

FIGURE 14.32
Express Tools (Text).

in one line at a time or redoing mtext completely. Simply activate the command, pick the lines of text you want to convert, and press Enter.

- **Arc-Aligned Text.** This command is used to tackle a problem that has never been addressed by AutoCAD prior to Express Tools' existence: how to create text that stretches around an arc. The procedure is to draw an arc and select this tool (or just type in `arctext`), then select the arc. The dialog box seen in Figure 14.33 appears.

Enter the actual text and adjust parameters such as fonts, size, and anything else you want (they all have explanatory tooltips if you float your mouse over them), and finally press OK when done. Your text should look similar to what is shown in Figure 14.34.

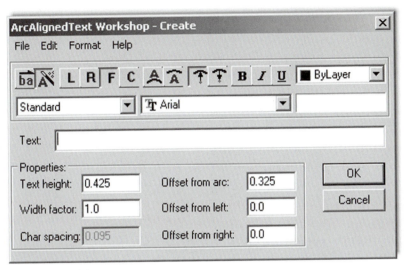

FIGURE 14.33
Arc-Aligned Text.

FIGURE 14.34
Arc-Aligned Text sample.

If the text is too squeezed or spread out, adjust the sizing. Note that the text can be over or under the arc. Also remember that you need to type in `arctext` to go back and edit it. Regular `ddedit` or double-clicking does not work here.

- **Enclose Text with Object.** This command encloses your text with rectangles, slots, or circles. Run through the menu options, setting offsets, shapes, and sizes.

Layout Tools

The third category from the top is Layout tools (Figure 14.35). Here, you find four menu choices, all having to do with Paper Space viewports. These commands are rarely used or are duplicates of existing commands, but do take a few minutes to briefly explore them. We do not go into further detail.

Dimension

The fourth category from the top is Dimension (Figure 14.36). Here, you find six relatively obscure dimstyle and leader tools. Take a few minutes to explore them, but we do not go into further detail.

Selection Tools

The fifth category from the top is Selection tools (Figure 14.37). Here, you find two relatively obscure commands that have to do with selection sets. We do not go into further detail here.

FIGURE 14.35
Express Tools (Layout tools).

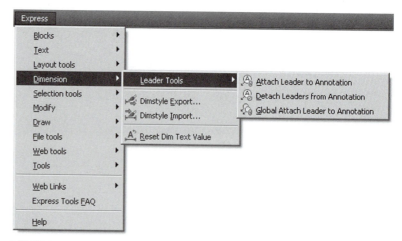

FIGURE 14.36
Express Tools (Dimension).

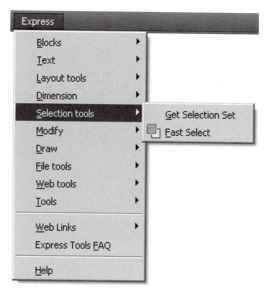

FIGURE 14.37
Express Tools (Selection tools).

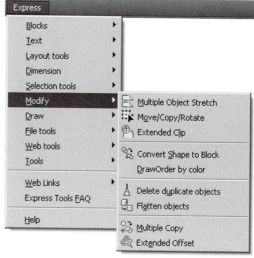

FIGURE 14.38
Express Tools (Modify).

Modify

The sixth category from the top is Modify (Figure 14.38). Here you find nine menu choices having to do with various move, copy, rotate, stretch, and other command variations. Several of them duplicate other commands (such as Multiple copy) or are of limited use (such as Move/Copy/Rotate). Do however explore all of them, but pay special attention to the ones described in more detail next.

- **Delete duplicate objects.** This command is basically the overkill command, which we cover in the next chapter.
- **Flatten objects.** If you go on to study 3D, this is a great tool for flattening (converting) a 3D image into a 2D one.

Draw

The seventh category from the top is called <u>D</u>raw (Figure 14.39). Here you will find just two menu choices, <u>B</u>reak-line Symbol and <u>S</u>uper Hatch…. We will cover both next.

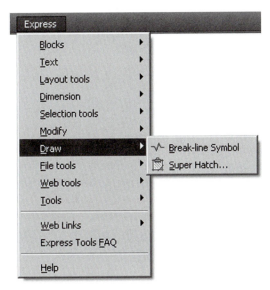

FIGURE 14.39
Express Tools (<u>D</u>raw).

- **Break-line Symbol.** Students have found this command one of the most useful among the Express Tools, because everyone it seems would like to avoid drawing this symbol from scratch. Here the process of making one is as easy as clicking two points and indicating a midpoint for the actual break line. Try it by selecting the command and specifying a first and second point along a line; then, using the MIDpoint OSNAP, place the break symbol in the middle. The Size option allows you to alter the break size.
- **Super Hatch**…. This command creates hatch patterns out of ordinary images by basically copying them many times over to fill an area. This looks interesting but really slows down the computer, as graphics are copied over and over. Give it a try but avoid selecting an overly complex image; you may have to wait a while then.

File tools

The eighth category from the top is <u>F</u>ile tools (Figure 14.40). Here, you find nine tools that deal mostly with file management. Much of what is here can be done manually in other ways, such as the numerous tools for opening, closing, and saving multiple drawings. We go over one tool in particular and do not go into depth with the rest.

- **Convert PLT to DWG.** This converts plot files to useable drawing files. PLT files are printer ready data files that allow a printer to print an AutoCAD drawing without AutoCAD being loaded. They are rarely used anymore, unless you take a file to a professional printer. If you find any stray PLT files lying around from the old days, you can use this express tool to convert them back to see what you've got.

Web Tools

The ninth category from the top is <u>W</u>eb tools (Figure 14.41). Here, you find three URL/hyperlink-related tools. AutoCAD hyperlinks have not yet been discussed, so we do not go into further detail here. You see a bit more on these tools when hyperlinks are introduced in the next chapter.

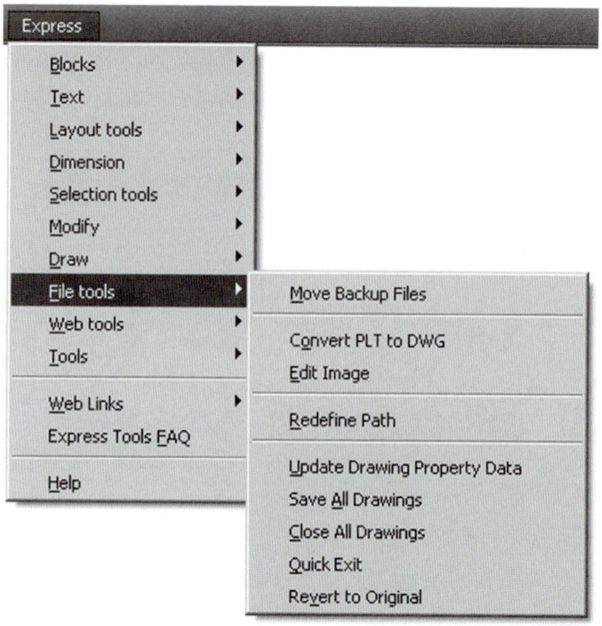

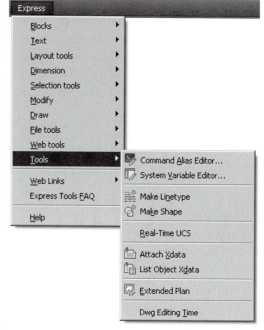

FIGURE 14.40
Express Tools (File tools).

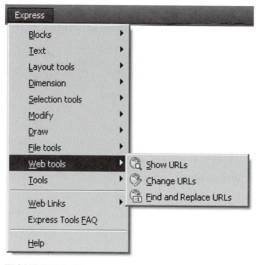

FIGURE 14.41
Express Tools (Web tools).

FIGURE 14.42
Express Tools (Tools).

Tools

The tenth and final category from the top is Tools (Figure 14.42). Here, you find nine tools that serve a variety of functions, from editing aliases and variables (outdated—the pgp file and the CUI take care of those) to making shapes and linetypes. Look over the commands on your own, but we do not go into further detail here.

Layer Express Tools

The Express Tools just covered are the most useful ones to the average designer and the ones typically used the most in many different professions. As you may have noticed, there is some overlap and redundancy between what is still in the Express Tools and what has been incorporated into regular AutoCAD. Autodesk does not like to throw much away.

The Express Tools are also missing something very important, perhaps the most important tools of all: the Layer tools. They were once part of the Express Tools gang, but their importance has been acknowledged by an upgrade to permanent status within regular AutoCAD menus. Although there are a bunch of them (many are redundant), we focus on the primary three that are of significant importance:

- Layer Freeze (command line: `layfrz`).
- Layer Isolate (command line: `layiso`).
- Layer Walk (command line: `laywalk`).

Besides typing, these tools are also accessible via the drop-down menus under **Format→Layer tools**, as seen in Figure 14.43.

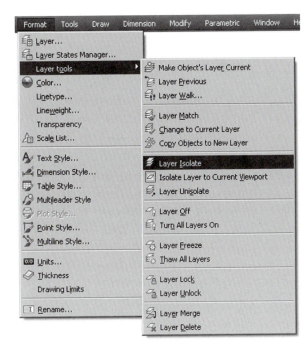

FIGURE 14.43
Layer tools.

LAYER FREEZE

In a complex drawing, layers may number in the dozens or even hundreds. Any tool that can simplify dealing with them is a welcome addition. Layer Freeze allows you to freeze a layer merely by selecting it. This bypasses the tedious process of finding out what layer an object is on (if not known), then freezing it manually via the LA dialog box or the toolbar. Simply type in `layfrz` or select **Format→Layer tools→Layer Freeze**.

- AutoCAD says: `Select an object on the layer to be frozen or [Settings/Undo]:`

Pick an object that represents the layer you want frozen and it disappears. This can be repeated as often as needed and is a good way to clear a lot of junk from a busy, dense drawing.

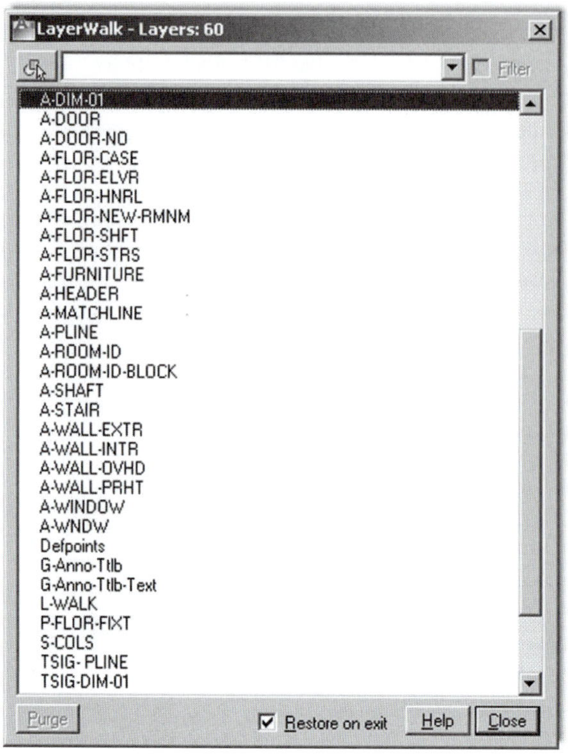

FIGURE 14.44
Layer Walk.

LAYER ISOLATE

Layer Isolate allows you to isolate a layer merely by selecting it. The command turns off (not freezes in this case) all the other layers, leaving the selected one by itself. This bypasses the tedious process of doing it manually. Simply type in layiso or select **Format→Layer tools→Layer Isolate**.

- AutoCAD says:
  ```
  Current setting: Hide layers, Viewports = Vpfreeze
  Select objects on the layer(s) to be isolated or [Settings]:
  ```

Pick an object that represents the layer you want isolated, press Enter, and all the others disappear. It is another good way to clear a dense drawing.

LAYER WALK

This is perhaps the most interesting of all the layer controls and a good way to shuffle through an entire drawing layer by layer. Layer Walk presents a list of layers and isolates one of them. Then, you use the arrow keys on your keyboard to select your way up or down the list. AutoCAD turns layers on or off in succession, allowing you to inspect each layer, one at a time (walk through them). To try this, open up any file that has a number of layers. Then, type in laywalk and press Enter or select **Format→Layer tools→Layer Walk**....

A dialog box appears, listing all the layers visible from layer 0 until the last one, as shown on a sample file in Figure 14.44. Click on layer 0; it highlights that layer and all the rest disappear. Use the arrow keys to cycle down through each layer, and notice how each one appears and disappears as its turn comes up in the listing.

This is a great tool to review all the layering on an unfamiliar drawing as well as one of your own to ensure accuracy. Notice that, along the way, many AutoCAD designers let their

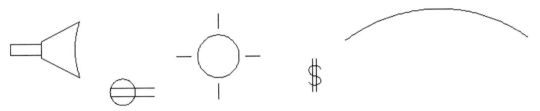

FIGURE 14.45
Various electrical symbols.

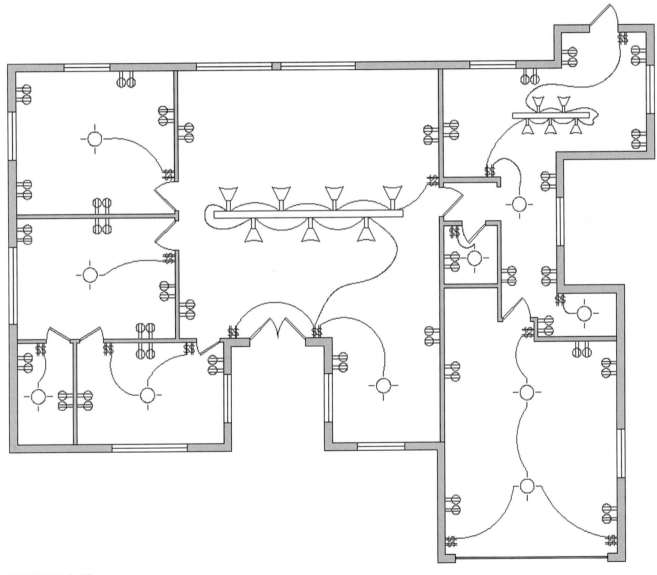

349

FIGURE 14.46
Electrical layout.

layering accuracy slip as the project gains complexity, and by the end, layers are sloppy, with objects residing where they are not supposed to be.

After inspecting all the layers, you can stop on any one of them for editing; if the Restore on exit selection box is checked, then all the layers come back on when you press Close.

Most of the other layer commands are redundant, as you can already perform these functions in other familiar ways (via the LA dialog box or toolbar), so we do not discuss them further, but review them at your own pace.

LEVEL 2 DRAWING PROJECT (4 OF 10): ARCHITECTURAL FLOOR PLAN

For Part 4 of the drawing project, you add in electrical wiring, switches, lights, panels, and outlets. This design is similar in principle, if not in complexity, to a real electrical design.

Step 1. You need the following layers (with suggested colors) for the electrical pieces. The pieces are monochrome in Figures 14.45 and 14.46.
  ○ E-Light (Green).
  ○ E-Outlet (Red).
  ○ E-Panel (White).
  ○ E-Switch (Magenta).
  ○ E-Wiring (Blue).
Step 2. Create the electrical symbols as shown next in Figure 14.45. Make sure the layers and colors are appropriate. Finally make a block out of each symbol, so you can just copy and rotate them into position. With switches and outlets, there are numerous types (dimmer, three-way, duplex outlet, GFI, etc.), so this only scratches the surface.
Step 3. Lay out the electrical symbols as seen in the floor plan in Figure 14.46. Some of the wiring runs are splines, and some are arcs.

SUMMARY

You should understand and know how to use the following concepts and commands before moving on to Chapter 15:

- Files tab
- Display tab
- Open and Save tab
- Plot and Publish tab
- System tab
- User Preferences tab
- Drafting tab
- 3D Modeling tab
- Selection tab
- Profiles tab
- Shortcuts
  ○ pgp file
- CUI
- Design Center
- Express Tools

REVIEW QUESTIONS

Answer the following based on what you learned in Chapter 14:

1. What is important under the Files tab?
2. What is important under the Display tab?
3. What is important under the Open and Save tab?
4. What is important under the Plot and Publish tab?
5. What is important under the System tab?

6. What is important under the User Preferences tab?
7. What is important under the Drafting tab?
8. What is important under the Selection tab?
9. What is important under the Profiles tab?
10. What is the `pgp` file? How do you access it?
11. Describe the overall purpose of the CUI.
12. Describe the purpose of the Design Center.
13. Describe the important Express Tools that are covered.

EXERCISES

1. Based on what you learned about the Options dialog box, set up your own profile. Be sure to include or at least consider all the variables and settings discussed. Save the profile as Profile_Your_Name. (Difficulty level: Easy; Time to completion: <10 minutes)
2. Based on what you learned about the `pgp` file, create your own based on whatever abbreviations you prefer. Be sure to utilize the commands shown in Figure 14.17 as well as add any of your own that you may want included. (Difficulty level: Easy; Time to completion: 20 minutes)
3. Based on what you learned about CUI, create your own toolbar that features the following command icons:
   - Line
   - Circle
   - Rectangle
   - Move
   - Copy
   - Rotate
   - Offset

   (Difficulty level: Easy; Time to completion: 15 minutes)
4. Review the Express Tools. Practice creating another arc-aligned text and break line. Open up any multilayer file and review the Layer tools: `layfrz`, `layiso`, and `laywalk`. (Difficulty level: Easy; Time to completion: 20 minutes)

Advanced Design and File Management Tools

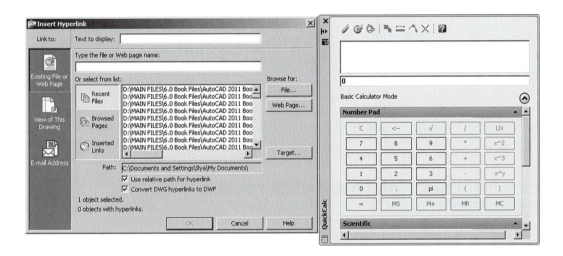

LEARNING OBJECTIVES

In this chapter, we continue to introduce a variety of advanced tools and discuss the following:

- Align
- Audit and Recover
- Blend
- Break and Join
- CAD Standards
- Calculator
- Defpoints
- Divide and Point Style
- Donut
- Draw Order
- eTransmit
- Filter
- Hyperlink

- Lengthen
- Object Tracking (OTRACK)
- Overkill
- Point and Node
- Publish
- Raster
- Revcloud
- Sheet sets
- Selection methods
- Stretch
- System variables
- Tables
- Tool palette
- UCS and crosshair rotation
- Window tiling
- Wipeout

By the end of the chapter, you will add significantly to your command repertoire.

Estimated time for completion of chapter: 3 hours.

15.1 INTRODUCTION TO ADVANCED DESIGN AND FILE MANAGEMENT TOOLS

This is easily the most diverse and unique chapter you will read in the book. Here, we really explore the "nooks and crannies" of AutoCAD and cover a wide range of very diverse commands. The reason these particular commands are grouped here is because this text was written from the start to simplify and isolate core topics and essential ideas, without cluttering the main themes with additional AutoCAD commands and concepts. Several more "big ideas" are left, such as xrefs, attributes, and isometric. Those are addressed in upcoming chapters. However, that leaves out a great deal of useful concepts and commands that did not quite fit in anywhere else.

These bite-sized chunks of information form the basis of this chapter. Many of these topics are just random commands, standing alone, not part of any other major concept, while some may be part of a larger family, but were not included earlier to avoid getting distracted from a much more important theme. All are useful to one extent or another and are necessary for a well-rounded knowledge of AutoCAD.

Obviously this is not an exhaustive list; some items have been left out due to space limitations, but for the most part the important ones are here. The list is in alphabetical order. Go through it at your own pace, noting how each concept or command can help you in your design work. Some are whimsical, some are rather odd, and surely one or two you will find you cannot do without!

15.2 ALIGN

This command does exactly what it advertises, aligns one object to another. In Figure 15.1, we have a rectangle drawn and rotated to some random angle and below it a straight line. While one can always measure the angle of the rectangle relative to the line and use the rotate command, there is a much easier and accurate way to align the rectangle to the line (or vice versa).

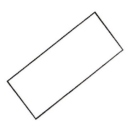

FIGURE 15.1
Rectangle to be aligned with
line.

Step 1. Type in `align` and press Enter. You can also use the cascading menus **Modify→3D Operations→Align**. There are no toolbar or Ribbon equivalents.
- ○ AutoCAD says: `Select objects:`
Step 2. Select the object to be moved into alignment (the rectangle). Press Enter.
- ○ AutoCAD says: `Specify first source point:`
Step 3. Pick the lower left corner of the rectangle using the ENDpoint OSNAP (point 1 in Figure 15.2).
- ○ AutoCAD says: `Specify first destination point:`
Step 4. Pick the left point of the line using the ENDpoint OSNAP (point 2 in Figure 15.2).
- ○ AutoCAD says: `Specify second source point:`
Step 5. Pick the lower right corner of the rectangle using the endpoint OSNAP (point 3 in Figure 15.2).
- ○ AutoCAD says: `Specify second destination point:`
Step 6. Pick the right point of the line using the ENDpoint OSNAP (point 4 in Figure 15.2).
- ○ AutoCAD says: `Specify third source point or <continue>:`
Step 7. You have no need for a third set of points, so press Enter.
- ○ AutoCAD says: `Scale objects based on alignment points? [Yes/No] <N>:`

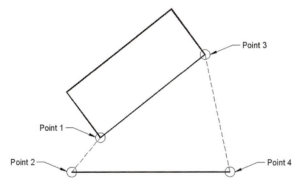

FIGURE 15.2
Alignment points.

There is something interesting you can do here. Not only can you align the rectangle to the line, but you can also scale it up to the line if needed by selecting `Yes`. We do not try this now but go back after learning the basic align command and try this variation of it. For now, just press Enter to accept the default `No` option. The rectangle is aligned to the line, as seen in Figure 15.3.

FIGURE 15.3
Alignment, final result.

15.3 AUDIT AND RECOVER

Although rare, AutoCAD does sometimes lock up and crash (usually a PC hardware or memory issue but can also be software related). In the event of a crash, the drawing is automatically saved but may be corrupted when you try to open it up again. Sometimes, drawings just cause AutoCAD to behave oddly for no specific reason (i.e., commands do not work as expected or just fail). In either case, the drawing needs to be audited or recovered.

Auditing is the lower-level check. AutoCAD evaluates the integrity of the drawing and corrects *some* errors. You simply type in `audit` and press Enter; alternatively use the cascading menus **Application menu→Drawing Utility→Audit**. There is no toolbar or Ribbon equivalent. AutoCAD asks about fixing any detected errors (Yes), cycles through drawing entities, and in a few seconds generates a log similar to what you see next (assuming nothing wrong was found or fixed, as seen here):

```
Auditing Header
Auditing Tables
Auditing Entities Pass 1
Pass 1 100 objects audited
Auditing Entities Pass 2
Pass 2 100 objects audited
Auditing Blocks
1 Blocks audited
Total errors found 0 fixed 0
Erased 0 objects
```

Of course, if something is found, fixes are reported. Sometimes, however, audit is not enough and you must recover the drawing. This is not as drastic as it sounds. The recover command is powerful and almost always brings the drawing back as a matter of routine. I can count on one hand the number of drawings that were lost forever due to corruption in 14 years of CAD design. And even in those cases, going to the previous night's backups restored the drawing, minus a few hours of lost work.

To use this command, open a blank drawing file, type in `recover`, and press Enter. Alternatively, you can use the cascading menus **File→Drawing Utilities→Recover**. There is no toolbar or Ribbon equivalent. You are prompted to browse and find the drawing. Once you select it and press Open, recover runs and generates its own log, as seen next. This particular example was also generated with a clean file. If there were items to fix, recover would have shown the actions taken.

```
Drawing recovery.
Drawing recovery log.
Scanning completed.
Validating objects in the handle table.
Valid objects 755 Invalid objects 0
Validating objects completed.
Salvaged database from drawing.
Auditing Header
Auditing Tables
Auditing Entities Pass 1
Pass 1 900 objects audited
Auditing Entities Pass 2
Pass 2 900 objects audited
Auditing Blocks
19 Blocks audited
Total errors found 0 fixed 0
Erased 0 objects
```

Following the log, you see the message box shown in Figure 15.4.

FIGURE 15.4
Recover message.

If something were indeed wrong (and then fixed), the drawing would have just opened as normal. Immediately save it.

15.4 BLEND

The blend command is a new one, just introduced in AutoCAD 2012. It allows you to join, or blend together, linework that is otherwise separate. This linework would typically be straight lines or curves, such as arcs or splines. To a lesser extent, some shapes (such as rectangles) can also be blended together, although that does not really yield useful results. In all cases, what the blend command does is create a spline between these shapes, curves, or lines. The spline shape can then be modified if need. To try it out, create two arcs as seen in the top part of Figure 15.5 and then follow the steps shown next.

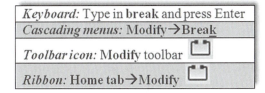

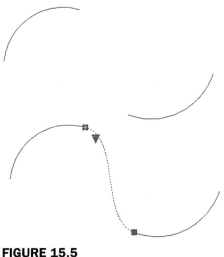

FIGURE 15.5
Blend.

Step 1. Start up the blend command via any of the preceding methods.
- AutoCAD says: `Continuity = Tangent`
  `Select first object or [CONtinuity]:`

Step 2. Pick the first arc, close to where the blend will start.
- AutoCAD says: `Select second object:`

Step 3. Pick the second arc, close to where the blend will end. You see a live preview of the new spline just as you are about to click. A new spline then is created to connect the arcs.

The bottom part of Figure 15.5 shows the results. The new blend has been selected, and the grip points and a down arrow are visible. The down arrow reveals a menu if clicked. This menu lets you choose between showing control vertices or fit points, which presents you with different methods for altering the blend. Experiment with both to see the effects.

15.5 BREAK AND JOIN

The break command can take a geometric object and put a break in it (remove a section, in other words) or, if applied at an intersection with another object, break that object at the intersection without actually removing any material. The join command is the exact opposite. It adds material to close the break, but works only in certain circumstances. Let us examine each command closer.

Break, Method 1

To try out the break command, open a new file and draw a line of any size or direction on your screen.

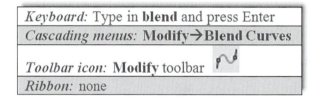

Keyboard: Type in **blend** and press Enter
Cascading menus: **Modify→Blend Curves**
Toolbar icon: **Modify** toolbar
Ribbon: none

Step 1. Start up the break command via any of the preceding methods. Be sure to turn off OSNAP.
  ○ AutoCAD says: `Select object:`
Step 2. Go ahead and click to select the line at some random spot. It turns dashed.
  ○ AutoCAD says: `Specify second break point or [First point]:`
Step 3. Pick another nearby point on the line, and a section between these two points is removed (Figure 15.6).

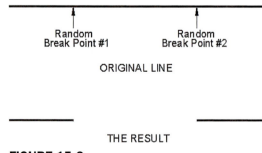

Random
Break Point #1

Random
Break Point #2

ORIGINAL LINE

THE RESULT

FIGURE 15.6
Break command, Method 1.

In this method, the first point of the break is somewhat random. When you selected the line, the first point was automatically chosen. However, as shown in the next method, you can select the line to be broken, then, using the First point option, specify exactly where to start the break. We use this to our advantage to create four lines out of two, as seen next.

Break, Method 2

Draw a horizontal line of any size, followed by a perpendicular vertical one that is of equal size. They should cross perfectly in the middle. You have a plus sign. One good trick to do this is to first draw a circle and just connect the four quadrant points with two lines, as seen in Figure 15.7.

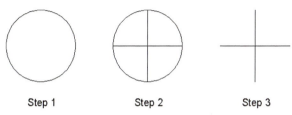

Step 1   Step 2   Step 3

FIGURE 15.7
Setup for break command, Method 2.

Step 1. Start the break command via any of the previous methods.
  ○ AutoCAD says: `Select object:`
Step 2. Go ahead and click to select the horizontal line at some random spot. It turns dashed.
  ○ AutoCAD says: `Specify second break point or [First point]:`
Step 3. Type in f for `First point`.
  ○ AutoCAD says: `Specify first break point:`
Step 4. Using OSNAPs select the midpoint of the horizontal line.
  ○ AutoCAD says: `Specify second break point:`
Step 5. Once again, using OSNAPs, select the same midpoint of the line.

359

Now, repeat these steps with the vertical line. The result is four lines where there were only two just moments earlier, as seen in Figure 15.8, when the grips are revealed. All four are broken at their intersection.

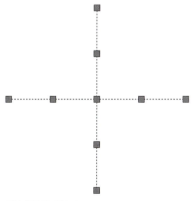

FIGURE 15.8
Break command, Method 2.

Join

The join command is included in this discussion because it is the exact opposite of break. It closes the gap between coplanar lines. Note the important disclaimer, however. The join command does not unite lines that are not on the same plane. This means that, if they are not pointing in the exact same direction and are not actually one after the other, then it does not work. To try this out this command, draw two coplanar lines (or use break to put a gap in), as seen in Figure 15.9. Note that join works on an unlimited number of coplanar lines; we are using only two in this example for simplicity.

Keyboard: Type in **join** and press Enter	
Cascading menus: Modify→Join	
Toolbar icon: Modify toolbar	→+
Ribbon: Home tab→Modify	→+

FIGURE 15.9
Setup for join command.

Step 1. Start the join command via any of the preceding methods.
  ○ AutoCAD says: `Select source object:`
Step 2. Select the first line.
  ○ AutoCAD says: `Select lines to join to source:`
Step 3. Select the second line.
  ○ AutoCAD says: `Select lines to join to source:`
Step 4. You are done with the two segments, so press Enter.
  ○ AutoCAD says: `1 line joined to source`

A solid line takes the place of the two segments.

15.6 CAD STANDARDS

This is a rather obscure procedure that may be valuable to a few organizations, and I have on occasion recommended it to prime contractors who often find their design files bouncing around from subcontractor to subcontractor. After all the respective additions to the drawings are finished, they are sent back to the main contractor. There it is found that the standards have deteriorated, as everyone did his or her own thing in regard to layers and text or dim styles. A way to identify and clean up the mess would be useful, hence the existence of this tool.

The CAD Standards procedures can be a bit confusing, and the process is not effortless nor guarantees perfect results. Depending on the damage, you may still need to do manual cleanup of layer properties, errant fonts, and the like. The best defense is to hold all others who may work on your drawings to a high standard from the start and outline what is acceptable and what is not. Obviously this can be challenging in just one office, never mind implementing it for designers at other, remote locations.

Here is the basic idea. Come up with one file that is in great shape. This can be a regular job that is singled out for its perfect or near perfect use of company standards and is of the highest quality and integrity. Then, do a Save As for this job under a .dws (drawing standards) extension. This is your standards file. All others are compared to it as they come in from subcontractors or just other designers.

To implement the standards, open up a job that just came in and is of questionable CAD quality. Type in `standards` and press Enter; alternatively use the cascading menu **Tools**→**CAD Standards**→**Check**…. There is no toolbar or Ribbon equivalent. The Configure Standards dialog box appears, as seen in Figure 15.10.

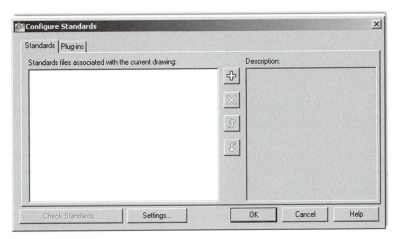

FIGURE 15.10
Configure Standards.

Next use the plus (+) sign to open up the saved dws file. The file appears in the window on the left with some descriptions on the right. The Check Standards button (lower left) also lights up, at which point you press it to execute the check. The Check Standards dialog box appears as seen in Figure 15.11.

FIGURE 15.11
Check Standards.

Next, you need to run through the fixes, item by item. Continue pressing the Fix button as AutoCAD runs through the fixes and matches up text, layers, and other items. When done, the Check Complete notice shown in Figure 15.12 displays. In this particular example, 74 problems were found but none fixed.

15.7 CALCULATOR

The calculator is a surprisingly powerful and versatile tool within AutoCAD, and one that is often overlooked by users. The function's power and versatility should come as no surprise,

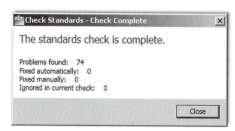

FIGURE 15.12
Check Complete.

as AutoCAD itself is highly mathematical software, a calculator in itself, performing complex operations "under the hood" as you draw, copy, and rotate objects.

The calculator function can perform all regular and scientific calculations you would expect from a handheld device but, in addition, can assist you in the construction of geometry and even allows you to use AutoLISP, a programming language. A full description of the calculator command can take up dozens of pages, so we explore only the basics here. However, we take a look at one example of construction assistance using the transparent line command cal and one example of the actual graphic calculator tool (QuickCalc) to do unit conversions. You are encouraged to read the Help files for a wealth of additional information.

First, a brief overview of the basic cal command. This was the original version of the calculator. It is a command line tool enabled by typing in cal and pressing Enter. AutoCAD then says: >> Expression:, at which point you can enter some math function using the familiar characters such as +, −, *, and /. Pressing Enter again yields an answer.

Not too long ago the limit of the calculator was +32,767 and −32,768 as the largest displayable final integer, but that restriction was lifted somewhere around the AutoCAD 2005 release, and you can now work with integer numbers between +2,147,483,647 and −2,147,483,648 (that is 2 to the 15th and 31st powers, respectively, if you *really* wanted to know where those seemingly random numbers came from).

Let us use the previously described cal command to assist in design, because crunching numbers alone can be done by your handheld or Windows calculator. Suppose you need to offset a line one half of some number, but you do not know what that value is (say one half of 17.625", for example). Instead of reaching for that calculator, let AutoCAD do both: calculate one half of 17.625 as well as offset the line, all in one command. To do this,

1. Draw a line of any reasonable size.
2. Start the offset command via any method.
3. You do not know the offset distance so let us use the calculator.
4. Type in 'cal (the apostrophe makes cal, or any command, temporarily transparent, meaning you can enter it without exiting out of another, in-progress command, which in this case is offset). Press Enter.
5. The >>>> Expression: prompt asks for the expression. Enter 17.625/2 and press Enter.
6. Your answer (8.8125) appears, and you may resume the offset command.
7. Continuing as usual, select the line and pick a direction to complete the offset and Esc to exit.

This was a very simple case of how the calculator assists in design tasks. It can do a lot more in that regard and you are encouraged to examine the Help files for additional examples.

The calculator function is not just a typed command line but also now comes in a graphic version. If you use this graphic version, you are using what is referred to as a QuickCal. It is almost a whole new command all to itself and has quite a bit more functionality.

Keyboard: Type in **quickcalc** and press Enter.	
Cascading menus: **Tools→Palettes→QuickCalc**	
Toolbar icon: **Standard** toolbar	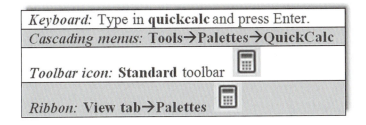
Ribbon: **View tab→Palettes**	

Start up the QuickCal command via any of the preceding methods and you see Figure 15.13.

FIGURE 15.13
QuickCalc.

The Number Pad and Scientific keypads are self-explanatory. If you ever used a handheld calculator, you should be right at home with this one. Variables, at the bottom, are not covered, but Units Conversion deserves a special mention. It is a *very* useful tool for converting between a wide variety of Length, Area, Volume, and Angular values. Simply select from the fields on the right (drop-down arrows appear) and enter the values to convert just below that. A converted value appears according to what you requested.

15.8 DEFPOINTS

Defpoints is not a command, but rather a concept that appears often and bears a short mention. *Defpoints* stands for definition points, and they appear in the drawing (and mysteriously in the Layers dialog box, where most users first notice them) as soon as you add any dimension to your drawing. What are they exactly? They are the tiny points where a dimension attaches itself to the object being dimensioned. The Defpoints layer does not go away easily and usually prompts questions from students as to its nature and intent. The layer fortunately does not plot, and you can pretty much disregard its existence if you choose. One benefit of it not plotting is that you can put any objects that you do not want to see on paper on it (such as viewports, as is discussed in Chapter 10). In fact, that is exactly what many users do, though a dedicated VP layer is recommended instead. In conclusion, do not worry about Defpoints and continue on.

15.9 DIVIDE AND POINT STYLE

Divide does exactly what it advertises; it divides an object into equally spaced intervals. Note however that the object is not actually broken into those segments (that is for a break command to do) but rather just divided by equally spaced points. Then, you can use these points as guides for further work. Let us give this a try.

Step 1. Draw a horizontal line of any size.
Step 2. Type in `divide` and press Enter. Alternatively you can use the cascading menu **Draw→Point→Divide**. There is no toolbar or Ribbon equivalent.
  ○ AutoCAD says: `Select object to divide:`
Step 3. Select the line.
  ○ AutoCAD says: `Enter the number of segments or [Block]:`
Step 4. Enter some value, 10, for example, and press Enter.

The line looks the same, with no apparent divisions. It seems as though the command has not worked. It actually worked just fine, and the line is divided into the ten segments using nine points, as expected. The problem is that you cannot see the points; they are obscured by the line, and we need to change the point style to something more visible. Using the cascading menus, select **Format→Point Style**. You will then see the dialog box shown in Figure 15.14.

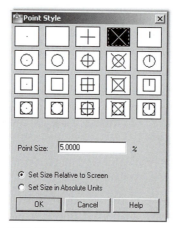

FIGURE 15.14
Point Style.

Select a point style that is not too complex; the X format in the top row (second from right) is a popular one. Press OK and the divided line appears as you were expecting it, divided into the ten visible segments (Figure 15.15).

So, how exactly can you use this to your benefit? Well, the Node OSNAP, which is discussed later in this chapter, allows you to connect to points. If you activate it, you can attach new

FIGURE 15.15
Divided line.

geometry, perhaps new lines, to those Xs. This is quite useful in basic construction and layout. Note an interesting quirk. If you zoom in or out, the points change size, but revert back to "standard" size upon typing in regen.

15.10 DONUT

Although this command sees only occasional use, it is still worth a mention. It creates exactly what it advertises: a donut of varying size. Broken down into its components, a donut is two circles of different size, one inside the other, and a solid hatch fill. You can technically create this shape with existing techniques, but the donut command just makes it easier. You have control over the sizes of both circles and, as a result, the size of the donut. The inner fill is always a solid hatch. Let us give it a try.

Step 1. Start the donut command by typing in donut and pressing Enter. Alternatively you can use the cascading menus **Draw→Donut**.
  ○ AutoCAD says: Specify inside diameter of donut <0.5000>:
Step 2. You can accept this by pressing Enter or type in another value (we accept the 0.5 in this example).
  ○ AutoCAD says: Specify outside diameter of donut <1.0000>:
Step 3. Once again, you can accept this value or type in your own (we accept the 1.0 in this example).
  ○ AutoCAD says: Specify center of donut or <exit>:
Step 4. Click anywhere on the screen and the donut appears, as seen in Figure 15.16.

FIGURE 15.16
Donut.

15.11 DRAW ORDER

Draw Order refers to the stacking order of objects positioned one on top of another. This feature is commonly used in most graphic design programs and has an occasional use in AutoCAD as well. The stacking order is usually not of major concern in day-to-day basic drafting. If items intersect, the one drawn last is on top; and in cases of two intersecting lines, the intersection point is one pixel in size, hardly worth a second thought.

Sometimes, however, such as when two or more hatch patterns are stacked, the order is critical. We use this feature later in this chapter to great effect when we discuss Wipeout. For now, just know where the Draw Order command is. It is found in the cascading menu under **Tools→Draw Order**, with options for Objects, Text, Hatches, and Dimensions.

Bring to Front and Send to Back send objects strictly to and from the top and bottom of the order, while Bring Above Objects and Bring Below Objects are more selective, moving items one step at a time. With these four tools, you can arrange any amount of objects in any order.

15.12 eTRANSMIT

The purpose of this command is to assist with emailing drawings. The problem is that, when drawings are sent out, items can be inadvertently omitted. These omissions may include something major, like associated xrefs, or something secondary, such as fonts and styles. As a project grows more complex it is easy to forget to send xrefs and even easier to forget about the unique fonts, styles, and ctb files that have been generated over time. The recipient of the email may have only some of these and what he or she downloads and reconstructs on that end may not be quite what you may have intended to be seen.

The eTransmit command is a go-getter of sorts. It sniffs out every associated file and everything else that may be relevant to recreate the file(s) in a new location as originally intended. Then, it packages everything together in a folder and sends it out. While veteran users have remarked that an experienced and aware designer needs to always have a clear idea of what to send (eTransmit is a recent invention, after all), it is still a valuable tool for novices and experts alike.

So, what is (and what is not) included in the eTransmit transmittal? Here is a partial list of what *is* included. You should know of most of these files. If not refer to Appendix C.

*.dwg. Root drawing file and any attached external references.
*.dwf. Design web format files that are attached externally to the root drawing or xrefs.
*.xls. Microsoft Excel spreadsheet files that are linked to data extraction tables.
*.ctb. Color-dependent plot style files used to control the appearance of the objects in the drawings of the transmittal set when plotting.
*.dwt. Drawing template file associated with a sheet set.
*.dst. Sheet set file.
*.dws. Drawing standards file associated to a drawing for standards checking.

Here is a partial list of what is *not* included. Do not worry if you do not recognize all these formats. You will know most of them by the end of Level 2.

*.arx, *.dbx, *.lsp, *.vlx, *.dvb, *.dll. Custom application files (ObjectARX, ObjectDBX, AutoLISP, Visual LISP, VBA, and .NET).
*.pat. Hatch pattern files.
*.lin. Linetype definition files.
*.mln. Multiline definition files.

To use eTransmit, save your work but keep the file open. Type in etransmit and press Enter. Alternatively you can use the cascading menu **File→eTransmit**…. There is no toolbar or Ribbon alternative. You are asked to save the file if it has not been saved yet. Then, the dialog box in Figure 15.17 appears.

The procedure is highly automated, but you do have to select a few parameters. The current drawing that is open is the one that is packaged together. Some of the included items are shown in the file tree. Expanding the + signs shows specific files (which you can deselect if desired). You also add more drawings to the transmittal using the Add File… button. If you are fine with just the current drawing (and what is included in it) then press the Transmittal Setups… button. You then see the dialog box shown in Figure 15.18.

Let us create a new setup as an exercise. Press the New… button and the dialog box in Figure 15.19 appears. Give the transmittal a name and press Continue. The dialog box in Figure 15.20 then appears.

All you really have to do here is visually check all the settings. Looking at the fields from the left, top to bottom, we see the Zip option for actually sending the files (recommended), followed by a variety of information related to formats, names, and locations. Unless you have a specific reason to change something, leave everything as default. On the right-hand

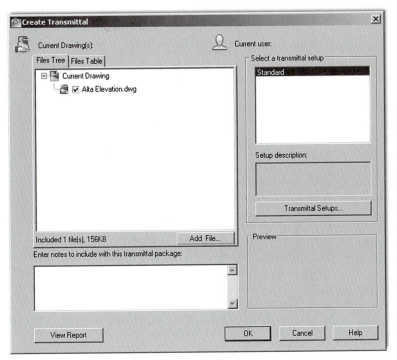

FIGURE 15.17
Create Transmittal.

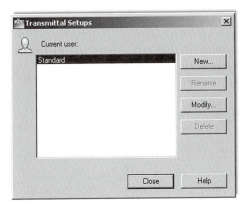

FIGURE 15.18
Transmittal Setups.

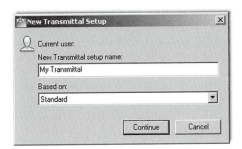

FIGURE 15.19
New Transmittal Setup.

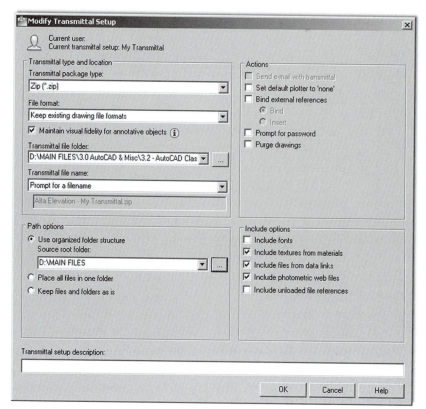

FIGURE 15.20
Modify Transmittal Setup.

side, top to bottom, leave everything under the Actions category as default (unchecked) as well as everything under the Include options category (middle three boxes checked).

When done press OK, then Close in the next dialog box (Figure 15.18), and finally OK in the main setup box (Figure 15.17). You are asked where to put the Zip file and for any last-minute name changes. Pressing Save drops the Zip file into its place, and it is ready for emailing.

15.13 FILTER

Filters were briefly discussed in the context of advanced layers (Chapter 12). While the ideas introduced here are roughly similar, the application is slightly different, so let us review them from the beginning.

Filters are nothing more than a method to isolate and select objects using their features or properties. So, for example, you can isolate and select all the circles by virtue of them being circles, and find green objects because they all have the color green in common. You can also add together properties and create requests like "Find me circles that are red only" and so on. This is a good tool for quickly finding stuff in a complex drawing. Once you find these objects, you can then change their properties, delete them, or do whatever else is needed.

To experiment with filters, let us first create a simple drawing that is just a collection of shapes. Draw the set of circles, rectangles, and arcs seen in Figure 15.21. Then, change the shapes' colors to blue or red.

Type in `filter` and press Enter. There is no cascading menu, toolbar, or Ribbon equivalent. Object Selection Filters appears, as shown in Figure 15.22. The procedure here is to first select the filter, then add it to the list, and finally apply it.

FIGURE 15.21
Basic shapes for filters.

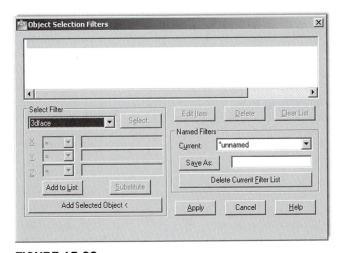

FIGURE 15.22
Object Selection Filters.

369

Select Arc from the Select Filter drop-down menu on the left. Then press the Add to List: button just below that. For good measure, let us also add the color red. Select the color from the same Select Filter menu. Notice, at that point, the Select button lights up. Go ahead and select the color red. Then, press Add to List:. Your filter should look like Figure 15.23.

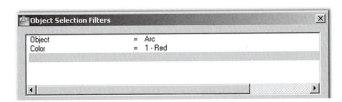

FIGURE 15.23
Objects and colors selected.

Now, press the Apply button. AutoCAD asks you to select objects. Select everything by typing in all. Every red arc in the drawing becomes dashed after the first Enter. Press Enter again and the grips come on, as seen in Figure 15.24. Now type in Erase and press Enter. Because they are selected, all the red arcs disappear. Filters can be saved for future use via the Save As: button and the listing can be modified or deleted via the Edit Item, Delete, and Clear List buttons.

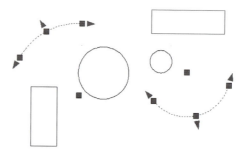

FIGURE 15.24
Segmented arcs selected by filter.

15.14 HYPERLINK

The hyperlink command attaches a hyperlink to either an object or text. During this process, you are asked to specify what this hyperlink opens when activated. It can be a website (usually) or a document (sometimes). In either case, it is a great command to quickly access additional useful information. Someone, for example, can see a drawing of a product and follow the link to a website of the manufacturer or call up an Excel file that lists all the specs of a design.

To try out the hyperlink features, draw a symbol for a chair (make a block) and a text website address, as seen in Figure 15.25.

WWW.GOOGLE.COM

FIGURE 15.25
Chair and web address.

We would like to link the chair (an *object*) to a furniture website (IKEA®, for example) and link the Google text (a *text string*) to the Google website.

Type in `hyperlink` and press Enter. Alternatively, you can use the cascading menu **Insert→ Hyperlink**. There is no toolbar or Ribbon equivalent. AutoCAD asks you to select objects. Go ahead and pick the chair block. The dialog box in Figure 15.26 then appears.

Enter the name of the desired website in the appropriate field. That is where it says Type the file or Web page name, not Text to display. Not much else needs to be done, so press OK. Repeat the process for the Google text.

Hover your mouse over the chair then the Google text and notice there is something new. A hyperlink symbol appears, with text prompting you to hold down the Ctrl key and click to follow the link (Figure 15.27). If your computer is currently connected to the internet, go ahead and do that in both cases (chair and Google website) to bring up the respective websites.

There is an interesting variation to this, as mentioned earlier. You need not attach a website to the link. You can also attach a regular document, so it is called up when the hyperlink is clicked on, as before. Pretty much anything can be attached if it can be found and the computer can run the application: Word, Excel, PowerPoint documents, pictures, videos—you name it.

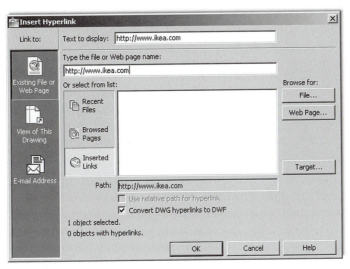

FIGURE 15.26
Insert Hyperlink.

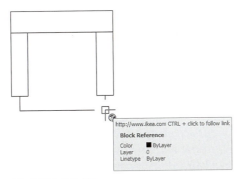

FIGURE 15.27
Hyperlink.

To do this, follow the same steps as before, but this time select the File… button under the Browse for: header on the right side of the dialog box. Find the file of interest, double-click on it, and OK the Hyperlink dialog box. That is it; the document is now attached to your object or text and supersedes any other links you may have.

As a final note, recall that Express Tools have a category called Web Tools. We skipped discussing them in Chapter 14 because the hyperlink command was not yet introduced. Take a few minutes now to look over what is available. These express commands include tools to Show, Change, and Find & Replace URLs. These are useful when you have multiple URLs in your drawing and need to manage them. The three tools are very straightforward to use and we do not spend any further time on them.

15.15 LENGTHEN

The lengthen command is quite a bit like the extend command in its effect but goes about doing it differently and is worth adding to your AutoCAD vocabulary. The command works by lengthening non-closed objects, such as lines or arcs (circles and rectangles do not qualify). You can lengthen the objects by change in size [DElta], by percentage added or subtracted [Percent], by total final size [Total], or dynamically "on the fly" [Dynamic]. All this applies to angles (of arcs) as well. To try it out, draw an arc and a line (Figure 15.28).

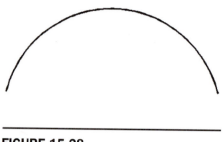

FIGURE 15.28
Arc and line for lengthen.

Step 1. Type in `lengthen` and press Enter. Alternatively, use the cascading menu **Modify→Lengthen**. There is no toolbar or Ribbon equivalent.

○ AutoCAD says: `Select an object or [DElta/Percent/Total/DYnamic]:`

Step 2. Pick the `Percent` choice by typing in p and pressing Enter.

○ AutoCAD says: `Enter percentage length <100.0000>:`

Step 3. Enter a value such as 150 and press Enter.

○ AutoCAD says: `Select an object to change or [Undo]:`

Step 4. Pick the line and it lengthens 150%. Press Esc to exit the command.

The process is essentially the same with the arc and other lengthening options. Repeat the command several times and run through each option.

15.16 OBJECT TRACKING (OTRACK)

Object snap tracking is a drawing aid that allows you to find various OSNAP points of an object and connect to them without actually drawing anything that touches that object, a sort of remote OSNAP. This function can be activated when the OTRACK button is pressed in at the bottom of the AutoCAD screen and is deactivated when the button is pressed back out. The best way to see the purpose and function of this command is to try it. Draw a series of rectangles or squares spaced some distance apart, as shown in Figure 15.29. You need not number them; that is only for identifying purposes.

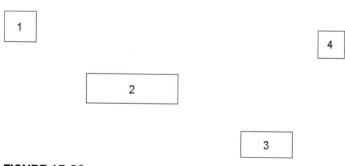

FIGURE 15.29
Rectangles for OTRACK.

What we want to do is connect a series of lines from (and to) the midpoints of each of these shapes via right angles (not directly). An example of this may be a cable diagram or an electrical schematic. One way would be to start a line from the midpoint of the right side of shape 1, stop somewhere randomly, and repeat the process from the top of shape 2. Then you can use trim, extend, or fillet to connect the lines. This, however, is cumbersome and slow.

A better way is to begin the line from the right side midpoint of shape 1 as before (turn on ORTHO also), but then activate OTRACK and, *without clicking*, position the mouse over the midpoint of the top of shape 2. A straight dashed *vertical* line appears. Follow that dashed line back up until the mouse meets the straight *horizontal* dashed line. Then and only then, click once and continue to draw the line to shape 2. This is shown in Figure 15.30. Notice what OTRACK does. It senses where that midpoint of shape 2 is and gives you a position to confirm and click before proceeding. Complete similar lines from the bottom of shape 2 to the top of shape 3 and then 4.

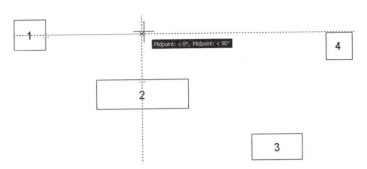

FIGURE 15.30
OTRACK in progress.

15.17 OVERKILL

This interesting command is technically part of the Express Tools under <u>M</u>odify→**Delete duplicate objects**, but it is more entertaining to type in its rather dramatic other name. What overkill does though is quite useful. It deletes duplicate objects that overlap each other. In the 2D flat world of AutoCAD, it can be quite challenging and time consuming to find these overlaps, especially if similar shapes and colors are involved. With overkill, you simply select the offending objects and AutoCAD figures out which to keep and which to delete, effectively merging them all into one.

To try it out, draw five lines, one on top of another. Although it may not be obvious to a novice, an experienced user notices these lines are a bit darker and "fuzzy," indicating the presence of an overlap; it is subtle but there. Now type in overkill and press Enter. AutoCAD asks you to select objects. Select all five lines using a Window or Crossing (not one at a time) and press Enter when done. The dialog box in Figure 15.31 appears.

373

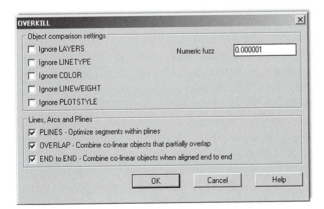

FIGURE 15.31
Overkill.

You can leave all the defaults as they are, pressing OK to have all the lines merge into one line. This is probably the simplest application of the overkill command. However, what if the lines are on different layers?

In this case, you have to check off Ignore LAYERS and the command merges the layers into one. Numeric fuzz works by comparing the distances between nearly overlapping objects and acts on them if the distance falls inside the fuzz value. Overkill has other useful features as well, so go through the command, referring to the Express Tools Help files for more information.

15.18 POINT AND NODE

You have already seen and used points when we discussed the divide command earlier in this chapter. In that example, points were created as a means to divide an object. To see them clearly, you needed to modify the form of the point, from a dot to something more visible, using **Format→Point Style**….

The point command can also be used to create points from scratch, as described next.

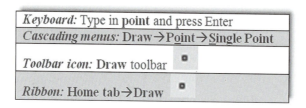

Keyboard: Type in **point** and press Enter
Cascading menus: Draw→Point→Single Point
Toolbar icon: Draw toolbar
Ribbon: Home tab→Draw

Step 1. Start the point command via any of the preceding methods.
  ○ AutoCAD says: `Current point modes: PDMODE=3 PDSIZE=0.0000`
    `Specify a point:`
Step 2. Click anywhere and a point appears.

As you may imagine, points by themselves are not too useful, but change them to an X and they can be used to create "anchors" to which you can attach geometry. For this, you need an OSNAP point that can do that. It is called Node and is one of the 13 OSNAPs. The symbol is a circle with an X superimposed on it. Go ahead and activate Node and practice drawing some lines that originate from the points you just created.

15.19 PUBLISH

The Publish command is a convenient way to print multiple sheets (layouts) all at once. This command depends on knowing the basics of Paper Space, so you may want to review Chapter 10 if needed. Assuming, however, that you have done so, the idea here is quite simple. When a file with multiple layouts is opened, the publish command sees them and allows you to print them all at once, as opposed to clicking on each tab one at a time and using the plot command. For this example, a file called Alta Elevation is opened. It features a design spread out over two layouts: a Plan View and an Elevation View. The publish command is then started via any of the following methods.

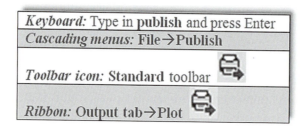

Keyboard: Type in **publish** and press Enter
Cascading menus: File→Publish
Toolbar icon: Standard toolbar
Ribbon: Output tab→Plot

The dialog box in Figure 15.32 then appears. Note that it can be expanded, if it is not, by pressing the Show details button on the lower left.

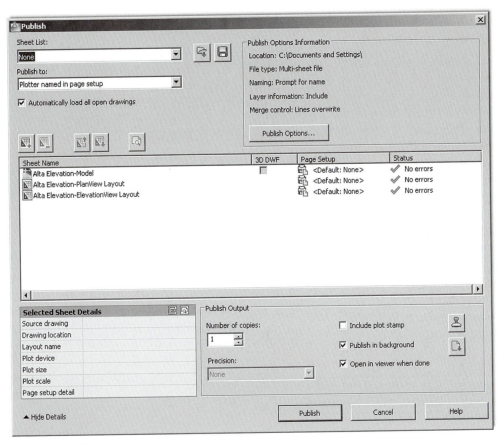

FIGURE 15.32
Publish dialog box.

The dialog box features some options to fine-tune publishing, such as adding additional sheets, plot stamp, deciding how many copies, and a few others. All you have to do at this point is just press the Publish button on the bottom and AutoCAD will print out one copy of every single layout the file contains. It is quick and easy and saves time. Go ahead and explore this very useful command on your own for additional features.

15.20 RASTER

Raster is a term you may hear in AutoCAD circles, and it is not so much a command as a concept in graphic design. A raster graphic is nothing more than a bitmap, which in turn is a grid of pixels viewable on any display medium, such as paper or a computer monitor. This method of presenting graphics stands in marked contrast to vector graphics, which are based on underlying mathematical formulas. Vectors can be scaled up or down with no loss of clarity, while bitmaps cannot. So raster images are basically pictures. These can easily be inserted into AutoCAD as covered in a previous chapter.

Raster graphics hold a unique place in AutoCAD history. They were needed to implement a massive worldwide conversion from paper-based hand-drawn designs to CAD files. When it became obvious that computer-aided design was the future, millions upon millions of hand-drawn documents, spanning decades of design work, were scanned into the computer, became backgrounds, and were patiently redrawn. Government and private industries around the industrial world collectively spent large sums of money and many work hours to transform these archives into viable CAD files for future use. Most of this work was done during a ten-year span (1988–1998), although some of it was started prior and some work is still ongoing. I recall putting in some significant screen time doing these conversions early in my career, and for a new AutoCAD user, this was excellent, if boring, practice: scan, draw, print, check, and repeat.

Raster scanning software can be quite sophisticated. At the time, we used regular large sheet scanners that simply pulled up the graphic on the screen. We then tugged at it until it was the right size (by comparing a drawn line to a scanned known dimension). Later, we acquired more sophisticated software that actually recognized shapes and attempted recreating these pieces of geometry to somewhat mixed results. Any smudges, imperfections, or odd shapes could confuse the system. Noise level (sensitivity) settings were used to control how much was picked up, and it was usually better to set it lower, so you could add what was missing rather than constantly erase accidentally picked up junk. As an AutoCAD user today, you may never need these tools, but it is good to know what they are just in case.

15.21 REVCLOUD

This is a command that thousands of architects were screaming for. Revision clouds, or *revclouds* for short, are used to indicate changes on a drawing. You can use an ellipse, rectangle, or circle just as well, but these shapes blend too easily with actual design work, and the unmistakable cloud shape is a better choice. Revclouds were once drawn using arcs, an extremely tedious process. Finally, some program-savvy designers wrote routines to automate the process. Autodesk then jumped on board and revclouds were born. Let us give it a try.

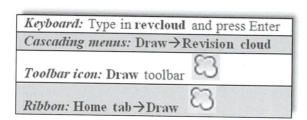

Keyboard: Type in **revcloud** and press Enter	
Cascading menus: Draw→Revision cloud	
Toolbar icon: Draw toolbar	
Ribbon: Home tab→Draw	

Step 1. Start up the revcloud command via any of the preceding methods.
  ○ AutoCAD says:
  ```
  Minimum arc length: 0.5000 Maximum arc length: 0.5000 Style:
  Normal
    Specify start point or [Arc length/Object/Style]<Object>:
  ```
Step 2. Click anywhere once.
  ○ AutoCAD says: `Guide crosshairs along cloud path...`
Step 3. Draw a circular shaped revcloud connecting the last arc to the first one (no extra clicking needed).
  ○ AutoCAD says: `Revision cloud finished.`

Your result should look somewhat similar to Figure 15.33 (you may have to zoom in or out to see the revcloud).

The command has several options. For example, if you do not connect the last arc to the first and instead just right-click, AutoCAD asks you: `Reverse direction [Yes/No] <No>:`. Here you can actually turn the revcloud "inside out," though this is rarely done. You can also change the size of the cloud's arcs by selecting the `Arc length` option, and even turn objects (like circle and rectangles) into revclouds using the `Object` option. Take the time to experiment with all of these.

FIGURE 15.33
Revcloud.

15.22 SHEET SETS

Like many new features in AutoCAD, these are not revolutions, rather but evolutions of previous ideas. This tool, introduced in AutoCAD 2005, was meant to combine some aspects of paper space layouts and the publish command with a few new twists, to create an "all under one roof" method of setting up the presentation and printing of a project. The idea is to set up sheet sets (1 Layout = 1 Sheet) that define exactly what is needed to plot out the entire job correctly. These sheets can come from different drawings and even different jobs and are grouped together by their respective categories. The idea is that they are now easier to handle and organize.

There has been some reluctance in the industry to use a tool that somewhat duplicates existing functionality (even if new features are added), but a few organizations, not surprisingly ones that produce jobs that span dozens of sheets, took to this sheet set idea. We provide a brief overview, and you can then decide if you want to pursue this further.

To create a sheet set, you need to set it up. Go to **File→New Sheet Set**… and you see what is shown in Figure 15.34. Select the first choice, an example sheet set, and press Next >. What is shown in Figure 15.35 appears.

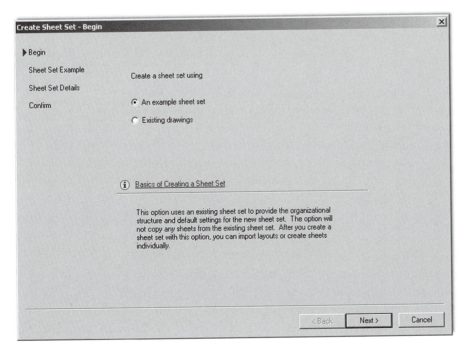

FIGURE 15.34
Create Sheet Set, Begin.

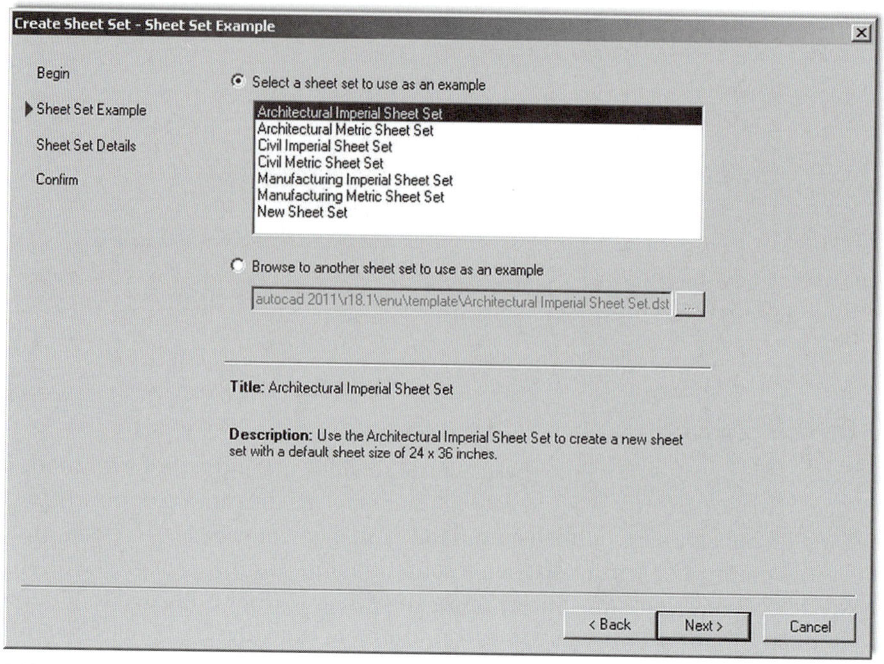

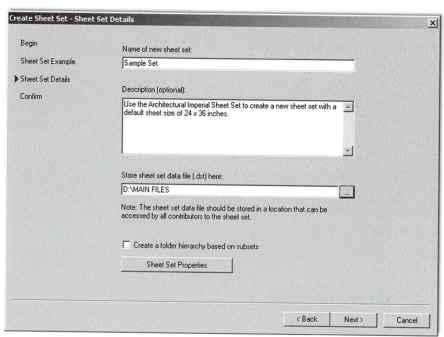

FIGURE 15.35
Create Sheet Set, Sheet Set Example.

Select the appropriate sheet set, Architectural Imperial for example, and press Next >.
Figure 15.36 appears. Here you can give the set a name and location. Press Next > and what
is shown in Figure 15.37 appears. The various sheet sets sorted by category are visible. Press
Finish and the Sheet Set Manager appears, as seen in Figure 15.38.

FIGURE 15.36
Create Sheet Set, Sheet Set Details.

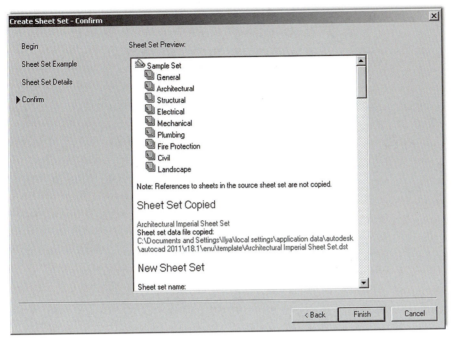

FIGURE 15.37
Create Sheet Set, Confirm.

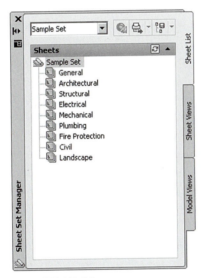

FIGURE 15.38
Sheet Set Manager.

This was all just to set up the basic sheets (placeholders). They are still blank and need drawings associated with them. Right-click on the Architectural folder and select Import Layouts as Sheet.... What is shown in Figure 15.39 appears.

You now have to browse for the right drawing associated with the Architectural layout. Continue down the categories until all are filled. This is just the basic idea of sheets sets. It is an extensive tool that can span many more pages of descriptions, and you should explore some of it on your own.

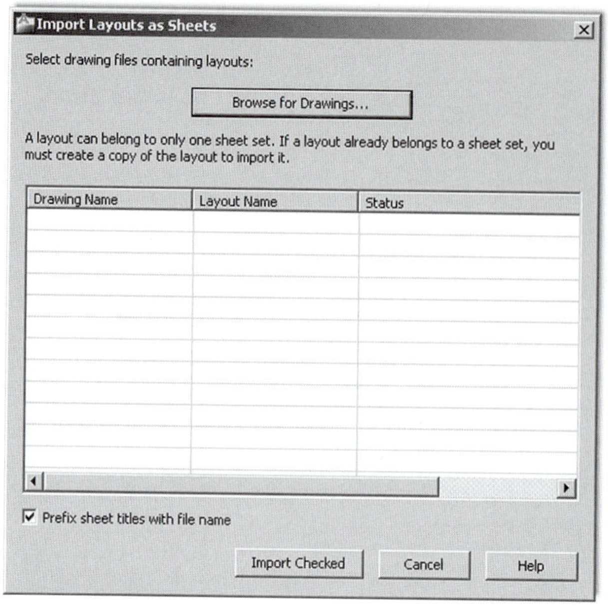

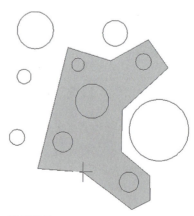

FIGURE 15.39
Import Layouts as Sheets.

15.23 SELECTION METHODS

What we outline here are a few other methods of selecting items that you may have not yet seen: the Fence (F), Window polygon (WP), and Crossing polygon (CP). All these are applicable when selecting objects for a number of operations such as erase, move, copy, and trim.

To try out the Window polygon:

Step 1. Draw a set of circles, as seen in Figure 15.40.

FIGURE 15.40
Window polygon.

Step 2. Begin the erase command via any method you prefer.
Step 3. When it is time to select objects, type in wp for Window polygon and press Enter.
○ AutoCAD says: First polygon point:
Step 4. Click anywhere.
○ AutoCAD says: Specify endpoint of line or [Undo]:

Step 5. Continue to "stitch" your way around the circles you wish to select, and right-click when done.

Notice the difference here between the standard Window selection tool (strictly rectangular in nature) and the flexible "any shape you want it" Window polygon (Figure 15.40). Its action of course is similar; everything it encloses is selected.

To try out the Crossing polygon:

Step 1. Draw another set of circles, as seen in Figure 15.41, or just undo your previous steps.

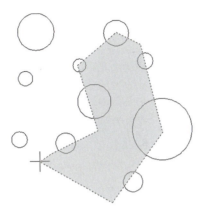

FIGURE 15.41
Crossing polygon.

Step 2. Begin the erase command via any method you prefer.
Step 3. When it is time to select objects, type in cp for Crossing polygon and press Enter.
  ○ AutoCAD says: First polygon point:
Step 4. Click anywhere.
  ○ AutoCAD says: Specify endpoint of line or [Undo]:
Step 5. Continue to "stitch" your way through the circles you wish to select, and right-click when done.

As you can see, the Crossing polygon (Figure 15.41) works the same exact way as the Window polygon, except anything it touches gets selected, much like how the regular Crossing operates.

The fence option works by having a drawn line be the cutting tool during a trim command and selects multiple lines to be trimmed. Draw a bunch of vertical lines, and one horizontal line, as seen in Figure 15.42. Then execute the trim command, but when it is time to select the lines to be trimmed, type in f for fence, and click twice to draw a line *across* the lines to be trimmed, finally pressing Enter to make them disappear.

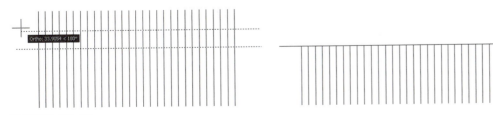

FIGURE 15.42
Fence command and the results.

In case you discovered it on your own, yes, you can accomplish the exact same thing using a regular Crossing to eliminate the lines once a cutting edge is selected. However, the fence command comes in handy when going around corners, as the flexible line-based approach can bend around any shape, while the rigid rectangular shaped crossing cannot. Try out both methods, as well as using fence around corners.

15.24 STRETCH

This is an easy command to learn and use, and it can work wonders for quick shape modifications. The stretch command uses the Crossing tool to literally stretch shapes. Create two rectangles, as seen in Figure 15.43.

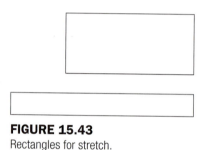

FIGURE 15.43
Rectangles for stretch.

We want to stretch the right sides of both rectangles some distance.

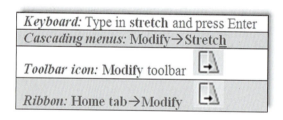

Keyboard: Type in **stretch** and press Enter	
Cascading menus: Modify→Stretch	
Toolbar icon: Modify toolbar	
Ribbon: Home tab→Modify	

Start the stretch command via any of the preceding methods.

Using a *Crossing* (this does not work with the Window selector), select the first quarter or so of the rectangles, as seen in Figure 15.44.

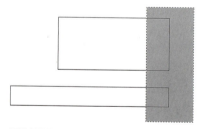

FIGURE 15.44
Starting the stretch command.

They become dashed. Press Enter and click anywhere to begin stretching them, as seen in Figure 15.45. You need to have Ortho on and all OSNAPs off.

Finally, click elsewhere and the rectangles have a new length. You did not have to explode and redraw them or use grips. You can use the Displacement option to key in specific X,

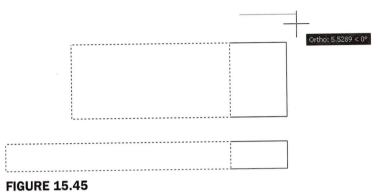

FIGURE 15.45
Stretch in progress.

Y, or Z distances or stretch to another OSNAP point. Note that stretch does not work on partially selected circles and it moves, not stretches, fully selected ones.

15.25 SYSTEM VARIABLES

There is not much to say about system variables except to make you aware of what they are, so if they come up in reading of other texts or the AutoCAD Help files, you know with what you are dealing. System variables are Yes/No, On/Off, or 0, 1, 2, 3 types of commands used to indicate a *preference or a setting*. AutoCAD 2012 has over 850 of them for every conceivable variable (something that changes) in the software, and listing them all is futile, although if you wish to see them all, type in setvar, press Enter, then type in ?, and press Enter twice.

Many system variables are set graphically via a dialog box, and users may not even be aware or know it is a system variable they are changing. Perusing the Help files and AutoCAD literature, especially on an advanced level, you find many references to them. Knowing system variables is important for those wishing to learn advanced customization (such as AutoLISP).

Here is a useful example that is typed in on the command line. Suppose you are mirroring text and the new set is backwards, as seen in Figure 15.46.

hello olləd

FIGURE 15.46
Mirrtext.

That is not quite right, as you want text to be readable even after mirroring, so some sort of system variable (SV) needs to be set. The responsible SV is called *mirrtext*, and it is of the On/Off variety represented by 0 and 1. Somehow it got set to 1, and needs to be 0. Type in mirrtext, press Enter, type in the new value of 0, and press Enter again. Try the mirror command again—it should work fine now.

As mentioned before, this is just one example; there are hundreds of others. Keep an eye out for them. After years of using AutoCAD, you will amass a formidable collection of useful SVs that can resolve some thorny problems for less experienced designers.

15.26 TABLES

If you have not yet used tables, here is a quick primer on them. They are pretty much what you might expect, a bunch of rows and columns, with a header on top, ready for you to enter information. You can specify the number of rows and columns and add or take them away at a later time as well. Tables are sort of like low-tech Excel spreadsheets and can be useful in certain situations (like when you need data displayed in a hurry). We cover some basics here and let you explore more on your own.

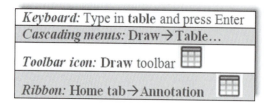

Start the table command via any of the preceding methods. You see Figure 15.47.

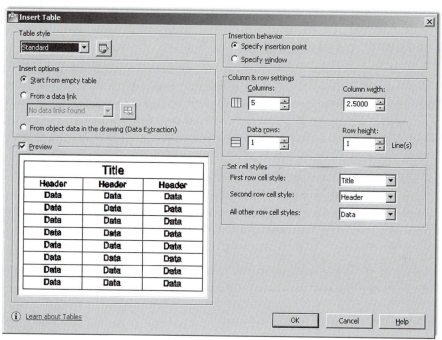

FIGURE 15.47
Insert Table.

Under the Column & row settings category, set your columns and rows as well as their respective widths and heights (we do a 10 × 10 here). When done press OK and you are asked to click where you want the table to go. When you position it, you see what is shown in Figure 15.48.

Here, you are asked to set up the header. Type in Sample Data Table and press OK. You see the completed table in Figure 15.49. You can now fill in the table with data by clicking in each individual cell and typing in values, as seen in Figure 15.50.

There is more to this command, such as sizing and modification of the table, as well as some ability to link up with Excel spreadsheets, so go ahead and explore it further. You can also set up styles of tables by first using the tablestyle command, which gives you the dialog box for "designing" a table (Figure 15.51).

Press New… and give your style a name. You then get the New Table Style dialog box (Figure 15.52). Here you can work with the general properties, text, and borders of the new table, similar to an Excel spreadsheet, and design a custom style that fits your needs.

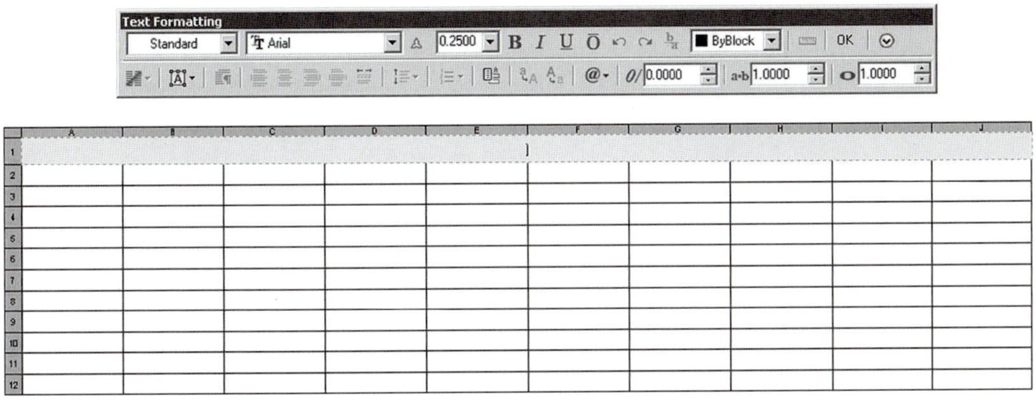

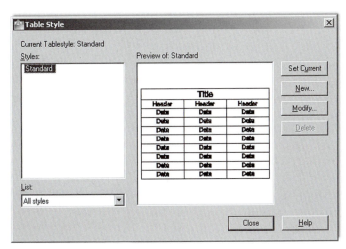

FIGURE 15.48
Insert header.

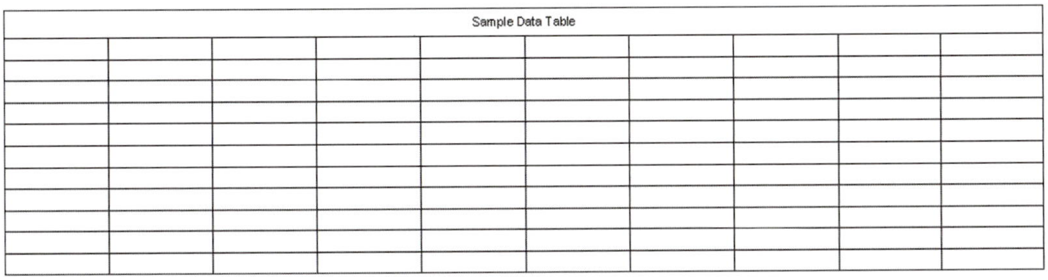

FIGURE 15.49
Fill in header.

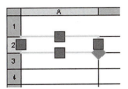

FIGURE 15.50
Fill in individual cells.

FIGURE 15.51
Table Style.

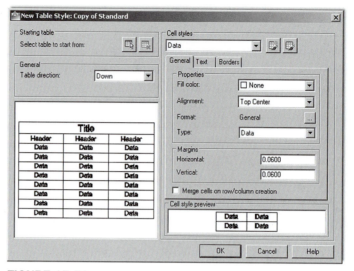

FIGURE 15.52
New Table Style.

15.27 TOOL PALETTE

This useful palette, shown in Figure 15.53, can be accessed by typing in `toolpalettes` and alternatively via cascading menu **Tools→Palettes→Tool Palettes** or by pressing Ctrl + 3. It may also be on your screen already, but hidden away off to the side. To see it, you have to hover the mouse over the palette's spine and it reveals itself. This "auto-hide" feature can be disabled via a right-click (on the spine) menu, seen in Figure 15.54.

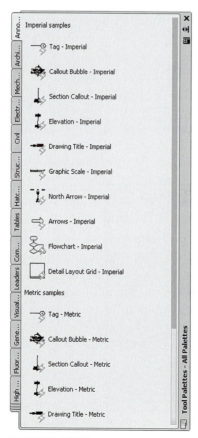

FIGURE 15.53
Tool Palettes, All Palettes.

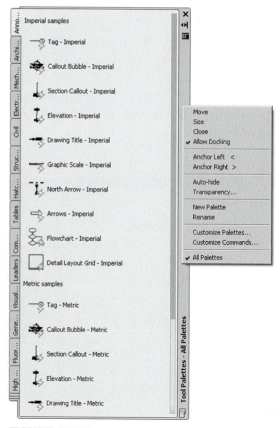

FIGURE 15.54
Tool Palettes, Spine menu.

The tool palette represents a *collection of commonly used commands and predefined blocks*. What you actually see when the palette comes up may be different from Figure 15.53, as it depends completely on what menu is pulled up or what tab is clicked. In all cases, to use this palette, just click on a command (or click and drag) and it is executed as though you typed or selected a toolbar icon. To access more menus, simply right-click on the gray "spine." As mentioned before, the menu in Figure 15.54 appears.

Even more tabs are available in this palette. Go to the bottom left of the tabs (where it looks like several tabs are bunched together) and left-click. Another extensive menu appears, as shown in Figure 15.55.

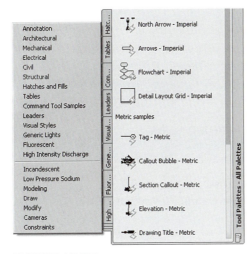

FIGURE 15.55
Tool Palettes, Tabs menu.

Take the time to go through each of the menu choices. The amount of available blocks is quite extensive.

Although we do not go into this further, you may also customize what tabs you have available by creating your own and adding custom commands and blocks to it. This is similar to creating new toolbars using the CUI, as covered in Chapter 14. You can also modify existing commands in the tabs; finally, the entire palette can be hidden, docked, and even made transparent, so it is always there but you can see what you are doing underneath it, a neat, if distracting, trick. Explore all the options; most are easy to figure out and self-explanatory.

15.28 UCS AND CROSSHAIR ROTATION

Rotating the crosshairs or the UCS icon is a very useful trick for working with tilted shapes. Not everything you work on will be perfectly horizontal and vertical. Some designs, or parts thereof, will be rotated at some angle, and it would be nice to align the crosshairs to them. This allows for easier drawing and better orientation for the designer. Commands such as Ortho can now be implemented at the new tilted angle.

Rotating the UCS icon is technically *not* the same as rotating just the crosshairs, but we do not split hairs (so to speak) over the conceptual differences, and we treat both methods as equivalent for now. There are two ways to rotate the Crosshairs/UCS icon: the legacy method prior to AutoCAD 2012 (using snap or the UCS command) and the new method via the UCS icon grips menu, which we cover first.

Method 1

In AutoCAD 2012, the UCS icon gained some new functionality in the form of grips. It seems like an odd place to add grips—the icon is not a construction object after all—but it works well. To demonstrate, draw a rectangle tilted at a 45° angle and click on the UCS icon. You see three grips appear. The outer ones can be clicked again to manually rotate the icon to any desired angle. The inner one can be used to move the icon around. Each of the grips reveals a menu as well. We ignore the inner grip for now and focus on the outer ones. Select the Rotate Around Z Axis choice (Figure 15.56) and AutoCAD prompts you to enter a rotation angle (also 45° in this case).

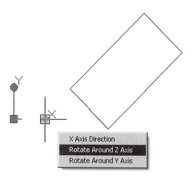

FIGURE 15.56
UCS grip menu.

Once you enter the value, the crosshairs rotate and turn a colorful green and red. See the rotation of the crosshairs in Figure 15.57. If you have Ortho on, you can now draw lines that are aligned to your rectangle.

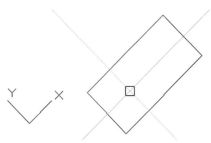

FIGURE 15.57
Rotated crosshairs.

Note that you can also achieve the same result by these following steps:

Step 1. Type in `ucs` and press Enter.
- ○ AutoCAD says: `Specify origin of UCS or [Face/NAmed/OBject/Previous/View/World/X/Y/Z/ZAxis] <World>:`

Step 2. Type in `z` and press Enter. You always want to rotate around the Z axis in 2D.
- ○ AutoCAD says: `Specify rotation angle about Z axis <90>`

Step 3. Type in 45° for this example. The UCS icon is rotated along with the crosshairs.

To return back to the usual orientation of the UCS/Crosshairs, you can type in `ucs`, press Enter, and type in `w` for `World`, or select World from the center grip menu, as seen in Figure 15.58.

Method 2

This legacy method does something different: It rotates the crosshairs but not the UCS icon. It is a minor variation on the same idea, but it is one you should be aware of. In 2D AutoCAD, this makes no practical difference, as stated earlier. In 3D, there is a more significant consequence, and the UCS method is used.

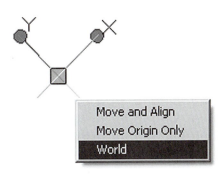

FIGURE 15.58
UCS World.

Step 1. Type in snap and press Enter.
Step 2. Type in ro and press Enter.
Step 3. For base point, type in 0,0.
Step 4. For angle of rotation type in per for perpendicular.
Step 5. Select an angled line.
Step 6. Turn off the SNAP command.

To get back to normal, repeat all six steps, except type in 0 at step 4.

15.29 WINDOW TILING

This is just a simple tool for allowing viewing of two drawings side by side in a convenient manner. Open up two drawings in one session of AutoCAD and press Restore Down (middle button) at the upper right for one of the drawings, so the two drawings are tiled one on top of the other. You see something similar to Figure 15.59. Then, select **Window→Tile Vertically** from the drop-down menu as seen in Figure 15.60. After the button is clicked on, you see what is shown in Figure 15.61. The active drawing is highlighted.

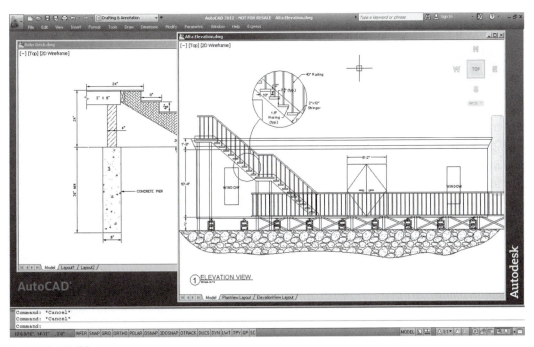

FIGURE 15.59
Window tiling, start.

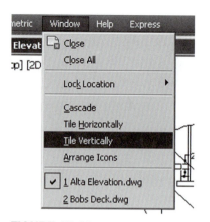

FIGURE 15.60
Window→Tile Vertically.

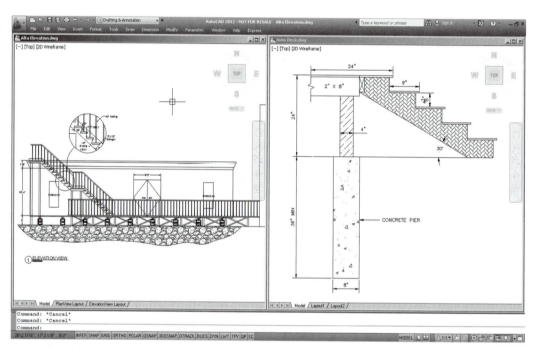

FIGURE 15.61
Window tiling, end.

Both drawings are lined up neatly on the screen. You can now click in and out of each and use copy and paste commands to transfer items if necessary. The same thing can of course be done for a horizontal tiling.

15.30 WIPEOUT

For reasons that may not be immediately apparent but are explained in detail, this command is one of the most important in 2D in producing drawings that not only possess extraordinary clarity but are easy to work with and modify. When used properly, wipeout is truly an "insiders'" secret, not known by beginner students, and one of the reasons why drawings done by AutoCAD experts have that intangible "something" that sets them apart from those done by less experienced designers.

First, let us define what wipeout is. It is simply a tool that creates a mask (or screen) that blocks out the viewer's ability to see other drawn geometry covered by the wipeout. To use it, draw a shape of some sort. Then type in wipeout, press Enter, and click your way around all or part of the shape. When you press Enter or right-click, the shape disappears behind the wipeout, as seen in Figure 15.62.

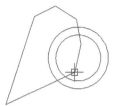

FIGURE 15.62
Creating a wipeout and the result.

Notice the wipeout itself is still visible (something to be addressed later), but the double circle shape *has* been partially blocked as expected. Practice using this command a few times just to become familiar with it.

Remember that a wipeout itself is an object. It can be moved, erased, copied, rotated, scaled, and just about anything else you can do to a regular geometric object. You can also click on the wipeout (click on the lines, not the blank center) to activate its grips. Then the grips can be used to reshape it into any shape whatsoever, as seen in Figure 15.63.

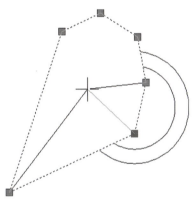

FIGURE 15.63
Wipeout grips.

391

So, what is so important about wipeout? Well, this command allows us to do something very special in 2D work. We can give our designs *depth*.

When you look around you in the real world, objects are generally not transparent. So, if you are looking at a bus in front of a building, part of the building is obscured by the bus. If another car were to park between you and the bus, it would obscure part of the bus, and so forth. In 3D, this is no problem at all. All objects are solid, and while they can be viewed in transparent wireframe form, a click of the button gives you rendered models that *create an instant illusion of depth*. This is important, as it matches our everyday experience and we quickly identify what we are looking at. In the image in Figure 15.64, the chair and the computer monitor both block a part of the computer desk. It is easy to tell what is in front of what in the rendered image on the left, but not as easy with the wireframe image on the right.

Let us try the same thing with the bus and car, as seen in Figure 15.65. Which is in front of which? You really cannot tell without shading the image, as in Figure 15.66.

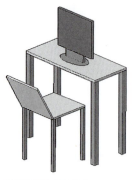

FIGURE 15.64
Solid and wireframe images.

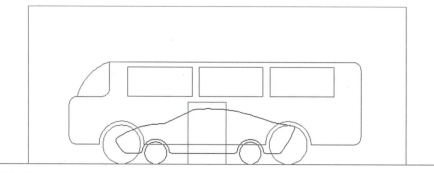

FIGURE 15.65
Bus, building, and car—wireframe.

FIGURE 15.66
Bus, building, and car—solid.

In 2D, though, we have a problem: We cannot shade in the conventional sense. Everything is drawn using lines and nothing is solid, even though it may be in real life. In a plan view of an architectural floor plan, this is not an issue, as we are looking straight down, but what about in elevation view, such as the previous images? A section through a room in elevation view may feature furniture obscuring walls and walls obscuring other items behind them. Or, in electrical engineering, we have equipment on a rack obscuring mounting holes. To provide realism and a sense of depth, we need to somehow obscure our view of items not seen. Of course, you can trim linework behind the object, but what if the object moves? You now have a gaping hole. We need a way to temporarily obscure and hide geometry that is tied directly to the object doing the hiding, so if it moves, the linework reappears. Wipeout is the secret to that trick.

In Figure 15.67, notice how the equipment bolted to the rack obscures part of the rack. Now, if the equipment is moved, such as in Figure 15.68 (notice the numbers to the left change from 20U – 22U to 8U – 10U), the wipeout moves with it and it acts like a solid object. The move is successfully done with no trimming necessary.

FIGURE 15.67
Electrical gear on a rack with a wipeout.

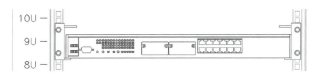

FIGURE 15.68
Electrical gear relocated.

What if you did not have a wipeout? Such is the case in Figure 15.69. Notice that the rack is visible underneath the equipment, giving no indication of depth. You would have to trim lines to achieve this.

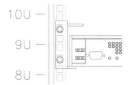

FIGURE 15.69
Same gear with no wipeout.

In summary, when a wipeout is attached to an object, it prevents anything behind it from being seen, exactly as in real life. So, a bookshelf with a wipeout attached blocks the wall behind it and electrical equipment blocks the mounting holes to which it is attached. Move the object and the wipeout moves, also blocking out whatever you happen to be blocking. This is the technique that the pros use to give drawings a uniquely realistic look in 2D design, and you should learn this as well. Let us go through some details.

The essential steps of making this technique work include the following:

Step 1. Create the drawing of what you are designing, such as the equipment rack in Figure 15.70.

FIGURE 15.70
Wipeout, Step 1.

Step 2. Using wipeout, click to trace the outline of the drawing point by point (just the four corners in this case).
Step 3. Using **Tools→Draw Order→Send to Back,** put the wipeout behind the design.
Step 4. Make a block out of the design and wipeout.

The object is now "solidified" and can be placed in front of other objects. If you create more objects in this manner, you can simply use the Draw Order tool to shuffle them around top to bottom as needed (as long as all of them are blocks). In this manner, you can create a rather impressive sense of depth and clarity to an otherwise busy drawing.

Take a close look at Figure 15.71. It is a larger section of the same cabinet shown in the previous figures. It could not have been done this way without extensive use of the wipeout command. Notice how it is easy to tell which equipment is in front and which is behind the rack. There are in fact several layers of objects: the equipment bolted to the front of the rack, then the rack itself, then the equipment in the back of the rack.

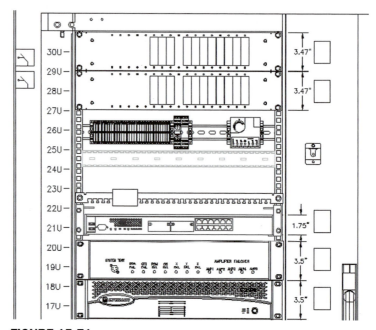

FIGURE 15.71
Full equipment rack with multiple wipeouts.

The final item to mention with wipeouts is that they themselves can be hidden (while their ability to hide objects remains in effect). To do this, simply type in wipeout, press Enter, and select f for Frames in the submenu. AutoCAD then says: Enter mode [ON/OFF] <ON>:. If you type in off, AutoCAD turns off all wipeouts on the screen, while preserving their effects on other objects.

LEVEL 2 DRAWING PROJECT (5 OF 10): ARCHITECTURAL FLOOR PLAN

For Part 5 of the drawing project, you add in carpeting and flooring to your floor plan. This is somewhat arbitrary and may or may not be the type of carpeting, tile, or flooring an architect would select on another design. You may of course use a different pattern if you wish. As always, be sure to use the proper layering.

Step 1. At a minimum use the following layer convention:
- A-Carpeting (Gray_9).
- A-Flooring (Color_32).

Step 2. Freeze all layers that are not relevant to the carpeting, leaving only walls, wall hatch, appliances, and windows. Draw a set of temporary lines to block off doorways so the hatch does not bleed from one room into another. Finally create the carpeting/flooring plan shown in Figure 15.72.

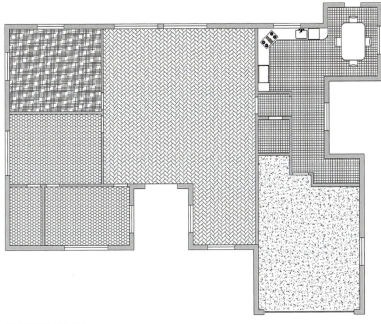

FIGURE 15.72
Carpeting and flooring layout.

SUMMARY

You should understand and know how to use the following concepts and commands before moving on to Chapter 16:

- Align
- Audit and Recover
- Blend
- Break and Join
- CAD Standards
- Calculator
- Defpoints
- Divide and Point Style
- Donut
- Draw Order
- eTransmit
- Filter
- Hyperlink
- Lengthen
- Object Tracking (OTRACK)
- Overkill
- Point and Node
- Publish
- Raster
- Revcloud
- Sheet sets
- Selection methods
- Stretch
- System variables
- Tables
- Tool palette

- UCS and crosshair rotation
- Window tiling
- Wipeout

REVIEW QUESTIONS

Answer the following based on what you learned in Chapter 15:

1. What are the purposes of the audit and recover commands? What is the difference?
2. For join to work, lines have to be what?
3. The defpoints layer appears when you create what?
4. Does the divide command actually cut an object into pieces?
5. What is the idea behind eTransmit?
6. To what can you attach a hyperlink? What can you call up using a hyperlink? How do you activate a hyperlink?
7. What is the idea behind overkill?
8. Can you make a revcloud out of a rectangle? A circle?
9. What three new selection methods are discussed?
10. Do you use a Crossing or a Window with the stretch command?
11. Name a system variable mentioned as an example.
12. Explain the importance of the wipeout command. What does it add to the drawing?

EXERCISES

1. Draw the following two rectangles according to the dimensions given. Pedit the width to 0.25" and shade each separately with a solid hatch, with color 9. Then use `align` to place the 10" box on top of the 4" box. (Difficulty level: Easy; Time to completion: <5 minutes)

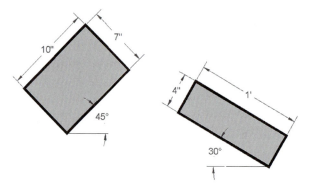

2. Draw the following set of lines and chamfer them to the dimensions shown. Then, pedit the result to a thickness of 0.25". (Difficulty level: Easy; Time to completion: <5 minutes)

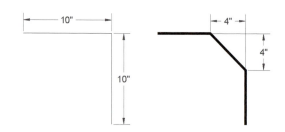

3. Draw the following line and divide it into seven sections using six points of the style shown. Then, pedit the line to a thickness of 0.25" and break it using the break command (First point option) at each of the points. Finally, offset each alternate section 0.5" up and 0.5" down, as shown. (Difficulty level: Easy; Time to completion: <5 minutes)

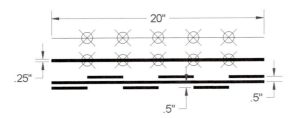

4. Draw the following rectangle, circle, and triangle according to the dimensions shown. Then use revcloud and the object option to turn each into a revcloud. Finally, pedit the new shapes to a thickness of 0.25". (Difficulty level: Easy; Time to completion: 5 minutes)

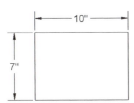
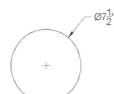
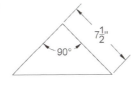

CHAPTER 16

Importing and Exporting Data

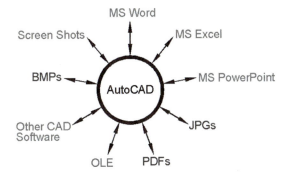

399

LEARNING OBJECTIVES

In this chapter, we learn how AutoCAD interacts with other software you are likely to use in the course of design work. We cover the following topics:

- AutoCAD and MS Word
- AutoCAD and MS Excel
- AutoCAD and MS PowerPoint
- AutoCAD and JPGs
- AutoCAD, PDFs, and screen shots
- AutoCAD and other CAD software
- Exporting and Save <u>A</u>s
- Importing and OLE

By the end of the chapter, you will smoothly import and export data among a variety of common office and design applications.

Estimated time for completion of chapter: 2 hours.

16.1 INTRODUCTION TO IMPORTING AND EXPORTING DATA

AutoCAD and the designers who use it usually do not work in a "software vacuum." They typically interact with not only AutoCAD but a number of other applications, some of them generic and used by many other professions, others unique to their own field. As such, there is often a need to either import data *into* or export data *out of* AutoCAD. Examples are numerous, such as an architect who wishes to transfer notes typed in MS Word into his project details page in AutoCAD or an engineer who needs to drop in an Excel spreadsheet. Just as often, AutoCAD drawings may need to be inserted into MS Word reports, PowerPoint presentations, and other applications. You may then need to create PDFs of your drawing or insert a PDF or an image into one.

Swapping files among the various software applications has been tricky in the past. Software companies are under no obligation to make such tasks flow smoothly, as it is not their business what else you may have on your PC (which may include programs from a direct competitor). Of course, such thinking scores no points with the customer, so developers do try to minimize compatibility issues whenever possible. Today, such tasks flow much smoother but still require knowledge of procedures that can be unique to each application.

The sheer quantity of software on the market is staggering, so we limit the discussion to

- Software or file types that are used or encountered by virtually everyone who uses a computer (Word, Excel, PowerPoint, JPG, and PDF).
- Software that has special relevance to AutoCAD users (such as other CAD programs).

We focus on two distinct tasks, how to *import* files *from* these applications and how to *export* AutoCAD files *into* them, analyzing each software pairing bi-directionally, and conclude with a brief overview of the export and insert features and OLE.

16.2 IMPORTING AND EXPORTING TO AND FROM MS OFFICE APPLICATIONS

We begin with MS Word, Excel, and PowerPoint, as those are probably the ones most important and relevant to the majority of users. Throughout, unless noted otherwise, we make extensive use of Copy and Paste. These Copy/Paste to clipboard tools are some of the most useful in computing, allowing you to shift around just about anything from one spot to another. The keyboard shortcuts are Ctrl + C for copy and Ctrl + V for paste or you can just right-click and select Copy and Paste.

Be aware that the "copy" is not AutoCAD's standard copy command, so do not use that. Yes, some students mix these up, so let us be clear from the start. Open up Word, Excel, and PowerPoint and keep them handy as we go through each one.

Word into AutoCAD

An example of this may be when you would like to bring in extensive typed construction notes onto one of the pages of the AutoCAD layout.

Go ahead and click and drag to highlight the text in Word first and right-click or Ctrl + C to copy to the clipboard. Go to the AutoCAD screen, but do *not* right-click and paste the text in. If you do this, it will appear as an embedded object, with a white background (seen in the top line of Figure 16.1), and will not be editable. Instead, create an *mtext* field of an appropriate size, and paste the copied text into it. The result is easily editable text that can be further formatted, as seen in the bottom line of Figure 16.1.

AutoCAD into Word

An example of this may be when you are preparing a report and would like to include some of the graphics spread out among the text to enhance clarity or support documentation.

This is a direct application of the Copy/Paste function *or* the PrtScn (print screen) key. With Copy/Paste you no longer have to change the background color of AutoCAD before insertion as you once did. Developers finally conceded to the fact that, yes, most paper is white and

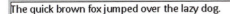

The quick brown fox jumped over the lazy dog.

The quick brown fox jumped over the lazy dog.

FIGURE 16.1
Insertion of MS Word text.

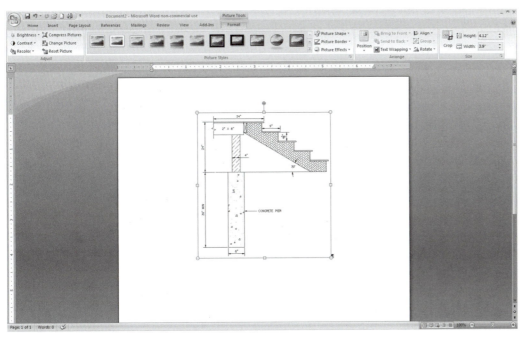

FIGURE 16.2
Insertion of an AutoCAD design into Word.

the black AutoCAD background just does not look right—an automatic color change is performed, but be careful with older AutoCAD releases.

Once inserted, some cropping is necessary to get rid of excess white space. Often the Print Screen button followed by pasting in is easier, and the 2007 Word and newer versions bring the cropping tool up right away when you click the pasted-in screen shot. Indeed, this was the initial method for importing screen shots throughout this textbook.

A typical view of this procedure is shown in Figure 16.2.

Excel into AutoCAD

An example of this may be when you would like to bring a spreadsheet into your electrical engineering drawing that details the type of connections and wiring used in a design.

Once again, you use Copy/Paste. In Excel, create a small group of cells, number them (adding borders if desired), then highlight all, right-click Copy, go into AutoCAD, and right-click Paste. As soon as you do this, the dialog box in Figure 16.3 appears.

This is the OLE (object linked embedded) Text Size dialog box, which allows you to alter the text height of your inserted spreadsheet. The exact size is not too important for now; you can easily scale objects up or down anytime after insertion, but the concept of OLE is important and is discussed again soon. For now, press OK, and you see what is shown in Figure 16.4 (zoom in if needed).

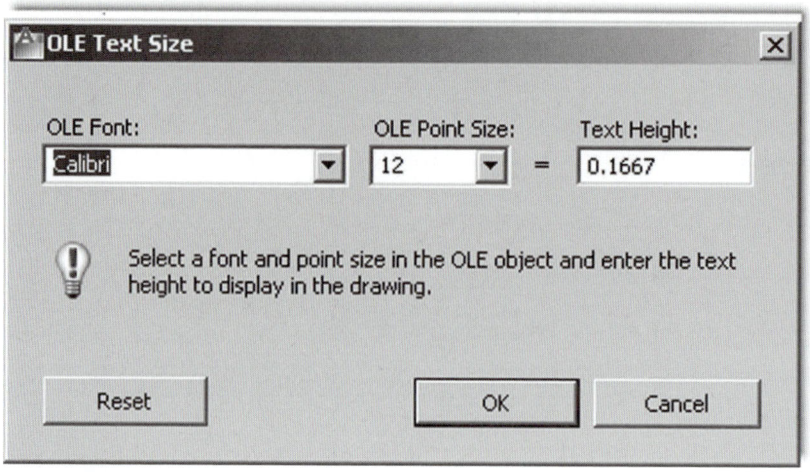

FIGURE 16.3
OLE Text Size.

1	2
3	4
5	6
7	8
9	10

FIGURE 16.4
Embedded Excel cells.

This is your basic Excel insertion. The cells are regular objects (very much like blocks), and you can erase, copy, and move them as needed. You can also scale them by clicking once and moving around the resulting grip points. You cannot mirror or rotate the cells, however.

A major question usually arises at this point from students. Are the inserted Excel cells linked to Excel, allowing for changes and updates? They are not in the true sense of the word. What you can do is double-click on these inserted cells and Excel is called up temporarily for editing purposes, but any changes made to the original file after accessing it *from* Excel (not AutoCAD) are not reflected in AutoCAD.

Having truly linked files is a very useful idea. There are some fancy Excel/AutoCAD interfaces in the industry where extensive Excel data is used to generate corresponding AutoCAD drawings, but this trick requires additional programming and extensive customization and is beyond the scope of this book.

AutoCAD into Excel

This action basically pastes an image into Excel with the same Copy/Paste procedure as for Word. This, however, is rarely done, as Excel is primarily for spreadsheet data; and unless you need a supporting drawing, there is just no overwhelming reason to do this.

PowerPoint into AutoCAD

This action is not very common, as PowerPoint is almost always the *destination* for text and graphics, not an intermediate step to be inserted into something else. However, if it is needed, you can insert slides into AutoCAD via regular Copy/Paste.

AutoCAD into PowerPoint

This makes a lot more sense, as you may want to insert a drawing into a PowerPoint presentation. Fortunately, PowerPoint behaves very much like Word in this particular case, and one Copy/Paste followed by some cropping is all it takes to insert an AutoCAD design, and no new techniques need to be learned.

16.3 SCREEN SHOTS

Screen shots or *screen captures* refer to capturing a snapshot of everything on your screen as seen by you, the user, and sent to memory, to be inserted somewhere. This is done by simply pressing the Print Screen button, which on most computers is called PrtScn, SysRq, or sometimes just F12. You can capture what is on your screen this way and insert it into Word or PowerPoint, then crop out what you do not need to see. While many computer users may know of the Print Screen function to actually print something, they may overlook this useful trick for inserting data into other applications.

16.4 JPG

This file compression format (used most often with photographs) can be inserted directly into AutoCAD and is a great idea if you would like a picture of what you drafted next to the drawing. Examples include an aerial photo of a site plan for architectural and civil applications and a photo of an engineering design after manufacture for as-built or record drawings. Inserting JPGs is straightforward process.

Step 1. Click on a file you want to insert and right-click to copy it.
Step 2. Then right-click and paste it into AutoCAD.
  ○ AutoCAD asks for an insertion point:
    `Specify insertion point <0,0>:`
Step 3. Click anywhere you like.
  ○ AutoCAD then asks for the image size:
    `Base image size: Width: 0.00, Height: 0.00 Inches`
    `Specify scale factor or [Unit] <1>:`
Step 4. Scale the JPG by moving the mouse to any size you wish (or type in a value).
  ○ AutoCAD then asks for the rotation angle:
    `Specify rotation angle <0>:`
Step 5. If you want the image rotated, enter a degree value; otherwise, press Enter.

The result of a JPG insertion is shown in Figure 16.5. It has also been selected (one click), revealing the Ribbon's Image Editor.

The image can be erased, copied, moved, rotated, mirrored, and just about anything else you can do to an element. You can also do some very basic image adjusting via the Ribbon or by *double-clicking* on the picture. The dialog box shown in Figure 16.6 appears, allowing you to adjust Brightness, Contrast, and Fade, as well as reset everything. Any changes apply only to that JPG, not copies of it (if other copies exist).

16.5 PDFS

The Portable Document Format (PDF) is of course Adobe's popular format for document exchange. AutoCAD drawings can be easily converted to PDFs by simply printing to them (in other words, selecting PDF as the printer, assuming Acrobat is installed). The Print function executes, and AutoCAD asks you where you want the file saved. Select the destination and your PDF file is created.

Another way to generate a PDF, especially if you need to customize the output, is via the Ribbon's **Output tab→Export** option. If you select the PDF option a Save As PDF dialog box

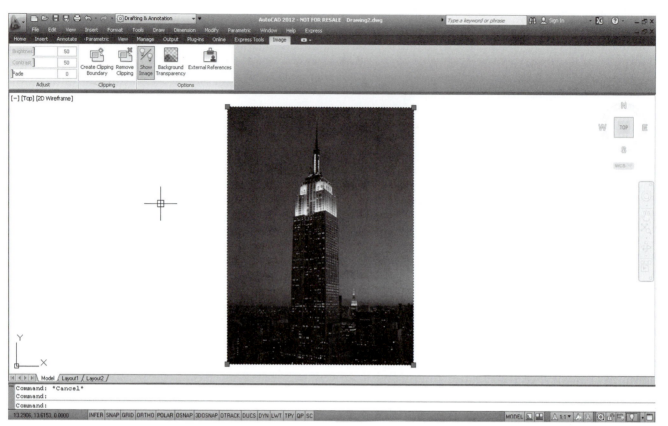

FIGURE 16.5
Embedded JPG.

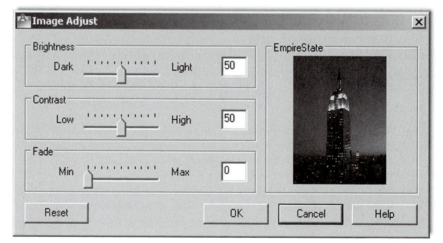

FIGURE 16.6
Image Adjust.

appears and allows some tinkering with options and output to finesse the results, as seen in Figure 16.7.

As of the 2010 release of AutoCAD, you can bring in a PDF file as an underlay. This means it comes into AutoCAD as a background, which you can see but not edit. This is not unlike raster images (mentioned in Chapter 15) and is quite useful if all you have of a design is a PDF. After insertion you can draw over it or use it as a supporting image, such as a key

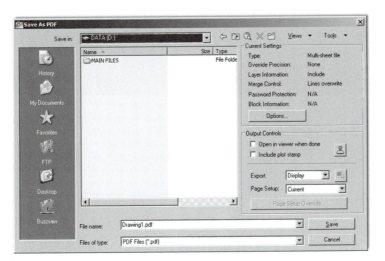

FIGURE 16.7
Save As PDF.

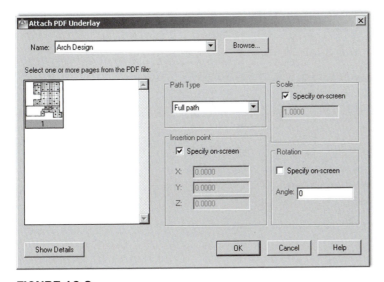

FIGURE 16.8
Attach PDF Underlay.

plan. To bring in the PDF, use the cascading menu **Insert→**PDF **Underlay**…. You then are prompted to look for the file you want to insert, and upon selection, you see the dialog box in Figure 16.8.

Once you press OK, the PDF inserts into the AutoCAD file and the box prompts you for an insertion point and scale factor. Once the image is embedded, you can click on it and the Ribbon changes to show you some of the image editing options, as seen in Figure 16.9, with Change to Monochrome and some Fade and Contrast adjustments among the more useful tools.

16.6 OTHER CAD SOFTWARE

Initially, as AutoCAD gained market dominance, it did not really attempt, nor had any incentive, to "play nice" with other CAD software on the market. Being the 2D industry leader allowed AutoCAD to get away with not accepting any other files in their native format. It was the company's way or the highway, but that slowly changed. One major concession in regards to importing and exporting was with MicroStation, AutoCAD's nemesis and only

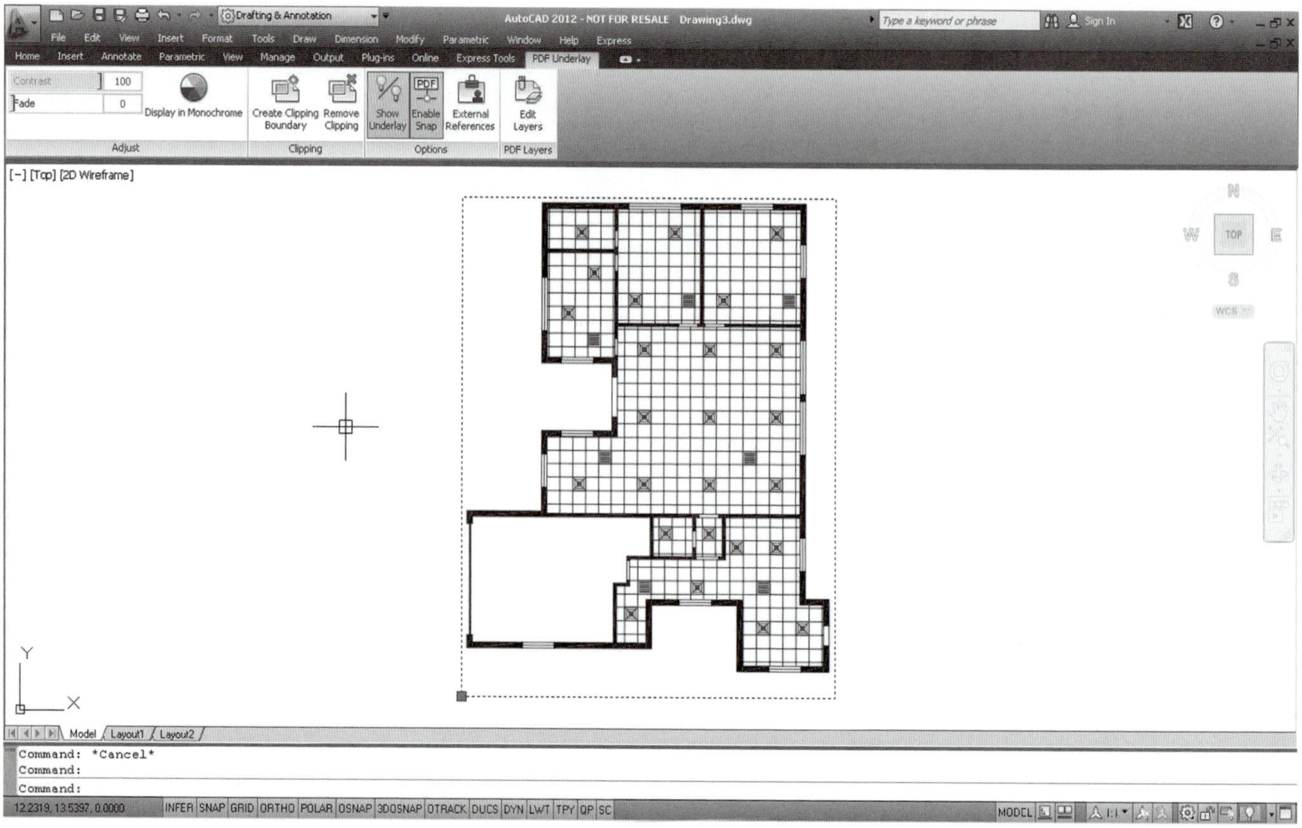

FIGURE 16.9
PDF Underlay Ribbon menu.

serious competitor. MicroStation always opened and generated AutoCAD files easily but not the other way around.

Finally, in recent releases, AutoCAD listed the `dgn` file format as something you can import and export. You were then able to bring in MicroStation files, more or less intact, as well as create ones from AutoCAD drawings. In AutoCAD 2009, the settings box was expanded and that was carried over into the current AutoCAD 2012, as seen in Figure 16.10. You can now compare layers and other features side by side and make adjustments so the target file is similar to the original.

To access this, use the cascading menu **File→Import**… and search for the MicroStation `dgn` file you are interested in importing. Another way around all of this is just to have the MicroStation user generate the AutoCAD file. You can also export `dwg` files directly to `dgn` as well by using the cascading menu **File→Export**… and selecting `dgn` as the target file. You then see the same Settings dialog box as in Figure 16.10. Of course, MicroStation accepts the AutoCAD file with no problem, so this may not be necessary. If you would like to learn a bit more about MicroStation in general, see Appendix B.

Let us expand the discussion to other CAD software. What should someone do who is using another CAD program and needs to exchange files with an AutoCAD user? The `dxf` file is the solution, as it is a format agreed upon by most CAD vendors as the go-between that can be understood by all applications for easy file sharing. AutoCAD easily opens a `dxf` or generates its own `dxf` files for any drawing. This is discussed momentarily.

A few observations on sharing files: Among architects, the issue is not much of a problem, as AutoCAD is the main application in this profession and competitors such as ArchiCAD easily open AutoCAD's files and send a `dxf` right back if needed. In engineering, however, AutoCAD has a lesser presence. While some electrical, mechanical, and civil engineers use AutoCAD,

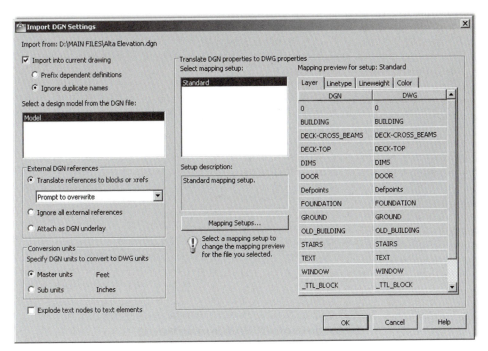

FIGURE 16.10
Import DGN Settings.

many others, especially industrial, rail, aerospace, automotive, and naval engineers, use 3D modeling and analysis software instead such as CATIA, NX, Pro/Engineer, or SolidWorks.

Until recently, these applications did not interact well with AutoCAD. Their respective "kernels" (the software's core architecture) are significantly different in design and intent from AutoCAD's ACIS 3D kernel. Their native and export file types (such as IGES and STEP) may also be unfamiliar to AutoCAD users. Some of the previously mentioned software *can* generate dxf files, but those files are "flattened" and what you get is a 2D snapshot of what used to be a 3D model. AutoCAD was also able to import images from CATIA and SolidWorks using the OLE command (to be discussed soon), but once again they were just flattened images.

Things improved somewhat with AutoCAD 2012. The 3D models can now be imported into AutoCAD with little loss in fidelity. Much more is said about this in the 3D version of this textbook.

All this can be somewhat confusing. Because so many different CAD packages are on the market, AutoCAD designers need to be aware and up to date on what others in their industry are using and think on their feet when it comes time to exchange files, as there usually *is* a way for most files types. Let us now take a look at the specifics of exporting, importing, and the OLE command to explore the available options.

16.7 EXPORTING AND THE SAVE AS FEATURE

As previously with the MicroStation example, exporting data can be done by selecting **File→Export…** from the cascading menu. What you see is the dialog box in Figure 16.11.

Your drawing can be exported to the following formats:

- **3D dwf.** This is to convert the drawing to web format for use on viewers.
- **Metafile.** This is to convert the drawing to a Windows Meta File, a Microsoft graphics format for use with both vectors and bitmaps.
- **ACIS.** ACIS is the 3D solid modeling kernel of AutoCAD as well as a format to which a drawing can be exported. Few other software currently use the ACIS kernel (SolidEdge comes to mind), as Parasolid is the industry standard, so this is a rarely used export format.

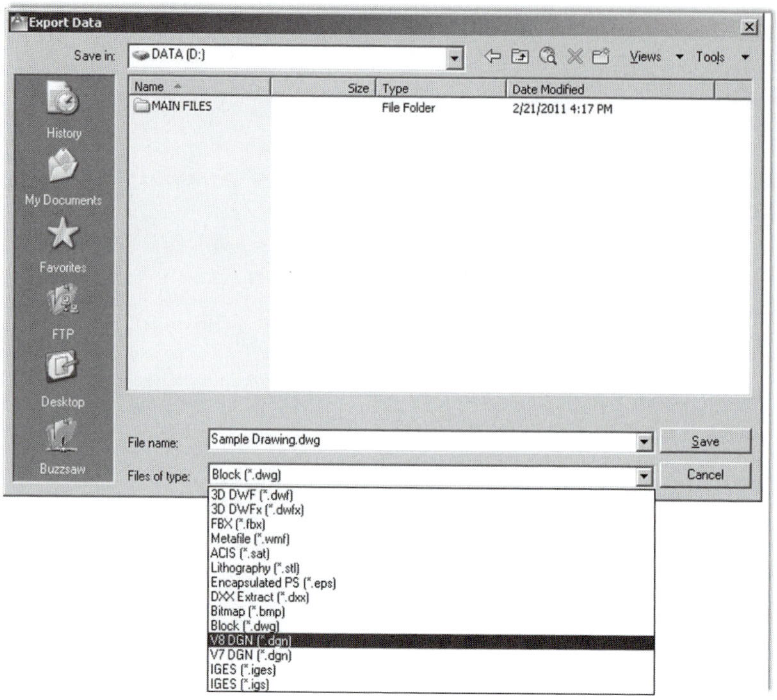

FIGURE 16.11
Export Data.

- **Lithography.** Stereolithography files (`*.stl`) are the industry standard for rapid prototyping and can be exported from most 3D CAD applications including AutoCAD. Basically, it is a file that uses a mesh of triangles to form the shell of your solid object, where each triangle shares common sides and vertices.
- **Encapsulated Post Script.** The `eps` is a standard format for importing and exporting PostScript language files in all environments, allowing a drawing to be embedded as an illustration.
- **DXX Extract.** DXX stands for Drawing Interchange Attribute. This is a `dxf` file with only block and attributes information and is not relevant to most users.
- **Bitmap.** This converts the drawing from vector form to bitmap form, with the resulting pixilation of the linework; it is not recommended.
- **Block.** This is another way to create the familiar block covered in Level 1.
- **V8 dgn.** MicroStation Version 8, as described previously.

The Save As dialog box can be accessed anytime you select **File→Save** or **File→Save As**… from the drop-down cascading menus. You can also type in `saveas` and press Enter. In either case, you get the dialog box shown in Figure 16.12.

The extensions presented here are described in detail in Appendix C, but in summary, you can save your drawing as an older version of AutoCAD (an important step that is automated via a setting in Options, as covered in Chapter 14) or as a `dxf` file, as described earlier. Note that the `dxf` files can also be created going back in time, all the way to Release 12 (that was a mostly DOS-based AutoCAD). The other extensions, `.dws` and `.dwt`, are used less often and are also mentioned in Appendix C.

16.8 INSERTING AND OLE

The final discussion of this chapter features the Insert menu option in general and the OLE concept or command in specific. The Insert cascading menu is shown in Figure 16.13.

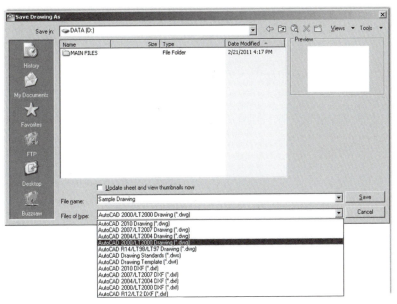

FIGURE 16.12
Save As dialog box.

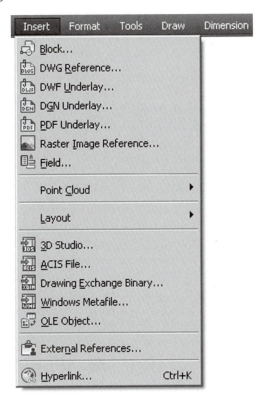

409

FIGURE 16.13
Insert cascading menu.

You may already be familiar with most of what is featured in this drop-down menu, or it is covered soon as a separate topic:

- Block insertion is the same as typing in `insert` (from Level 1).
- Hyperlinks and raster images are covered in Chapter 15.
- External references is all of Chapter 17.
- Layouts is a Paper Space topic from Chapter 10.

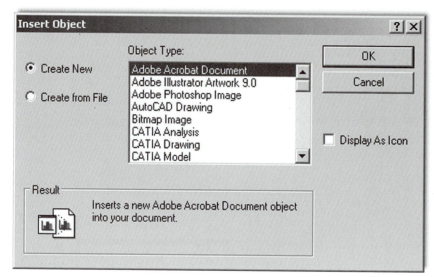

FIGURE 16.14
Insert Object.

Other items we do not cover (and you rarely need) are

- Fields.
- 3D Studio, ACIS, Drawing Exchange Binary, and the like.

Let us then focus on OLE (object linked embedded). OLE is a way to embed (insert) nonnative files into AutoCAD on a provisional basis, meaning these files are not part of AutoCAD (they cannot be anyway, as they are of non-AutoCAD formats), but they are merely dropped in and are visually present. In principle, you can drop in just about anything, with just about any results. Some files sit nicely, others do not appear, and still others may even cause a crash. Let us take a look at the dialog box.

Select <u>O</u>LE Object… from the Insert menu and the dialog box in Figure 16.14 appears.

These, in theory, are the various software file items that can be inserted into AutoCAD. Looking through it, top to bottom, you can see that a wide variety of Adobe, Microsoft, and even Dassault (they own CATIA and SolidWorks) files can be inserted. You can also, in principle, insert video and audio clips (one can envision music files playing upon the opening of an AutoCAD drawing) and even Flash animations. Simply pick what you want to insert and select Create from File. Some of the more useful OLE insertions are related to the Microsoft products listed, but much of that topic was already covered in this chapter. To insert a file, select Create from File and browse for the file in which you are interested.

LEVEL 2 DRAWING PROJECT (6 OF 10): ARCHITECTURAL FLOOR PLAN

For Part 6 of the drawing project, you complete the internal design of the house by adding a reflected ceiling plan (RCP) and a heating, ventilating, air conditioning (HVAC) system that will be made up of diffusers and a few return air ducts. Note that a residential home is not likely to have an RCP, which is so named because it looks exactly as if you held a mirror in your hand and looked down at it to see the ceiling. RCPs are mostly part of commercial space design and include what you generally see in a typical office: white ceiling tiles, lights, and HVAC diffusers. Here, for educational purposes, we will add this ceiling into the residential design.

Step 1. Create some appropriate layers, such as
  ○ M-RCP (Gray_8).

410

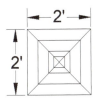

Diffuser   Return Air Duct

FIGURE 16.15
Diffuser and return air duct symbols.

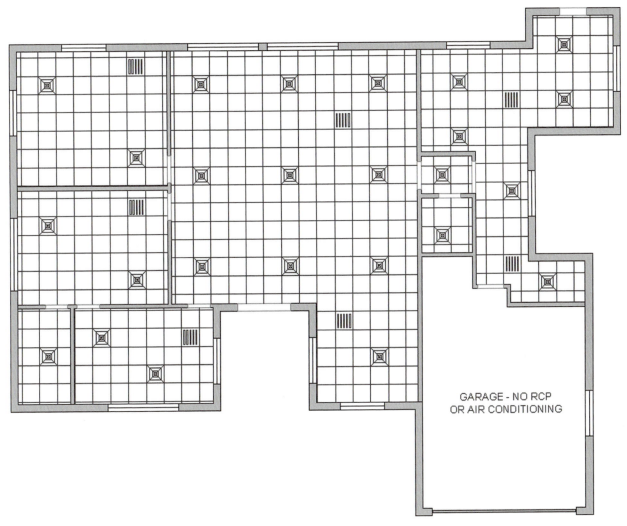

GARAGE - NO RCP
OR AIR CONDITIONING

FIGURE 16.16
RCP and HVAC layout.

- ○ M-Diffuser (Yellow).
- ○ M-Ret_Air (Cyan).

Step 2. Create some HVAC symbols, as shown in Figure 16.15. Make blocks out of them for easier handling. Be sure to be on that object's layer before making a block.

Step 3. Create the RCP plan by drawing an arbitrary vertical and horizontal "starter" line in the middle of each room and offsetting 2 feet in either direction until you reach the walls. Then trim as necessary.

Step 4. Add in the diffuser and return air duct symbols as shown in the plan (Figure 16.16).

SUMMARY

You should understand and know how to use the following concepts and commands before moving on to Chapter 17:

- Data transfer from Word to AutoCAD
- Data transfer from AutoCAD to Word
- Data transfer from Excel to AutoCAD
- Data transfer from AutoCAD to PowerPoint
- Screen shots, JPGs, and PDFs
- Interaction with other CAD software
- Exporting and Save As features
- Inserting and OLE

REVIEW QUESTIONS

Answer the following based on what you learned in Chapter 16:

1. What is the correct way to bring Word text into AutoCAD?
2. What is the procedure for bringing AutoCAD drawings into Word?
3. What is the correct way to bring Excel data into AutoCAD?
4. What is the procedure for bringing AutoCAD drawings into PowerPoint?
5. What is the procedure for importing JPGs and generating PDFs?
6. Describe the basic ideas of interacting with other CAD software.
7. What are Exporting and Save As?
8. What is OLE?

EXERCISE

1. At your own pace, review everything presented in this chapter and do the following:
   - Type any short paragraph in MS Word and import it into AutoCAD correctly, using Copy and Paste.
   - Type in several columns of data in MS Excel, then import it into AutoCAD. Adjust it by scaling it up and down.
   - Import any image from AutoCAD into MS Word using
     - ○ The Print Screen function.
     - ○ Copy and Paste.
   - Export any drawing to a dxf format.
   - Bring in any JPG (picture) using Copy and Paste. Adjust it in AutoCAD.

(Difficulty level: Easy; Time to completion: 15 minutes)

External References (Xrefs)

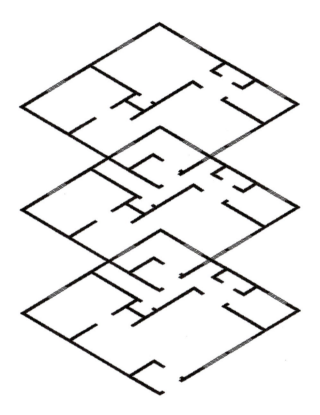

LEARNING OBJECTIVES

In this chapter we introduce and thoroughly cover the concept of external references (xrefs). We specifically introduce the following topics:

- The primary reasons for using xrefs
- Loading xrefs
- Unloading xrefs
- Binding xrefs
- Updating and editing xrefs
- Layers in xrefs
- Multiple xrefs

By the end of the chapter, you will be well versed in applying this critical concept to your design work.

Estimated time for completion of chapter: 1–2 hours.

17.1 INTRODUCTION TO XREFS

External references, or xrefs for short, is a critically important topic for architectural AutoCAD users and very useful knowledge for all others. Along with Paper Space (Chapter 10), xref forms the core of advanced AutoCAD knowledge. Like many advanced topics, you need to understand fundamentally what it is and when to use it. The "button pushing" or the mechanics of how to make it work is the easy part. In this introduction, let us explain what an xref is and why there is a need for this rather clever concept, which has existed since AutoCAD's early days.

What Is an Xref?

As defined, an xref is a file that is electronically attached (referenced) to another, new, file. It then appears in that new file as a fully visible background against which you can position new design work. This xref is not really part of the file, but it is merely "electronically paper clipped" to it and cannot be modified in a traditional way; it only serves as a background. The word *xref* is also an action verb, as in "I need to xref that file in."

Why Do We Need an Xref? What Is the Benefit?

The preceding paragraph was just a strict definition and may not have yet illuminated the real reason for xrefs. To see why we have this concept in AutoCAD let us propose the following scenario.

You are designing a tall office building. It is boxy in shape (similar to the former World Trade Center towers), and as such, the exterior does not change from floor to floor. The interior, however, does. The first floor is a lobby, the second is a restaurant, the third is storage, and floors 4 through 30 are office space. In short, for a designer, once the exterior is done, the rest is all interior-space work. Therefore, going from floor to floor, the exterior design file can be xrefed in and just "hang out," providing a background for placement of interior walls and other items. This dramatically reduces the size of each file as a copy of the exterior is merely referenced and never part of the interior files. Got 100 floors to do? No problem; one `xref` file of the exterior is all that is needed if those floors do not change.

- **Xref benefit 1. Reduce file size by attaching a core drawing to multiple files.** If this were all there was, xref would be a neat trick and would have surely fallen out of favor as computer speed, power, and disk storage space increased over the years. However, the preceding was just a warm-up. The main reason for the xref concept is *design changes*. If you need to change the shape of the exterior walls, you need to do it only *once*, and the change propagates through all files using that xref. Think about the implications for a moment. In the old hand-drafting days, each floor had to be drawn separately. Imagine now a last-minute change to the exterior design. Hundreds of sheets would have been updated by hand. Score one for AutoCAD.
- **Xref benefit 2. Critical file security.** Need another benefit? How about security? The `xref` file can be placed in a folder that is accessible only to the structural engineer and lead architect, not the interior designer or electrical engineer. Granted, this "need to know" basis for access may not be necessary, but it is good to have a way of keeping someone from shifting structural walls to accommodate the carpeting.
- **Xref benefit 3. Automated design change updates to core drawings.** Ideally, you are now sold on the benefits of an xref. So how is this actually done in practice? The xref is typically not just the exterior walls but also columns and the interior core (elevators and stairs), as those also tend to not change from floor to floor. The designer creates these files, naming them something descriptive such as `Exterior_Walls_Ref` or `Columns_Core`. Next, the

xrefs are attached together and then to the first floor file or to the first floor file individually; more on that later.

The first floor file is the active file of course and is named `Floor_One_Arch`. At first, it is blank, showing just the xref. Then, the interior design is completed. The designer opens another blank file, attaches the same xref or set of xrefs, and names this file `Floor_Two_Arch` and so on. Xrefs can be attached, detached, refreshed, and bound to the main drawing file. We cover all this shortly. We also cover some xref-related topics, such as layering and methods to edit xrefs in place.

17.2 USING XREFS

To demonstrate the features of the xref command, we first need to have a building exterior file with which to work. You can select one of your own or one assigned by your instructor if taking a class. Alternatively (and for good practice), draw the floor plan in Exercise 1 at end the end of the chapter. We use that floor plan for the rest of the xref discussion. In Figure 17.1, the plan is rotated horizontally and the left corner moved to point 0,0. All the dimensions are frozen and the file is saved as `ExteriorWalls.dwg`.

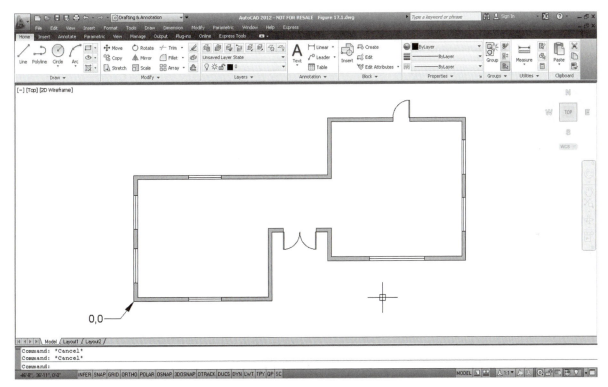

FIGURE 17.1
The xref file.

As of now, nothing is special about this file. We still need to xref it to the working file.

Open up a blank file and save it as `Floor_One_Layout.dwg`. Then type in `xref` and press Enter. Alternatively, you can use the cascading menu **Insert→External References**.... The palette in Figure 17.2 appears.

At the upper left is a white rectangle with a paper clip (hence the earlier paper clip reference). That is the Attach drawing button. You can click on the down arrow to reveal further attachment options (Image, dwf, dgn), but we do not need those. Simply click on the Paper icon and the

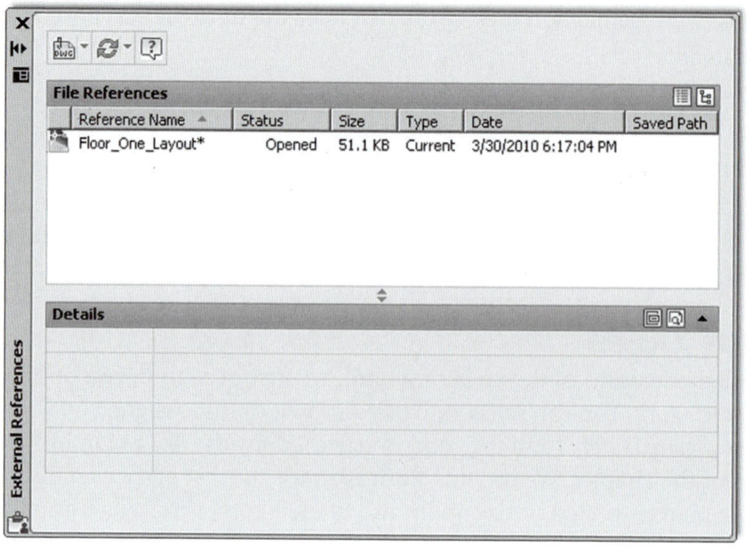

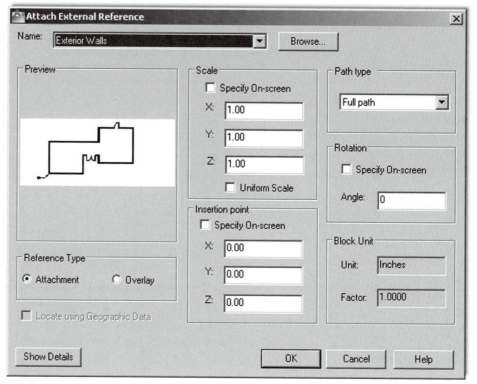

FIGURE 17.2
The xref palette.

Select Reference File browsing window opens. Browse to find the xref file (`ExteriorWalls.dwg`) and click on Open. The External Reference insertion box appears, as shown in Figure 17.3.

416

FIGURE 17.3
Attach External Reference.

Examine what is featured here, most of which you should be familiar with (such as Insertion Point, Scale, and Rotation). Uncheck Insertion Point; the fields turn from gray to white. We want the xref to insert at 0,0,0. Finally, press OK.

The xref appears with its lower left corner at 0,0,0. Zoom out to see the entire drawing if not apparently visible. Notice the new addition to the xref palette, as seen in Figure 17.4. The `ExteriorWalls` file is visible and indicated as Loaded with the appropriate date, time, and path.

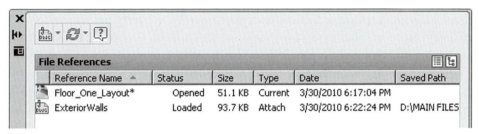

FIGURE 17.4
Xref loaded.

You can now proceed to add new design work to the file, as the xref procedure is complete. Let us skip ahead and explore some of the other available options, now that the xref is loaded. By right-clicking on the `ExteriorWalls` paper icon, you get access to the important menu in Figure 17.5.

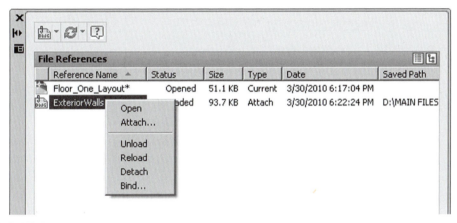

FIGURE 17.5
Xref menu.

417

Xref Menu

- **Open.** This option allows you to open the original xref if you wish to make changes.
- **Attach….** This option allows you to attach another xref to your working file with the same procedure as outlined earlier in the chapter.
- **Unload.** This option removes the xref from your working file but preserves the last known path, so if you wish to bring the xref back, you need not browse for it again but just click on Reload.
- **Reload.** The option just referred to for bringing back an unloaded xref.
- **Detach.** This option is more permanent and removes the xref from the working file, requiring you to browse and find it if reattachment is called for again.
- **Bind.** This option allows you to attach a permanent copy of the xref to your working file and severs all ties with the original. The xref then becomes a block; and if it is exploded, it completely fuses together with the working file. This action is potentially dangerous and is rarely used, as you are now stuck with the xref permanently. Two possible uses include permanent drawing archiving and emailing the drawing set (so as to not forget the xrefs), though eTransmit can take care of that.

17.3 LAYERS IN XREFS

A very logical problem creeps up when using xrefs. The `xref` file itself has layers, often many of them. There may also be several xrefs in one active drawing. And, of course, the drawing itself may have numerous layers of its own. It is almost guaranteed that, on a job of even medium complexity, layer names can and will be duplicated. This is not allowed in AutoCAD, and some method is needed to separate layers in the xref from mixing and being confused with layers in other xrefs or the main design drawing. The solution is quite simple. In the same manner that people have first and last names to tell them apart (especially helpful if the first names are the same), so do layers when xrefs are involved.

All layers that are part of a certain xref have that xref's name preceding the layer name with a small vertical bar (and no spaces) between them. All layers that are native to the working file of course have no such prefix.

This simple method ensures no direct duplication by having the xref file name act as the "first name" and the layer name as the "last name." An example is shown in Figure 17.6. New layers have been added in the host file. Notice the small vertical bar in the xref layer name and the inclusion of the `xref` file name.

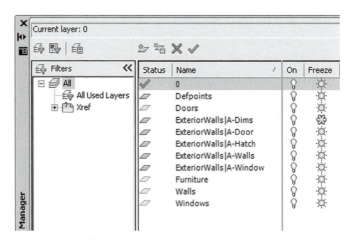

FIGURE 17.6
Xref layering.

When an xref is bound to the drawing, the layer's vertical bar changes to a 0 set of symbols, indicating they are now bound and part of the working drawing, as seen in Figure 17.7. You should be familiar with this if you have ever examined a bound set.

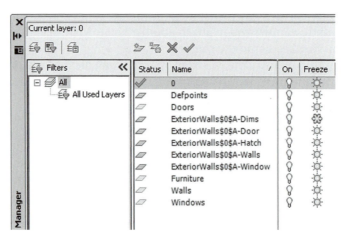

FIGURE 17.7
Xref bound layering.

17.4 EDITING AND RELOADING XREFS

At one time xrefs could not be changed in place. This means you had to open the original `xref` file to edit it and could not do this from the main working file. This made sense, as you generally want to limit access to the original xref, so it cannot be casually edited.

All things (and software) change, and a number of releases ago, AutoCAD started allowing xrefs to be edited in place using a technique similar to editing blocks (somewhat diminishing the security benefit of xrefs, as mentioned earlier). Let us give it a try. Double-click on any of the lines making up the xref while in the main file. The Reference Edit box appears, as seen in Figure 17.8.

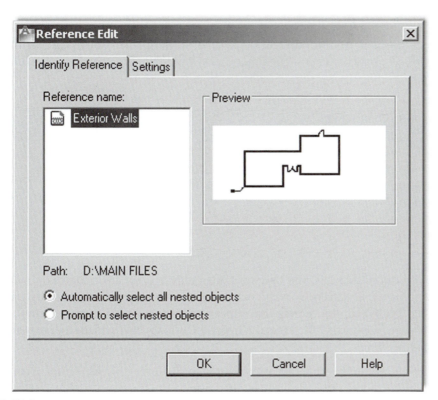

FIGURE 17.8
Reference Edit.

The xref you clicked on is listed under Reference name. Press OK and you are taken into Reference Edit mode. You may notice a few new features or items of interest:

- The xref has remained brightly lit up while everything that is on the main drawing file has faded down. This is a feature of Ref Edit. You can now clearly tell apart the xref geometry from the rest. The intensity of the fade can be adjusted under **Option→Display→Reference Edit fading intensity**.
- The xref can now be edited in place; go ahead and change a few items around.
- A new toolbar, called *Refedit*, appears (shown in Figure 17.9).

FIGURE 17.9
Refedit toolbar.

When you finish editing the xref, press the last button all the way to the right, Save References Edits. A warning icon appears; press OK and you are done. The xref has been edited in place.

Just as a reminder, this procedure is only for editing the xref while you have the main working file open, and what you just did can also be done in a far simpler manner by just opening the xref as its own file (although with Ref Edit you have the advantage of seeing the internal design while editing). In fact, when changes are extensive, the designer may just open the xref itself and work on it all day, without resorting to Ref Edit. Every time the designer saves his or her work, the changes become permanent and accessible to any other designers who may be working on a job that uses that xref.

A good question may come up at this point. How do other designers know that the xref is changed so they can adjust their interior work accordingly? Many releases ago, the only way to know was for the person working on the `xref` file to simply tell others via a visit to their cubicle or office, phone, or email. The other users would then press Reload All References, as seen in Figure 17.10.

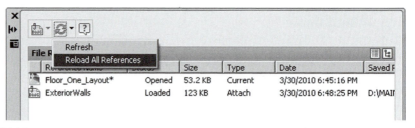

FIGURE 17.10
Reload xref.

Of course you can still do this, and designers often refresh every few minutes on their own if they know someone is actively working on the xref. However, a new tool appeared recently. It is simply a "blurb" announcement that appears in the lower right corner of AutoCAD's screen, as shown in Figure 17.11. Press the hyperlink and the xref reloads and refreshes automatically.

One final brief topic in this chapter concerns how to chain together xrefs if there are more than one of them. Some methods are suggested next.

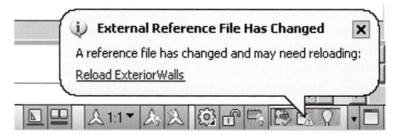

FIGURE 17.11
Reload xref announcement.

17.5 MULTIPLE XREFS

As mentioned earlier in the chapter, many complex drawings use more than one xref. Typically, you may find the exterior walls, columns, and the core as three separate xrefs. The ceiling grid (RCP) can also be xrefed in for extensive ceiling work, as can the demolition plan and many other combinations.

There are essentially two options. Let us say you have five xrefs total to work with. You can attach them one to another and then to a master working file (daisy chain) or one at a time as shown in Figures 17.12 and 17.13.

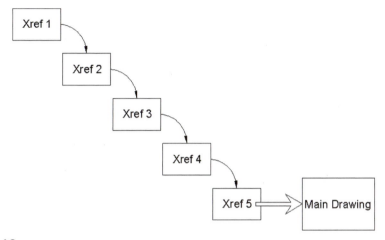

FIGURE 17.12
Multiple xref attach, Method 1.

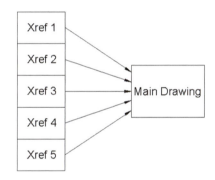

FIGURE 17.13
Multiple xref attach, Method 2.

Which method you use is up to you (there may be company policy or a convention in use at your job). Each method has minor pros and cons, and you should expect both to be used in the industry.

One final word on when *not* to xref, as this feature gets abused occasionally. Xrefing of title blocks, furniture, and text is not what xref is for. It is intended for major pieces of infrastructure not secondary objects. The logic behind xref in the first place is to attach objects that are not likely to change and can serve as the backdrop. Use the tool wisely.

17.6 RIBBON AND XREFS

Some xref functions can be accessed via the Ribbon, and we discuss them separately in this section. You can find them under the Insert tab→Reference panel, as seen in Figure 17.14.

One important control is the xref fading slider you see toward the bottom of the tab. This controls the darkness of the xref. Some users prefer the xref to be dark, similar to the surrounding design, while some prefer it very light, so they clearly know what is an xref and what is the design.

Finally, the Ribbon switches to a dedicated xref tab if you click on the xref, as seen in Figure 17.15, although there is much duplication of the previous tab's commands.

421

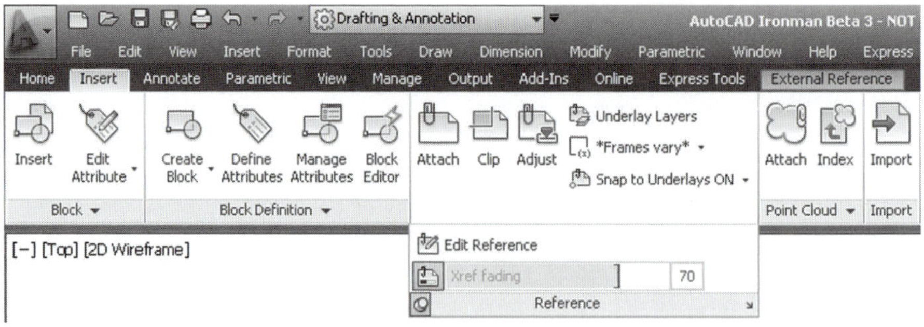

FIGURE 17.14
Xref via Ribbon, 1.

FIGURE 17.15
Xref via Ribbon, 2.

LEVEL 2 DRAWING PROJECT (7 OF 10): ARCHITECTURAL FLOOR PLAN

For Part 7 of the drawing project, we will shift to the outside of the building and create some landscaping in the form of trees, shrubs, and drive/walkways. You will need some Chapter 11 tools for the greenery.

Step 1. Create some appropriate layers as shown next. Freeze all layers that are not relevant to the task at hand.
  ○ L-Driveway (Gray_8).
  ○ L-Trees_Shrubs (Color_108).
  ○ L-Walkway (Gray_8).
Step 2. Create the landscaping plan as seen in Figure 17.16. You need to create plines for the borders of the driveway and the walkways. Then, after adding the hatch patterns (Sand and Brick), erase some or all of the plines. For the trees and shrubs you need to call upon the xline and spline.

SUMMARY

You should understand and know how to use the following concepts and commands before moving on to Chapter 18:

- Xref command
  ○ Inserting
  ○ Detaching
  ○ Unloading
  ○ Reloading
  ○ Binding
- Layers in xref
- Nesting xrefs

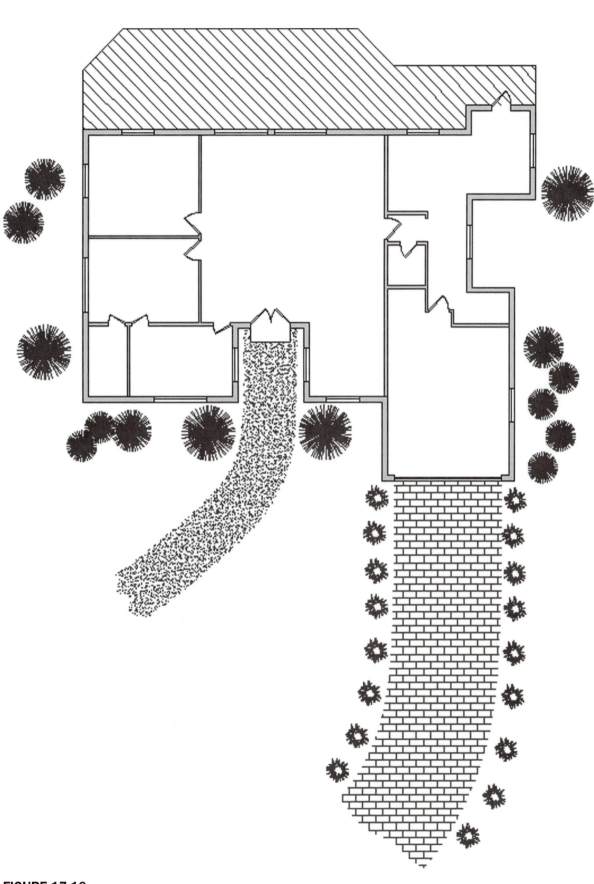

FIGURE 17.16
Landscaping plan.

REVIEW QUESTIONS

Answer the following based on what you learned in Chapter 17:

1. What are several benefits of xref?
2. How is an xref different from a block?
3. How do you attach an xref?
4. How do you detach, reload, and bind an xref?
5. What do layers in an xrefed drawing look like?
6. What do layers look like once you bind the xref?
7. What are some nesting options for xref?

EXERCISE

1. Draw the following floor plan to practice the xref command. (Difficulty level: Easy/Moderate; Time to completion: 45–60 minutes)

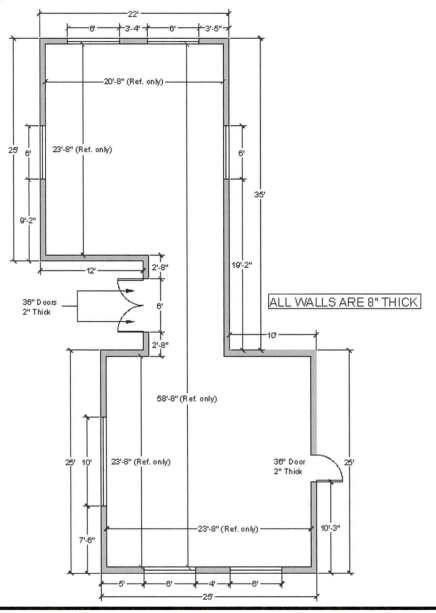

Spotlight On:
Electrical Engineering

Electrical engineering is the branch of engineering that deals with electricity and its applications. This can be on a large scale, such as power generation and transmission (Figure 1), or small scale, such as design of chips and consumer electronics (Figure 2), sometimes referred to as *electronics engineering*. The profession appeared in the mid-1800s, as major advancements in the understanding of electrical phenomena were put forward by Ohm, Tesla, Faraday, Maxwell, and others. Today, electrical engineering is a wide-ranging profession vital to our society. Few human-made objects have no electrical components, and electrical engineering figures prominently in virtually every product or device you see before you. Electrical engineers held about 318,700 jobs in 2009, making this the second largest branch of the U.S. engineering community (behind software engineering).

Just some of electrical engineering's specialties include power, controls, electronics, microelectronics, signal processing, telecommunication, and computer engineering. Electrical engineers need a mastery of a wide variety of engineering sciences, such as basic and advanced circuit theory, electromagnetics, embedded systems, signal processing, controls, solid state physics, and computer science and programming, among others.

Education for electrical engineers starts out similar to that for the other engineering disciplines. In the Unites States, all engineers typically have to attend a four-year ABET-accredited school for their entry level degree, a Bachelor of Science. While there, all students go through a somewhat similar program in their first two years, regardless of future specialization. Classes taken include extensive math, physics, and some chemistry courses, followed by statics, dynamics, mechanics, thermodynamics, fluid dynamics, and material science. In their final

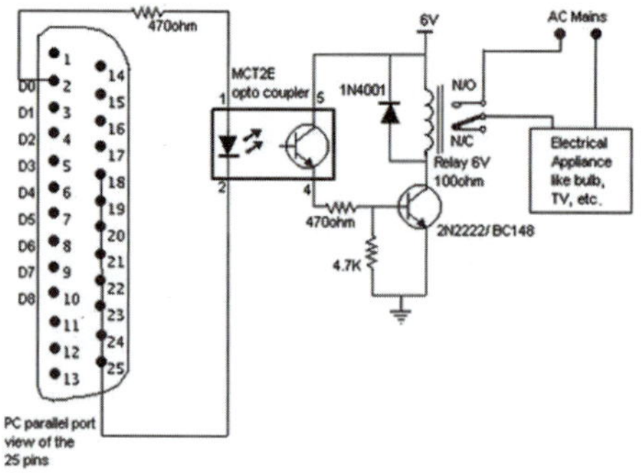

FIGURE 1
Electrical engineering on a large scale.

FIGURE 2
Electrical engineering on
a micro scale.

two years, engineers specialize by taking courses relevant to their chosen field. For electrical engineering, this includes courses in most or all of the sciences just listed.

Upon graduation, electrical engineers can immediately enter the workforce or go on to graduate school. Though not required, some engineers choose to pursue a Professional Engineer (P.E.) license. The process first involves passing a Fundamentals of Engineering (F.E.) exam, followed by several years of work experience under a registered P.E., and finally sitting for the P.E. exam itself.

In 2009, electrical engineers received starting salaries averaging $60,300 with a bachelor's degree and $71,500 with a master's degree. This of course depends highly on market demand and location. A master's degree is highly desirable and required for many management spots. The job outlook for electrical engineers is excellent, as the profession is extremely diversified; and by selecting a specialization carefully, an engineer can join one of many growing fields, such as controls, nano-engineering, robotics, and other such career paths.

So how do electrical engineers use AutoCAD and what can you expect? Industry wide, AutoCAD enjoys a significant amount of use for the simple reason that most electrical work is 2D in nature and plays into AutoCAD's strength. AutoCAD layering is far simpler in electrical engineering than in architecture. There is no AIA standard to follow, only an internal company standard, so there is room to improvise. As always, the layer names should be clear as to what they contain.

In electrical engineering, you will not likely need advanced concepts such as Paper Space and xrefs, but often you need to build attributes that hold information on connectors and wires. You also need to have a library of standard electrical symbols handy, so nothing needs to be drawn from scratch every time. The key in electrical engineering drafting is accuracy (lines straight and connected), as the actual CAD work is quite simple. A related field, network engineering, is similar. Here, you may use plines or splines that indicate cables going into routers and switches. Shown in Figure 3 is an example of an electrical schematic. Notice the extensive use of symbols for capacitors, diodes, transformers, and switches.

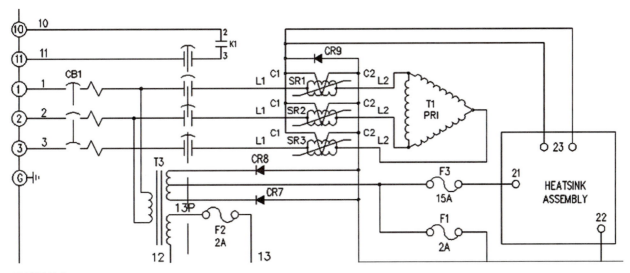

FIGURE 3
Electrical schematic.

Figure 4 is an example of some racked equipment in a network/electrical engineering drawing. There is nothing complex in this image, just careful drafting. You saw this example earlier when wipeout was discussed.

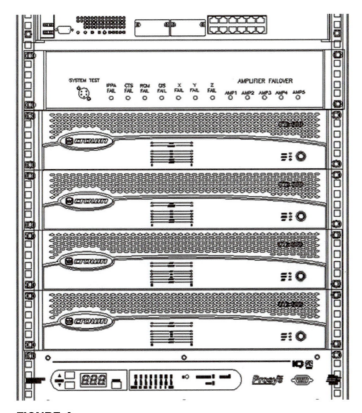

FIGURE 4
Racked equipment.

CHAPTER **18**

Attributes

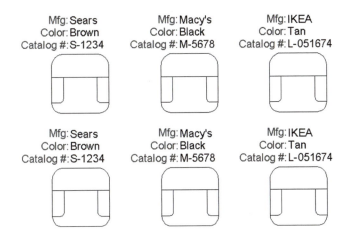

429

LEARNING OBJECTIVES

In this chapter, we introduce and thoroughly cover the concept of attributes.
We specifically discuss

- The purpose of attributes
- Defining attributes
- Editing attributes and properties
- Extracting data from attributes
- Invisible attributes

By the end of the chapter, you will be well versed in applying this critical concept to your design work.

Estimated time for completion of chapter: 1–2 hours.

18.1 INTRODUCTION TO ATTRIBUTES

Attributes is an important advanced topic. An *attribute* is defined in AutoCAD as information inside a block. The key words here are *information* and *block*. You can have a block without information—you have done this many times before; it is called a *block* or *wblock* but not an attribute. You can also have information in the form of text sitting in your drawing. That is just text or mtext but also not an attribute. However, put the two concepts together and you have what we are now about to explore: essentially an "intelligent" block, one that has useful embedded information.

Up and Running with AutoCAD® 2012: 2D and 3D Drawing and Modeling.
© 2012 Elsevier Inc. All rights reserved.

What is the reason for creating attributes in the first place? Why would you want information inside of blocks, and what can you do with this information? The information itself can be anything of value to the designer or client. For an office chair in a corporate floor plan, one may want to know its make, model, color, and catalog number. Then, if there are a variety of chair types for this office, the designer can keep track of all of them. Another example is a list of connections going into an electrical panel. It would be nice to know what and how many. All this information can be entered into the attribute and displayed on the drawing if you wish.

However, this is only half the story and displaying information is not the main reason for creating attributes. As a matter of fact, this information is often completely hidden so as not to clutter up the drawing. So, what is the main reason? The key is that this information can be extracted out of the attributes and sent to a spreadsheet (among other destinations). Now you have a truly useful system in place, where extensive information is hidden unobtrusively in the drawing, yet can be accessed if needed at the touch of a few buttons, and put into an Excel file or other very useful form for cost analysis or just plain old keeping track and organizational purposes.

You may have already heard of these tools in use or envisioned a need for them in architecture with such items as doors and windows schedules. The software keeps track of each door and window in the drawing and links the information to a database that is updated as you add or remove the objects from your design. While basic AutoCAD does not do this automatically, other add-on software does. Underneath it all is the concept of embedded attributes, similar to what we discuss here.

You should have a good understanding of the preceding few paragraphs. As always, the button pushing is far easier once the concept is crystal clear. We now go ahead and do the following:

- Create a simple design.
- Create a basic attribute definition.
- Make a block out of the simple design *and* the attribute definition.
- Extract information out of our new attribute.

The information is extracted via a multistep process to two locations, an Excel spreadsheet and a table. In both cases, the information is in an organized tabular form and, while not linked bi-directionally, is an exact reflection of drawing conditions. Let us give it a try.

18.2 CREATING THE DESIGN

We need nothing complicated drawn to learn the basics of attributes. Go ahead and draw a simple chair made up of several rectangles and rounded corners, as shown in Figure 18.1. Make the chair 30" × 30" and approximate the rest of the dimensions relative to those.

We would like to embed the following information in it—manufacturer, color, and catalog number—and put these three items just above the chair. Let us first add some text, a header of sorts, indicating what the three categories are, as seen in Figure 18.2. Use regular text of any appropriate size, font and color.

This is not yet the attribute definition, merely the name of the categories, so we know what we are looking at. You do not have to do this, but it helps clarify things initially, as the attributes are just plain *X*s. You now have a simple design and the start of some useful information. We are now ready to make an attribute definition.

18.3 CREATING THE ATTRIBUTE DEFINITIONS

Here, we define the new attributes that contain the actual information. The idea for now is to simply assign them all a value of *X* and change the *X* to meaningful data later in the process,

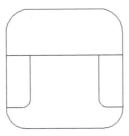

FIGURE 18.1
Basic chair design.

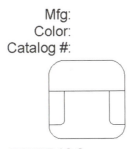

Mfg:
Color:
Catalog #:

FIGURE 18.2
Basic chair design with
text.

so we are creating generic definitions, which allow us to create multiple sets of data from one
template.

431

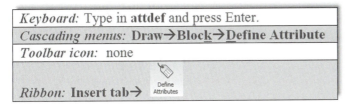

Keyboard: Type in **attdef** and press Enter.	
Cascading menus: **Draw→Block→Define Attribute**	
Toolbar icon: none	
Ribbon: **Insert tab→**	Define Attributes

Start the attribute definition via any of the preceding methods. The dialog box in Figure 18.3
appears.

We are interested in only the upper right of the dialog box for now. Under the Attribute
category, let us focus on Tag:, Prompt:, and Default:.

- For Tag, type in Mfg for Manufacturer (this is the actual category).
- For Prompt, type in Please enter manufacturer (this is just a reminder for the user).
- For Default, type in X (this is the generic placeholder) as seen in Figure 18.4.

The rest of the information in the dialog box can be left as is for now, although we discuss a
feature under Mode (upper left) later in the chapter. One thing you can adjust however is the
text height, matching up the attribute size to the size of the text over the chair. The text is 4"
in height, so you can enter that value into the field.

Press OK and your attribute definition appears attached to your crosshairs. Go ahead and place
it next to the Mfg: text. Note that it is in capital letters by default. Go ahead and go through
the same procedure for the remaining two categories, changing the Tag value and slightly
modifying the Prompt. The default remains as X. Note also that you cannot have blank spaces
in the tag, hence the CATALOG_# underscore. The final results are shown in Figure 18.5.

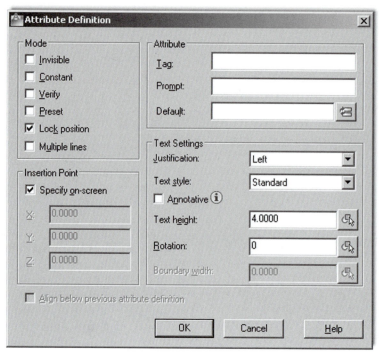

FIGURE 18.3
Attribute Definition.

FIGURE 18.4
Attribute fields.

FIGURE 18.5
Attribute fields completed.

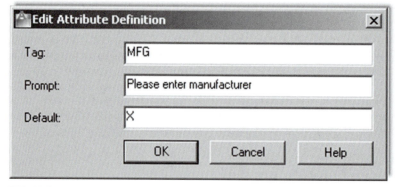

FIGURE 18.6
Edit Attribute Definition.

We are almost done creating the attribute definitions. Note the differences between the text and the attribute. Double-click on the original text, and you get the standard text editing field. Double-click on the attribute definition and you get a different, Edit Attribute Definition, field, as shown in Figure 18.6.

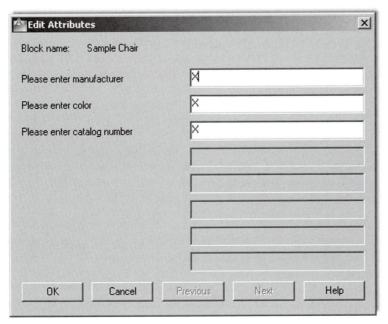

FIGURE 18.7
Edit Attributes.

FIGURE 18.8
Edit Attribute.

433

At this point, it may not be obvious why we have two identical sets of categories next to each other or why we entered the *X*. After the next step, what we are doing should become much clearer.

18.4 CREATING THE ATTRIBUTE BLOCK

This next step should be familiar: Create a regular block named `Sample Chair` via any method you prefer. Give it a name, select the objects (both the chair and all the text), and pick an insertion point. Finally, click on OK, and the dialog box shown in Figure 18.7 appears.

This Edit Attributes dialog box allows you to enter specific values for those placeholders (marked as *X*), but we do *not* do that at this point. Simply press OK and your new attribute is officially born (Figure 18.8). Notice all the previous categories are now just *X* and ready for customization. This is the process for creating a generic attribute. In the example of the office chairs, you create several copies (the different types of chairs) and customize the information for each. Then, copy and distribute the types as needed on the floor plan, as we do next.

18.5 ATTRIBUTE PROPERTIES AND EDITING

Now that you have a block with attributes, you need to learn how to edit its fields to add or delete information from an attribute as well as change its color, font, and just about any other property. Some of these techniques overlap each other and duplicate effort, so we need to clearly state what is used for what.

First of all, if you want to change only the content of an attribute (in other words, what it says), you need to get back to the dialog box shown in Figure 18.7. It is basic and simple, but does not change fonts and colors. Also, the only way to get to it is by typing. The command, attedit, takes care of the content. Indeed, this is the way you should "populate," or fill in, the fields once we start copying the chair. Try it right now and edit the chair to say Sears for the manufacturer, Brown for color, and S-1234 for catalog number.

Now, what if you wanted to change an attribute's properties, as discussed earlier? Well, the method here changes slightly, as you need to deal with the Enhanced Attribute Editor. Although a full command matrix is shown next, the easiest way to get to it is just to double-click on the attribute.

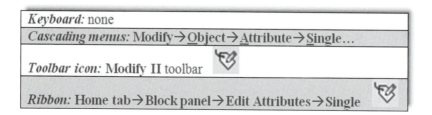

Keyboard: none	
Cascading menus: Modify→Object→Attribute→Single…	
Toolbar icon: Modify II toolbar	
Ribbon: Home tab→Block panel→Edit Attributes→Single	

What you get after either double-clicking or using the previous methods is Figure 18.9. Here, you can do quite a lot. This Editor not only allows you to fill in the fields but also a variety of properties (colors, layers, fonts, etc.). Click through the three tabs, Attribute, Text Options, and Properties, to explore all the options. Here are some additional pointers on attributes.

Exploding Attributes

Attributes can be easily exploded, but this action destroys the blocks and turns them into what you see in Figure 18.9, where you have just generic placeholders unattached to a

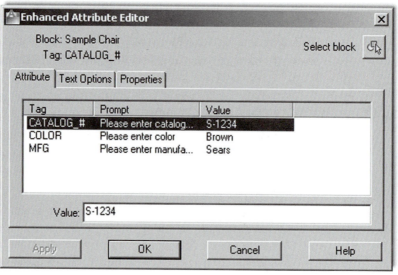

FIGURE 18.9
Enhanced Attribute Editor.

physical design. You can change a few fields and reassemble the block, which is a perfectly valid technique, especially if the attribute is giving you a hard time. Be careful though; exploding complex attributes can suddenly fill your screen with an enormous amount of data, so keep one finger on that Undo button.

Inserting Attributes

As you learned in this chapter, attributes are blocks; therefore, you can insert them as needed, as an alternative to copying them. When you insert a block that contains an attribute, you are prompted to fill in the X placeholders at the command line. Indeed, this is how title blocks are often filled out when a generic template block is inserted into a specific job file.

Ideally, everything makes sense now that you see the big picture of what was done. Review the chapter up to now if not. Our next step is to learn how to extract the useful information. For this you need to duplicate the attribute to simulate having a number of them in a real drawing. Then, go ahead and change some of the values. The final result is shown in Figure 18.10.

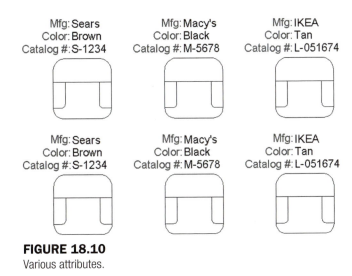

FIGURE 18.10
Various attributes.

18.6 ATTRIBUTE EXTRACTION

Extracting the information is a multistep process, necessarily so because of the number of options offered along the way. Our target extraction is both an Excel spreadsheet and a table embedded in AutoCAD. Remember of course that this is a basic generic extraction and more complex spreadsheets and tables can be created to hold such data, which is left to you to explore.

Step (Page) 1. Begin.

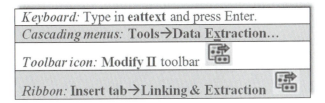

Start up the data extraction process via any of the preceding methods. The Data Extraction – Begin (Page 1 of 8) dialog box shown in Figure 18.11 appears. You are going to create a new data extraction, so leave the default

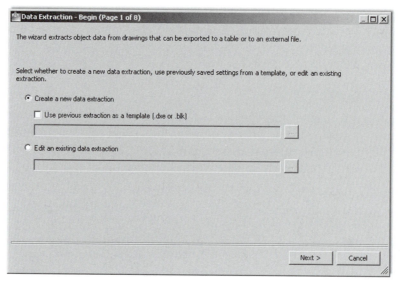

FIGURE 18.11
Data Extraction – Begin (Page 1 of 8).

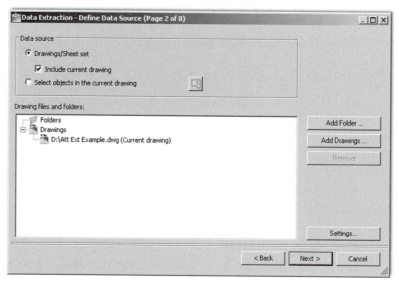

FIGURE 18.12
Data Extraction – Define Data Source (Page 2 of 8).

top choice but uncheck the Use previous… box. Then, press Next > and you are taken to a File→Save Data Extraction As dialog box. Save your file as `Sample.dxe` anywhere you will find it easily on your drive. You are then taken to the next page.

Step (Page) 2. Define Data Source (Figure 18.12). Here, we are interested in the data source, or location of the file to be looked at. Leave the default choice with the Include current drawing box checked off. Press Next >.

Step (Page) 3. Select Objects (Figure 18.13). Here, we are interested in selecting from what we want to extract the data. In our case, there is only one choice, the sample chair, as it is the only available block (as seen under the type column). You can also click on Display blocks only, under the Display options, to isolate the chair. In either case, select it and press Next >.

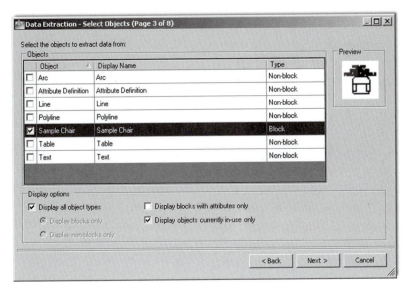

FIGURE 18.13
Data Extraction – Select Objects (Page 3 of 8).

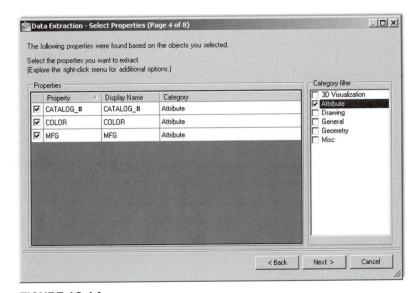

FIGURE 18.14
Data Extraction – Select Properties (Page 4 of 8).

Step (Page) 4. Select Properties (Figure 18.14). Here, you initially see many property choices. Use the category filter on the right to uncheck all the choices except Attribute. You then easily see the three properties on the left that we created earlier. Check off all three and press Next >.

Step (Page) 5. Refine Data (Figure 18.15). This step is for you to refine the data if needed. It contains some interesting features, such as Link External Data… (to tie in this output to another Excel file) as well as some tools for adjusting and tweaking the columns. Much of this and more can be done in Excel itself after extraction is completed, but it can be useful to adjust the look and content of the table extractions, as tables are less flexible. If you want to change nothing, press Next > for the final few steps.

Step (Page) 6. Choose Output (Figure 18.16). Here, we are interested in our destination output. Although the Excel spreadsheet is the one that gets all the attention,

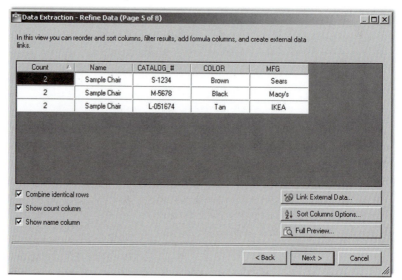

FIGURE 18.15
Data Extraction – Refine Data (Page 5 of 8).

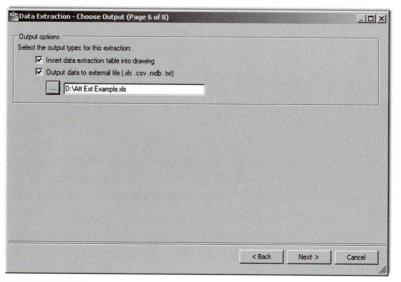

FIGURE 18.16
Data Extraction – Choose Output (Page 6 of 8).

you can also extract to a table that will be inserted into the drawing itself. Tables were first introduced in Chapter 15, so review that section if necessary. Check off both boxes if you have not done so yet, and finally select the location of the Excel file by using the three dot browse button. When ready, press Next >.

Step (Page) 7. Table Style (Figure 18.17). We do not do anything to the table, leaving it as is, but you should review the options available and press Next > when done.

Step (Page) 8. Finish (Figure 18.18). On the final page, there is little to do except read what AutoCAD is telling you and press Finish. You are asked for an insertion point and notice the table attached to the mouse. You may have to zoom in to see it better. Finally, when you can see it, click to place the table. The table is inserted into your drawing, as shown in Figure 18.19.

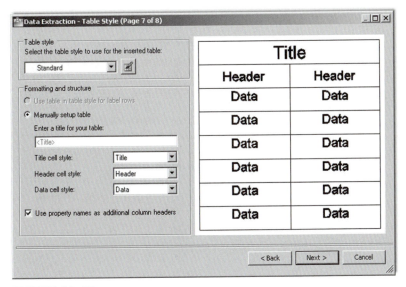

FIGURE 18.17
Data Extraction – Table Style (Page 7 of 8).

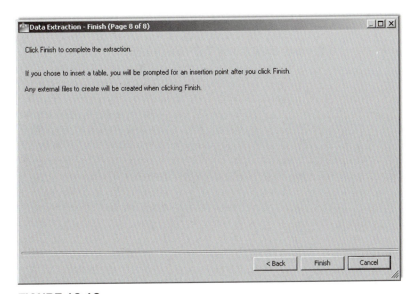

FIGURE 18.18
Data Extraction – Finish (Page 8 of 8).

Count	Name	CATALOG_#	COLOR	MFG
2	Sample Chair	L-051674	Tan	IKEA
2	Sample Chair	M-5678	Black	Macy's
2	Sample Chair	S-1234	Brown	Sears

FIGURE 18.19
Table Insertion (AutoCAD).

It is not fancy, but it gets the information across. The main output, however, is the Excel file. Look for it wherever you may have dropped it off and double-click on it to start Excel. Then, after some cleanup (cell borders, column width, fonts, colors, etc.), select the cells and Copy/ Paste them into AutoCAD, as you did in Chapter 16. It looks like Figure 18.20.

Count	Name	CATALOG #	COLOR	MFG
2	Sample Chair	L-051674	Tan	IKEA
2	Sample Chair	M-5678	Black	Macy's
2	Sample Chair	S-1234	Brown	Sears

FIGURE 18.20
Table insertion (Excel).

FIGURE 18.21
Invisible option.

This in essence is attribute data extraction. You should recognize the value in what you did and the potential it holds for easy data access and accounting.

There are other uses for attributes. One common use is formatting a full title sheet and block with generic attributes and saving it as a company standard. In this case, the text is the fixed permanent categories, such as Date, Sheet, Drawn by, and other fields, and the attribute definitions are the variables, such as the actual date, sheet number, and the initials of whoever created the drawing. Then, everything is blocked out and saved for future use, with the Xs representing to-be-filled-in variables. As the need comes up for a new title block, it is inserted as a block into the design's Paper Space with the variables filled out on the spot.

We conclude the attributes chapter by briefly going over an important option in the attdef command dialog box.

18.7 INVISIBLE ATTRIBUTES

This option appears as the first check-off box in the upper left of the attdef dialog box and is the most important of all the ones in that column. It is used to make attributes invisible upon creation.

Now, why would you do this? As mentioned at the beginning of this chapter, this is a relatively common way of creating attributes for the simple reason that you do not need to see the text and the attribute information. If you know what you created, then there is no need to clutter up the drawing with that list over and over again. This is especially true if there is a great deal of information (electrical termination panels come to mind) and revealing all would create a dense blanket of text on the drawing, obscuring everything else.

To use the invisible option, simply start the attribute creation process with attdef, check off the Invisible option, then proceed as outlined previously. The attribute of course remains visible until the block is created and disappears afterward. To see what you have, there is no problem: Simply double-click on the block or type in attedit, as outlined previously.

As you can see, being invisible for an attribute is no problem at all, as you can easily access it for editing if needed, and it is unobtrusively out of the way until then, the best of both worlds. The only downside of course is that you do not know that the attribute is there if

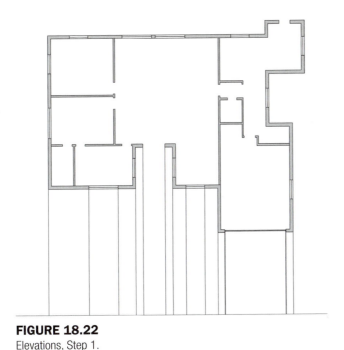

FIGURE 18.22
Elevations, Step 1.

you were not the one who set it up. It is always a good habit to double-click on blocks just to see what pops up on an unfamiliar drawing, and indeed, this is a standard investigative procedure in these cases.

441

Figure 18.21 shows the Invisible option checked off in the `attdef` dialog box.

LEVEL 2 DRAWING PROJECT (8 OF 10): ARCHITECTURAL FLOOR PLAN

For Part 8 of the drawing project, we finish up the design by creating an exterior elevation. Elevations are the "head on" views of a building. They are projected from overhead or plan views and are meant to give the client and builder a visual of the architect's design. These elevations feature walls, doors, windows, any other exterior design features (such as columns or fireplace chimneys), and of course the roof design.

Step 1. You need no new layers for this. Open up the floor plan, freeze all layers except the walls and windows, and draw guidelines straight down from each bend or feature in the wall (such as a door or window), as seen in Figure 18.22.

Step 2. Raise the elevation and add the doors and windows according to the provided dimensions, as seen in Figure 18.23. For simplicity, some details, such as gutters, are missing.

Step 3. Complete the elevation by adding exterior wall hatch and some other finishing touches. Figure 18.24 shows the final result. Repeat the same procedure for at least one other side of the house or complete all four.

SUMMARY

You should understand and know how to use the following concepts before moving on to Chapter 19:

- Attribute definitions
- Attribute blocks

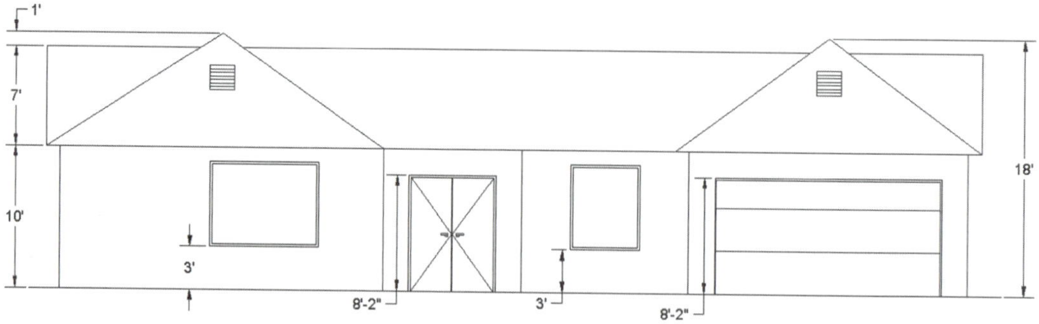

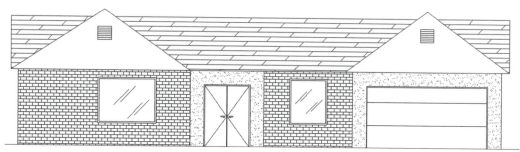

FIGURE 18.23
Elevations, Step 2.

FIGURE 18.24
Elevations, Step 3.

442

- Editing fields and properties
- Attribute extraction
- Invisible attributes

REVIEW QUESTIONS

Answer the following based on what you learned in Chapter 18:

1. What is an attribute and how do you create one?
2. How do you edit attributes?
3. How do you extract data from attributes? Into what form?
4. How do you create an invisible attribute? Why can this be important?

EXERCISE

1A. Draw the following doors:

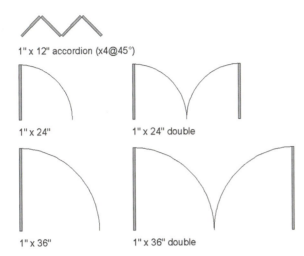

1" x 12" accordion (x4@45°)

1" x 24"

1" x 24" double

1" x 36"

1" x 36" double

1B. Next, add text attributes to each, as shown by the sample that follows for the 1" × 12" accordion:
   - *Type:* Accordion Double.
   - *Size:* 1" × 12".
   - *Color:* Natural Wood.
   - *Mfg:* XYZ Door Company.
   - *Stock #:* 654-A-1x12.
1C. Create similar attribute data for the remaining four doors. Vary the colors and the stock numbers.
1D. Make a block out of each one and fill in the attribute information.
1E. Copy the entire set once for a total of ten doors.
1F. Extract the data to Excel.
(Difficulty level: Intermediate; Time to completion: 30 minutes)

443

CHAPTER 19

Advanced Output and Pen Settings

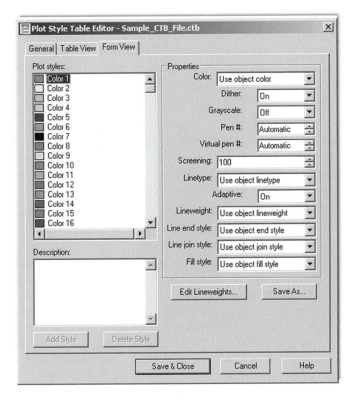

445

LEARNING OBJECTIVES

In this chapter, we introduce and thoroughly cover the concept of advanced output, which includes the concept of the ctb pen settings file. We specifically discuss

- ctb files and their purpose
- Editing lineweights
- Screening
- The lwt option

By the end of the chapter, you will be well versed in applying this critical concept to your design work.

Estimated time for completion of chapter: 1 hour.

19.1 INTRODUCTION TO ADVANCED OUTPUT AND PEN SETTINGS

This chapter is the last of three that talk about some aspect of printing and output. Although general and advanced printing was already explored in Chapters 9 and 10, a crucial piece of information was omitted. This information concerns pen settings and other advanced output techniques, essential knowledge for professional-looking output. The reasons we are doing this are outlined next.

Most designs are rather complex when everything has been added to the drawing. Lines cross other lines and text is everywhere. The eye needs to be able to quickly and easily focus on what is essential, for example, the outlines of the walls of a floor plan or the outline of the mechanical part among the many other elements in the design.

This is analogous to a mix of a rock song. Not every instrument can be the loudest; something has to give, and a balance needs to be found so the nuances and color of the music can stand out. Usually, when the singer sings, other instruments have to be turned down a bit, but when a guitar solos, its volume gets turned up, and so on. Otherwise it would be a loud, obnoxious mess with all sounds competing for attention.

Such is the situation with a pencil and paper drawing. All lines cannot be the same thickness and darkness. Such a monochrome mess would be confusing to look at. Draftspersons have known this for as long as the profession has been around and, as a result, have numerous pencils at their disposal ranging from lightest (hard lead) to darkest (soft lead) and everything in between. They use the darkest for the primary design and the lighter leads for all else.

AutoCAD has a similar philosophy. Lines must have differences in thickness for some to stand out more than others. But, how do we show these differences, now that we no longer use pencil leads? The answer is to use color. Colors are then assigned line thicknesses, and assuming of course a consistent and intelligent pattern of use on the part of the designer, differences in line thickness can easily be incorporated into the design by assigning certain colors to the primary geometry. As you can see, colors play a dual role. They allow you to tell geometries apart not only on the screen but also on paper. The challenge then is to set up a table that assigns colors to certain line thicknesses, and most important, stick to using these colors properly and consistently.

19.2 SETTING STANDARDS

The thing you need to determine before anything gets touched in AutoCAD is what colors will be used to represent the most important geometry. You actually did this back in Chapter 3, when layers were introduced. This of course is something done once by the CAD manager and usually is applied to all drawings. All we are really doing is establishing a "pecking order," so to speak, for colors. This order corresponds from the most important (hence thickest) to the least important (the thinnest) pen setting. There are no hard rules here as far as *what* colors to use; it is more or less arbitrary, although convention favors some colors over others.

Let us say (based on common convention) that green is the A-Walls layer color in our architecture plan and is assigned the thickest lineweight. The A-Windows layer is typically red and the A-Doors layer yellow. Both those objects are almost as important as the walls but can be set a bit thinner. The A-Furniture and A-Appliances layer is magenta and thinner still. A-Text and A-Dims layers are cyan and thinner yet. Finally in our hypothetical example is hatch. Those dense patterns really need to be light, so as not to overwhelm the drawing, so we assign them a gray color and the thinnest lineweight of all.

Although this is just an arbitrary example, it is also exactly the thought process you need to go through to have a rough idea of how things will look. Once this is generally documented, you can move on to the actual settings.

19.3 THE ctb FILE

The ctb file is the file that needs to be modified to set the pen settings. It is where you assign pen thickness to a color and where we spend our time for the remainder of the chapter. We open the ctb file, give it a name, modify settings, and save. Then, as you set up plots, you assign the ctb file to those plots, which is a permanent, "do only once" action.

There are two ways to set up the ctb file: either by accessing it in the Page Setup Manager or as a separate action, which is what we do. Select **Tools→Wizards→Add Plot Style Table…** from the drop-down menu. The dialog box shown in Figure 19.1 appears.

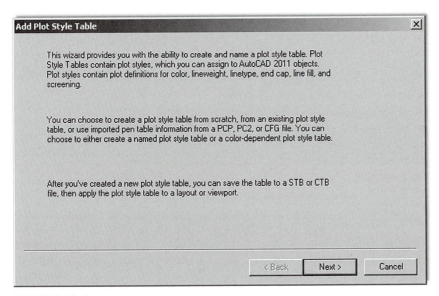

FIGURE 19.1
Add Plot Style Table.

Press Next > and you are taken to the next page of this dialog box (Figure 19.2), where you select the first option, Start from scratch. This begins a new line setting procedure from the default base values.

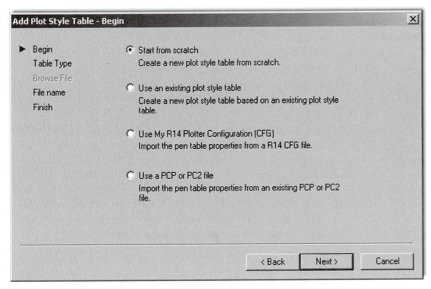

FIGURE 19.2
Add Plot Style Table – Begin.

Press Next > and you are taken to the next page of this dialog box (Figure 19.3); select the Color-Dependent Plot Style Table choice. This is the only choice we look at; the named plot styles are not covered and are rarely used.

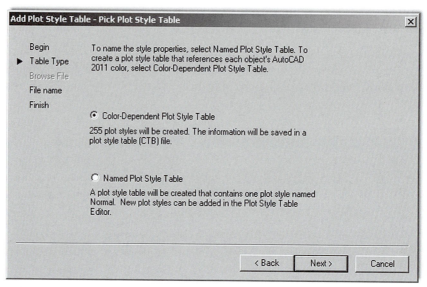

FIGURE 19.3
Add Plot Style Table – Pick Plot Style Table.

Press Next > and you are taken to the next page of this dialog box (Figure 19.4). Give the new ctb file a descriptive name, like `StandardCTB`, `ColorCTB`, or anything else that will be recognized. Here, it is `Sample_CTB_File`.

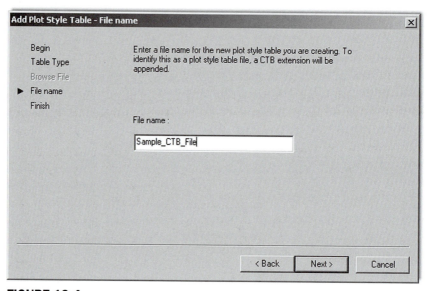

FIGURE 19.4
Add Plot Style Table – File Name.

Press Next > for the final step (Figure 19.5). Here, we are interested in the Plot Style Table Editor… to actually set the lineweights. Press that button and you are taken to our final destination, the actual Plot Style Table Editor, shown in Figure 19.6.

Here is where we set the colors, weights, and a few other tricks. Stay in the default Form View (referring to the tabs on top), as we do not need the information in any other format.

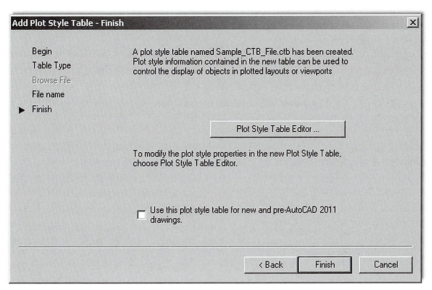

FIGURE 19.5
Add Plot Style Table – Finish.

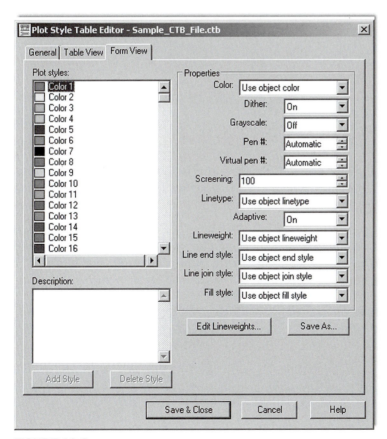

FIGURE 19.6
Plot Style Table Editor for Sample_CTB_File.

Familiarize yourself with this editor. On the left is the palette of the available 255 colors, and on the right is a variety of effects and settings you can impart on whatever color or colors you select from the palette. All the tools are there; you just need to proceed carefully.

Step 1

The first thing you need to do is to set all the colors in the color palette to *black*. As the default settings are now, each color is interpreted literally, which means that if you are sending your design to a color plotter, all the linework will come out in color, exactly as seen on the screen. This is a very desirable outcome in certain situations but not in most (we talk about this more later on). For now, we need to monochrome everything so we can work on line thicknesses.

- Select all the colors (1–255) in the palette by clicking your mouse in the empty white area just to the right of the colors, holding down the button, and sweeping top to bottom. Holding down the Shift key and clicking the first and last entry accomplishes the same thing.
- Then, as all the colors are highlighted blue, select the very first drop-down choice in Properties, called Color:, and select Black (Figure 19.7). All colors are assigned black ink.
- Clear the palette choices by clicking randomly anywhere in the white empty area of the left side of the palette, and test out your changes by clicking random colors. They should all be black.

FIGURE 19.7
Selecting black for all colors.

Step 2

Next we need to set the actual thicknesses. This is done in the Lineweight: drop-down menu, about halfway down the list on the right. Currently, it says Use object lineweight, which simply means nothing is set specifically and all lineweights are the same (the default value is 0.1900 mm).

Here, we need to discuss a major point you must understand in working with this feature. The actual line sizes are of secondary importance to the *relative differences* between the sizes. This means that you need not worry too much about how big 0.2 mm or 0.15 mm is—although that is the range of sizes that happen to look just right on paper. What is more important is that there is a large enough difference between the choices (0.2 mm vs. 0.15 mm as opposed to 0.2 mm vs. 0.18 mm), so the eye can notice this difference. You can just as easily use 0.25 mm and 0.3 mm, but as you get up in the higher numbers, all lines end up being overly thick, so it is best to stick to the range of 0.25 mm on the upper end and 0.05 mm on the lower end.

Let us put this all to use and set some values:

- Click on and select color 3 (green); set Lineweight to 0.2500 mm.
- Click on and select color 1 (red); set Lineweight to 0.1900 mm.
- Click on and select color 2 (yellow); set Lineweight to 0.1900 mm.
- Click on and select color 4 (cyan); set Lineweight to 0.1500 mm.
- Click on and select color 6 (magenta); set Lineweight to 0.1500 mm.
- Click on and select color 9 (gray); set Lineweight to 0.1000 mm.

Now, to put it all together, recall the Level 1 apartment drawing. The A-Walls layer was colored green and appears darkest (0.2500 mm) as required. The A-Doors and A-Windows layers were yellow and red, respectively, and are slightly lighter (0.1900 mm), followed by the A-Text and the A-Appliances (or A-Furniture) layers, colored cyan and magenta, respectively, lighter still (0.1500 mm). Last were the gray-colored hatch patterns coming in at (0.1000 mm).

So, will these particular lineweight choices look good? The drawing will certainly look more professional and features will stand apart from each other. The exact values and colors are all determined by the architect, engineer, or designer, as this person inspects the results and tweaks the lineweights as needed. This is an extensive trial and error procedure and not all individuals select the same settings; however, in any professional office, these techniques must be used regardless of exactly what specific settings are set. Nothing screams amateur more than a monochrome drawing with equal lineweights.

Step 3

To complete the procedure, press Save & Close and Finish in the Add Plot Style Table – Finish dialog box. Then, open a drawing, such as the Level 1 floor plan, if available. Type in `plot` and set the drawing up for plotting or printing as outlined in Chapter 10, being sure this time to select `Sample.ctb` instead of `monochrome.ctb` in the appropriate field.

19.4 ADDITIONAL CTB FILE FEATURES

We briefly cover two other useful features about which you should know. The first one was already alluded to earlier. When you selected all the colors and turned them to black, you could have left a few out. These colors would then have been plotted in color. So this action is certainly not an "all color" or "all black" deal. You can mix and match, and what is typically done in a piping or electrical plan (for example) is that the entire floor or site plan is monochrome, but the pipes or cables and wiring are left in vivid red, blue, or green (yellow and cyan do not show up well on white paper). The intent of the design is much clearer. It is a great effect, although it can obviously be derailed by not having a color capable printer or plotter.

The second feature has to do with the screening effect. What this does is lighten any color of your choosing so less ink is put on the paper during printing. It is literally "faded out" and is a very useful trick, as we will see with an example. The color most often chosen to be screened is color 9 (gray), as it is a natural for this effect and already looks faded, although any color can be used.

To screen a color, select it in the editor as outlined just previously, and go to the Screening field (as seen in Figure 19.8). Then, use the arrows or type in a value. Typically, the screening effect is most useful around 30–40%. Then proceed as before with saving and plotting.

451

FIGURE 19.8
35% screening of color 9.

So, what is the effect like? Figure 19.9 shows two solid fills on the same solid white color: one with no screening (left), meaning it is 100%, and one with the 35% screening (right) as viewed through the plot preview window.

It is a great trick to use. Here are some ideas where you can do this:

- In a key plan, where the area of work is grayed out and screened, yet the underlying architecture is still visible.

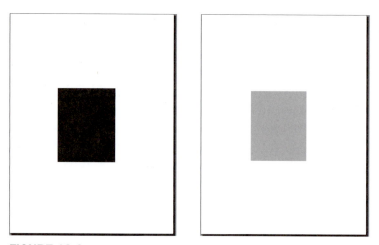

FIGURE 19.9
Plot preview without and with 35% screening.

- In an area of work on a demolition plan, where the area to be demolished for a redesign is grayed out, with all the features still visible.
- As a rendering tool, to create contrasting shading.
- As a rendering tool, to give a sense of depth or solidness to exterior elevations.

and many more applications….

19.5 THE lwt OPTION

An additional Lineweight (lwt) option is available to you, and it is one that is often noticed first because of its visibility. It is found all the way on the right among your bottom of the screen menu choices, as shown in Figure 19.10.

| INFER | SNAP | GRID | ORTHO | POLAR | OSNAP | 3DOSNAP | OTRACK | DUCS | DYN | LWT | TPY | QP | SC |

FIGURE 19.10
lwt (Lineweight) option.

This is *not* the same thing as what we covered earlier in this chapter. The ctb file and setting the lineweights in that manner gets you only plotted results. This means that on-screen all the lines remain the same size, regardless of the settings. The results of the ctb settings can only be seen on paper.

In contrast, the lwt option allows you to set line thickness that is actually visible on-screen. It is a rather interesting effect. Once set, the lines remain at that thickness regardless of how much you zoom in or out. Opinions on this effect vary among AutoCAD users. Some think it is very useful and accurately reflects what will be printed, while others see it as a distraction and are perfectly fine remembering what colors are set to what line thickness via the ctb method.

It is up to you to experiment and see where you stand. To set the line thickness using this option, you need to go to the Layers dialog box, create a new layer (usually with a color), and set the value under the Lineweight header as seen in Figure 19.11 (with 0.50 mm used for clarity). The result is seen in Figure 19.12.

After setting this layer as current, draw something. Notice that the line is thin, as if nothing happened. Now press in the LWT button; the line suddenly turns thicker, as seen in Figure 19.13. This is exactly how it will print. Note however that ctb settings supersede the lwt

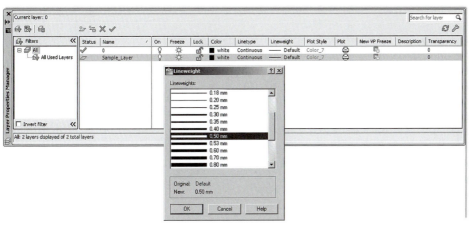

FIGURE 19.11
Lineweight settings.

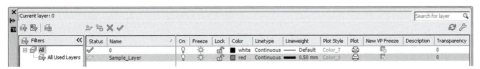

FIGURE 19.12
Lineweight set to 0.50 mm.

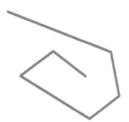

FIGURE 19.13
Line drawn with the
0.50 mm lwt setting.

settings, and you would leave the ctb settings as default if you are inclined to use this lwt feature instead.

LEVEL 2 DRAWING PROJECT (9 OF 10): ARCHITECTURAL FLOOR PLAN

For Part 9 of the drawing project, you do two things. You first set up a master title block that contains attributes. It is reused numerous times when we get to set up full Paper Space layout tabs in Part 10 in the next chapter. Then, use what you learned in this chapter to create a useable ctb file.

Step 1. Apply attributes to create an intelligent title block. You can use a modified version of the title block first created in Chapter 10. Start out with that basic template, insert it into a Paper Space tab, and add some extra lines to the lower right part of the title block (see Figure 19.14). Then, following what you learned in Chapter 18, add text for the following categories, followed by the attributes themselves. Make sure you are drawing on a title block layer (color: white).
  ○ Drawn by:

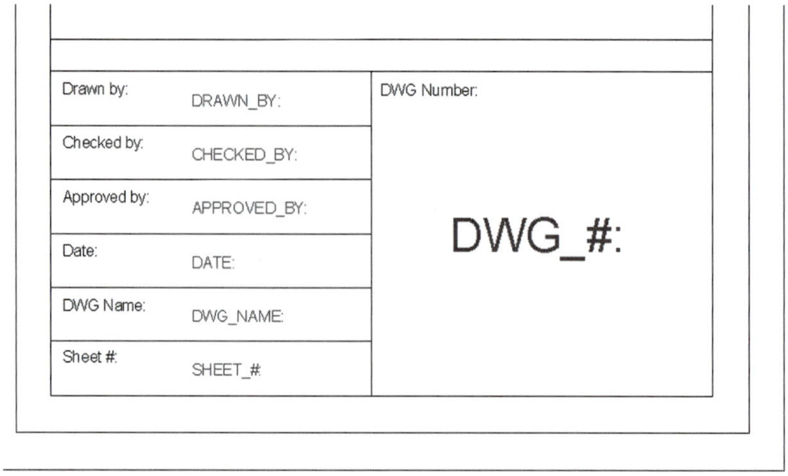

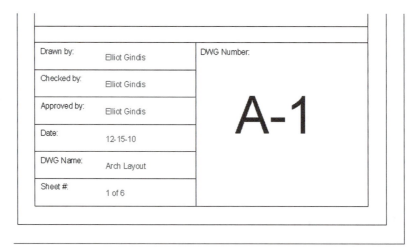

FIGURE 19.14
Title block with attributes.

      ○ Checked by:
      ○ Approved by:
      ○ Date:
      ○ DWG Name:
      ○ Sheet #:
      ○ DWG Number:

Step 2. When you have filled everything in, make a block of the entire title block and either fill in the Xs right away or leave them as variables until you set up the rest of the Paper Space layout in the next chapter. Figure19.15 shows the title block filled in with information after it was made into a block.

FIGURE 19.15
Title block completed.

Step 3. The next step is to assign proper lineweights to the ctb file of the architectural design. Assume the standard we have been following and make green the darkest line, followed by yellow and red, then all the others. The exact values are up to you.

SUMMARY

You should understand and know how to use the following concepts before moving on to Chapter 20:

- ctb file
- Editing lineweights
- Screening
- `lwt`

REVIEW QUESTIONS

Answer the following based on what you learned in Chapter 19:

1. What is the ctb file?
2. Describe the process of setting up a ctb file.
3. What is screening?
4. What is lwt?

EXERCISE

1. For practice, create a brand-new ctb file. Call it `Ch_19_Sample_File`. Monochrome all colors and assign the following thicknesses to the listed colors as seen next:
   - Color 1, red, 0.2500.
   - Color 2, yellow, 0.1900.
   - Color 3, green, 0.1500.
   - Color 4, cyan, 0.3000.
   - Color 5, blue, use object lineweight.
   - Color 6, magenta, 0.0900.
   - Color 7, white, 0.1000.
   - Color 9, gray, use object lineweight.
   Screen color 9 to 30%.

   (Difficulty level: Easy; Time to completion: <10 minutes)

CHAPTER 20

Isometric Drawing

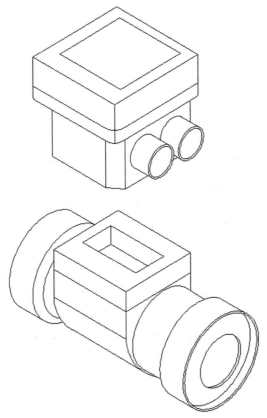

LEARNING OBJECTIVES

In this chapter, we introduce isometric perspective and how to use it. We cover

- What isometric perspective is
- When to use it
- When not to use it
- Setting isometric perspective
- Changing planes (F5)
- Ellipses in isometric
- Text in isometric

Up and Running with AutoCAD® 2012: 2D and 3D Drawing and Modeling.

In the course of the chapter, you will create several isometric designs, including a computer desk and mechanical devices.

Estimated time for completion of chapter: 1 hour.

20.1 INTRODUCTION TO ISOMETRIC PERSPECTIVE

By strict definition, *isometric perspective* means representing a three-dimensional object in two dimensions. As applied to AutoCAD, it means we draw a 3D object without resorting to the complexities of 3D space, sort of "faking it" in 2D, by angling horizontal lines 30° (the accepted isometric drafting standard, though there is also Axonometric, which is 45°). This creates the illusion of 3D and is good enough to get an idea across; indeed, that is how it is done with paper and pencil. In our case, we sometimes leave the horizontal lines flat, as with the box in Figure 20.1, and sometimes angle all of them 30°, as with the computer table exercise at the end of the chapter.

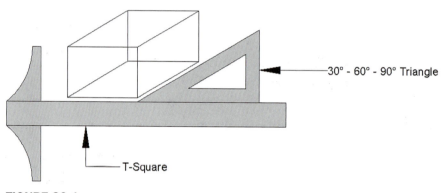

FIGURE 20.1
T-square and triangle.

Why Use Isometric Perspective Instead of 3D?

There are a few reasons. One is that it is much easier to learn isometric drawing than to develop proficiency in real 3D. So, for someone who only needs a quick 3D sketch once in a while just to accent an otherwise excellent 2D presentation, this perfectly fits the bill. Another reason is that 3D takes up more time and computing resources, something that is not always available, and there is also some difficulty in combining 2D and 3D drawings on one sheet. Many students do not go on to take AutoCAD 3D right away (nor should they), and isometric fills that temporary void quite well. For those who do go on, isometric drawing serves as an excellent introduction to 3D and should be reviewed before starting the 3D instruction.

When *Not* to Use Isometric Perspective

Isometric drawing is inappropriate when precision 3D drawings are required for design, testing, and manufacture. Remember, isometric perspective is only for visualization. These are *not* true 3D objects and have no real depth, only the illusion of depth. As a result, true measurements and the use of the offset command are not possible in the traditional sense.

20.2 BASIC TECHNIQUE

There is nothing inherently special about drawing an isometric design. Draw a straight (Ortho) horizontal line, then rotate it 30° counterclockwise. Next attach a perfectly straight line to one of the tips using ENDpoint. Copy the original horizontal line from where it meets the vertical to the other end (using ENDpoint as well). You are on your way to drawing an isometric box. It is the same as taking a T-square in hand drafting and putting a 30° −60° −90° triangle on it, as illustrated in Figure 20.1.

That was not hard, but it was tedious. There is of course a way to have AutoCAD preset the crosshairs so you are always drawing in isometric mode. To explain it, we need to introduce the idea of planes, so here is a short introduction to the plane concept.

In isometric drawing, you are always drawing in one of three available planes: top, right, or left (see Figure 20.2). Because you are using a 2D pointing device (a mouse), which exists only in a 2D world, you need to be able to easily move from one plane to the other to be able to draw on it. To toggle between planes, press the F5 key. Your crosshairs line up accordingly, and now you just draw using the line command. Here is one final rule: You almost always have Ortho on while in isometric mode, otherwise, the lines are not straight and rotating them 30° is meaningless. Let us put it all together and draw an isometric box. After you are able to do this, more advanced isometric drawings are surprisingly easy.

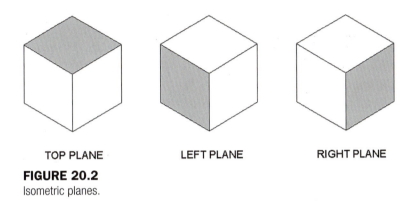

TOP PLANE   LEFT PLANE   RIGHT PLANE

FIGURE 20.2
Isometric planes.

Step 1. Open a blank file.

Step 2. From the cascading menu select: **Tools→Drafting Settings…** and the dialog box seen in Figure 20.3 appears.

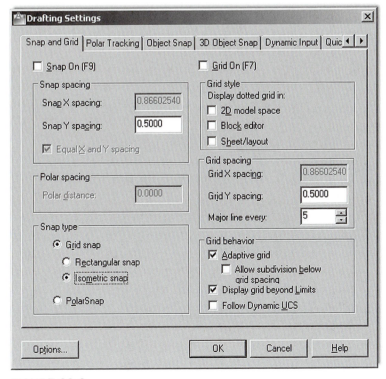

FIGURE 20.3
Drafting Settings with Isometric snap selected.

Step 3. Pick the Snap and Grid tab, go to the Snap type category (lower left), pick Isometric snap, and press OK. Figure 20.3 is a screen shot of the Drafting Settings dialog box with Isometric snap selected.

Step 4. Make sure Ortho is activated and the crosshairs are set at 100% for easier visualization of planes. Notice also how the crosshairs are positioned now that you are in isometric mode.

Step 5. Draw straight lines (as they should be with Ortho) of any size and continue drawing until you have the front face of a box. Very important: Do not try to connect the final line to the first line; you will never get it exactly right. Instead overshoot and click after passing it by, as shown in Figure 20.4.

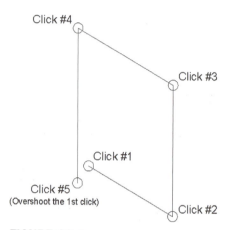

FIGURE 20.4
Isometric drawing, Step 5.

Step 6. Now, use the fillet command to fillet the line between Clicks 4 and 5 and Clicks 1 and 2. What is shown in Figure 20.5 is the result.

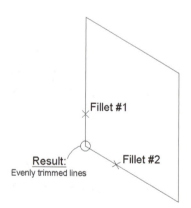

FIGURE 20.5
Isometric drawing, Step 6.

Step 7. Now, press F5. The crosshairs shift to another plane and you can draw another surface.

Step 8. You can press F5 again for the final surface or use the copy command. Be sure to use the ENDpoint OSNAP to cleanly connect lines. Figure 20.6 shows what you get after a few lines.

460

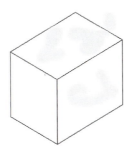

FIGURE 20.6
Isometric drawing,
Step 8.

Step 9. And the final result, after a few more lines, is seen in Figure 20.7.

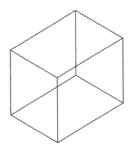

FIGURE 20.7
Isometric drawing,
Step 9.

461

What makes isometric drawing easy is that the techniques do not change for more complex objects. There may be more lines, but it is the same thing over and over.

Here are some additional pointers:

- To draw circles in isometric perspective use ellipses. This command is introduced in Chapter 2 and used to make the bathtub in the apartment drawing at the end of Chapter 4. We say a few more words about using ellipses in isometric design here.
- Always stay in Ortho mode unless you need to deliberately angle a line at some odd angle. You see an example of this with the back of the monitor in the upcoming computer desk drawing.
- Text can be angled up or down by 30° to align with the isoplanes. We discuss this shortly.
- Have fun with this. Isometric drawing is a semi-artistic endeavor and not as rigid as drafting. Eyeball distances but try to make the drawing look good. Do not be afraid to throw in some hatch patterns, as is done on the computer desk drawing. Remember though, if you need more precision, you have to go to the real 3D.

20.3 ELLIPSES IN ISOMETRIC DRAWING

A circle angled away from you is an ellipse, and these shapes find wide application in isometric drawing. You need ellipses for the base of the monitor in the upcoming computer desk drawing as well as the power button and extensive end-of-chapter exercises.

There is no surefire way to create the perfect ellipse; it always requires a bit of adjusting to get right and a bit of a creative eye to do so. Reproducing what was first shown in Chapter 2, we have Figure 20.8.

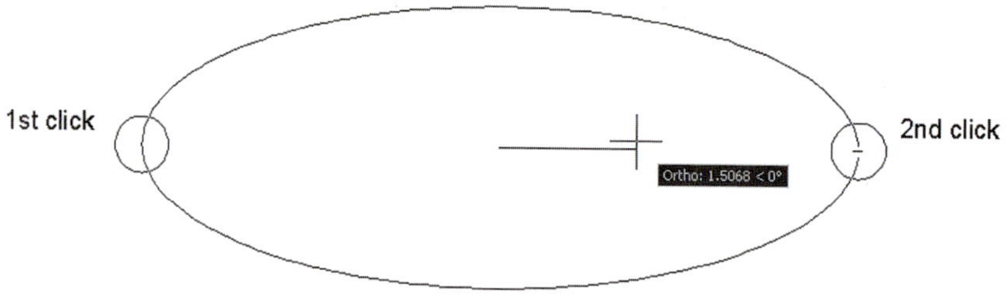

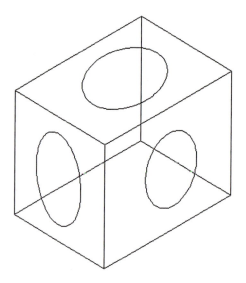

FIGURE 20.8
Basic ellipse.

Once the ellipse is positioned and is the right size, you can rotate it until it looks correct and in the same plane as the rest of the isometric geometry. Figure 20.9 is an example of some ellipses in an isometric box.

FIGURE 20.9
Isometric box with ellipses.

20.4 TEXT AND DIMENSIONS IN ISOMETRIC DRAWING

Text can easily be aligned with isoplanes by rotating it plus or minus 30°, or other values depending on the plane, as shown in Figure 20.10.

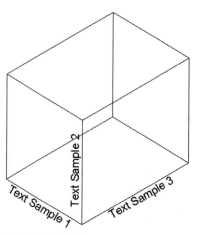

FIGURE 20.10
Text in isometric drawing.

Although not ideal, you can dimension in isometric drawing, using **Dimension→Aligned** and vertical dimensions, as seen in Figure 20.11.

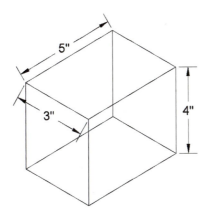

FIGURE 20.11
Dimensions in isometric drawing.

LEVEL 2 DRAWING PROJECT (10 OF 10): ARCHITECTURAL FLOOR PLAN

For Part 10 of the drawing project, you completely finish the design by adding the Paper Space layouts. You need to put together a lot of what you learned in Levels 1 and 2.

Step 1. In Paper Space, create the following seven layouts:
- ○ Architectural Layout.
- ○ Furniture Layout.
- ○ Carpeting Layout.
- ○ Electrical Layout.
- ○ RCP_HVAC Layout.
- ○ Landscaping Layout.
- ○ Elevations.

Step 2. Insert a title block into each Paper Space layout. Label the title block as appropriate for that layout. Make sure all the attribute fields are filled in and page numbers are incremented.

Step 3. Create an mview window for each layout (VP layer, color: white), centering and zooming each layout to the proper scale using the techniques you learned in Chapter 10. The scale will be either ¼", ³/₈" or not to scale. Freeze the VP layer when you are done with each of the layouts.

Step 4. In each layout, use advanced layer settings to freeze all layers that are not necessary for that particular view. You will likely need to create as many layer settings as there are Paper Space layouts, as each one shows something different.

Step 5. Add labels and any other title block information you may need.

Step 6. If your computer is connected to a printer or plotter, plot out all of the layouts on D-sized paper. Make sure all of the following are addressed:
- ○ Paper Space layouts are correct.
- ○ Proper scale is used in each mview window.
- ○ Proper layers show in each mview window.
- ○ The title block is correct for each layout.
- ○ The ctb file is correct and there are varying lineweights.

All the layout plans are shown in Figures 20.12 through 20.18. They will be in full color, when you see them on your screen.

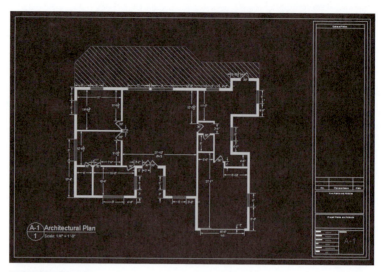

FIGURE 20.12
Architectural plan.

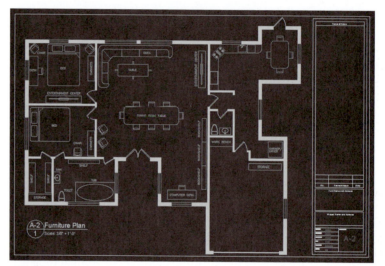

FIGURE 20.13
Furniture plan.

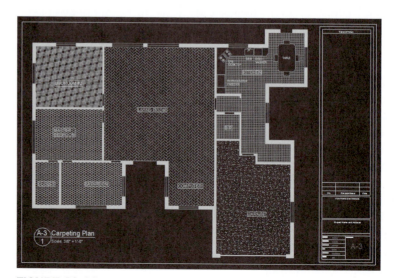

FIGURE 20.14
Carpeting plan.

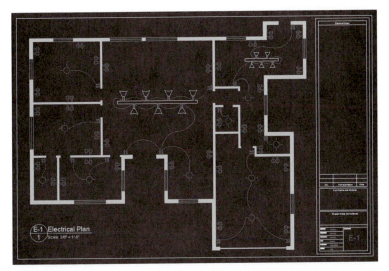

FIGURE 20.15
Electrical plan.

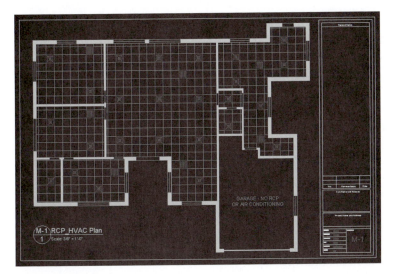

FIGURE 20.16
RCP_HVAC plan.

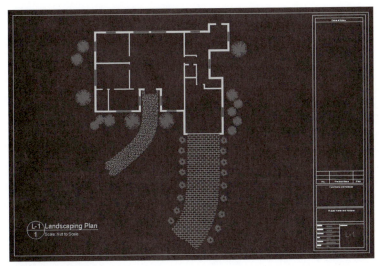

FIGURE 20.17
Landscaping plan.

FIGURE 20.18
Elevations.

SUMMARY

You should understand and know how to use the following concepts at the conclusion of this chapter:

- Isometric drawing
  - What is it?
  - When to use it
  - When not to use it
  - Isometric settings
- Planes
  - What are the three planes?
  - Switching between them (F5)
- Ellipses in isometric drawing
- Text and dimensions in isometric drawing

REVIEW QUESTIONS

Answer the following based on what you learned in Chapter 20:

1. List some instances when isometric drawing is appropriate.
2. List some instances where 3D would be better than isometric drawing.
3. Name the three planes that exist in isometric perspective.
4. In what dialog box is Isometric snap found?
5. What F key switches the cursor from plane to plane?
6. What is almost always on when drawing in isometric?
7. What is the equivalent of a circle in isometric drawing?

EXERCISES

1. Draw the following computer desk. You need your basic isometric drawing knowledge and construction techniques developed in this chapter. Also make use of the ellipse command and hatch the desk itself. All sizes are approximate and may be estimated. (Difficulty level: Easy/Moderate; Time to completion: 20–30 minutes)

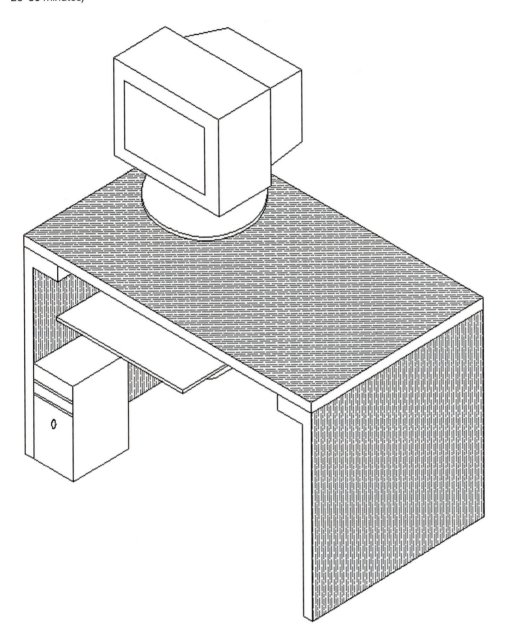

2. Draw the following mechanical device. You need your basic isometric knowledge and construction techniques developed in this chapter. Also make extensive use of the ellipse command. All sizes are approximate and may be estimated. (Difficulty level: Moderate; Time to completion: 30 minutes)

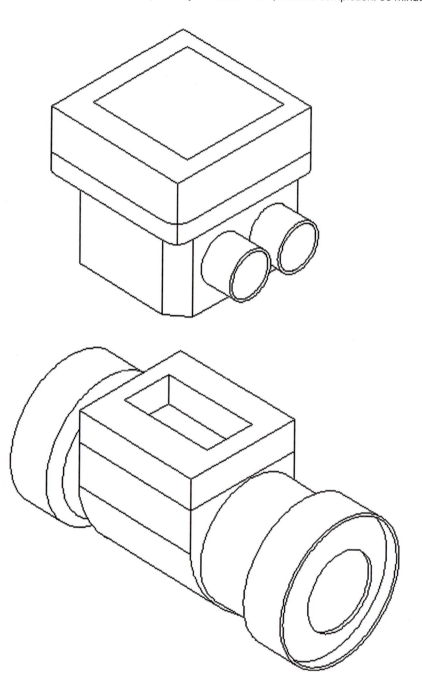

3. Draw the following architectural detail. All sizes are approximate and may be estimated. (Difficulty level: Easy; Time to completion: 10–15 minutes)

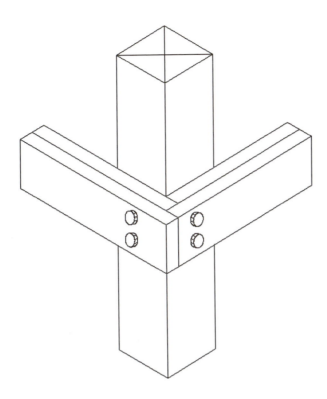

4. Draw the following architectural detail. All sizes are approximate and may be estimated. (Difficulty level: Easy; Time to completion: 15–20 minutes)

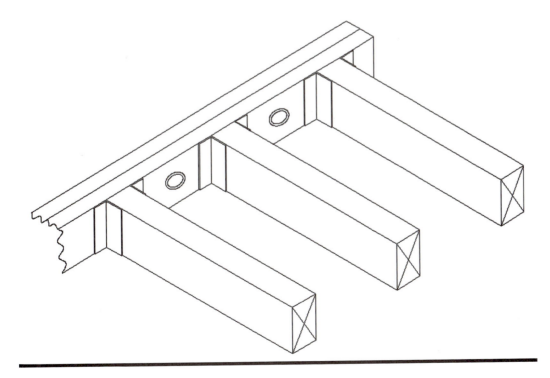

Answers to Review Questions

CHAPTER 11 REVIEW QUESTIONS

1. What are the two major properties a *pline* has and a line does not?

Property 1: A pline's segments are joined together. Property 2: A pline can have thickness.

2. What is the key command to add *thickness* to a pline?

Pedit.

3. What is the key command to create a *pline* out of regular lines?

Pedit.

4. What option gives the pline *curvature*? What option undoes the curvature?

Spline option. Decurve option.

5. What two major things happen when you *explode* a pline?

It loses its thickness and is split into segments.

6. What other *options* for pline were mentioned?

Making a pline with variable thickness, adding an arc, and various other vertex options.

7. Define an *xline* and specify a unique property it has.

An xline is a continuous line that has no beginning or end.

8. How is a ray the same and how is it different from an xline?
   A ray has no end, just like an xline, but it does have a beginning, unlike an xline.

9. Describe a *spline* and list its advantages.
   A spline is a NURBS curve and is used to create organic curvature, something that cannot be done any other way.

10. List some uses for a spline in a drawing situation.
    Gradient lines, cables, electrical wires, pathways, landscaping, and many other uses.

11. What is an mline? List some uses.
    An mline is a multiline. It can be used for everything from architectural walls to highway design.

12. What are some of the items likely to be changed using *mlstyle*?
    Number of lines making up the mline, distances between those lines, linetypes making up the lines, colors of the lines and fills, end caps, and other details.

13. What are some of the uses of *mledit*?
    Mainly to edit intersections that are made using mlines; also to create vertexes and make cuts in mlines.

14. Describe the *sketch* command. What is a *record increment*? What size is best?
    The sketch command allows you to draw freehand. A record increment is the smallest size piece of line that makes up the sketch pathway; .01 is an optimal size.

CHAPTER 12 REVIEW QUESTIONS

1. Describe what is meant by *layer management*. Why is it needed?
   Layer management describes a set of tools and concepts that, taken together, manage a large amount of layers, so they can be grouped together under some common theme and controlled as one unit (frozen, locked, etc.).

2. Describe the basic premise of a *script*.
   A script automates a series of input commands and saves the process under a name so it can be recalled and put to use automating a repetitive task.

3. What can a *script file* do for layer management?
   A script file can freeze and thaw a large amount of layers by simply executing the script. This was the original way to do layer management.

4. Describe what *Layer State Manager* does.
   It saves layer settings under a descriptive name, so this setting can be recalled to easily freeze and thaw (among other actions) large amounts of layers all at once.

5. What are the fundamental steps in using *Layer State Manager*?
   Once the layers are set up as desired, the setting is saved in the LSM.

6. What is *layer filtering*? What are the two types?
   Layer filtering sorts layers based on common traits (such as color or names), so they can all be worked with at once. Property filter and Group filter.

CHAPTER 13 REVIEW QUESTIONS

1. Describe the important features found under the *Lines* tab.
   Nothing critical is found here, but the colors of the dimension and extension lines are changed.

2. Describe the important features found under the *Symbols and Arrows* tab.
   Here, the arrowheads are changed to an architectural tick.

3. Describe the important features found under the *Text* tab.
   Here, the text style and color are changed, with a box added around the text.

4. Describe the important features found under the *Fit* tab.
   Here, the most important feature is the setting of the overall scale.

5. Describe the important features found under the *Primary Units* tab.
   Here, everything under Linear dimensions, Zero suppression, and Angular dimensions is discussed.

6. Describe the important features found under the *Alternate Units* tab.
   Here, the approach to setting alternate units is discussed, including Zero suppression and placement.

7. Describe the important features found under the *Tolerances* tab.
   The types of tolerances are covered.

8. Describe the purpose of *geometric constraints.*
   To constrain the positioning of drawn geometry.

9. What seven *geometric constraints* are discussed?
   Perpendicular, Parallel, Horizontal, Vertical, Concentric, Tangent, and Coincident.

10. Describe the purpose of *dimensional constraints.*
    To constrain the sizing of drawn dimensions and the geometry they represent.

11. What six *dimensional constraints* are discussed?
    Horizontal, Vertical, Aligned, Radius, Diameter, and Angle.

CHAPTER 14 REVIEW QUESTIONS

1. What is important under the *Files* tab?
   It is recommended that nothing be changed under this tab.

2. What is important under the *Display* tab?
   Here, you can turn off some unnecessary screen tools like scroll bars; change the background colors of the screen, command line, and other features; clean up the Paper Space layouts; and change the crosshair size.

3. What is important under the *Open and Save* tab?
   Here, you can save a drawing down to an older release of AutoCAD, delete Automatic save and backup files, and set security options.

4. What is important under the *Plot and Publish* tab?
   Here, you can set and modify plot stamp settings.

5. What is important under the *System* tab?
   It is recommended that nothing be changed under this tab.

6. What is important under the *User Preferences* tab?
   Here, you can set right-click customization.

7. What is important under the *Drafting* tab?
   Here, you can modify the look of the AutoSnap and Aperture boxes.

8. What is important under the *Selection* tab?
   Here, you can modify the pickbox and grip size as well as set the visual effects settings.

9. What is important under the *Profiles* tab?
   Here, you can save your settings and preferences under a name for future recall.

10. What is the `pgp` file? How do you access it?

The pgp file holds AutoCAD shortcuts. It can be accessed via Tools→Customize→Edit Program Parameters (acad.pgp) in the drop-down menus.

11. Describe the overall purpose of the *CUI*.

The Customize User Interface is used to modify the AutoCAD environment with changes to toolbars, palettes, shortcuts, macros, and much more.

12. Describe the purpose of the *Design Center*.

The Design Center is useful for bringing entities such as layers and blocks into a drawing from another location as well as downloading symbol library blocks from the Autodesk Seek website.

13. Describe the important *Express Tools* that are covered.

Some of the more useful Express Tools include Copy Nested Objects, Explode Attributes to Text, Convert Text to Mtext, Arc-Aligned Text, Enclose Text with Object, Delete Duplicate Objects, Flatten Objects, Break Line Symbol, Super Hatch, Convert PLT to DWG, and Layer Express Tools such as LAYFRZ, LAYISO and LAYWALK.

CHAPTER 15 REVIEW QUESTIONS

1. What are the purposes of the *audit* and *recover* commands? What is the difference?

Both commands are used to check up on the integrity of an AutoCAD drawing in cases of errors or a drawing not opening. Audit is the lower-level check for a drawing that is already opened but not running right. Recover is a higher-level check and may be used to try to open a corrupted drawing.

2. For *join* to work, lines have to be what?

They have to be coplanar.

3. The *defpoints* layer appears when you create what?

Dimensions.

4. Does the *divide* command actually cut an object into pieces?

No, it only inserts points to create sections, but the object is still intact.

5. What is the idea behind *eTransmit*?

The eTransmit command allows you to send a drawing via email along with all associated fonts, linetypes, xrefs, etc.

6. To what can you attach a *hyperlink*? What can you call up using a hyperlink? How do you activate a hyperlink?

To an object or text. Either a website or a document. Ctrl + click.

7. What is the idea behind *overkill*?

Overkill allows you to remove multiple stacked lines.

8. Can you make a *revcloud* out of a rectangle? A circle?

Yes, via the Object option.

9. What three new *selection methods* are discussed?

Window polygon, Crossing polygon, and fence.

10. Do you use a Crossing or a Window with the *stretch* command?

A Crossing.

11. Name a system variable mentioned as an example.

Mirrtext.

12. Explain the importance of the *wipeout* command. What does it add to the drawing?
  The wipeout command hides objects by covering them up. It can be used to add depth to a drawing.

CHAPTER 16 REVIEW QUESTIONS

1. What is the correct way to bring *Word* text into AutoCAD?
  Paste copied text from Word into mtext.

2. What is the procedure for bringing *AutoCAD* drawings into Word?
  For most applications, a Copy/Paste suffices. PrtScn can be used as well.

3. What is the correct way to bring *Excel* data into AutoCAD?
  Copy/Paste into an OLE.

4. What is the procedure for bringing AutoCAD drawings into *PowerPoint*?
  Copy/Paste or PrtScn works.

5. What is the procedure for importing JPGs and generating PDFs?
  The JPGs can be Copied/Pasted in, then adjusted. To generate a PDF, select it as the output printer or via the Ribbons output tab.

CHAPTER 17 REVIEW QUESTIONS

1. What are several *benefits* of xref?
  Reduce file size by attaching a core drawing to multiple files, critical file security, and automated design change updates to core drawings.

2. How is an xref different from a block?
  An xref is attached to a drawing and is never truly a part of it. A block is inserted into a drawing and does become part of it.

3. How do you *attach* an xref?
  Xref command, browse for the xref file, specify insertion point, and then OK.

4. How do you *detach, reload,* and *bind* an xref?
  By opening up a menu via a right-click on the drawing name in the xref palette.

5. What do *layers* in an xrefed drawing look like?
  They are preceded by the name of the xref itself.

6. What do layers look like once you bind the xref?
  They contain a 0 in the name.

7. What are some *nesting* options for xref?
  One xref attached to another then to another (etc.) and then to the main drawing, or all attached to one drawing simultaneously.

CHAPTER 18 REVIEW QUESTIONS

1. What is an *attribute* and how do you create one?
  An attribute is information inside of a block. To create one you use the attdef command and fill in the required fields.

2. How do you *edit* attributes?
  Double-click on the attribute.

3. How do you *extract data* from attributes? Into what form?

Data extraction is via the eattext command. It can be extracted into an Excel spreadsheet or an AutoCAD table.

4. How do you create an *invisible* attribute? Why can this be important?

By checking off the invisible button during the initial creation of the attribute. This is important so that the information is not visible in the drawing but can still be used.

CHAPTER 19 REVIEW QUESTIONS

1. What is the ctb file?

This is the file that stores information on line thickness and other settings.

2. Describe the process of *setting up* a ctb file.

You do this through a multipage Add Plot Style dialog box.

3. What is *screening*?

Screening fades the intensity of any color.

4. What is lwt?

Lineweights.

CHAPTER 20 REVIEW QUESTIONS

1. List some instances when isometric drawing is *appropriate*.

If you do not yet have 3D skills or need only a quick and simple 3D-looking model. Also if computing resources are limited or you do not want to spend time combining 2D and 3D.

2. List some instances where 3D would be better than isometric drawing.

In any instances where a true 3D model is needed to convey the design intent, including for purposes of shading and rendering. Also when more accuracy is needed.

3. Name the three planes that exist in isometric perspective.

Top, Left, and Right.

4. In what dialog box is *Isometric snap* found?

ddosnap, Snap and Grid tab.

5. What *F key* switches the cursor from plane to plane?

F5.

6. What is almost always on when drawing in isometric?

Ortho.

7. What is the equivalent of a circle in isometric drawing?

Ellipse.

Chapters 21–30

Welcome to the fascinating world of 3D computer-aided design. Ideally, you have completely mastered everything in Levels 1 and 2 and are ready to learn a new set of skills. 3D is a different world in AutoCAD, yet it builds on what you learned in 2D; really what we are doing is just adding another dimension (literally).

3D has come a long way in AutoCAD. From humble beginnings in the 1980s (which left users joking that it is not really 3D but 2.5D), through steady improvements in the 1990s, to a major retooling a few years ago with release 2007, AutoCAD's 3D capabilities have grown exponentially, and it is now quite capable 3D software.

Note however that AutoCAD still does not do engineering modeling and analysis and cannot compete with CATIA, SolidWorks, SolidEdge, Pro/Engineer, NX, Inventor, and others, nor was it ever designed to. At its heart AutoCAD is still *drafting* software first and foremost and is *not* meant for complex modeling and simulation. Autodesk has other products, such as Inventor and Revit, which venture into the world of parametric modeling while AutoCAD is firmly in the world of visualization. If you need to test a design for structural failure or send the geometry to a CNC mill for manufacturing, AutoCAD is not the software for this.

This of course is just fine for thousands of architects and engineers worldwide who use AutoCAD's 3D capabilities for visualization and create superb 3D designs. Some typical uses include 3D architectural models of residential and commercial interiors and exteriors, furniture design, small part design, electrical riser equipment diagrams, and others applications that require a 3D view of the object(s) to visualize, promote, display, and sell the design or concept. Often, these models are exported to specialized rendering software, such as 3ds Max, Form-Z, or Rhino, to obtain photorealistic images. Creating these underlying models is what we focus on here.

Our study of 3D is loosely broken into three major sections. The first part, which starts with Chapter 21, is an introduction to the basics. We cover all the fundamental tools needed to create what can be referred to as *flat design*, meaning objects without major curvature. This broadly refers to architectural models, such as simple 3D floor plans that feature walls, door, windows, and furniture. For the most part, these are elements that feature essentially straight edges. We learn basic tools like Extrusion, 3D Rotate and Mirror, and Booleans.

The second part is an extension of this knowledge into what can be thought of as *curved design*, meaning objects with curvature. This broadly refers to mechanical models, such as gears, shafts, bearings, springs, and shape-formed piping, to name a few examples. There is of course a great deal of overlap between architecture and mechanical design in terms of what is required of AutoCAD, and you really need to learn everything regardless of what field you are in. Here, we cover tools such as Revolve, Loft, Shell, and Taper.

The third part is all about advanced tools, such as Meshes, Lighting, and Rendering, which round out your study of 3D and add realism in the form of lighting and shading as well as some rendering and backgrounds to your design.

A word of caution: Be sure to have truly mastered 2D before proceeding. All the 3D chapters are written with the assumption that you are an experienced 2D user, and no explanations are forthcoming when we use 2D tools to begin 3D projects. Also, review isometric drawing (Chapter 20). That seemingly unrelated topic is actually a precursor to 3D, as you will soon see, and you should go back and review it.

3D Basics

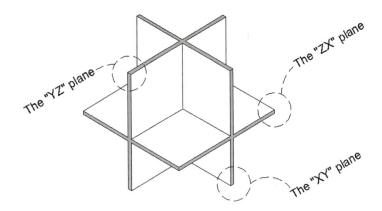

LEARNING OBJECTIVES

In this chapter, we introduce the basics of 3D theory and operating in 3D space. Specific topics include the following:

- Axes, planes, and faces
- 3D Workspace—Ribbon, toolbars, and options
- Entering and exiting 3D
- Projecting into 3D
- 3D dynamic views
- Extrude
- Visual Styles—Hide; Realistic and Conceptual Shade
- ViewCube and Navigation Bar

By the end of the chapter, you will understand basic 3D theory and be able to navigate in and out of 3D, navigate inside 3D, as well extrude, hide, and shade your designs.

Estimated time for completion of chapter: 1 hour.

21.1 AXES, PLANES, AND FACES

So, what exactly does *drawing in 3D* mean? It is the ability to give depth to objects, or to expand them into the "third dimension" from a flat plane. This concept should be intuitively obvious—after all, we live in a 3D world. Everything has not just a length and width but also a depth (or height).

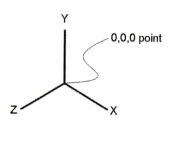
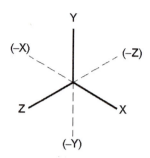

Positive X, Y, and Z axes   Positive and negative X, Y, and Z axes

FIGURE 21.1
The X, Y, and Z axes.

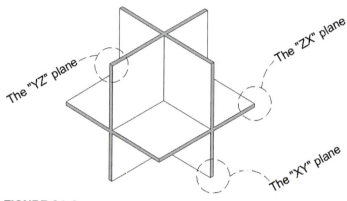

FIGURE 21.2
The XY, YZ, and ZX planes.

It turns out that, as you were learning AutoCAD in Levels 1 and 2, you were in 3D all along. It is just that you did not see or use this vertical third dimension; therefore, everything appeared flat, similar to not seeing the height of a tall building if you are flying directly above it. All this changes as we discuss this hidden dimension and learn how to project into it. For this, we need to start at the very beginning and learn about axes, planes, and faces, because by definition, using the third dimension involves using a third, and previously ignored, Z axis.

Concept 1. There exists in the Cartesian coordinate system a total of three axes: X, Y, and Z. These axes intersect each other at the 0,0,0 point, as seen in Figure 21.1(left), and by definition can be positive or negative, as represented by solid and dashed lines in Figure 21.1(right).

Concept 2. A plane is defined as an intersection of two axes. Therefore, the X, Y, and Z axes can define three unique planes: the XY, YZ, and ZX planes, as seen in Figure 21.2.

Concept 3. A total of six faces of an imaginary cube can then be formed using the three planes. This can be easily seen if we move the planes out and connect them edge to edge, as in Figure 21.3. For our purposes right now, planes and faces are really the same thing, and we refer only to planes from here on forth (though the term *faces* does get some use in related concepts).

The preceding discussion on axes and planes is very important; make sure you completely understand everything thus far. We use both axes and planes shortly.

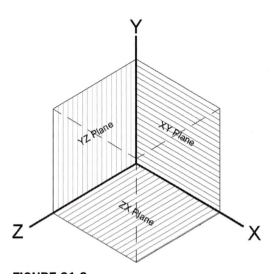

FIGURE 21.3
The XY, YZ, and ZX faces (planes) of a cube.

FIGURE 21.4
3D Basics and 3D
Modeling workspaces.

481

21.2 3D WORKSPACES, RIBBON, TOOLBARS, AND 3D OPTIONS

Before we can start to work in 3D, we need to set up the appropriate tools, so as to not be distracted with trying to find them later on when we need them. Just as in 2D, you can work in 3D via the Ribbon, toolbars, and the occasional typing. Also, quite a few 3D palettes locate many tools in one spot. You often use a combination of all these tools depending on the situation.

AutoCAD simplifies your work in 3D by allowing you to switch to a 3D workspace. Here you find the Ribbon altered, with many of the needed 3D tools in one place. You have two choices, 3D Basics and 3D Modeling, as seen in Figure 21.4. If you have older versions of AutoCAD installed on your computer, you may see other 3D choices (migrated from AutoCAD 2010 or 2009), but ignore them and choose 3D Modeling.

You see your AutoCAD screen change to 3D mode. Note the difference in the Ribbon. It now has a full complement of 3D tools available under multiple tabs (Solid, Surface, Mesh, Render, etc.) while retaining some essential 2D tools. Also be sure to restore the menu bar at the top of the screen (see Chapter 1 if you do not recall how). Finally, shut down any toolbars that may have been present, as we bring up some new 3D ones, as described next.

It is *very useful* to have 3D toolbars on your screen in addition to the Ribbon while you are learning 3D. Despite the overlapping redundancy between the Ribbon and toolbars, a few commands are still found only in the toolbars, so we build a collection of them as we go. If you prefer the Ribbon, you can start eliminating the toolbars after getting acquainted with them as they are introduced.

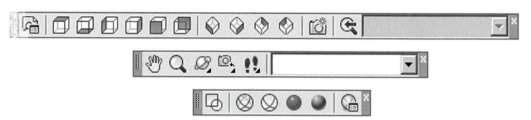

FIGURE 21.5
Essential 3D toolbars.

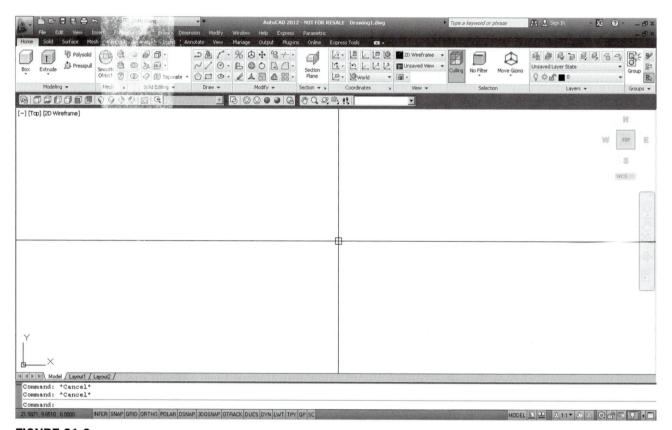

FIGURE 21.6
3D Modeling workspace.

Using **Tools→Toolbars→AutoCAD** in the cascading menus, bring up the toolbars seen in Figure 21.5. They are View, 3D Navigation, and Visual Styles. Arrange them around the screen area. Your screen should look close to, if not exactly like, Figure 21.6.

It is worth noting that, although you have less and less typing to do in 3D, a few commands are still available only via this method. Also, do not forget, you still have the cascading menus with full 3D tools as an alternative to all of these in many cases. This preceding covers all four primary methods of interacting with AutoCAD in 3D. As you can see, except for more of an emphasis on graphical inputs (Ribbon and toolbars), the methods remain the same as 2D.

A few words about 3D and the Options dialog box, first examined in Chapter 14. It is highly recommended that your crosshairs are size 100 (all the way across) in 3D, as this helps a lot with visualizing what you are doing and where in 3D space you are. You can change this under the Display tab. Just as important, you need your UCS icon. In 2D, you may have been

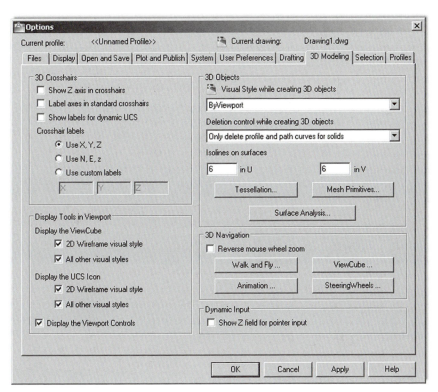

FIGURE 21.7
Options, 3D Modeling tab.

tempted to turn it off (it was one of the tips early in Level 1), but here you need it to know where you are in 3D space, and you really cannot work without it. Finally, let us take a look at the 3D Modeling tab we skipped in Chapter 14. It is shown in Figure 21.7.

For now, the only suggestion is to make sure the Show Z axis in crosshairs option is unchecked (it may already be). It is found all the way at the top left. The 3D concepts are generally much clearer when only two axes are present in the crosshairs. We return to this Options dialog box later in our 3D studies to tweak a few more settings, mostly those at the bottom right (Walk and Fly, ViewCube, etc.). Click on OK and let us jump right into 3D in the next section.

21.3 ENTERING AND EXITING 3D

The key to starting out learning 3D is to get into 3D mode and reveal the third axis. The gateway to *entering* 3D is any of the isometric views, though we use the *SW Isometric View* most often. The way to *exit* 3D is any of the flat views (front, back, left, right, etc.), although we use the *Top View* most often. Let us give this a try using the View toolbar, as shown in Figure 21.8.

Press the SW Isometric View icon and the screen switches to what you see in Figure 21.9. Examine what happened after going into SW Isometric. The Z axis, previously unseen, is revealed by rotation of the UCS icon and the associated planes. Note how this is reflected in the new position, or shape, of the crosshairs. Note that your UCS icon may be in the center of the screen. To make it go back into its lower left corner, as seen in Figure 21.9, just type in ucsicon, press Enter, then type in n for Noorigin.

Go ahead and press the Top View icon and go back to 2D. Now that you understand the two essential views, go ahead and press all the other icons to see what view they give you. Icons

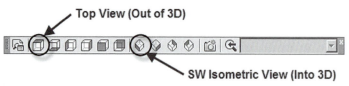

FIGURE 21.8
Top and SW Isometric Views.

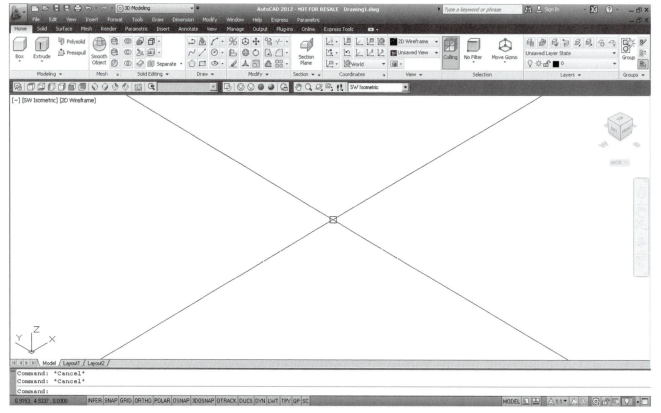

FIGURE 21.9
3D SW Isometric View.

2–7 from the left are all 2D views, while icons 8–11 are all 3D. When done, return to the Top 2D view.

Now, try the exact same thing via the Ribbon under the Home tab→View, as seen in Figure 21.10. You have the same basic symbols, so you should know how to proceed.

Finally let us try the same thing via the cascading menu, as seen in Figure 21.11. The path is **View→3D Views→SW Isometric** (or **Top**). Once again, the symbols are the same, so you should know how to proceed.

Note that in the SW Isometric View, your UCS icon may stay in the middle of the screen. You generally want it tucked away in the lower left corner, just as it was in 2D drafting. To do this, type in `UCSICON` and press Enter.

- AutoCAD says: `Enter an option [ON/OFF/All/Noorigin/ORigin/Properties]`
  `<ON>:`

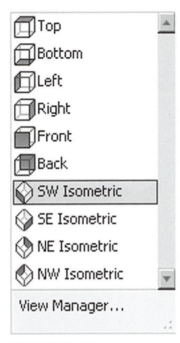

FIGURE 21.10
3D SW Isometric View (Ribbon).

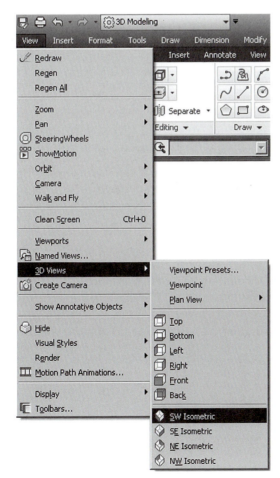

FIGURE 21.11
3D SW Isometric View (cascading menus).

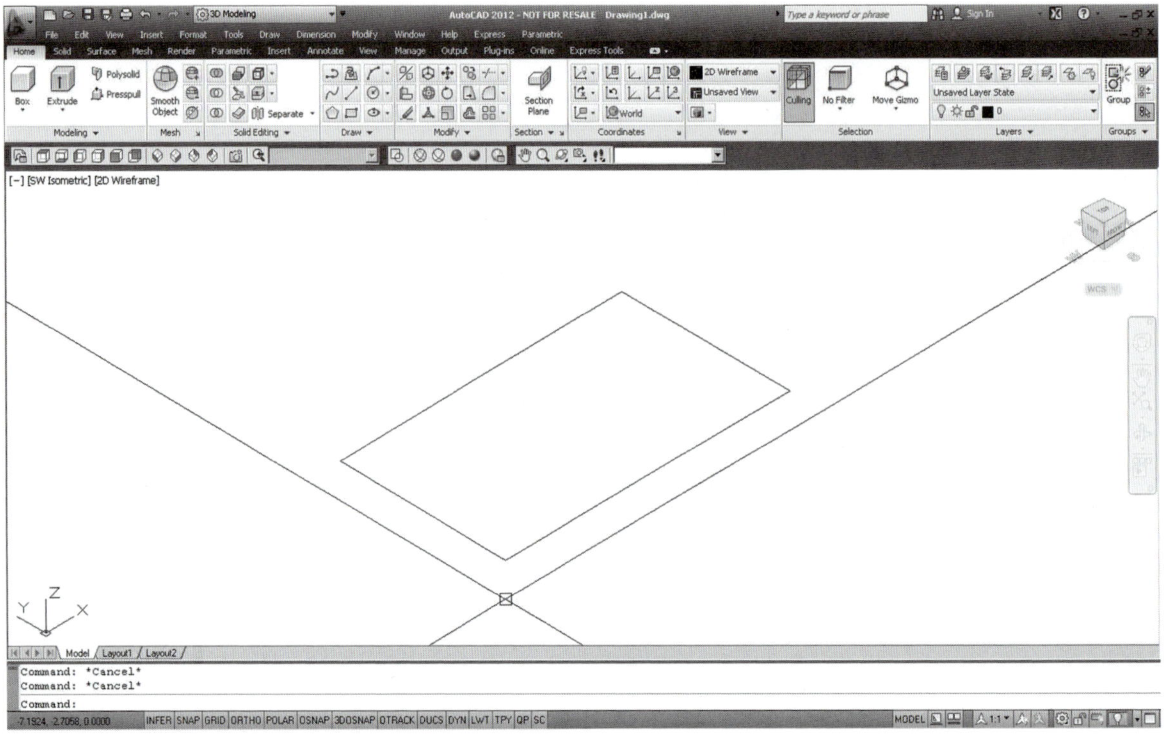

FIGURE 21.12
Rectangle in 3D.

Type in n for Noorigin, press Enter, and the icon returns to the corner and does not interfere with viewing design work. In later chapters, we ask it to move out of the corner when we need the icon to be aligned with objects we are working on, but for now this is not necessary.

You now know how to get into 3D (SW View) and back to 2D (Top View). So, in 3D, go ahead and draw a rectangle that is 10" × 6". You can do this using the rectangle command or individual lines; at this point it does not matter. The result is shown in Figure 21.12.

Here, we come to a roadblock of sorts. Say you would like to draw four sides and a top to create a full 3D box. How do you do this? Although you are in 3D, you have no way yet of actually projecting objects in 3D. This is the subject of the next section.

21.4 PROJECTING INTO 3D

Your input device (universally a mouse these days) is by its nature a 2D device. It works by going forward-back and left-right on your desk or mouse pad but not straight up. This is obviously a bit of a problem for projecting and drawing into the third dimension. Industry CAD software designers and researchers noticed this many years ago and came up with a simple and elegant solution that was integrated into numerous software packages over the years.

When you need to go into 3D, instead of raising your mouse into the air, you simply switch from the "flat" plane you are on to a vertical plane. The effect is immediate; you can now draw "up" relative to you, the observer. To go back to flat, you switch the planes right back. This is shown graphically in Figure 21.13.

You already did this when learning isometric drawing by pressing F5 and cycling through a total of three planes: top, left, and right. We do something similar here in 3D. It is why isometric drawing was important to review. In a way, you already knew some 3D concepts after finishing that chapter.

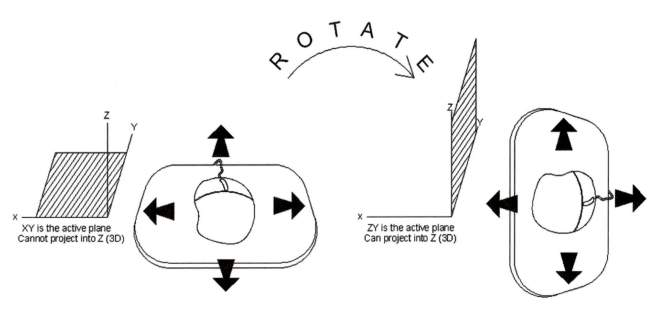

FIGURE 21.13
Rotation of planes.

FIGURE 21.14
UCS toolbar.

So, how do you actually rotate these planes? Rotating planes is equivalent to *rotating the UCS icon*. We introduce the various ways you can rotate the UCS icon via the familiar command matrix. You can do this via typing, cascading menus, toolbars, or the Ribbon. If you wish to continue using the toolbars, you will need to add the UCS toolbar to your collection (Figure 21.14). We discuss most of this toolbar in Chapter 28, but for now need only three of the icons, as shown next.

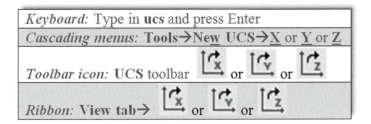

To perform the UCS rotation, use any of the methods shown in the preceding command matrix. However there is a slight difference in procedure, depending on what method you choose. Before you start, first read all the way through the steps to get a full understanding!

Step 1(a). If you *typed* in ucs and pressed Enter, then,
   ○ AutoCAD says: `Specify origin of UCS or [Face/NAmed/OBject/Previous/ View/World/X/Y/Z/ZAxis]<World>:`

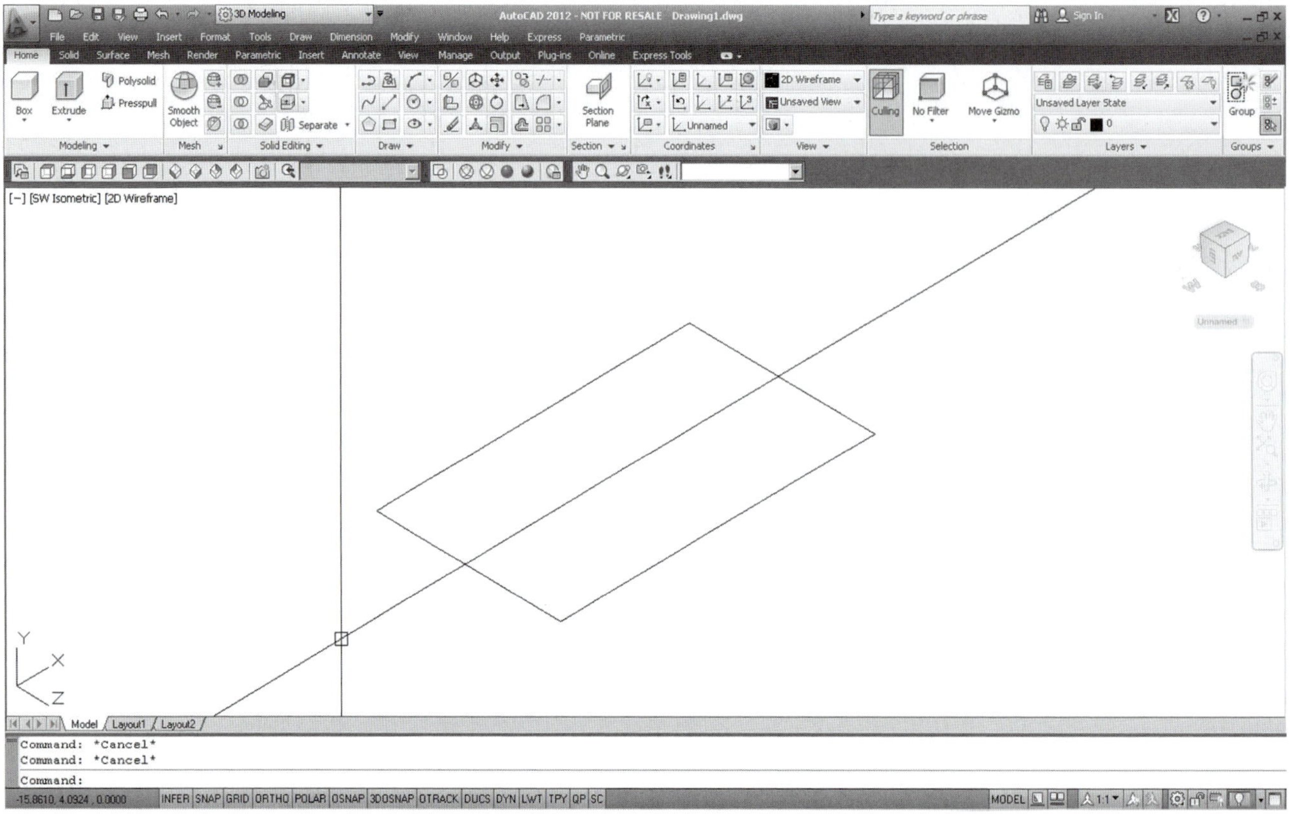

FIGURE 21.15
UCS rotation around the X axis.

You have to choose an axis in Step 2, before proceeding to Step 3, however….

Step 1(b). If you used the *icons, cascading menus,* or the *Ribbon,* then the process of choosing an axis is part of the command right away (as you picked either an X, Y, or Z), and you can skip Step 2.

Step 2. Pick the axis around which you want to rotate: X, Y, or Z.

A very important skill needs to be learned right at the point of Step 1 or Step 2. Around which axis do we need to rotate the UCS icon to be able to draw "up" into the third dimension: X, Y, or Z? It is something you need to be able to quickly identify. Fortunately, in this case, you have a choice of two out of the three axes: either the X or the Y. Rotating around the Z axis does nothing useful (in this example) and just spins the UCS axes around like the blades of a helicopter. Do you see why? So type in either choice X or Y and press Enter, or if using the other methods, choose X or Y right away. This example proceeds with the X axis being chosen.

Step 3. After the X axis is selected,
  ○ AutoCAD says: `Specify rotation angle about X axis <90>:`
Step 4. You can just press Enter, as 90° (the default) is exactly what we need.

You should see what is shown in Figure 21.15. The X axis is the "hinge" around which we pivoted and the crosshairs now point in the direction we need to go, which is straight up, relative to you.

Now go ahead; using standard drafting techniques, draw a straight line starting from one of the rectangle's corners and continuing up to any reasonable distance away (using Ortho and

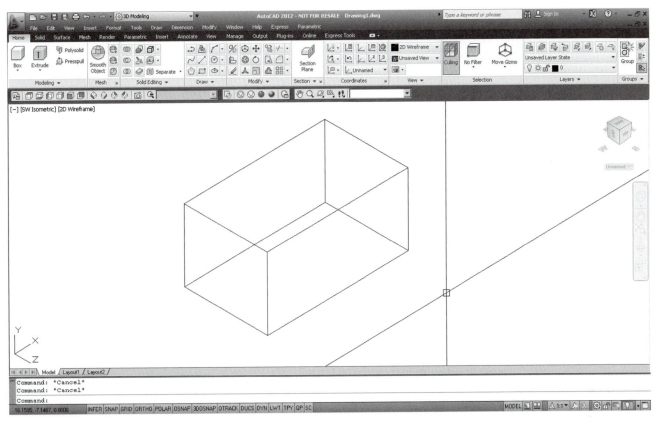

FIGURE 21.16
3D box.

FIGURE 21.17
3D UCS icon reset to
World View.

OSNAP points, of course). Copy it to the other three corners and copy the bottom four lines (or rectangle) to the top, thereby creating a box. You should have what is shown in Figure 21.16. Note that you may have to use grips to "trim" or "extend" the lines if necessary to create a neatly drawn box.

To bring your UCS icon back to the previous familiar state, as originally seen (and shown again in Figure 21.17) you can do any of the following.

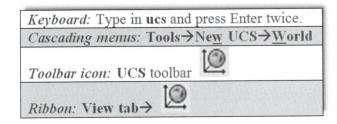

Keyboard: Type in **ucs** and press Enter twice.
Cascading menus: **Tools→New UCS→World**
Toolbar icon: UCS toolbar
Ribbon: **View tab→**

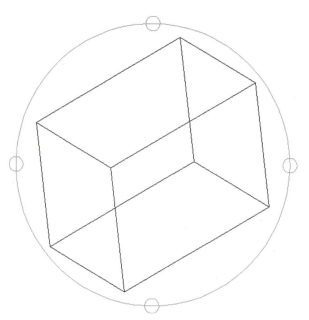

FIGURE 21.18
3D Free Orbit in use.

This is called restoring the icon to "World View." Practice rotating the UCS icon around all three axes. Notice how the axes act as "hinges" for rotations. You need to recognize which "hinge" to rotate around later on as we do actual object rotations.

Next, we need to learn some 3D dynamic viewing options, such as Orbit.

21.5 3D DYNAMIC VIEWS

In 3D, you need to constantly spin the object around to get a good look at it from all sides. AutoCAD has powerful 3D Orbit tools to easily look at your design in real time as opposed to just preset views. Remember that, even though we say that we are rotating the object, really we are not, but rather *rotating our view of it*. To begin the 3D orbit, use any of the following methods.

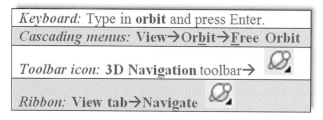

As soon as you click on the Orbit icon, a green circle appears (it does not appear if you type the command, but the effect is the same) and you are able to do the orbit by clicking down, holding the left mouse button, and moving the mouse around as seen in Figure 21.18. When done just press Esc, and the new view is permanent. To restore the familiar SW view, just press that icon.

You can also do a continuous orbit (the icon looks similar but has a small arrow) that rotates the view nonstop until you exit the command. This is more entertaining than useful but does look impressive in a presentation.

What we have done thus far is create a *wireframe* model. This is a common term in computer-aided design (not just AutoCAD) and simply means the model is not shaded or rendered

FIGURE 21.19
Modeling toolbar.

and resembles a wire that has been shaped into something (a frame), hence *wireframe*. Although we introduced a lot of useful concepts here, such as axes, planes, 3D views, 3D orbit, and UCS rotation, an easier and faster way exists to create 3D models, ones that can also be shaded. Wireframes created so far cannot be shaded in AutoCAD, as they are essentially "hollow" pieces and contain no surfaces (although surfaces can be added as we see much later in the course). Creating solids is the subject of the next section, so go ahead and erase the box; we will create it again.

21.6 EXTRUDE

Extrusion is the method most often used to quickly and easily create solid objects and, as such, is perhaps one of the most important commands in 3D. If you prefer toolbars, you will need a new toolbar, called *Modeling* (Figure 21.19) to access the extrude command via that method.

We begin the same way as before, by drawing a 10" × 6" rectangle, but this time use only the rectangle command, not individual lines, then proceed as described next.

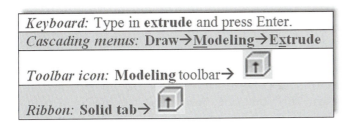

Keyboard: Type in **extrude** and press Enter.

Cascading menus: **Draw→Modeling→Extrude**

Toolbar icon: **Modeling** toolbar→

Ribbon: **Solid tab→**

Step 1. Start the extrude command via any of the preceding methods.
  ○ AutoCAD says: `Current wire frame density: ISOLINES=4`
      `Select objects to extrude:`
Step 2. Pick the rectangle and press Enter.
  ○ AutoCAD says: `Specify height of extrusion or [Direction/Path/Taper angle]:`
Step 3. You can actually now move your mouse up and down and create a thickness in real time, but it is more practical to type in a value, so go ahead and type in 4 for the thickness. You should see what is shown in Figure 21.20.

This new, extruded box may at first seem quite similar to the previous one you tried doing. That is because wireframe is not just a *method of construction* but also a *method of presentation*. There are some significant differences, however. This box is a real solid model, not just a collection of "wires" spliced together. We can now hide and shade it as described in the next section.

21.7 VISUAL STYLES: HIDE AND SHADE

Visual Styles is an important set of tools that allow you to view your design in a variety of useful ways. We explore two variations of Hide and two distinct versions of Shade: realistic and conceptual. We also look at a few other settings, such as Sketchy and X-Ray. Those can

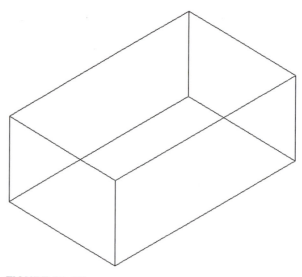

FIGURE 21.20
Extruded box.

FIGURE 21.21
Render toolbar.

be changed easily via the Colors… button under the Display tab in Options…, as detailed in Level 2.

The hide command simply hides wireframe linework that you would not see with a solid object. There are two slightly different variations of it. The first type of hide is temporary while you are viewing a stationary design. The hidden view reverts to wireframe if you try to rotate it or regenerate it. This can be done by simply typing in hide and pressing Enter, via the Hide icon on the Render toolbar (see Figure 21.21), or finally via the cascading menu **View→Hide**.

The other variety of hide is more permanent and the design can be rotated in 3D while staying fully hidden. That is accomplished via the 3D Hidden Visual Style icon on the Visual Styles toolbar (first shown in Figure 21.6).

To return to wireframe, you can no longer just regenerate but must press the 2D Wireframe icon on the same Visual Styles toolbar. All of the previous is summarized in the command matrix shown next. Go ahead and try all the ways to hide your 3D box before moving on.

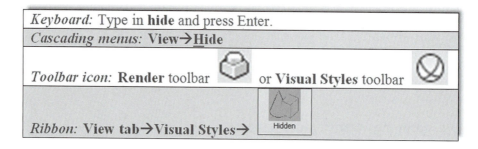

The result of hiding via the keyboard hide entry, cascading menus, and the Render toolbar's Hide icon is shown in Figure 21.22. Note how you can no longer zoom and pan. You can

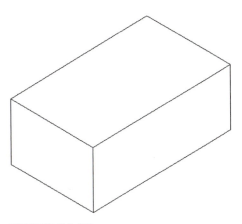

FIGURE 21.22
Hidden box, Method 1.

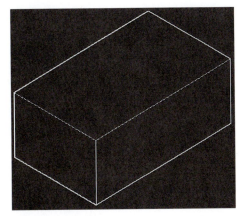

FIGURE 21.23
Hidden box, Method 2.

still do a Free Orbit, but the hidden lines reappear, so this is useful to just view the design and maybe run off a print.

The result of using the 3D Hidden Visual Style icon as well as the Hidden choice in the Ribbon is shown in Figure 21.23. Note how the background color of the screen changes (which can be changed, as mentioned). Also note how you can now zoom and pan. This version of hide is good for continued work on the design while in hidden mode.

Let us now give shading a try. As already mentioned, it comes in two versions, conceptual and realistic. Which you use depends on a variety of shape and background color settings. First, we focus on the default shade, which is the Realistic Visual Style. Start shade via any of the following methods.

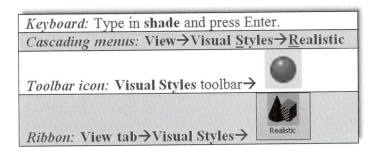

Keyboard: Type in **shade** and press Enter.
Cascading menus: **View→Visual Styles→Realistic**
Toolbar icon: **Visual Styles** toolbar→
Ribbon: **View** tab→**Visual Styles**→

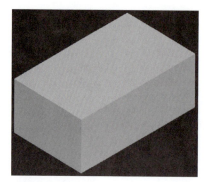

FIGURE 21.24
Realistic shading box.

FIGURE 21.25
Conceptual shading box.

494

The result of this shading is shown in Figure 21.24.

The other style of shade is set as follows. Note that typing shade is no longer applicable, as that leads to the realistic version of shade being applied by default.

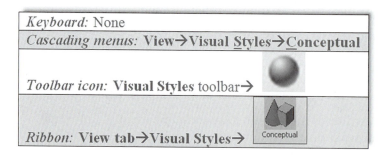

Keyboard: None	
Cascading menus: **View→Visual Styles→Conceptual**	
Toolbar icon: **Visual Styles** toolbar→	
Ribbon: **View tab→Visual Styles→**	Conceptual

The result of this shading is shown in Figure 21.25.

You probably did not notice too much of a difference between the two types of shading. That is because the boxes are not really colored themselves and also they appeared against the default background, altering the appearance of their own coloring. To appreciate the differences between the two types of shading, observe the following two sets of colored and shaded shapes (Figures 21.26 and 21.27).

These shapes are set against a white then a black background with the realistic and conceptual shading on, and similar colors added to all the shapes. Notice how the softer shading on the

Realistic   Conceptual

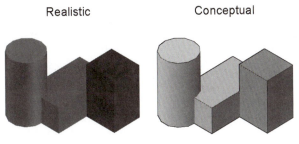

FIGURE 21.26
Realistic vs. conceptual shading (white background).

Realistic   Conceptual

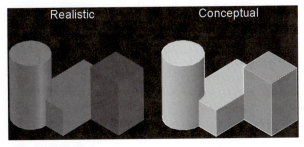

FIGURE 21.27
Realistic vs. conceptual shading (black background).

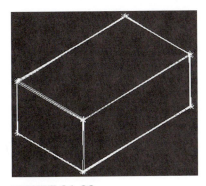

FIGURE 21.28
Sketchy visual style.

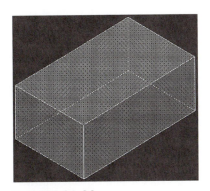

FIGURE 21.29
X-ray visual style.

right gives a more pleasing picture of the model. For this reason, the Conceptual Visual Style shading is used more often in presenting designs both on-screen and on paper. It is also highly recommended that you set your background to white or black in 3D for a more accurate representation of colors. To return to the unshaded image, press the 2D Wireframe option or icon.

Two more types of presentation views are quickly mentioned: sketchy and X-ray. Both can be accessed through the Ribbon's View tab→Visual Styles or through the cascading menu's **View→Visual Styles**. Try them both out on your box. Figure 21.28 shows the sketchy visual style. It presents a "hand-drawn" rendering of your design and may find some use in a more artistic representation. X-ray makes the shaded shape transparent and is seen in Figure 21.29.

All the views and settings thus far described can be accessed via a (new for AutoCAD 2012) set of drop-down menus at the very upper left of the drawing area—shown with a black background for clarity. The first category from the left, [-], is the Viewport Controls. They present various choices for viewports and whether or not to show the ViewCube, SteeringWheels, and Navigation Bar (to be discussed in the next section). The menu is shown in Figure 21.30. Just to the right of that, under [SW Isometric], are the View Controls. They allow yet another way to set and manage the various views. The menu is shown in Figure 21.31. Finally the [2D Wireframe] menu contains the Visual Style Controls. It allows you to set the visual styles, and the menu is shown in Figure 21.32.

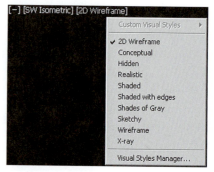

FIGURE 21.30
Viewport Controls.

FIGURE 21.31
View Controls.

FIGURE 21.32
Visual Style Controls.

After the final discussion regarding the ViewCube, SteeringWheels, and the Navigation Bar in the next section, you should be familiar with almost everything in these menus.

21.8 VIEW CUBE AND NAVIGATION BAR

You almost certainly have been looking at the ViewCube and Navigation Bar since Chapter 1. The top view of this cube and its friend the Navigation Bar hang out in the upper right corner of your screen, and have done so since the beginning, as seen in the 2D top view in Figure 21.33. SteeringWheels (an advanced tool not discussed here) is also seen in Figure 21.33 and is brought up via the first button on the Navigation Bar.

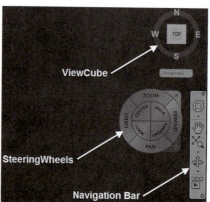

FIGURE 21.33
ViewCube, SteeringWheels, and Navigation Bar.

496

With your box shaded, switch to SW Isometric view. Now that you are in 3D, the bar stays the same, but the cube looks like what is shown in Figure 21.34.

FIGURE 21.34
ViewCube in 3D.

This cube is primarily useful in 3D navigation, and that is why there was no mention of it in the previous 2D chapters. It first appeared in AutoCAD 2009 and is used to dynamically rotate to the graphically shown views, such as top, left, or front. You can click on the faces of the cube, its edges, or its corners; and of course you can just manually rotate it. Whatever you click on "lights up," and your design moves accordingly. Additionally you can press the N, S, E, and W buttons. Go ahead and try it out, returning to the SW Isometric View afterward.

The ViewCube presents no completely new ideas but rather combines a host of various tools in one convenient location. Much of it is equivalent to the toolbar view icons, though you now have the ability to also view a design from a "corner" or an "edge."

The buttons on the Navigation Bar should also be generally familiar at this point, except for the top button, Full Navigation Wheel, and the bottom button, ShowMotion (also an advanced topic, not covered in his chapter). The rest of the buttons from the top going down are Pan, Zoom Extents, and the various Orbits. Review the functionality of each button.

One last item to mention with the navigation cube is its ability to put you into perspective view. We cover perspective view in later chapters, but you can try it out briefly now. Hover your mouse around the top left area of the cube and a tiny blue house appears, as seen in Figure 21.35.

FIGURE 21.35
Additional menu button.

Click on the house icon and your shaded cube switches to a perspective view (Figure 21.36).

To exit out of this view, right-click on the house and you see the menu in Figure 21.37. Select and click on Parallel and you exit the perspective view.

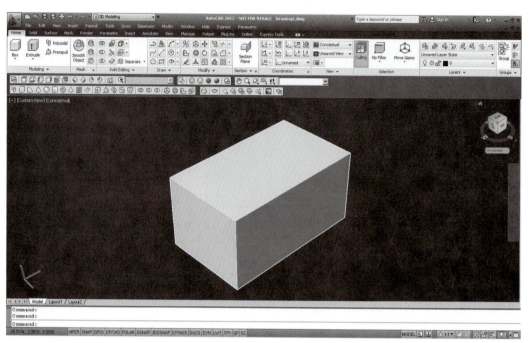

FIGURE 21.36
Perspective view.

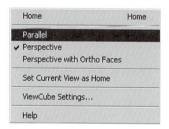

FIGURE 21.37
Restoring the Parallel view.

This is it for the basics. It seems we introduced an endless collection of new concepts, but really it was just six "big ideas" as listed next.

- The concepts of axes and planes.
- How to get in and out of 3D.
- How to rotate planes.
- How to orbit around your design.
- How to extrude the design.
- How to shade the design.

Review everything carefully; your success in the rest of 3D depends on understanding these basic concepts.

SUMMARY

You should understand and know how to use the following concepts and commands before moving on to Chapter 22:

- Axes: X, Y, and Z
- Planes: XY, ZX, and YZ
- Faces

- 3D toolbars: View, 3D Navigation, and Visual Styles
- SW Isometric View
- Top view
- UCS rotation
- Free Orbit
- Extrude
- Hide
- Shade: realistic and conceptual views
- ViewCube and Navigation Bar

REVIEW QUESTIONS

Answer the following based on what you learned in Chapter 21:

1. What three axes are discussed?
2. What three planes are discussed?
3. What 3D toolbars were introduced in this chapter?
4. What command is used most often to create 3D objects?
5. How do you rotate the UCS icon?
6. What does hide do?
7. What does shade do? What are the two types of shade?
8. What is the purpose of the Navigation Bar and the ViewCube?

EXERCISES

499

1. In 3D SW Isometric Conceptual Visual Style (CVS), set your units to Architectural and draw a 10" × 5" rectangle. Extrude this rectangle to 8'. This is one of several ways to draw a wall. We need this in the coming chapters. Zoom out to extents if necessary and alter the color if you wish. The result is shown next. (Difficulty level: Easy; Time to completion: <5 minutes)

2. In the 3D SW Isometric CVS, set your units to Architectural and draw a 4" × 3" box. Extrude it to 6". Now rotate your UCS around the Y axis by 30° and around the X axis by 30°. Then, draw the same rectangle with the same extrusion. Change the colors if you wish. This is what you should get. (Difficulty level: Easy; Time to completion: <5 minutes)

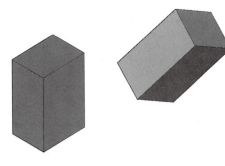

3. In the 3D SW Isometric CVS, set your units to Architectural and draw a 5" circle, extruding it to 25". Now rotate your UCS around the X axis by 90° and draw the same extruded shape. Then, restore the UCS position, and rotate it around the Y axis by 90°. Draw the same shape again. You should have the three cylinders shown in the left image. Move two of them center to center with the third one as shown in the right image. (Difficulty level: Easy, Time to completion: 5–7 minutes)

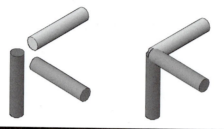

CHAPTER 22

Primitives

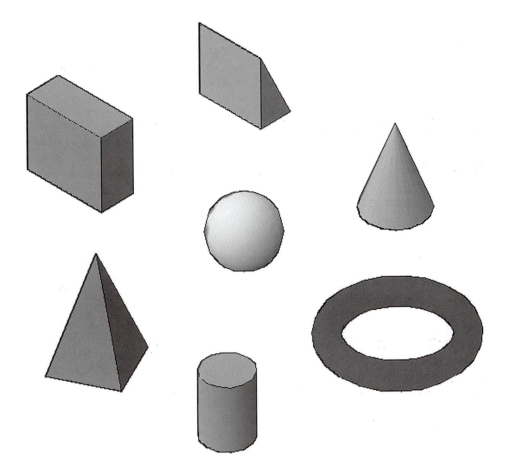

LEARNING OBJECTIVES

In this chapter, we introduce the concept of AutoCAD primitive shapes, known as just *primitives*, and discuss the following:

- Box
- Wedge
- Cone
- Sphere
- Cylinder
- Torus
- Pyramid

Up and Running with AutoCAD® 2012: 2D and 3D Drawing and Modeling.

By the end of the chapter, you will understand basic primitive shapes and be able to apply them in design work as necessary.

Estimated time for completion of chapter: 1 hour.

22.1 INTRODUCTION TO PRIMITIVES

Before we jump into more complex extruded-shape manipulations, we take a brief detour with this chapter and go over a very basic set of tools that you may find useful in certain situations. Called *primitives*, they are nothing more than ready-to-go shapes built into AutoCAD. Think of them as Lego® pieces for computer-aided design; all you need to do is specify a few dimensional values and place them in your drawing as needed. The available primitives are

- Box
- Wedge
- Cone
- Sphere
- Cylinder
- Torus
- Pyramid

As with most commands you learned in AutoCAD, there are four ways to access these tools:

1. *Typing in* the desired primitive and pressing Enter.
2. *Cascading menus* under **Draw→Modeling**.
3. Using the *Modeling toolbar*; icons 2–8 from the left.
4. Using the *Ribbon*; Solid tab.

Once you select the primitive via one of these methods, simply enter the shape's dimensional values as prompted (such as height, width, angle) or just click or move the mouse for random values. It is best to have the Conceptual Visual Style shading turned on, as the shapes become clearer, as well as staying in the 3D SW Isometric View, to get the full effect. Hold off with trying the other shapes on the Modeling toolbar, as we go over them in advanced chapters. Finally, all primitives can be erased, moved, copied, and rotated in space just like any other solid shape.

Box

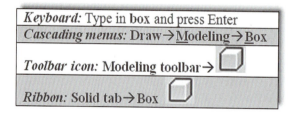

Keyboard: Type in box and press Enter
Cascading menus: Draw→Modeling→Box
Toolbar icon: Modeling toolbar→
Ribbon: Solid tab→Box

Step 1. To try this first primitive, use any of the preceding methods.
  ○ AutoCAD says: Specify first corner or [Center]:
Step 2. Click anywhere on the screen.
  ○ AutoCAD says: Specify other corner or [Cube/Length]:
Step 3. A 2D solid rectangle is drawn based on the movements of your mouse, or you can enter a value by pressing L for length. Click randomly or enter a value.
  ○ AutoCAD says: Specify height or [2Point] <7.2545>:
Step 4. Drag your mouse up (or enter a value) and the shape is automatically extruded into a cube, as seen in Figure 22.1.

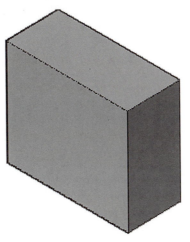

FIGURE 22.1
Box primitive.

Wedge

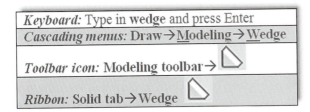

Keyboard: Type in wedge and press Enter
Cascading menus: Draw→Modeling→Wedge
Toolbar icon: Modeling toolbar→
Ribbon: Solid tab→Wedge

503

Step 1. To try this primitive, use any of the preceding methods.
  ○ AutoCAD says: `Specify first corner or [Center]:`
Step 2. Click anywhere on the screen.
  ○ AutoCAD says: `Specify other corner or [Cube/Length]:`
Step 3. A 2D solid rectangle (the base of the wedge) is drawn based on the movements of your mouse, or you can enter a value by pressing L for length. Click randomly or enter a value.
  ○ AutoCAD says: `Specify height or [2Point] <2.0000>:`
Step 4. Drag your mouse up (or enter a value) and the wedge shape is automatically extruded, as seen in Figure 22.2.

Cone

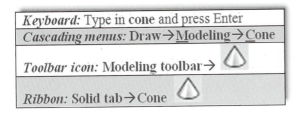

Keyboard: Type in cone and press Enter
Cascading menus: Draw→Modeling→Cone
Toolbar icon: Modeling toolbar→
Ribbon: Solid tab→Cone

Step 1. To try this primitive, use any of the preceding methods.
  ○ AutoCAD says: `Specify center point of base or [3P/2P/Ttr/ Elliptical]:`
Step 2. Click anywhere on the screen or use the 3P, 2P Ttr features, as in the 2D world.
  ○ AutoCAD says: `Specify base radius or [Diameter] <7.1596>:`

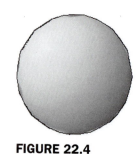

FIGURE 22.4
Sphere primitive.

FIGURE 22.3
Cone primitive.

FIGURE 22.2
Wedge primitive.

Step 3. A 2D solid circle (the base of the cone) is drawn based on the movements of your mouse, or you can enter a value by typing a radius or pressing d for diameter. Click randomly or enter a value.
  ○ AutoCAD says: `Specify height or [2Point/Axis endpoint/Top radius] <6.9580>:`
Step 4. Drag your mouse up (or enter a value) and the cone shape is automatically extruded, as seen in Figure 22.3. The command has some interesting suboptions, such as `Top radius`, which can chop off the pointy top of the cone. Experiment with these options.

Sphere

Keyboard: Type in **sphere** and press Enter
Cascading menus: Draw→<u>M</u>odeling→Sphere
Toolbar icon: Modeling toolbar→ 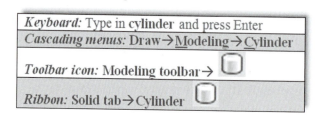
Ribbon: Solid tab→Sphere

Step 1. To try this primitive, use any of the preceding methods.
  ○ AutoCAD says: `Specify center point or [3P/2P/Ttr]:`
Step 2. Click anywhere on the screen or use the 3P, 2P Ttr features, as in the 2D world.
  ○ AutoCAD says: `Specify radius or [Diameter] <5.7587>:`
Step 3. A solid sphere is drawn based on the movements of your mouse, or you can enter a value by typing a radius or pressing d for diameter. The final shape is shown in Figure 22.4.

Cylinder

Keyboard: Type in **cylinder** and press Enter
Cascading menus: Draw→<u>M</u>odeling→Cylinder
Toolbar icon: Modeling toolbar→
Ribbon: Solid tab→Cylinder

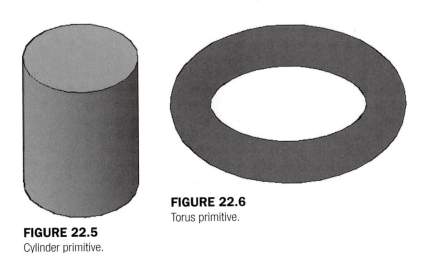

FIGURE 22.6
Torus primitive.

FIGURE 22.5
Cylinder primitive.

Step 1. To try this primitive, use any of the preceding methods.
- AutoCAD says: `Specify center point of base or [3P/2P/Ttr/ Elliptical]:`
Step 2. Click anywhere on the screen or use the 3P, 2P Ttr features, as in the 2D world.
- AutoCAD says: `Specify base radius or [Diameter] <7.9122>:`
Step 3. A solid circle (the base of the cylinder) is drawn based on the movements of your mouse, or you can enter a value by typing a radius or pressing d for diameter.
- AutoCAD says: `Specify height or [2Point/Axis endpoint] <22.8198>:`
Step 4. Drag your mouse up (or enter a value) and the cylinder shape is automatically extruded, as seen in Figure 22.5.

Torus

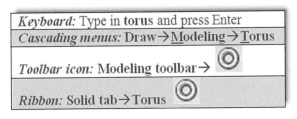

Keyboard: Type in **torus** and press Enter
Cascading menus: Draw→Modeling→Torus
Toolbar icon: Modeling toolbar→
Ribbon: Solid tab→Torus

Step 1. A torus is basically a donut. To try this primitive, use any of the preceding methods.
- AutoCAD says: `Specify center point or [3P/2P/Ttr]:`
Step 2. Click anywhere on the screen or use the 3P, 2P Ttr features, as in the 2D world.
- AutoCAD says: `Specify radius or [Diameter] <6.3128>:`
Step 3. A hoop is drawn based on the movements of your mouse or you can enter a value by typing a radius or pressing d for diameter.
- AutoCAD says: `Specify tube radius or [2Point/Diameter] <6.2433>:`
Step 4. Drag your mouse (or enter a value) and the torus shape is automatically extruded, as seen in Figure 22.6.

Pyramid

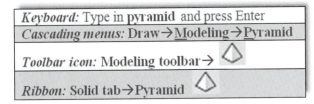

Keyboard: Type in **pyramid** and press Enter
Cascading menus: Draw→Modeling→Pyramid
Toolbar icon: Modeling toolbar→
Ribbon: Solid tab→Pyramid

FIGURE 22.7
Pyramid primitive.

Step 1. To try this primitive, use any of the preceding methods.
  ○ AutoCAD says: `4 sides Circumscribed`
    `Specify center point of base or [Edge/Sides]:`
Step 2. You can then click anywhere, or use the `Sides` option to give the pyramid more than four sides, a useful feature.
  ○ AutoCAD says: `Specify base radius or [Inscribed] <22.2294>:`
Step 3. A pyramid base is drawn based on the movements of your mouse, or you can enter a value. The inscribed feature works in the same manner as it was used with the polygon command.
  ○ AutoCAD says: `Specify height or [2Point/Axis endpoint/Top radius]`
    `<38.8857>:`
Step 4. Drag your mouse (or enter a value) and the pyramid shape is automatically extruded, as seen in Figure 22.7. Just like with the cone, you can vary the top radius of the shape.

22.2 APPLYING PRIMITIVES

So, how useful are these primitives in real-life applications? The best way to describe this is "extremely useful in isolated cases and not used much in most others." For example, if you need a shape not quickly done any other way, such as a sphere or torus (you can easily extrude the other shapes), then primitives are indispensable. But, since most of the 3D objects you will design are not simple shapes, primitives are rarely used. It is still a good skill to have and an easy one to learn.

Go over each shape several times, add some color, and be sure to practice 3D orbit to get various views of the objects, as well as switching between wireframe and shaded. This chapter marks the end of 3D introductory reading. Flat design begins in earnest next, and the following chapters pick up in speed and level of difficulty.

SUMMARY

You should understand and know how to use the following concepts and commands before moving on to Chapter 23:

- Box
- Wedge
- Cone
- Sphere
- Cylinder
- Torus
- Pyramid

REVIEW QUESTIONS

Answer the following based on what you learned in Chapter 22:

1. Name the seven primitives covered in this chapter.
2. What four methods are used to create primitives?
3. What two primitives cannot be created any other way?

EXERCISES

1. Create each primitive shape once via only mouse clicking and once using keyed-in values. (Difficulty level: Easy; Time to completion: 5–10 minutes)
2. Create the following layout using primitives and the rectangle command. All sizes and positioning are approximate. (Difficulty level: Easy; Time to completion: 10 minutes)

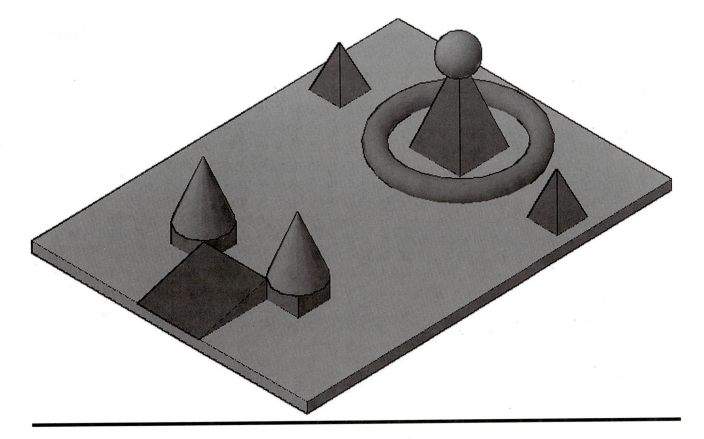

Object Manipulation

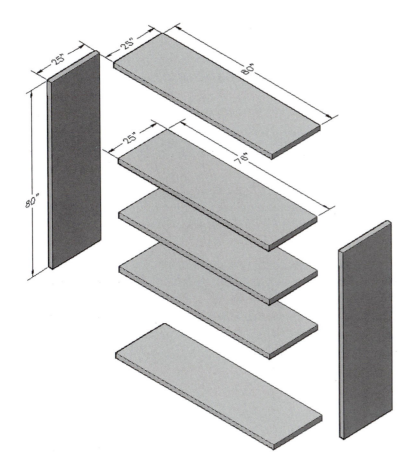

509

By the end of the chapter, you will be able to create flat designs of moderate complexity. We introduce multiple small projects throughout this and upcoming chapters to put each new concept to use right away.

Estimated time for completion of chapter: 2 hours.

23.1 INTRODUCTION TO OBJECT MANIPULATION

It was already mentioned that extrude is perhaps the single most important command in 3D AutoCAD. Using it, you can create a wide array of objects that are flat in nature, such as building walls, doors, and some furniture, just to name a few items. However, unless you are designing a perfect box of a house, you need to rotate and mirror these objects into position as well as array objects and add fillets or chamfers for more realistic edges. All these tools and more are covered in this chapter. We go through each one and do a few sample exercises. These tools, together with Boolean operations (introduced in the following chapter), are virtually all you need to model a basic architectural design of a building in 3D.

Rotate3D

Although you may not have yet tried it, you can use the regular rotate command in 3D. You can rotate the object on whichever plane you are on. The rotational axis is perpendicular to the plane and the net result is the object "spinning in place." However you may often want to rotate the object around another axis, for example, make a flat plate "stand up." Recall from the first chapter that you can use the UCS command to simply rotate your active plane and draw new objects that way, but for existing ones you need to use the 3D version of rotate to bring them into position. Rotate3D gives you this ability and is an important design tool.

This command comes in two versions: rotate3D and 3Drotate. The difference is that the former is command driven and the latter uses a graphical "gizmo" to facilitate rotation. We try both, starting with rotate3D. Draw a flat 10" × 10" plate and extrude it to 1" thickness, as seen in Figure 23.1, with the Conceptual Visual Style shading on and color added.

Notice the colorful UCS icon at the bottom of the figure. It is very prominent on purpose, because it is the focus of our attention while rotating in 3D. What we want to do right now

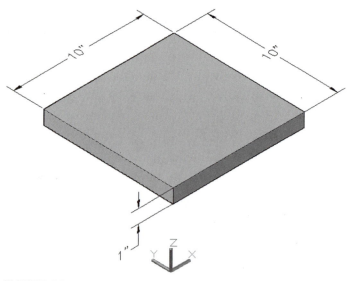

FIGURE 23.1
Flat plate.

is rotate the plate so it stands up on its side like a vertical wall. What axis do we pick as a "hinge" around which the plate rotates? Fortunately, we have a choice of two: the X and the Y axes. Rotating around the Z does nothing for us in this case. Notice the similarities of this discussion to the one on UCS rotation in Chapter 21; it is essentially the same idea.

So now that we know what we want to do, let us give the command a try. There is no Ribbon, cascading menu, or toolbar equivalent, so do the following:

Step 1. Type in rotate3D and press Enter.
  - ○ AutoCAD says: Current positive angle: ANGDIR=counterclockwise
    ANGBASE=0
    Select objects:
Step 2. Select the plate and press Enter.
  - ○ AutoCAD says: Specify first point on axis or define axis by [Object/
    Last/View/Xaxis/Yaxis/Zaxis/2points]:
Step 3. Here, you need to choose which axis to rotate around. Let us go with X. Type that in and press Enter.
  - ○ AutoCAD says: Specify a point on the X axis <0,0,0>:
Step 4. Here, you need to pick the point on the axis of rotation, which should be on the object itself. Using OSNAPs pick the lower left corner point of the plate.
  - ○ AutoCAD says: Specify rotation angle or [Reference]:
Step 5. Here, you can enter any numerical value, but let us stay with 90° for now. Type in 90 and press Enter. The plate rotates 90° around the X axis and stands on its end, as in Figure 23.2.

In the same exact manner, practice rotating the plate around the other axes. From where it is right now in Figure 23.2, a rotation around the Z axis makes it face the lower left side of the page. A rotation around the X axis makes it lay down flat again, and a rotation around the Y axis produces no visible results, aside from a possible shift in position, as the plate is a perfect 10" × 10" square.

3Drotate (Gizmo)

An alternative approach to rotating in 3D is a relatively new tool called a *gizmo*. The net result is exactly the same as with the typed command just covered, but the approach is more graphical in nature. You also have toolbar, Ribbon, and cascading menu approaches available, as seen here in the command matrix.

FIGURE 23.2
Plate 3D rotation around
the X axis.

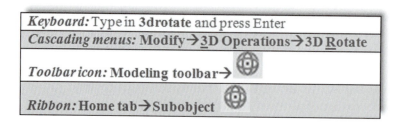

Keyboard: Type in **3drotate** and press Enter	
Cascading menus: Modify→<u>3</u>D Operations→3D <u>R</u>otate	
Toolbar icon: **Modeling toolbar**→	
Ribbon: Home tab→Subobject	

When you start up the command via any of the preceding methods you are asked to select the object in question and you see the gizmo shown in Figure 23.3.

The purpose behind this tool is to be able to easily visualize rotational motion. Following the command line directions, first pick a base point, then the circular band that best represents the rotational axis of motion you want. Finally enter an angle and the shape is rotated. This all closely matches the previous method and detailed steps are not given, as they are essentially the same. A snapshot of the gizmo in action is shown in Figure 23.4. Practice both approaches and stick with the one you prefer.

Be sure to understand these procedures thoroughly. Next, we put everything together and draw our first real 3D project, a chair, and then move on to mirror3D.

3D MODELING EXERCISE: CHAIR

Create the model of a chair based on the dimensions given in Figure 23.5. The full assembly, at a slightly different angle, is shown in Figure 23.6. Specifically you need to

- Create four 2" × 2" rectangles extruded 30" for the legs
- Create two 24" × 24" rectangles extruded 2" for the seat and backrest
- Rotate3D one of the rectangles 30° from vertical for the reclining backrest
- Make sure everything fits together perfectly, adding shading and color of your choice

Mirror3D

Mirror3D is similar in approach to rotate3D. Much like before, you can use the regular 2D mirror command, but this results in mirroring in the current plane only. What we need is the ability to mirror between *any* planes. As you may have guessed, the key here is selecting the correct plane (as opposed to axis) to mirror over. Just as a reminder, we have three axes (X, Y, and Z) and three unique planes (XY, ZX, and YZ). Unlike the axes, you do not see the planes;

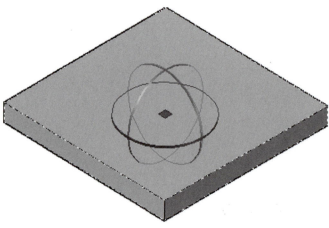

FIGURE 23.3
Gizmo tool.

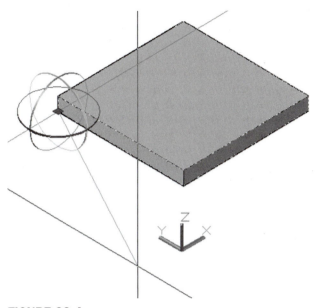

FIGURE 23.4
Gizmo tool rotating an object.

513

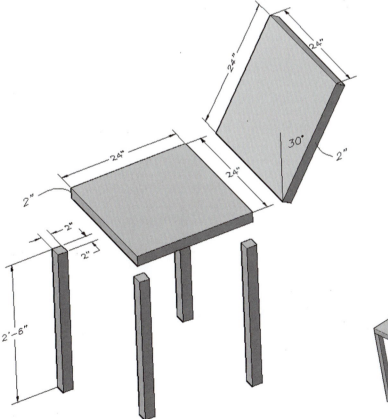

FIGURE 23.5
Drawing project, 3D chair.

FIGURE 23.6
3D chair, full assembly.

it is something you must envision in your mind. Think of them as a thin sheet of glass that slices through parts. Whatever is on the other side is the mirror image. Erase your chair and let us start from a clean screen. Draw the previous 10" × 10" flat plate extruded to 1" (Figure 23.7).

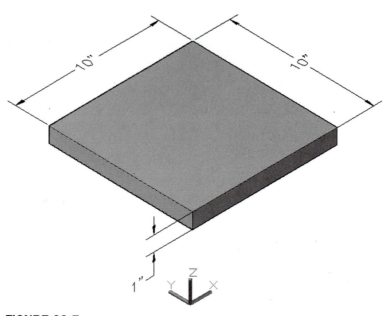

FIGURE 23.7
Flat plate for mirror3D.

514

We would like to mirror it so there is another plate next to the darker shaded right side. Before proceeding, identify the correct plane. Remember a plane is not an axis; rather it is formed by *two axes* put together, so your choices once again are XY, ZX, and YZ. The correct answer is ZX. Do you see why? It is important to be able to quickly recognize these planes. Let us now run through the full mirroring procedure. Mirror3D can also be written as 3Dmirror, with no effect on the command. No gizmo is associated with it, but there is a Ribbon icon, as seen here in the command matrix.

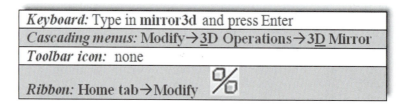

Keyboard: Type in **mirror3d** and press Enter		
Cascading menus: Modify→3D Operations→3D Mirror		
Toolbar icon: none		
Ribbon: Home tab→Modify		

Step 1. Start up the command via any of the preceding methods.
  ○ AutoCAD says: `Select objects:`
Step 2. Select the plate and press Enter.
  ○ AutoCAD says:
    `Specify first point of mirror plane (3 points) or`
    `[Object/Last/Zaxis/View/XY/YZ/ZX/3points] <3points>:`
Step 3. It is here that you need to select the correct plane. As discussed before, type in ZX and press Enter.
  ○ AutoCAD says: `Specify point on ZX plane <0,0,0>:`

Step 4. Pick a point somewhere on the mirroring plane using OSNAPs.
- ○ AutoCAD says: `Delete source objects? [Yes/No] <N>:`

Step 5. This is similar to the 2D version of the mirror command. Press Enter for the default. The mirrored image appears as shown in Figure 23.8.

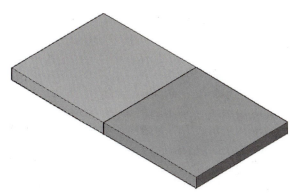

FIGURE 23.8
Flat plate mirrored over the ZX plane.

Go ahead and mirror the two plates over the remaining planes. The procedure is exactly the same, except you type in YZ for the first mirror plane and XY for the second. The expected results are shown in Figures 23.9 and 23.10, respectively.

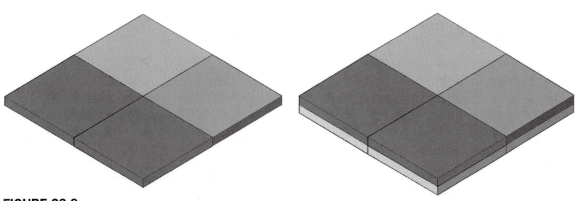

FIGURE 23.9
Flat plates mirrored over the YZ plane.

FIGURE 23.10
Flat plates mirrored over the XY plane.

3D MODELING EXERCISE: BOOKSHELF

Create the model of a bookshelf based on the dimensions given in Figure 23.11. The full assembly is shown in Figure 23.12. Specifically you need to:

- Create four 25" × 80" rectangles extruded 2" for the top, bottom, and side shelves.
- Create three 25" × 76" rectangles extruded 2" for the inner shelves.
- Rotate3D any parts that need to be rotated according to design requirements.
- Keep in mind that the distances between inner shelves can be a random value.
- Make sure everything fits together perfectly, adding shading and color of your choice.

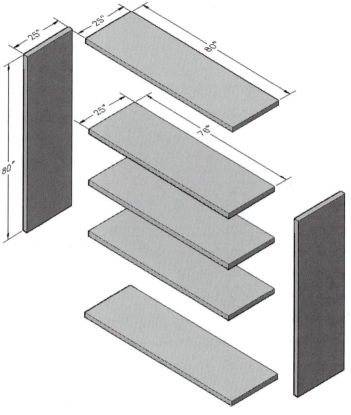

FIGURE 23.11
Drawing project, 3D bookshelf.

FIGURE 23.12
3D bookshelf, full
assembly.

3Darray

The 3D Array is quite similar in principle to the regular 2D one. There is, for starters, both a rectangular and polar form, just as in 2D. With the rectangular array, you just add a 3D component (a level) to the familiar column and row. For the polar array, the one big difference is two points that define the axis of rotation instead of just one point for the center. Another difference is the curious fact that 3Darray is all command line driven; there is no dialog box.

Let us go ahead and try out the command, starting with the *rectangular array*. Clear your screen and draw a 2" × 3" rectangle extruded to 4", as seen in Figure 23.13. Add shading and a color of your choice.

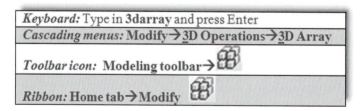

Step 1. Start up 3Darray via any of the preceding methods.
- ○ AutoCAD says: `Select objects:`

Step 2. Select the cube and press Enter.
- ○ AutoCAD says: `Enter the type of array [Rectangular/Polar]<R>:`

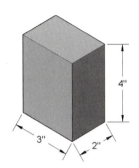

FIGURE 23.13
Extruded block for 3D
rectangular array.

Step 3. Press Enter to accept the default `Rectangular` array.
- ○ AutoCAD says: `Enter the number of rows (---) <1>:`

Step 4. Enter a small value, such as 5, and press Enter.
- ○ AutoCAD says: `Enter the number of columns (|||) <1>:`

Step 5. Enter a small value, such as 5, and press Enter.
- ○ AutoCAD says: `Enter the number of levels (...) <1>:`

Step 6. Enter a small value, such as 5, and press Enter.
- ○ AutoCAD says: `Specify the distance between rows (---):`

Step 7. Here, be sure to enter a number large enough to allow for space between the rows, 10 in this case.
- ○ AutoCAD says: `Specify the distance between columns (|||):`

Step 8. Be sure to enter a number large enough to allow for space between the columns, 10 in this case.
- ○ AutoCAD says: `Specify the distance between levels (...):`

Step 9. Be sure to enter a number large enough to allow for space between the levels, 10 in this case.

The result of the array with the 125 elements is shown in Figure 23.14. This exact type of array may not be used that often but is still valuable to know. Next, we go over the vastly more useful polar version.

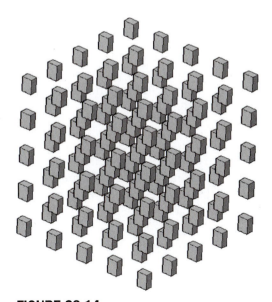

FIGURE 23.14
3D rectangular array.

To demonstrate the *polar array*, we draw an axle with spokes. To create an axle, draw a circle 5" in diameter and extrude it to 36". Then draw a small spoke, say a 1" circle extruded to 10", and position it carefully in a perpendicular manner on the tip of the axle. You likely have to rotate3D the spoke into position then attach it using quadrant or center points. After adding shading and color, the result is as shown in Figure 23.15.

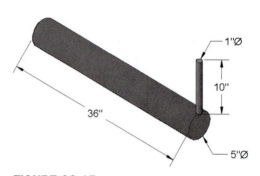

FIGURE 23.15
Axle and spoke for 3D polar array.

Step 1. Start up 3Darray via any of the same methods shown earlier in the matrix for the rectangular array.
  ○ AutoCAD says: `Select objects:`
Step 2. Select the spoke and press Enter.
  ○ AutoCAD says: `Enter the type of array [Rectangular/Polar]<R>:`
Step 3. Press p to accept the `polar` array.
  ○ AutoCAD says: `Enter the number of items in the array:`
Step 4. You can enter any value, but let us stick to 10 in this case.
  ○ AutoCAD says: `Specify the angle to fill (+=ccw, −=cw) <360>:`
Step 5. Press Enter to accept the default 360°, as we would like to go all the way around.
  ○ AutoCAD says: `Rotate arrayed objects? [Yes/No] <Y>:`
Step 6. Press Enter for `Yes`, as we want the spokes to rotate.
  ○ AutoCAD says: `Specify center point of array:`
Step 7. Here, you need to specify the center of rotation. Pick the center of the axle using the Center OSNAP.
  ○ AutoCAD says: `Specify second point on axis of rotation:`
Step 8. Here, you need to pick the second center point using OSNAPs on the other side of the axle. These two points together indicate the axis of rotation for the spokes. The result is shown in Figure 23.16.

3Dscale

This 3D version of the scale command is not much different from the regular scale (which you can also use in 3D). There is no icon or cascading menu for this command; you have to type it or use the Ribbon. Draw a basic 3D shaded cube as you did just before for 3Darray. It is reproduced again with a different color (Figure 23.17).

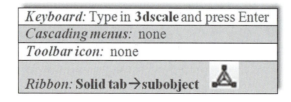

Keyboard: Type in **3dscale** and press Enter	
Cascading menus: none	
Toolbar icon: none	
Ribbon: **Solid tab → subobject**	

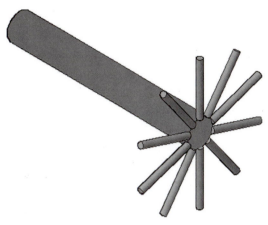

FIGURE 23.16
Completed 3D polar array.

FIGURE 23.17
Extruded block for
3D scale.

519

Step 1. Start up the 3Dscale command via one of the two preceding methods.
  ○ AutoCAD says: `Select objects:`
Step 2. Select the cube and press Enter.
  ○ AutoCAD says: `Specify base point:`
Step 3. You also see a scaling gizmo appear, which snaps to whatever base point you select
    (Figure 23.18).
  ○ AutoCAD then says: `Pick a scale axis or plane:`
Step 4. Pick the `XY` plane. You can then move the mouse to scale the object or type in a
    value, as seen in Figure 23.19.

3Dmove

Just as 3Dscale was somewhat of a duplication of the regular scale command, so is 3Dmove.
However, there is a difference, as this command allows for *constrained* movement of an object
along any selected axis or plane, regardless of your UCS orientation or whether or not Ortho is
on. In that regard, it is quite useful, because just randomly moving an object in 3D space (with
no constraint, such as Ortho) places the object literally anywhere, and usually nowhere near
where you think it is in relation to other objects. The 3D space is quite deceptive. For proof
of that, try moving something then using 3D Orbit to glance at the objects from a different
perspective. You may be surprised where it really is. Let us try the 3Dmove command on the
same random cube you just used.

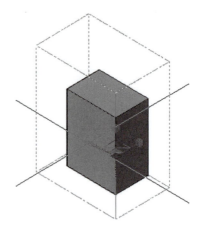

FIGURE 23.18
3Dscale gizmo.

FIGURE 23.19
3Dscale in progress.

Keyboard: Type in **3dmove** and press Enter
Cascading menus: Modify→3D Operations→3D Move
Toolbar icon: Modeling toolbar→
Ribbon: Solid tab→subobject

Step 1. Start up the 3Dmove command via any of the preceding methods.
   ○ AutoCAD says: `Select objects:`
Step 2. Select the cube and press Enter.
   ○ AutoCAD says: `Specify base point or [Displacement] <Displacement>:`
Step 3. A 3Dmove gizmo appears, but do *not* pick a base point. Instead, use the mouse to carefully select an axis. As soon as you do, the axis turns gold in color. The cube is constrained (even if Ortho is not on), and you can now slide the cube up and down that axis, as seen in Figure 23.20.

The procedure is the same for the Y and Z axes. If you select a plane, however, you *do* need to have Ortho on to be constrained to the chosen plane.

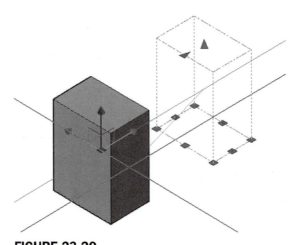

FIGURE 23.20
3Dmove in progress (X axis).

Fillets and Chamfers in 3D

Recall the operation of the fillet command (with a radius of zero) in 2D AutoCAD. You could use the command to trim two intersecting lines or close a gap between those lines. None of that makes any sense when working with 3D models, which are, by definition, already closed shapes. However, recall also that in 2D we gave fillets radiuses. We use this feature in 3D by adding radiuses or rounded corners on solid objects. In a way we now use the original and more correct form of the fillet command, which by definition has always been a rounded corner.

There are two ways to work with the fillet command in 3D. One way is by using the 2D approach, that is, typing, toolbar, cascading menu, or Ribbon. The other is via the dedicated Fillet tool available as part of the Solid Editing toolbar. The result is the same, but the approach is different. Let us try the 2D version first.

Draw a 12" × 8" box and extrude it to 10", as seen in Figure 23.21.

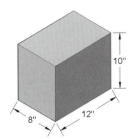

FIGURE 23.21
Extruded block for fillet.

Step 1. As first described in Chapter 1, start up fillet via any method you choose.
  ○ AutoCAD says: `Current settings: Mode = TRIM, Radius = 0.0000`
    `Select first object or [Undo/Polyline/Radius/Trim/Multiple]:`
Step 2. Press r for `Radius`.
  ○ AutoCAD says: `Specify fillet radius <0.0000>:`
Step 3. Enter a reasonable value such as 1" and press Enter.
  ○ AutoCAD says: `Select first object or [Undo/Polyline/Radius/Trim/`
    `Multiple]:`
Step 4. Select one of the cube edges facing you.
  ○ AutoCAD says: `Enter fillet radius <1.0000>:`
Step 5. Confirm this by pressing Enter and continue selecting the remaining two edges. Each time,
  ○ AutoCAD says: `Select an edge or [Chain/Radius]:`
Step 6. After all three edges are selected, you see them dashed. Finally press Enter.
  ○ AutoCAD says: `3 edge(s) selected for fillet.`

The result is shown in Figure 23.22.

FIGURE 23.22
3D fillet.

The color of the fillets may be different from the rest of the object, as seen in Figure 23.22. This depends on what layer is active as you perform the fillet. In this case it was 0, and the color defaults to the standard gray (shown in dark gray, here), although the block itself is colored (shown here in medium- and light gray).

The 3D version of fillet is a bit more dynamic in nature. Create another 12" × 10" × 8" box and use the commands seen in the following matrix. There is no typing equivalent.

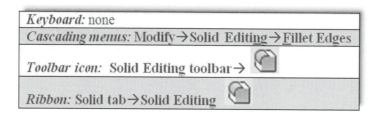

○ AutoCAD says: `Radius = 1.0000`
   `Select an edge or [Chain/Radius]:`

Select the edge to be filleted and it becomes shaded. You may change the radius or press Enter to accept. You then see a red triangle (a grip). Click on it and drag your mouse back and forth, noting how the fillet changes dynamically, as seen in Figure 23.23.

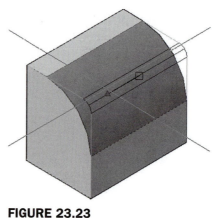

FIGURE 23.23
Dynamic filleting in 3D.

Click again and press Enter to confirm the final fillet. While this looks impressive, in most cases, you will likely want to specify an exact fillet size, so the dynamic filleting may not find much use in real design work.

Chamfering is quite similar, so we do not run through the specific steps, but you can also do it through the standard 2D approach or via the 3D dynamic chamfer, as shown with the command matrix.

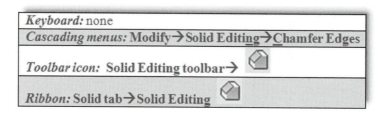

Once again the edge to be chamfered is shaded and you must enter sizes when prompted or press Enter to accept. You then see two triangular grips this time (Figure 23.24). Each can be moved to dynamically expand the sides of the chamfer. As with the fillet, you may want to just type in the value.

Review everything covered in this chapter. In the next chapter, we cover Boolean operations, the final key to flat design.

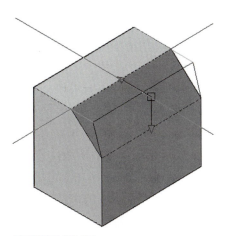

FIGURE 23.24
Dynamic chamfering in 3D.

523

SUMMARY

You should understand and know how to use the following concepts and commands before moving on to Chapter 24:

- Rotate3D
- 3Drotate (gizmo)
- Mirror3D
- 3Darray
    - Polar
    - Rectangular
- 3Dscale
- 3Dmove
- Fillets in 3D
- Chamfers in 3D

REVIEW QUESTIONS

Answer the following based on what you learned in Chapter 23:

1. Describe the rotate3D procedure. Do you need axes or planes for this command?
2. How is 3Drotate different from rotate3D?
3. Describe the mirror3D procedure. Do you need axes or planes for this command?
4. Describe the 3Darray procedure, both rectangular and polar.
5. Describe 3Dscale and 3Dmove.
6. Describe fillets and chamfers in 3D.

EXERCISES

1. Create the following table using the rectangle, circle, extrude, fillet, mirror3D, and a few other commands. The exact size of the table can be anything you want, but if you need some values to get started, the tabletop is 50" × 30" and the bottom is 20" × 20". All fillets and thicknesses are 2". (Difficulty level: Easy; Time to completion: 10 minutes)

2. Create the following buggy frame. You need the circle, extrude, 3Darray, donut, copy, and some of the UCS rotation commands. Once again, sizing is not what is important here, but you can make the two main axles 72" in length and base everything else off of that. (Difficulty level: Moderate; Time to completion: 15 minutes)

3. Create the following helicopter model using the tools you learned in this chapter. While this design will not win any awards, it also should take you no more than 30 minutes to do and really gets you using recently learned commands such as extrude, fillet, rotate3D, and 3Darray. Start out by creating a basic blade profile (an airfoil) using pline. Then, extrude it and 3Darray the resulting blade around a rotor. Create the body using extrude and fillet. Attach a tail using basic extrude and scale down a copy of the main rotor for the tail rotor, placing everything into position. Finally, shade and color the model. (Difficulty level: Moderate; Time to completion: 30 minutes)

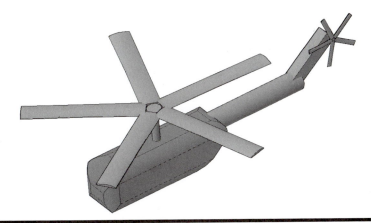

Spotlight On: Aerospace Engineering

Aerospace engineering is a highly specialized branch of mechanical engineering that evolved relatively recently in our history, as humankind took to flight, first in the atmosphere, then into space. Formally defined, it is the science behind the design and construction of aircraft (aeronautical engineering) and spacecraft (astronautical engineering). These engineers conceptualize, build, and test everything from gliders to missiles, jet fighters, and space rockets. Many also work in naval design, as many of the principles guiding the flow of air around an aircraft (a fluid) readily transfer to the flow of water (also a fluid) around a ship or submarine.

FIGURE 1

FIGURE 2

Aerospace engineers need a mastery of a wide variety of engineering sciences, such as fluid dynamics, structures, control systems, aeroelasticity, thermodynamics, materials, and electrical engineering. It is perhaps one of the more diverse engineering disciplines and specialization is necessary. Three broad areas of specialization include aeronautics, propulsion, and space vehicles.

Education for aerospace engineers starts out similar to that for the other engineering disciplines. In the Unites States, all engineers typically have to attend a four-year ABET-accredited school for their entry level degree, a Bachelor of Science. While there, all students go through a somewhat similar program in their first two years, regardless of future specialization. Classes taken include extensive math, physics, and some chemistry courses, followed by statics, dynamics, mechanics, thermodynamics, fluid dynamics, and material science. In their final two years, engineers specialize by taking courses relevant to their chosen field. For aerospace engineering, this includes aerodynamics, structures, controls, propulsion, and orbital mechanics, among others.

Upon graduation, aerospace engineers can immediately enter the workforce or go on to graduate school. Although not required and not common in the aerospace industry, some engineers choose to pursue a Professional Engineer (P.E.) license. This would typically be in related mechanical engineering, as there is no aerospace P.E. exam. The process first involves passing a Fundamentals of Engineering (F.E.) exam, followed by several years of work experience, and finally the P.E. exam.

Aerospace engineers can generally expect somewhat higher starting salaries than civil, mechanical, and industrial engineering graduates but on par with electrical and lower than petroleum engineering and some computer science degrees. A master's degree is usually the path to higher starting pay, which with a bachelor's, currently averages around $55,000/year nationwide. This of course depends highly on locality, market demand, and even GPA on graduation.

How do aerospace engineers use AutoCAD and what can you expect? Industry wide, AutoCAD use is not widespread due to the nature of the profession, which relies on more sophisticated software to model parts directly in 3D. The industry standard is CATIA, though NX and Pro/Engineer are also used. Aerospace engineering also relies heavily on software for initial testing or design validation. Examples include finite element analysis (FEA) and software such as NASTRAN or ANSYS to model loads, deflections, and even heat propagation. Computational fluid dynamics (CFD) and software such as Fluent are used to model the flow of air around the design at various speeds and temperatures (also applicable to the flow of water). See Appendix B if you would like to learn more about these techniques.

AutoCAD, however, can still be found for small part design, aerospace electrical schematics, and overall system layouts. Smaller companies that have little to no modeling, testing, and manufacturing needs also may use AutoCAD due to the associated cost savings. Many a home kit airplane designer or builder turned to AutoCAD to produce drawings or just to document ideas. The general trend is to move away from AutoCAD as design requirements become more sophisticated and testing and manufacturing are involved.

AutoCAD layering is far simpler in aerospace engineering than in architecture. There is no AIA standard to follow, only an internal company standard, so there is room to improvise. As always the layer names should be clear as to what they contain. Figure 3 shows an aerospace application of AutoCAD for a set of wings for a small unmanned aerial vehicle (UAV).

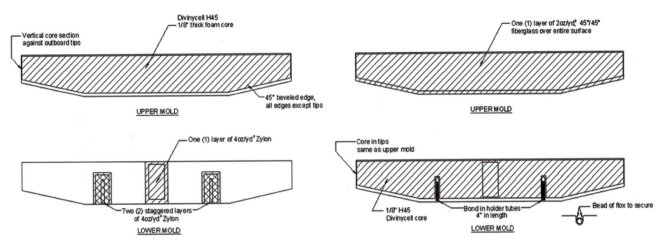

FIGURE 3
Wings for a small unmanned aerial vehicle drawn with AutoCAD.
(Image courtesy of DynaWerks LLC., www.DynaWerks.com)

Figure 4 shows a typical design of a small general aviation aircraft wing, with overall dimensions and specs.

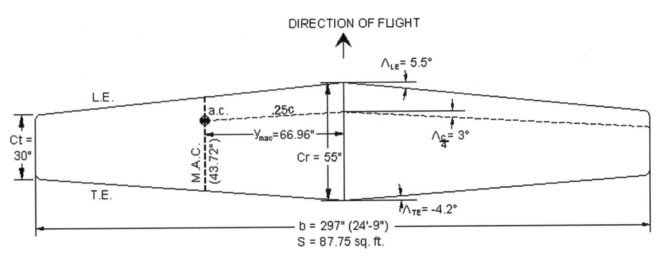

FIGURE 4
Wing specifications for a small, general aviation aircraft drawn with AutoCAD.

Finally, Figure 5 shows a side profile of a small private plane (a Lancair Legacy®) drawn in AutoCAD. This side profile study is used for sizing; the vertical and horizontal numbers are inch position from the tip of the propeller cone and the ground.

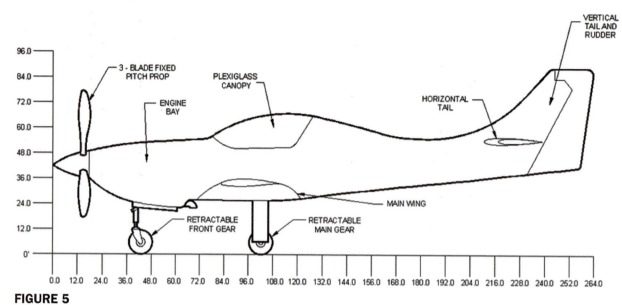

FIGURE 5
A small, general aviation aircraft (Lancair Legacy) drawn with AutoCAD.

Boolean Operations

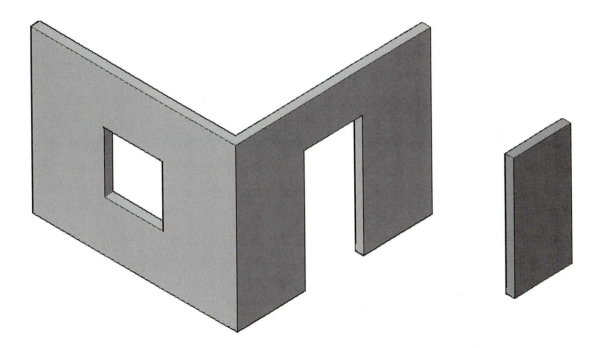

By the end of the chapter, you will be able to create designs that feature openings and holes as well as complex unions and intersections.

Estimated time for completion of chapter: 1–2 hours.

24.1 INTRODUCTION TO BOOLEAN OPERATIONS

We have thus far been able to put together a few fairly simple objects using extrude and the 3D manipulation commands (mirror3D, rotate3D, etc.). What we still miss is the ability for objects to interact with each other, such as being fused together (e.g., two adjoining walls) or subtracted from each other (e.g., a hole in a wall representing a window).

This functionality, referred to as *Boolean operations*, is very important for advanced 3D modeling in both architectural and mechanical design. We cover the three Boolean operations of union, subtract, and intersect. The functionality of each can be represented by two overlapping circles (representing two solids), as shown in Figure 24.1.

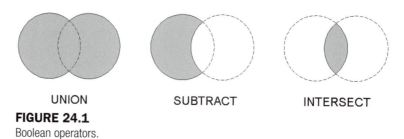

UNION   SUBTRACT   INTERSECT

FIGURE 24.1
Boolean operators.

These circles also form the AutoCAD icons for these operations. As you can see from the shaded areas, the solids can be

- Joined together completely.
- Subtracted, leaving only one part of the solid (either one can be chosen).
- Intersected where only the intersecting geometry remains.

All three Boolean operators can be accessed via typing, cascading menus, a toolbar (Modeling toolbar, which you should already have up), and of course the Ribbon.

Union

The union operation fuses solids or regions together into one object. This is an irreversible operation, so be careful before you implement it. Objects do not actually have to be touching to be joined together. Union is not as commonly used as subtract but does find some use when joining together walls or mechanical assemblies.

Let us give union a try by first creating two solid cubes of any size and positioning them to intersect each other, as seen in Figure 24.2. Notice that their grips have been activated, and you can see that the two cubes are distinct and separate entities, even though they are touching.

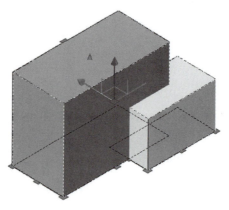

FIGURE 24.2
Two solid cubes before union.

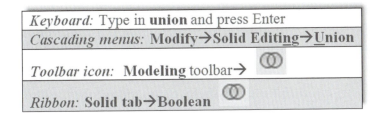

Step 1. Start the union command via any of the preceding methods.
  ○ AutoCAD says: `Select objects:`
Step 2. Select both cubes (they turn dashed) and press Enter. Both cubes now are fused, as seen in Figure 24.3. Notice how the grips reflect one entity, not two. Go ahead and move the objects; they move as one. This becomes even clearer if we switch to wireframe (Figure 24.4), with a color change for additional clarity.

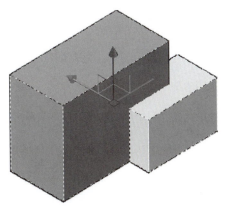

FIGURE 24.3
One solid cube after union (shaded view).

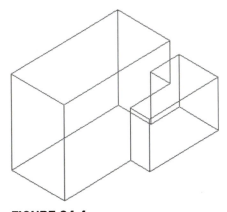

FIGURE 24.4
One solid cube after union (wireframe).

Subtract

This is a frequently used command, and it is quite indispensable in 3D work. As you may have guessed from the name, it involves the process of subtracting one object from another. In AutoCAD, this is the only way to take away material to create an empty space. Examples of this include the creation of doors and windows in 3D architectural floor plans and bolt holes in mechanical design, just to name a few. Let us set up just such a situation with a flat plate and a circle, which are then extruded and subtracted to form a hole.

Staying in wireframe mode (for clarity), draw a 10" × 10" square, extruding it to 2", as shown in Figure 24.5.

Now, draw a guideline from corner to corner (use endpoints) of the top surface of the extruded rectangle and a circle of 2" radius based on the midpoint of the guideline, as shown in Figure 24.6. Note that switching to the wireframe view and adding temporary guidelines is a common technique in 3D.

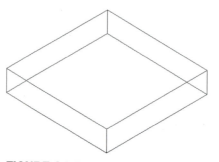

FIGURE 24.5
10" × 10" × 2" block.

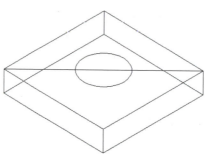

FIGURE 24.6
Circle and guideline positioned.

You can erase the guideline and extrude the circle a total of −2" (yes, that is a negative sign), as seen in Figure 24.7. Although it seems like we are done, we are not; the "hole" is not yet a hole.

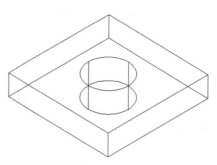

FIGURE 24.7
Circle extruded.

The final step is to subtract the extruded circle from the extruded rectangle, leaving a hole in its place.

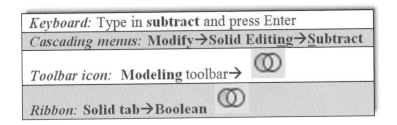

Keyboard: Type in **subtract** and press Enter
Cascading menus: **Modify→Solid Editing→Subtract**
Toolbar icon: **Modeling** toolbar→
Ribbon: **Solid tab→Boolean**

Step 1. Start up the subtract command via any of the preceding methods.
- ○ AutoCAD says:
  ```
  Select solids, surfaces, and regions to subtract from...
  Select objects
  ```
Step 2. Here, you have to be careful. You must first select the material that you want to keep. In this case, it is the extruded square. Go ahead and select it and press Enter.
- ○ AutoCAD says: `Select solids and regions to subtract...`
  ```
  Select objects:
  ```
Step 3. Select the material you want to get rid of to make the hole. That of course is the extruded circle.

The subtraction operation is performed. Although you may not be able to immediately see what you did, a simple switch to the shaded mode reveals the hole, as seen in Figure 24.8.

FIGURE 24.8
Completed subtraction operation.

Be sure you thoroughly understand the essence of the subtract command; as mentioned earlier, we use it often. The key is to visualize what material you want removed and where and create a shape to exactly reflect that. Remember again that the first selection is always for the *material you want to keep*; the second is for the *material you want to remove*.

Intersect

The final Boolean command creates a new shape out of the overlapping parts of other shapes and can be used to create some interesting objects. Let us create a simple intersection. Use the previous rectangle with the hole and copy it to the side (with Ortho on) to create two shapes that intersect each other. Figure 24.9 shows you what you get.

Notice the overlap between the shapes. This is what is kept after the intersection command is executed.

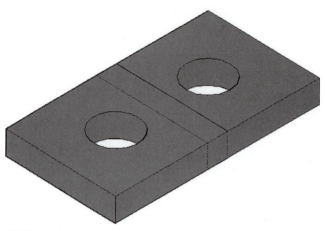

FIGURE 24.9
Intersected shapes.

Keyboard: Type in **intersect** and press Enter	
Cascading menus: **Modify→Solid Editing→Intersect**	
Toolbar icon: **Modeling** toolbar→ 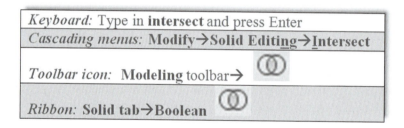	
Ribbon: **Solid tab→Boolean**	

Step 1. Start intersect via any of the preceding methods.
- AutoCAD says: `Select objects:`

Step 2. Select both pieces and press Enter. The result is shown in Figure 24.10; only the intersecting part remains.

FIGURE 24.10
Completed intersect
operation.

3D Modeling Exercise: Building Wall with Door and Window

This next practice drawing is a very important one. We are creating a basic 3D model of two building walls, complete with a door and a window opening. Although this model is quite simple, it contains many of the important elements needed to create an entire building in 3D, the differences being only the amount of pieces present and some minor refinements. All the essentials are here, however, so follow all the steps closely.

Step 1. Draw two walls that are 8' high and 10' long, with a thickness of 5". One way to do this is to first draw a rectangle that is 5" × 10' and extrude it to 8'. Then, copy the structure and reattach the copy at 90° to the original wall. Figure 24.11 shows what you should see in wireframe mode.

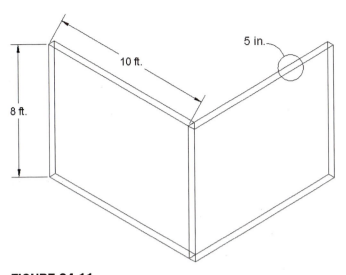

FIGURE 24.11
3D modeling exercise, Step 1.

Step 2. Use the union command to fuse the walls together. Figure 24.12 shows another, slightly rotated, view of the walls in wireframe after the application of union. Notice where they intersect; they are clearly in one piece, not two separate panels.

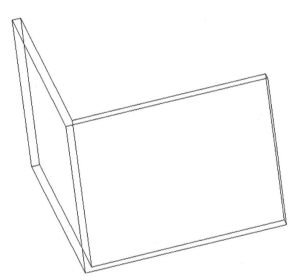

FIGURE 24.12
3D modeling exercise, Step 2.

Step 3. Now create a 36" × 36" window (off to the side) as well as a 36" wide × 72" high door (also off to the side). Both should be 5" thick, as that is the overall thickness of the walls. Rotate3D the objects into position if needed. The result is shown in Figure 24.13.

Step 4. Let us think carefully about what to do next. The window is actually a hole in the wall, so we need not reuse it. All we need to do is position it in the right place and subtract. The door is a slightly different story. We need one copy of it to make the opening in the first place, and one copy to then install into the new opening (and rotate into an open position). So it is a good idea to make a copy of the door right now, because the process of subtracting deletes the original. Do not forget this, or you will end up redoing the door after using the original to create the opening.

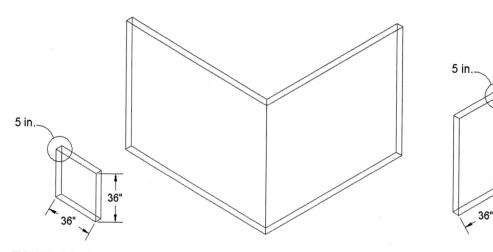

FIGURE 24.13
3D modeling exercise, Step 3.

Step 5. After copying the door, insert it carefully into the wall, making sure it is lined up correctly (midpoint of bottom front of door to midpoint of bottom front of wall). Then, insert the window in a similar manner and raise it up (using Ortho) to roughly the halfway point in the wall. You may have your own method of doing this as well, but either way, Figure 24.14 shows what you should have as a result.

Step 6. Go ahead and subtract out the window and the door. Figure 24.15 shows what you see in wireframe mode. Figure 24.16 shows the same thing in shaded mode with color added.

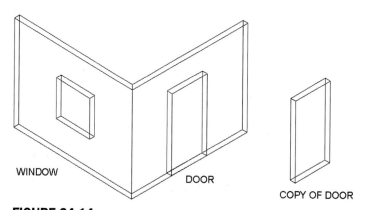

FIGURE 24.14
3D modeling exercise, Steps 4 and 5.

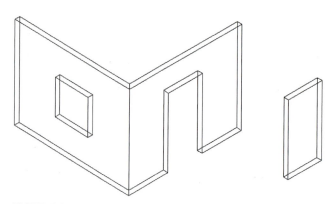

FIGURE 24.15
3D modeling exercise, Step 6 (wireframe).

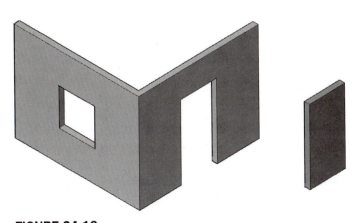

FIGURE 24.16
3D modeling exercise, Step 6 (solid).

Step 7. Finally, let us put the copy of the door back where it belongs and give it a −30°
rotation (yes, negative). The final result is in Figure 24.17.

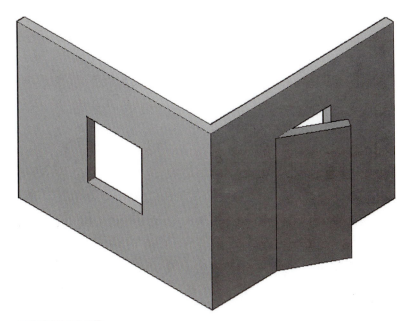

FIGURE 24.17
3D modeling exercise, Step 7.

This chapter concludes 3D "flat design." These basic 3D skills go a long way in creating a
realistic model of a building or even a simple mechanical device. In the coming chapters,
we add additional tools and refinement, starting with curved design. For now be sure to
completely understand everything and practice everything learned thus far.

SUMMARY

You should understand and know how to use the following concepts and commands before
moving on to Chapter 25:

- Union
- Subtract
- Intersect

REVIEW QUESTIONS

Answer the following based on what you learned in Chapter 24:

1. What are the three Boolean operations?
2. In what toolbar are they found?
3. Is union a permanent setting?
4. Why is it a good idea to make a copy of something that is being subtracted?

EXERCISES

1. Using your new Boolean skills, create this washer and bolt assembly. Make the washer 3" square and the hole (as well as the bolt) .75" in diameter, and size the hex bolt top off of that. You need the circle, extrude, and polygon commands at the very least. (Difficulty level: Easy; Time to completion: 10 minutes)

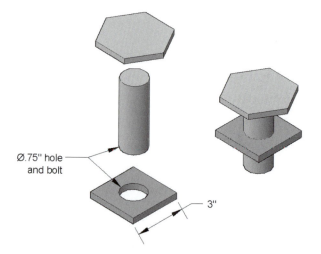

Ø.75" hole and bolt

3"

2. Create the following design of a bed. It consists of a thick panel, a mattress, two identical headboards (with cutouts), and two pillows angled at 30°. Sizing is entirely up to you; draw a reasonable-looking base panel and size the rest off of that. You need the rectangle, line, arc, pedit, extrude, fillet, mirror3D, and subtract commands at the very least, as well as some UCS rotations. (Difficulty level: Moderate; Time to completion: 30 minutes)

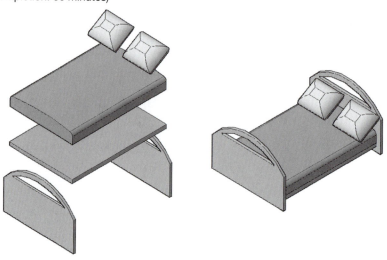

3. Create the following design of a drilled, vented brake rotor. Create a circle of the given size, extrude to 1", add the center axle cutout, and arrange the drilled holes as shown on the left, arraying them eight times. Extrude and subtract as needed. Shade and color to finish off the model as seen on the right. (Difficulty level: Moderate; Time to completion: 15 minutes)

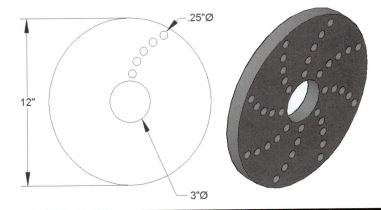

CHAPTER 25

Solid Modeling

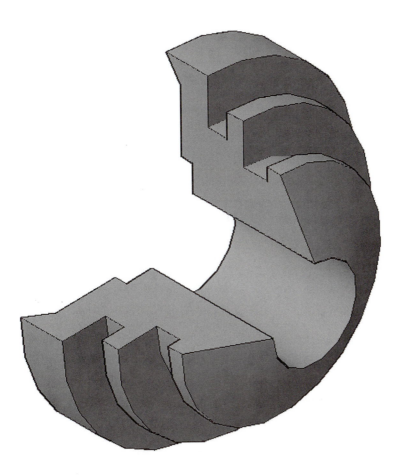

LEARNING OBJECTIVES

In this chapter, we introduce a completely new set of tools in solid modeling, which are used for "curved design." These dramatically expand the range of design work that you can do. We specifically look at

- Revolve
- Shell
- Taper
- Loft
- Path extrusion
- Sweep

By the end of the chapter, you will be able to create designs that were not possible with previous tools, with much of this new knowledge applicable to engineering design.

Estimated time for completion of chapter: 2–3 hours.

25.1 INTRODUCTION TO SOLID MODELING

With this chapter, we head into a new direction and begin talking about some of the advanced solid modeling tools that make up what can be referred to as *curved design*. While these tools are not necessarily more difficult to master, they do open up a new range of modeling possibilities, including some rather complex 3D designs. We introduce *revolve*, *shell*, *taper*, *loft*, *path extrusion*, and *sweep*. Although some of these commands, such as taper and shell, do not involve curves, they are still included with this advanced tools group.

Revolve

As essential as extrude was to flat design, such is the importance of revolve to curved design, and it is the first command we cover. Revolve easily creates complex revolved shapes based on the concept of *solid or surface of revolution*, and this command has no other equivalent in AutoCAD.

The basic idea is the following. We would like to draw a 2D outline of an object and revolve this outline or "profile" around an axis. If the profile is closed, the result is a *solid of revolution*; if the profile is open (has gaps) or lines are used instead of polylines, the result is a *surface of revolution*. This process can create some unique shapes that are impossible to make any other way, but the entire concept rests on the designer's ability to visualize the profile on which the model is built.

In past releases of AutoCAD, the revolve command was a bit picky, and you had to use a closed polyline for the revolution to work, no exceptions. These restrictions have been lifted and you can create a profile using the regular line command and even leave gaps in it; although, once again, just be aware that this creates surfaces, not solids, as is shown later. Your axis of revolution can be anywhere on or near the object, and you need not revolve all the way around but can use any degree value in between if necessary.

Let us put all this together with some examples. In 2D, create a profile approximately similar to what is shown in Figure 25.1, using a *polyline* to ensure a solid. We try it with regular lines later on.

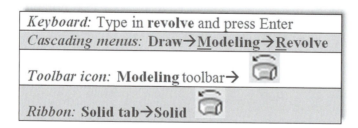

Keyboard: Type in **revolve** and press Enter
Cascading menus: Draw→Modeling→Revolve
Toolbar icon: **Modeling** toolbar→
Ribbon: **Solid tab→Solid**

Step 1. Start revolve via any of the preceding methods.
   ○ AutoCAD says: `Current wire frame density: ISOLINES=4`
      `Select objects to revolve:`
Step 2. Select the profile and press Enter.
   ○ AutoCAD says: `Specify axis start point or define axis by`
      `[Object/X/Y/Z]`
Step 3. Here, AutoCAD is asking for an axis of revolution for the profile. There are many ways this can be addressed, and the axis can be on the object or some distance away

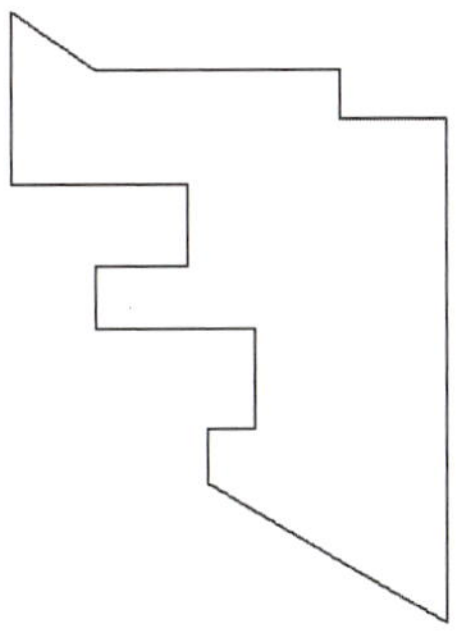

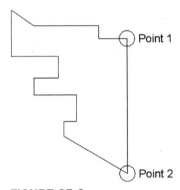

FIGURE 25.1
Polyline profile for
revolve.

from it (causing the revolved solid to have a hole about the axis). For this example,
we pick two points at the top right and bottom right of the profile, as seen in Figure
25.2. Using the ENDpoint OSNAP, click on these two locations.

Point 1

Point 2

FIGURE 25.2
Selecting rotation axis.

Step 4. After the first click,
  ○ AutoCAD says: Specify axis endpoint:
  After the second click,
  ○ AutoCAD says: Specify angle of revolution or [STart angle]<360>:
Step 5. Here, you can ask for a full 360° revolution by just pressing Enter, but instead type
  in 270, so we can see a cutout of the part. If you are in 3D, you can also just trace out
  the revolution angle with movements of your mouse. For now though, you see a 2D
  revolved surface, as in Figure 25.3.

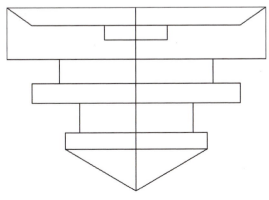

FIGURE 25.3
Revolved profile in 2D wireframe.

To see the full effect of what you have done, switch to the familiar SW Isometric view and also shade and color the solid. The far more dramatic result is seen in Figure 25.4.

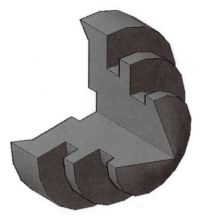

FIGURE 25.4
Completed revolution, shaded and colored.

For some other ideas, undo your last step, and try locating the axis of revolution some distance away from the object (Figure 25.5). This creates a hole in the center of the object on execution of revolve, as seen in Figure 25.6.

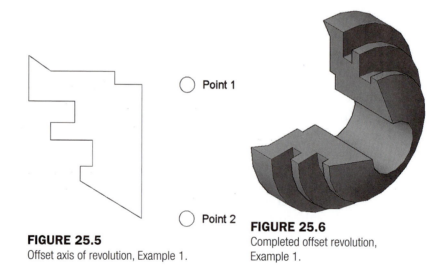

◯ Point 1

◯ Point 2

FIGURE 25.5
Offset axis of revolution, Example 1.

FIGURE 25.6
Completed offset revolution, Example 1.

The axis of revolution can actually be anywhere. In Figure 25.7, it is above the profile and at an angle, creating a completely different solid of revolution, as seen in Figure 25.8.

Next let us try the same profile but with a piece missing (Figure 25.9). The profile no longer is a closed polyline, and in this case, the revolve command creates a surface, not a solid, a difference you can easily see. The same effect is achieved using plain old lines. Even if they are connected in a loop, AutoCAD still defaults to the surface revolution. Figure 25.10 shows the result.

Try the following example of revolution, a 3D cup. Draw the profile in Figure 25.11, pedit, and revolve 360°. Then, switch to 3D, and shade, color, and rotate if needed. You now have your first piece of glassware for a future 3D house, as seen in Figure 25.12.

Point 2

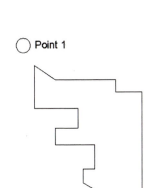

Point 1

FIGURE 25.7
Offset axis of revolution, Example 2.

FIGURE 25.8
Completed offset revolution,
Example 2.

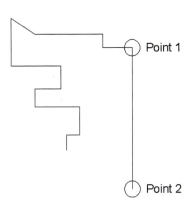

Point 1

Point 2

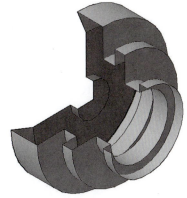

FIGURE 25.9
Surface of revolution profile.

FIGURE 25.10
Completed surface revolution.

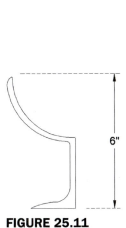

6"

FIGURE 25.11
Cup profile.

FIGURE 25.12
Completed cup.

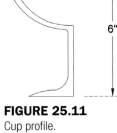

As you can see, revolve is a very versatile, powerful command and is really limited only by your imagination and visualization skills. Be sure to go through each example presented thus far. None of the profiles or solids has to look exactly like the ones shown, and indeed no dimensions are provided (except for the cup height), but be sure to understand the underlying principles and what is being done to produce what you see.

Shell

Shell is one of several AutoCAD 3D commands "borrowed" from engineering solid modeling software due to their usefulness. The command creates a "shell" of a predetermined thickness out of a solid object. The shell is of an even thickness all around and can have as many open surfaces as needed.

Let us demonstrate by example. Draw a 10" × 8" rectangle and extrude it out to 4", as seen in Figure 25.13.

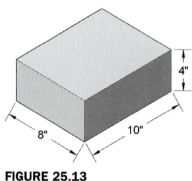

FIGURE 25.13
Basic block for shell.

To activate the command, you cannot type in shell (that is an OS command), but you can use other methods, as seen next in the command matrix.

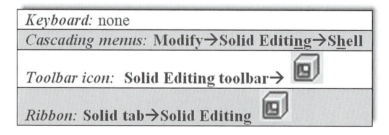

Keyboard: none
Cascading menus: Modify→Solid Editing→Shell
Toolbar icon: Solid Editing toolbar→
Ribbon: Solid tab→Solid Editing

Step 1. Start the shell command via any of the preceding methods.
- ○ AutoCAD gives you a lengthy response:
  ```
  Solids editing automatic checking: SOLIDCHECK=1
  Enter a solids editing option [Face/Edge/Body/Undo/eXit] <eXit>:
  _body
  Enter a body editing option
  [Imprint/seParate solids/Shell/cLean/Check/Undo/eXit] <eXit>:
  _shell
  Select a 3D solid:
  ```
Step 2. Select the extruded block.
- ○ AutoCAD says: Remove faces or [Undo/Add/ALL]:
Step 3. Carefully pick the upper surface of the extruded block. Be careful not to select an edge in this case but an actual flat surface. You may select more than one surface, and

it is something you should try the next time you practice the command. For now, just press Enter after selecting the top surface.

○ AutoCAD says: `Enter the shell offset distance:`

Step 4. Here, you need to enter a valid shell size (thickness). Some complications can result at this point if your shape has other modifications to it, such as fillets. Then, if the thickness of the walls is too thin, you can "eat through" the material, causing errors or actual holes in the shape. This particular example should proceed smoothly. Enter .5" for the thickness.

○ AutoCAD says: `Solid validation started.`
`Solid validation completed.`
`Enter a body editing option`

The shell command executes, and the result is as seen in Figure 25.14.

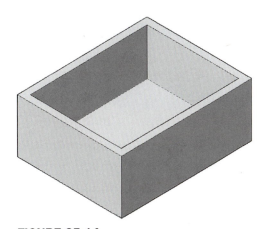

FIGURE 25.14
Completed shell.

As suggested earlier, try removing more than one face. The result is an "open" box, as seen in Figure 25.15. You can remove all four sides if you wish and end up with a flat plate, although that somewhat defeats the whole point of shell.

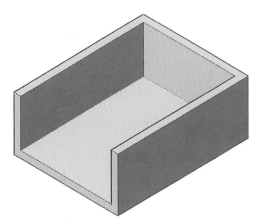

FIGURE 25.15
Completed shell (two faces removed).

Taper

Taper is a useful command that does exactly what it advertises: It tapers (or creates a gradient) on one or more sides of an object. A simple example, and one we actually model soon, is a plastic wastebasket commonly found in any office. If you have one nearby, take a look at it. Its base is a slightly smaller rectangle than the top rim. The reason for tapering such items is because of the nature of plastic molds and the manufacturing process. Another useful result is that now these wastebaskets are easily stackable one on top of another.

Let us practice this command on a tall rectangular box. Draw a 20" × 15" rectangle and extrude it to 30", as seen in Figure 25.16.

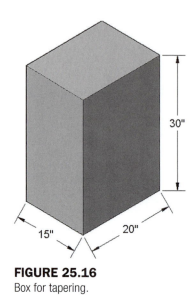

FIGURE 25.16
Box for tapering.

The taper command is another one that cannot be typed in, and the other input methods must be used.

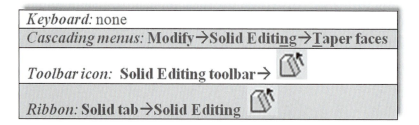

Keyboard: none

Cascading menus: **Modify→Solid Editing→Taper faces**

Toolbar icon: **Solid Editing toolbar→**

Ribbon: **Solid tab→Solid Editing**

Step 1. Start the taper command via any of the preceding methods.
  ○ AutoCAD again gives you a lengthy response:
```
_solidedit
Solids editing automatic checking: SOLIDCHECK=1
Enter a solids editing option [Face/Edge/Body/Undo/eXit] <eXit>:
_face
Enter a face editing option
[Extrude/Move/Rotate/Offset/Taper/Delete/Copy/coLor/mAterial/
Undo/eXit] <eXit>:
_taper
```
Step 2. Here, the first task is to select all the faces you want to taper. This can be as few as one face or as many as all of them. Be careful though; by *face*, we mean the ones that will taper. Therefore, the top and bottom of the box are not valid choices, only the sides. Also, note that it is easier to select faces when you are in wireframe view. In shaded view, you need to rotate around the object to get all the faces; in wireframe, you can (carefully) pick "through" the shape. When you pick a face it becomes dashed.

○ AutoCAD says: `Select faces or [Undo/Remove/ALL]: 1 face found.`

Step 3. When done press Enter.

○ AutoCAD says: `Specify the base point:`

Step 4. Here, you need to consider what is meant by *base point*. It is the start of the taper. You would need to pick either the top or the bottom of the block. Since this is a wastebasket, which tapers in as you go down, we pick the midpoint of one of the lines representing the *top face or surface*. If you choose a similar point on the bottom face or surface, then the block tapers up, like a pyramid. Go ahead and select the base point as discussed.

○ AutoCAD says: `Specify another point along the axis of tapering:`

Step 5. This of course is the opposite of where the first point was, so pick the midpoint of the *bottom face or surface*. In summary, you are giving AutoCAD what amounts to a straight line that tells it in what direction the taper is going.

○ AutoCAD says: `Specify the taper angle:`

Step 6. This final request asks for the amount of taper in degrees. Enter a small value, perhaps 4 or 5, and press Enter. The block tapers evenly on all sides. The result in shade mode is shown in Figure 25.17.

FIGURE 25.17
Completed taper.

551

3D Modeling Exercise: Wastebasket

Let us put extrude, taper, fillet, and shell together into one exercise.

Step 1. Clear your screen of previous work, and make sure you are in the SW Isometric view. Starting from the beginning, draw a 22" × 18" rectangle and extrude it to 34". The result is a block of material similar to what we just created when we practiced the taper command. It is shown shaded and with a different color in Figure 25.18.

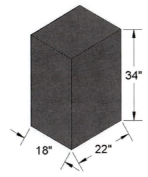

FIGURE 25.18
Wastebasket, Step 1.

Step 2. Taper the block on all four sides to 4°, as seen in Figure 25.19.

FIGURE 25.19
Wastebasket, Step 2.

Step 3. Fillet the side edges with a 3" radius, and the bottom with a 1.5" radius, as seen in Figure 25.20.

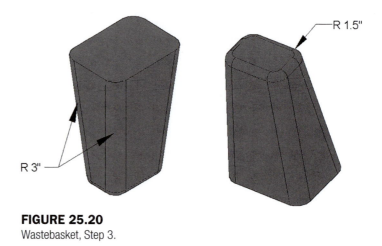

FIGURE 25.20
Wastebasket, Step 3.

Step 4. Shell the wastebasket to a 0.35" thickness, as seen in Figure 25.21.

FIGURE 25.21
Wastebasket, Step 4.

Loft

Loft is a recent addition to the AutoCAD toolbox, added along with many other 3D enhancements in AutoCAD 2007. It is another command "borrowed" from the engineering solid modeling world and is one of the more useful and versatile.

Despite some rather complex surface and part modeling programming "under the hood" with loft, as far as the AutoCAD end user is concerned, the command works on a simple principle. It connects together open or closed profiles. All you have to do is pick those profiles in the order you want them lofted and AutoCAD "connects the dots," so to speak, and creates either a surface (if the profile is open) or a solid (if the profile is closed). Take the very simple example of a classic wooden barrel, as shown in Figure 25.22.

FIGURE 25.22
Barrel.

While in this particular case you can use revolve to model this object, this is only because it is symmetric about the center axis. We often do not have such a convenience, so we need to default to loft instead. The barrel's lengthwise profile can be modeled as three circles, one after another, with the first and third representing the top and bottom, and the second (slightly larger one) representing the fatter middle part, as seen in Figure 25.23.

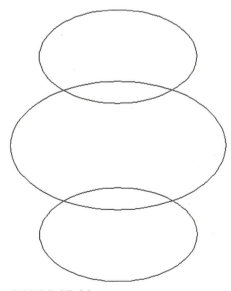

FIGURE 25.23
Three-ring profile of a barrel.

These rings are drawn to a random size, but both outer rings are an equal distance from the larger center one, ensuring our barrel is not lopsided. You can easily set up these 3D profiles by drawing two sizes of circles in 2D Top View, with each having the same center point. Then, switch to a side view and "lift" the smaller circle upward (or larger one downward). Finally a quick mirroring completes the profile. You should be able to do this type of 3D work with little effort at this point in the course.

Let us now run the loft command and get a barrel.

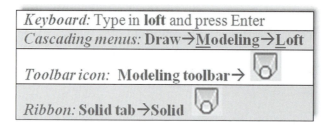

> *Keyboard:* Type in **loft** and press Enter
> *Cascading menus:* **Draw→Modeling→Loft**
> *Toolbar icon:* **Modeling toolbar→**
> *Ribbon:* **Solid tab→Solid**

Step 1. Start the command via any of the preceding methods.
  ○ AutoCAD says: `Current wire frame density: ISOLINES=4,`
    `Closed profiles creation mode=Solid`
    `Select cross sections in lofting order or [POint/Join`
    `multiple edges/Mode]`

Step 2. Select the hoops, one after another, from top to bottom (or vice versa). Each one becomes dashed as you select it. Selecting the hoops out of order does not necessarily stop the command but gives you unexpected results.
  ○ AutoCAD says: `1 found, 3 total`

Step 3. Press Enter when all three are selected.
  ○ AutoCAD says: `Enter an option [Guides/Path/Cross sections only/`
    `Settings]`
    `<Cross sections only>:`

Step 4. Press Enter. The loft is performed and the barrel is visible if you are in the shaded mode. You also see a prominent down arrow, which when pressed reveals a menu to fine-tune the loft (Figure 25.24). Experiment with the various choices to see the

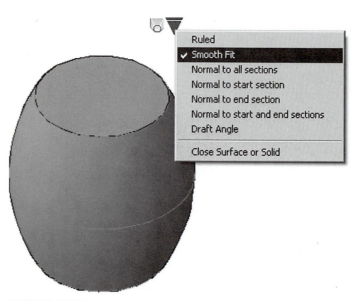

FIGURE 25.24
Loft settings menu.

effect on the overall shape. In almost all cases, we want a smooth fit (the default), so after experimenting restore that choice.

The final result is shown again in Figure 25.25. The down arrow is gone but can be brought back by a single mouse-click on the shape.

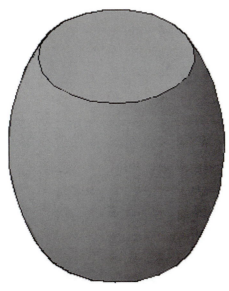

FIGURE 25.25
Loft completed.

While this was a very simple example, the essentials of loft are all there. Because the circles by definition are closed profiles, we get a solid out of this. If, for example, we had three arcs instead (in the same positions), loft would yield a surface, as seen in Figure 25.26. Go ahead and try this by trimming the circles. We use this surfacing ability later on to generate some rather complex surfaces quickly and easily.

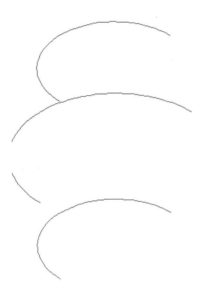

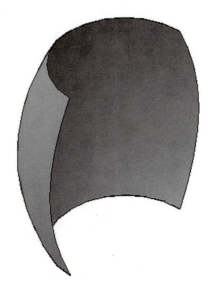

FIGURE 25.26
Loft using arcs.

The true power of loft shows when you need to smoothly merge profiles of varying shapes, sizes, and locations. AutoCAD's Help files have an extensive list of items that qualify as cross sections, so be sure to take a look. Sometimes you may hit a snag in performing loft due to an unusual combination of elements, but as long as they are not on the same plane (two plates on a dinner table is an example of objects on the same plane), you should be able to complete the operation. Some examples shown next include lofting a solid between a rectangle and a circle some distance away (Figure 25.27) and lofting a surface between two arcs some distance apart (Figure 25.28).

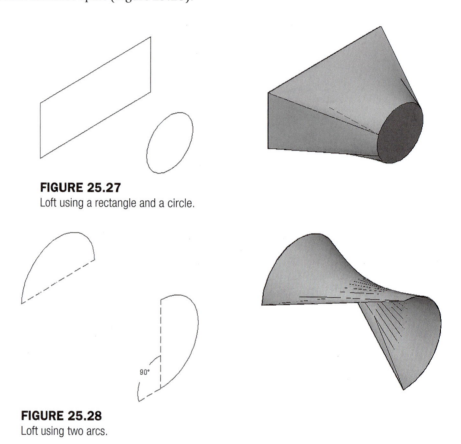

FIGURE 25.27
Loft using a rectangle and a circle.

90°

FIGURE 25.28
Loft using two arcs.

3D Modeling Exercise: Mechanical Drill Bit

This is a good example of a loft application. We create three profiles, position them, and loft between them, creating a complex shape.

Step 1. Draw a circle and one tooth, similar to what is found on a gear. Then, array it 12 times around the circle. Finally, do a little bit of cleanup trimming and pedit the entire structure. This is your basic profile, as seen in Figure 25.29. You need not draw the exact same thing; something close is just fine.

Step 2. Now copy the profile for a total of three items. Make each new copy 1.5 times larger than the previous one using scale, as seen in Figure 25.30.

Step 3. Carefully place the three profiles one on top of another using their center points and rotate profiles 2 and 3 (the medium and the largest ones) 10° and 20°, respectively, as seen in Figure 25.31.

Step 4. Now, switch to the 3D SW Isometric view and separate the three profiles evenly in size order, as seen next. Be careful, you may have to 3Drotate all the profiles to get them to "stand upright." Also, be sure to check the side profiles to make sure they are spaced evenly. Use Ortho, and when done, 3Dorbit around the profile to check that they are located just right. You should see what is shown in Figure 25.32.

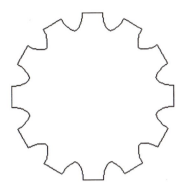

FIGURE 25.29
Drill bit, Step 1.

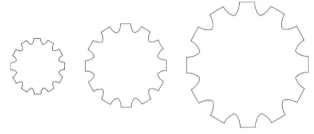

FIGURE 25.30
Drill bit, Step 2.

557

FIGURE 25.31
Drill bit, Step 3.

FIGURE 25.32
Drill bit, Step 4.

Step 5. Finally, you are ready to execute loft. Figure 25.33 shows the result.

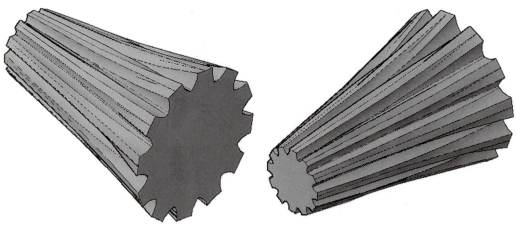

FIGURE 25.33
Drill bit, Step 5.

As you can see from this and the previous examples, loft is a very versatile and powerful command. You can represent virtually any object as long as you are able to visualize and correctly draw a suitable profile.

Path Extrusion

Path extrusion and sweep, which follow next, are commands with a similar mission but different approaches. Path extrusion is the older command and still quite useful in certain situations, while the newer sweep just expands and simplifies the same idea.

The central idea here is to allow extrusion along a given path, which can be curved and otherwise nonlinear (recall that the basic extrude command was linear in nature). Path extrusion is technically not a new command, as the option to select a path while extruding was actually always there in the command's submenus, so all we are doing here is finally trying it out.

Draw a circle of any diameter or radius. Then draw an L-shaped set of lines, as shown in Figure 25.34, and add a small fillet to the intersection of the lines; finally pedit the lines and arc.

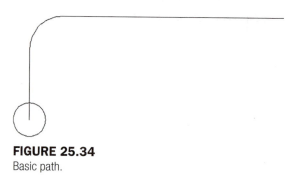

FIGURE 25.34
Basic path.

This line is the path, and the circle is the profile we would like to extrude. There is one more step before you can do the path extrusion, however. This command requires that the profile (circle) be turned perpendicular to the path (line). Use 3Drotate to rotate the circle 90°. It will look like what is shown in Figure 25.35. Now switch to 3D. You may need to 3Drotate the pieces into an upright position, as seen in Figure 25.36.

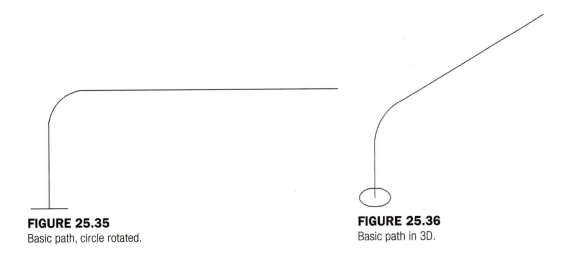

FIGURE 25.35
Basic path, circle rotated.

FIGURE 25.36
Basic path in 3D.

We can now do the path extrusion. There is no separate path extrusion command, rather it is a submenu of the regular extrude.

Step 1. Start up extrude via any preferred method.
  ○ AutoCAD says: `Current wire frame density: ISOLINES=4`
    `Select objects to extrude:`
Step 2. Select the circle and press Enter.
  ○ AutoCAD says: `Specify height of extrusion or [Direction/Path/Taper angle]`
Step 3. Select p for `Path` and press Enter.
  ○ AutoCAD says: `Select extrusion path or [Taper angle]`
Step 4. Click to select the path itself. You see what is shown in Figure 25.37, in wireframe.

The extrusion is then shaded and colored to give the final result, as seen in Figure 25.38. This view is also switched to SE Isometric to show the other side of the pipe.

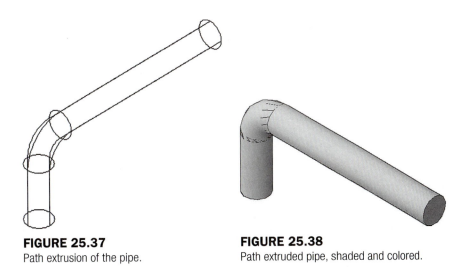

FIGURE 25.37
Path extrusion of the pipe.

FIGURE 25.38
Path extruded pipe, shaded and colored.

Figure 25.39 shows the same path extruded pipe with a hollow center, to actually make it a pipe. This was done by going back to the original profile and simply adding another (smaller) circle profile and extruding it along the same path. Then a subtraction was done, subtracting one extrusion from the other. As an exercise, go ahead and recreate this yourself.

FIGURE 25.39
Path extruded pipe with hollow center.

Sweep

As already mentioned, sweep is a variation on the same theme but with some enhancements. Once again, we have a profile and a path, but instead of carefully lining up the circle to the path, we can select almost *any* profile, *anywhere* on the screen, and click almost *any* path to generate a sweep. Therefore, it is a much more flexible command. Figure 25.40 shows an example with a rectangle used to create a sweep along essentially the same type of path. Notice the profile is nowhere near the path. Go ahead and set this up via the following steps.

FIGURE 25.40
Rectangle and sweep path.

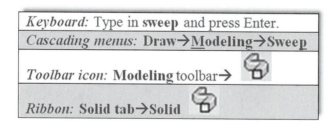

| *Keyboard:* Type in **sweep** and press Enter. |
| *Cascading menus:* **Draw→Modeling→Sweep** |
| *Toolbar icon:* **Modeling** toolbar→ |
| *Ribbon:* **Solid** tab→Solid |

Step 1. Start the sweep command via any of the preceding methods.
  ○ AutoCAD says: `Current wire frame density: ISOLINES=4`
    `Select objects to sweep:`
Step 2. Pick the rectangle and press Enter.
  ○ AutoCAD says: `Select sweep path or [Alignment/Base point/Scale/`
    `Twist]:`
Step 3. Select the path and the sweep is produced, as seen shaded and colored in the SE Isometric view in Figure 25.41.

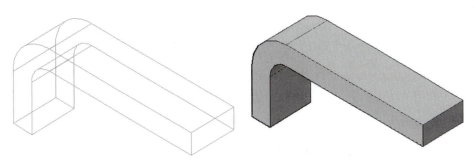

FIGURE 25.41
Rectangle sweep.

There are some limits to this of course, and those have to do mostly with the size of the profile or path. If the profile does not physically fit the path, AutoCAD tells you that it is unable to produce the sweep. Note that the sweeping action destroys the original profile, so if you think you may need it again, keep a copy of it. This is an important point in creating multiple sweeps later on.

A major advantage of this command over path extrusion is that a sweep can be produced even if the profile is on the same plane as the path, as seen with an arc and a circle in Figure 25.42.

FIGURE 25.42
Circle sweep.

The sweep command is not bothered by the similar plane. It just takes whatever profile is selected and applies it perpendicularly to whatever path you choose.

This concludes the advanced 3D tools. Practice and learn to apply everything thus far. We have an additional chapter with a few more tools to complete your 3D construction knowledge, but the ones presented in this chapter are the primary ones.

Drawing Challenge: Helical Coil

Using what you know so far, draw the helical coil shown in Figure 25.43. You have *all* the tools at your disposal at this point. Think carefully, and do not worry about size or how many coils to do, only the shape. The answer is featured at the end of the next chapter, but do not look it up right away; give it your best shot first. Also, do not be tempted to use AutoCAD's helix tool, which will be covered in the next chapter. It is relatively new (introduced in AutoCAD 2007), and the results do not look like what is shown here anyway. There is a reason why you should try to figure this out, and it is mentioned along with the answer in the next chapter. Good luck.

562

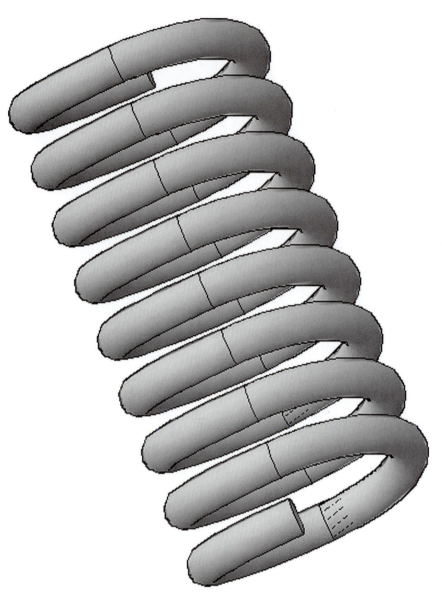

FIGURE 25.43
Helical coil drawing challenge.

SUMMARY

You should understand and know how to use the following concepts and commands before moving on to Chapter 26:

- Revolve
- Shell
- Taper
- Loft
- Path extrusion
- Sweep

REVIEW QUESTIONS

Answer the following based on what you learned in Chapter 25:

1. Describe the revolve command.
2. Does the profile need to be open or closed? Describe the effect of each.
3. Describe the shell command.
4. How many surfaces can you select with shell?
5. Describe the taper command.
6. Describe the loft command.
7. Describe the path extrusion command.
8. Describe the sweep command.
9. Why is sweep better than path extrusion?

EXERCISES

1. Use the following profile and axis of rotation (the line below the profile) to revolve it 360° to create the model of a basic tire and rim. Sizing and exact profile shape are up to you, and you need various drawing and editing commands, including pedit, prior to revolve. (Difficulty level: Moderate; Time to completion: 20 minutes)

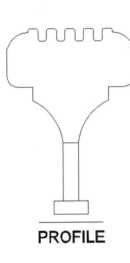

PROFILE

2. Create four 4.5" × 6" rectangles, then explode, trim, and pedit them together into an outline as shown in the profile on the left. Then, extrude them to a height of 5.5" and shade and color them. Finally use the shell command (0.2" thickness) to remove the faces, as shown in the profile on the right. (Difficulty level: Moderate; Time to completion: 10 minutes)

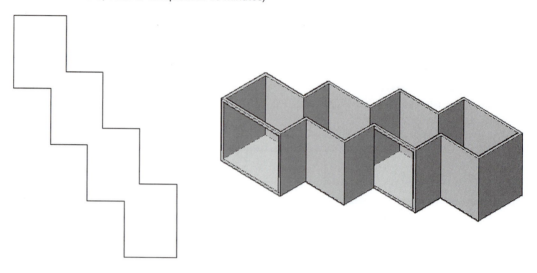

3. Create the following basket using the rectangle (20" × 10"), extrude (10"), taper (10°), fillet (1.5"), and shell (0.25") commands. Then, add the handle using arc and sweep (1" diameter circle). (Difficulty level: Moderate; Time to completion: 15–20 minutes)

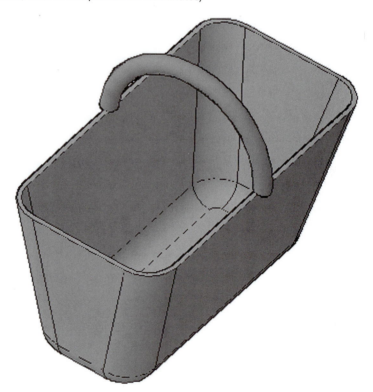

4. Create the following bracket and pole model. An exploded view as well as a breakdown of individual parts are shown for clarity. The sizing of the parts is up to you. (Difficulty level: Moderate; Time to completion: 30 minutes)

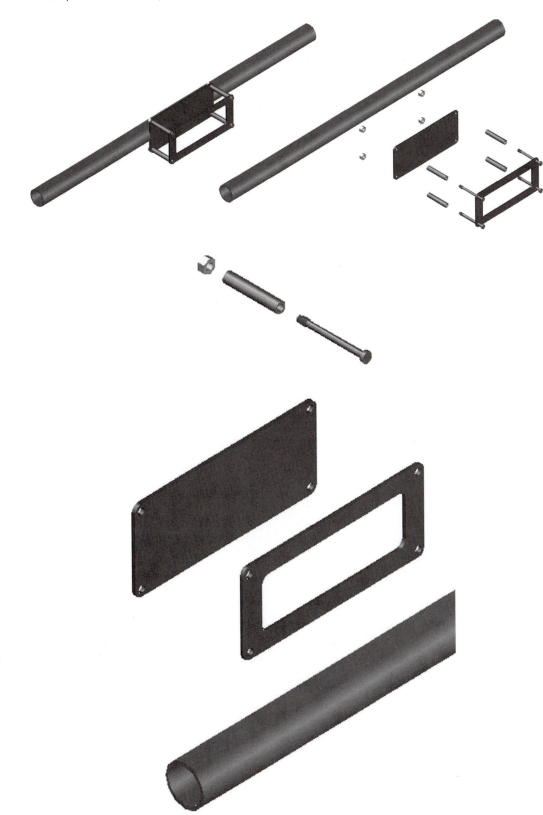

CHAPTER 26

Advanced Solids, Faces, and Edges

LEARNING OBJECTIVES

In this chapter, we continue to introduce additional advanced tools, including

- Polysolid
- Helix
- 3D path array
- 3Dalign
- Faces
  - Move
  - Offset
  - Delete
  - Rotate
  - Copy
  - Color
- Edges
  - Copy
  - Color

Up and Running with AutoCAD® 2012: 2D and 3D Drawing and Modeling.

By the end of the chapter, you will have the complete 3D object design tool set in your hands.

Estimated time for completion of chapter: 2 hours.

26.1 INTRODUCTION TO ADVANCED SOLIDS, FACES, AND EDGES

With this chapter, we continue to introduce advanced 3D tools. While they are not the primary tools you use in day-to-day 3D work, they are still needed to round out your knowledge. Specifically, we add polysolid, helix, 3D path array, and 3Dalign to the list and focus on working with faces (move, offset, delete, rotate, copy, and color) as well as edges (copy and color).

Polysolid

After trying out polysolid, you may never create a wall via the extrude command again. A polysolid is analogous to a polyline. What you get is continuous sections of extruded rectangles similar to continuous sections of a pline in 2D. It is a great tool for quickly constructing 3D walls of a building. Once you begin creating the polysolid, you can continue to "stitch" forward one section at a time by just pointing the mouse in the direction you want to go and entering a distance (usually with Ortho on).

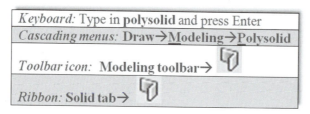

Keyboard: Type in **polysolid** and press Enter

Cascading menus: **Draw→Modeling→Polysolid**

Toolbar icon: **Modeling toolbar→**

Ribbon: **Solid tab→**

Step 1. Start polysolid via any of the preceding methods.

    ○ AutoCAD says:

```
_Polysolid Height = 4.0000, Width = 0.2500, Justification = Center
    Specify start point or [Object/Height/Width/Justify] <Object>:
```

Step 2. These are the default values for the height and width, and they are OK for now. Go into the SW Isometric view and activate Ortho. Then, click anywhere on the screen.

    ○ AutoCAD says: `Specify next point or [Arc/Undo]:`

Step 3. A connected polysolid is drawn every time you click, so do several clicks and press Enter when done, then shade and color the polysolid. You should have something similar to what is shown in Figure 26.1 on your screen.

To create a section of a wall similar to what you did in Chapter 24, you need to change the height and width of the polysolid by selecting h for `height` or w for `width` in the suboptions.

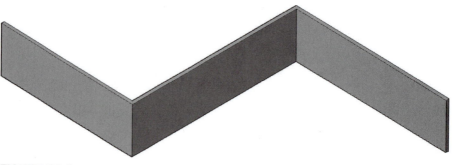

FIGURE 26.1
Basic polysolid.

For *height*,

* AutoCAD says: `Specify height <4.0000>:`

Enter 72 for the height.

For *width*,

* AutoCAD says: `Specify width <0.2500>:`

Enter 5 for the width.

Polysolid has two more useful features. The first occurs if you select the `Arc` suboption while clicking around; you can create an arc section, as seen in Figure 26.2.

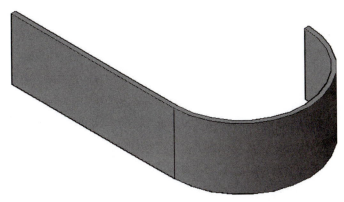

FIGURE 26.2
Basic polysolid with arc.

The second very useful feature is the ability of the polysolid command to convert regular 2D geometric objects into polysolids using the `Object` suboption. These objects can be rectangles, arcs, circles, and others, as seen in Figures 26.3 and 26.4. Simply begin the polysolid command, press o for `Object`, and select the shape. The height is dictated by the `Height` setting, which can be easily changed.

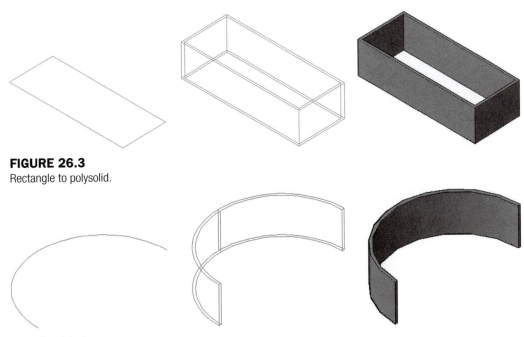

FIGURE 26.3
Rectangle to polysolid.

FIGURE 26.4
Arc to polysolid.

Helix

After apparently receiving many requests, called *wish list* entries, on this topic, Autodesk finally included the helix command. It pretty much does what it advertises, creating a basic 3D helix coil, as seen in Figure 26.5.

FIGURE 26.5
Basic helix coil.

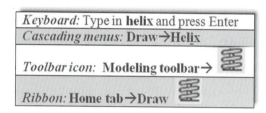

Keyboard: Type in **helix** and press Enter	
Cascading menus: **Draw→Helix**	
Toolbar icon: **Modeling toolbar→**	
Ribbon: **Home tab→Draw**	

Step 1. Start helix via any of the preceding methods.
  ○ AutoCAD says: `Number of turns = 3.0000 Twist = CCW`
    `Specify center point of base:`
Step 2. You are asked here to specify where you want the helix coil to be. Generally, you click on any random point to just create the helix. Afterward, you can move it into a specific position according to design needs.
  ○ AutoCAD says: `Specify base radius or [Diameter] <1.0000>:`
Step 3. This is the overall radius or diameter of the bottom part of the coil. Enter any value; 10 is used in this example.
  ○ AutoCAD says: `Specify top radius or [Diameter] <10.0000>:`
Step 4. The top radius or diameter of the coil does not necessarily have to be the same as the bottom; this way your coil can taper in or expand if needed. For this example, however, we choose the top and bottom to be the same size (10).
  ○ AutoCAD says: `Specify helix height or [Axis endpoint/Turns/turn`
    `Height/tWist] <1.0000>:`
Step 5. Although you can manually pull the helix into shape at this point, a better approach is to indicate the number of turns and helix height. Type in `t` for `Turns` and press Enter.
  ○ AutoCAD says: `Enter number of turns <3.0000>:`
Step 6. Enter any reasonable value; 10 is used for this example.
  ○ AutoCAD says: `Specify helix height or [Axis endpoint/Turns/turn`
    `Height/tWist] <1.0000>:`
Step 7. Now type in the height you wish the helix to have; 40 is used in this example.

The final result is seen on the left-hand side of Figure 26.6. Now, how would you give the helix a thickness, like a real spring? We recently covered the answer to that: the sweep command. Create a small circle profile positioned near the helix. Using the helix as a path, create a sweep of the circle profile as seen in the rest of Figure 26.6.

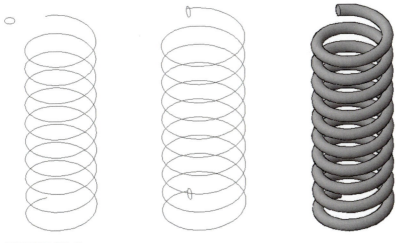

FIGURE 26.6
Completed helix coil with sweep.

While this is an easy way to model the coil as presented in the challenge drawing of the previous chapter, resist the temptation to do so and search for other ways.

3D Path Array

The array command was first introduced in Chapter 8. Having gone through a bit of a redesign, array is now command line driven and features a new third variation, the path array. This allows you to select an object and a path. Multiple instances of that object are then created along the path. The same exact thing can be done in 3D.

To try this out, we create a spiral staircase. Draw a small step consisting of a 20" × 8" rectangle extruded to a thickness of 1", as seen in Figure 26.7. Do not forget to add color and shade.

FIGURE 26.7
20" × 8" × 1" step for path array.

Then, using your new helix skills, create a helix of any reasonable size. The one used in this example is 15 × 15, with a height of 50. The step is rotated 90° to align with the bottom of the helix, as seen in Figure 26.8.

Finally, run through the path array command (it is the same for 3D as it is for 2D). Just as a reminder, you are asked to select the object and indicate this is a path array. Then, type in 50

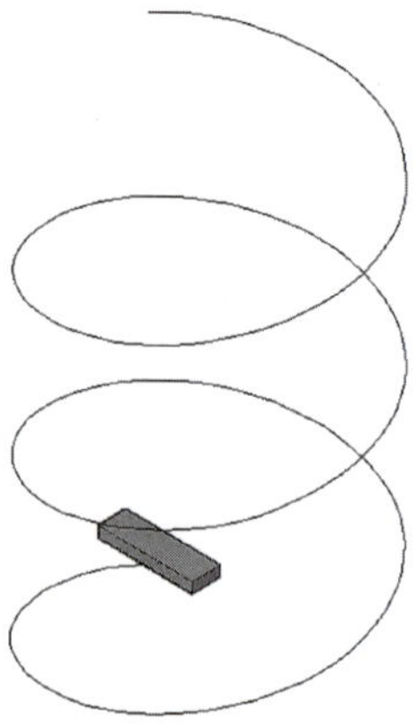

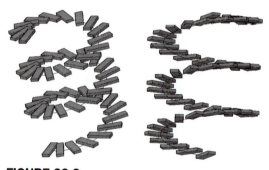

FIGURE 26.8
Helix for path array.

when asked how many objects. Finally press Enter twice to complete the 3D path array command. Your results should look similar to Figure 26.9, which shows the new staircase from two different views. Note that the path is still visible but can of course be erased if desired.

FIGURE 26.9
Completed spiral staircase (two views).

While this was just a "quick and dirty" example, you can easily see how some time spent crafting a proper set of stairs, handrails, and a frame yields a graceful and impressive design of a spiral staircase with comparatively modest effort.

3Dalign

As you may recall from Level 2, aligning objects in 2D can be somewhat tricky, and the align command is quite useful in quickly and easily mating up geometric pieces. In 3D, it can be even more of a challenge to align objects, especially if they are out of alignment in all three axes. Therefore, 3Dalign is almost mandatory to accomplish this. Go ahead and create and position the following two blocks as described next.

In Figure 26.10, on the left (Block 1) is a 25" × 15" block, extruded to 15". It is even with the X, Y, and Z axes, as seen in the standard SW Isometric view, while a similar copied Block 2 on the right has been rotated 15° along the X and Y axes and 30° along the Z axis. We need to align Face 1 with Face 2 by moving Block 2 (since it is the rotated one) and bringing it around to have its face mate up with the face of the correctly aligned Block 1.

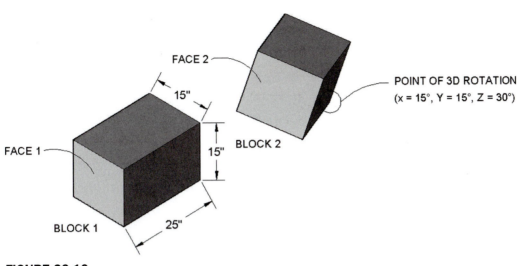

FIGURE 26.10
Blocks for 3Dalign.

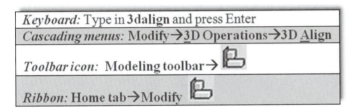

Step 1. Start the 3Dalign command via any of the preceding methods.
 ○ AutoCAD says: `Select objects:`
Step 2. Select just the misaligned Block 2 (you could, in theory, pick Block 1 as well) and press Enter.
 ○ AutoCAD says: `Specify source plane and orientation ...`
   `Specify base point or [Copy]:`
Step 3. Here, you need to click on and choose three consecutive source points on Block 2's face that you want used as points of attachment to Block 1. Several combinations are possible, so let us arbitrarily choose the points shown in Figure 26.11. As you select these three points,
 ○ AutoCAD says: `Specify second point or [Continue] <C>:`
   `Specify third point or [Continue] <C>:`
   `Specify destination plane and orientation ...`
   `Specify first destination point:`
Step 4. Now, you are ready to specify the destination points. Here you have to be careful and think about what those points actually are, as random choices produce random results. Observe that, in this example, Block 2 needs to be rotated 180° for its greenish yellow face to match up to Block 1's greenish yellow face. Therefore, the destination points need to be mirror images of the source points, not the exact same ones.

The order we picked the source points was Top Left→Top Right→Bottom Right. As we swing Block 2 around, the corresponding points on Block 1 are Top Right→Top Left→Bottom Left. Do you see why? Visualize this on a nearby object, one that you can pick up, locate the points on, and turn around: perhaps a cell phone or a book. Let us now select those points, as seen in Figure 26.12. As you are doing this,

 ○ AutoCAD says: `Specify first destination point:`
   `Specify second destination point or [eXit] <X>:`
   `Specify third destination point or [eXit] <X>:`

Step 5. Select the points carefully; Block 2 may be in the way, so you need to trust your OSNAPs (or just stay in wireframe mode to see the points). The final result is shown in Figure 26.13.

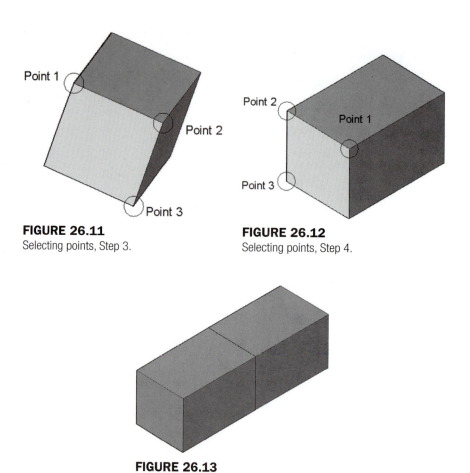

FIGURE 26.11
Selecting points, Step 3.

FIGURE 26.12
Selecting points, Step 4.

FIGURE 26.13
3Dalign completed.

26.2 WORKING WITH FACES

Suppose you created a hole in a block via extrusion and subtraction, as seen in Figure 26.14. How would you delete that hole if it is no longer needed, without redrawing or otherwise damaging the original block? What if you needed to move the hole from its center position

FIGURE 26.14
Block with a hole.

to another area of the block? Or, what if you just needed to change its diameter? All this and more are addressed by the various faces operations, such as move face, copy face, offset face, delete face, rotate face, and even color face.

Being able to easily modify the position, geometry, and sizing of 3D features is a fundamental requirement of all solid modeling programs. The parametric modeling software on the market approaches this in various ways, some better than others. The Spatial Technologies ACIS kernel that powers AutoCAD's 3D modeler is not quite on the level of the Parasolid, Granite, or CAA that powers the high-end NX, Pro/Engineer, and CATIA software, respectively (see Appendix H for more information on kernels), but it does a decent job working with embedded 3D features using the various faces tools. Let us cover them all, one at a time, as they are very important in effective 3D design.

Delete Faces

Everything you create in AutoCAD 3D has faces. They may be the faces of an object, such as the extruded hole or actual surfaces of larger objects. When AutoCAD refers to faces, it is those objects or surfaces that can be moved or copied. In some cases, you can interpret this as moving or copying an entire feature (such as the extruded hole in the upcoming examples) or, in other cases, an actual surface of an object.

Delete faces is probably the easiest of the face commands to start with. It is meant to be used to delete full features, such as an extruded cylinder (hole), not surfaces. To try it out, draw a random size block with a hole in it, as seen in Figure 26.15.

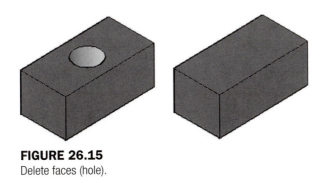

FIGURE 26.15
Delete faces (hole).

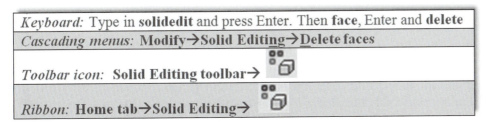

Keyboard: Type in **solidedit** and press Enter. Then **face**, Enter and **delete**

Cascading menus: **Modify→Solid Editing→Delete faces**

Toolbar icon: **Solid Editing toolbar→**

Ribbon: **Home tab→Solid Editing→**

Step 1. Start up delete face (a solid editing command) via any of the preceding methods.
- ○ AutoCAD brings up a lengthy solid editing menu:
```
_solidedit
Solids editing automatic checking: SOLIDCHECK=1
Enter a solids editing option [Face/Edge/Body/Undo/eXit] <eXit>:
_face
Enter a face editing option
[Extrude/Move/Rotate/Offset/Taper/Delete/Copy/coLor/mAterial/
Undo/eXit] <eXit>:
_color
Select faces or [Undo/Remove]:
```

Step 2. Click to select the hole. Be careful to pick only the hole not surrounding geometry.
- AutoCAD says: 1 face found.

Step 3. Press Enter.
- AutoCAD says:
  Solid validation started.
  Solid validation completed.
  Enter a face editing option.

Step 4. The hole is removed, as seen in Figure 26.12. Press Esc to return to the command line.

The delete faces command can also be used to delete unwanted fillets in 3D. Simply run the command, select the fillet, and it reverts to a "0" radius (sharp corner), as seen in the sequence of Figure 26.16.

FIGURE 26.16
Delete faces (fillet).

Move Faces

The move faces command moves the hole to another location. Undo the face deletion you just practiced and let us try to move the hole around.

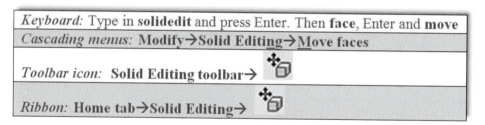

Keyboard: Type in **solidedit** and press Enter. Then **face**, Enter and **move**	
Cascading menus: **Modify→Solid Editing→Move faces**	
Toolbar icon: **Solid Editing toolbar→**	
Ribbon: **Home tab→Solid Editing→**	

Step 1. Start up move face via any of the preceding methods.

Step 2. AutoCAD shows the same lengthy solid editing menu (not reproduced here again) and asks you to select the faces (the hole). From there, the command is exactly like a regular move command. It asks for a base point to move *from* and a second point to move *to*. Be sure to have Ortho on when doing this. Upon completion, the hole is moved, as seen in Figure 26.17.

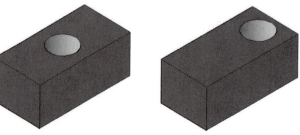

FIGURE 26.17
Move faces, before and after.

Copy Faces

Copy faces copies any element you select in the same exact manner that move faces works. It is however more useful for a slightly different application involving actual faces as opposed to elements like a hole. To try out both approaches, leave your hole in its new position at the edge of the block and do the following.

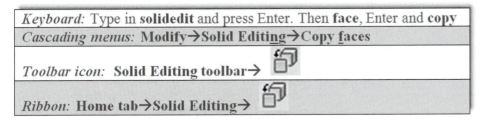

Keyboard: Type in **solidedit** and press Enter. Then **face**, Enter and **copy**
Cascading menus: **Modify→Solid Editing→Copy faces**
Toolbar icon: **Solid Editing toolbar→**
Ribbon: **Home tab→Solid Editing→**

Step 1. Start up copy face via any of the preceding methods.
Step 2. You are prompted for a base point to copy *from* and a second point to copy *to*. However, you have to be careful in how you apply this command. If you copy the extruded hole, the new hole is not a solid and cannot be subtracted. What you get is shown in Figure 26.18.

FIGURE 26.18
Copy faces applied to hole.

577

What happened was that just the edge was copied, and we discuss how to deal with this properly when we discuss edges. Meanwhile, what exactly is copy faces good for? Well, remember that you can move and copy not just features but also surfaces. We copy a surface off to the side and extrude it into a shape, as shown next in Figures 26.19 and 26.20. This can be useful because you need not create new profiles from scratch but can reuse complex geometry.

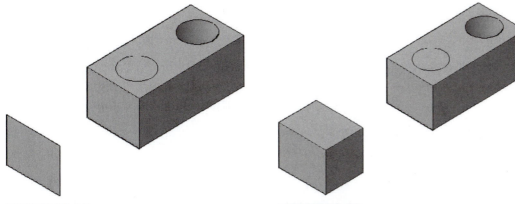

FIGURE 26.19
Copy faces applied to surface.

FIGURE 26.20
Copied face extruded.

Offset Faces

Offset faces can accomplish essentially the same thing as extrude. While you are used to extruding 2D objects into the third dimension, extrude and offset faces can also be used to lengthen or shorten what is already a 3D object by extruding or offsetting the face you pick, and the solid material stretches or shrinks to meet the new position of this face. The approach is essentially the same, so we focus on offset faces.

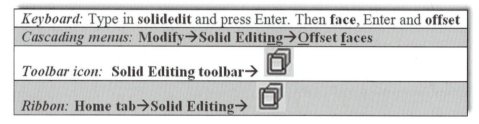

Keyboard: Type in **solidedit** and press Enter. Then **face**, Enter and **offset**
Cascading menus: **Modify→Solid Editing→Offset faces**
Toolbar icon: **Solid Editing toolbar→**
Ribbon: **Home tab→Solid Editing→**

Step 1. Start up offset face via any of the preceding methods.
Step 2. AutoCAD once again runs through the lengthy solid editing menu and asks you to select a face. Pick the face you want to offset and press Enter. AutoCAD then asks you to specify the offset distance. Enter some value, and you see the result, similar to what is shown in Figure 26.21.

Offset faces can also come in handy when attempting to change the size of an embedded feature, such as the hole's diameter in this case. Applying offset faces to the hole, we get the result seen in Figure 26.22 when adding an offset of 3 to it.

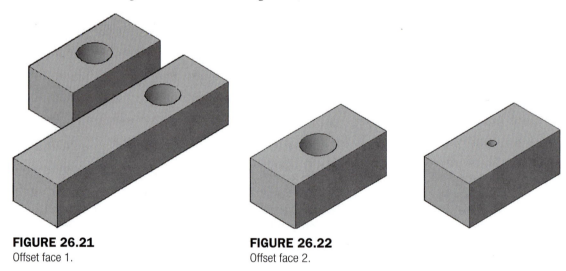

FIGURE 26.21
Offset face 1.

FIGURE 26.22
Offset face 2.

Rotate Faces

Rotate faces is primarily for rotating embedded 3D features. Since it is hard to see the rotation of a circle, let us draw a new part, consisting of a random size plate with a cutout, as seen in Figure 26.23. Next to the cutout is a small extrusion for reference purposes. We would like to rotate this cutout 90°.

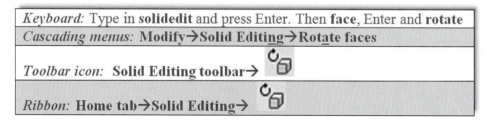

Keyboard: Type in **solidedit** and press Enter. Then **face**, Enter and **rotate**
Cascading menus: **Modify→Solid Editing→Rotate faces**
Toolbar icon: **Solid Editing toolbar→**
Ribbon: **Home tab→Solid Editing→**

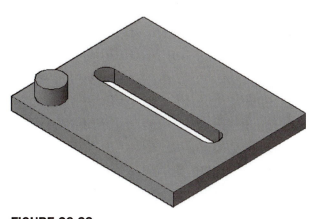

FIGURE 26.23
Plate with groove and extrusion.

Step 1. Start up rotate face via any of the preceding methods.
Step 2. AutoCAD once again gives you a lengthy solid editing menu (not reproduced here) and asks you to pick the faces for rotation. Carefully select both straight and curved faces and press Enter.
- ○ AutoCAD says: Specify an axis point or [Axis by object/View/Xaxis/ Yaxis/Zaxis] <2points>:
Step 3. You are free to choose any axis for the rotation or use two points to specify a new one. For this example, the Z axis is used.
- ○ AutoCAD says: Specify the origin of the rotation <0,0,0>:
Step 4. The origin of the rotation is up to you to choose. For this example though, OTRACK is used to select the exact center of the cutout.
- ○ AutoCAD says: Specify a rotation angle or [Reference]:
Step 5. Once again, this is up to you, but 90° is used for this example.
- ○ AutoCAD rotates the cutout and says:
  Solid validation started.
  Solid validation completed.
  Enter a face editing option.

The result is seen in Figure 26.24.

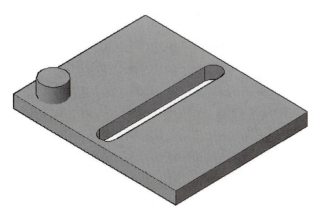

FIGURE 26.24
Groove rotated via rotate face.

Color Faces

You may have noticed that many faces are colored for clarity in this chapter. This is easy to do and is a useful trick to make the faces stand out (we do this with edges as well later on).

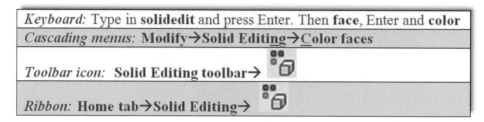

Keyboard: Type in **solidedit** and press Enter. Then **face**, Enter and **color**

Cascading menus: **Modify→Solid Editing→Color faces**

Toolbar icon: **Solid Editing toolbar→**

Ribbon: **Home tab→Solid Editing→**

Step 1. Start up color face via any of the preceding methods.
Step 2. Select a face, and you see the familiar color dialog box from Chapter 3.
Step 3. Select a color and the face changes.

26.3 WORKING WITH EDGES

Working with edges is a short topic, as there are only two items to cover: color edges and copy edges.

Color Edges

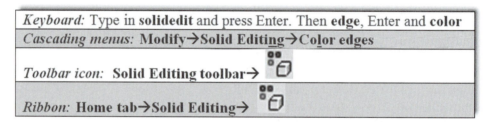

Keyboard: Type in **solidedit** and press Enter. Then **edge**, Enter and **color**

Cascading menus: **Modify→Solid Editing→Color edges**

Toolbar icon: **Solid Editing toolbar→**

Ribbon: **Home tab→Solid Editing→**

This command is very similar to color faces. You may not see the new colored edge as easily since it is just that, an edge, basically a thin line.

Step 1. Start up color edge via any of the preceding methods.
Step 2. Select an edge, and you see the familiar color dialog box from Chapter 3.
Step 3. Select a color and the edge changes.

Copy Edges

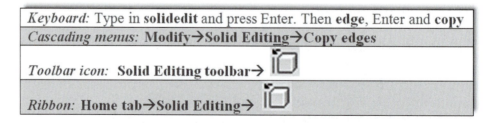

Keyboard: Type in **solidedit** and press Enter. Then **edge**, Enter and **copy**

Cascading menus: **Modify→Solid Editing→Copy edges**

Toolbar icon: **Solid Editing toolbar→**

Ribbon: **Home tab→Solid Editing→**

Recall that, when copy faces was attempted on an extruded hole, there was a problem, as we were unable to subtract the copied hole. Copy edges gets around this problem by copying only the original circle, which can now be extruded. With a simple shape, such as a circle, you could also just redraw it and extrude; but with more complex shapes, this copy edge technique is quite useful. An example follows graphically via Figure 26.25. Go ahead and try this command out on your own.

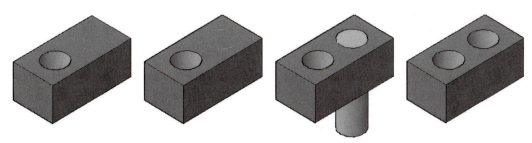

FIGURE 26.25
Edge copied, extruded, and subtracted.

Practice all the face and edge editing commands we just covered. We conclude the chapter with an explanation on how the helix was created by hand.

Helical Coil Explained

So how was helical coil drawn in the previous chapter? Have you made an earnest attempt in creating it without using the helix tool? Here is the answer. You have to visualize what a helix really is: a bunch of circles connected together. All we need to do is figure out a way to connect them and the rest is easy. Here is one possible way. Draw a circle and "split it in two" by drawing a line across it quadrant to quadrant, trimming off one half and mirroring the half right back, as seen in Figure 26.26 (you can also use the break command).

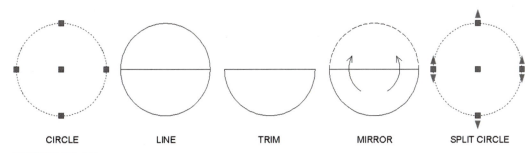

CIRCLE   LINE   TRIM   MIRROR   SPLIT CIRCLE

FIGURE 26.26
Creating the helical coil, Part 1.

Now go into 3D and use rotate3D to raise one half of the circle 15° up (or down) around the Y axis, as seen in Figure 26.27.

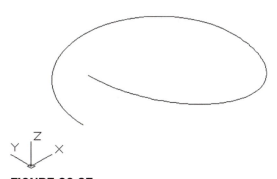

FIGURE 26.27
Creating the helical coil, Part 2.

At this point, it is just a matter of connecting a bunch of these segments, as seen in Figure 26.28. The final step is to use the sweep command to create the thickness for the coil, as seen in Figure 26.29 (remember to use many circles, as one gets used up for each section). Union the entire coil and assign color and shading. The final coil is shown in Figure 26.30.

FIGURE 26.28
Creating the helical coil, Part 3.

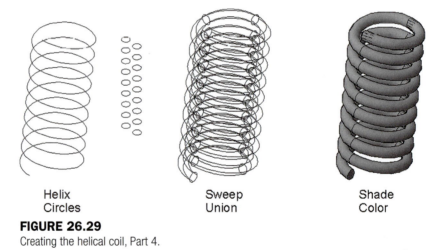

| Helix | Sweep | Shade |
| Circles | Union | Color |

FIGURE 26.29
Creating the helical coil, Part 4.

So why were you asked to create this helix when AutoCAD has an automated command for this? Well, mainly because it is a good exercise in critical thinking in a situation that has no easy answer (at least prior to the introduction of the helix command in AutoCAD 2007). Very often situations arise in which you have to depict something complex and no ready-to-use tools are available—after all, every possible design scenario cannot be accounted for by AutoCAD. You have to stop, think of what you do have available, and see how you can use these tools creatively to construct something new.

Everything used in the construction of this helix is familiar to you, but the combination of commands is novel. That is the key, finding ways to accomplish a task using the tools you already have. It is a virtual certainty you will run into drawing situations where you need to think out of the box. Ideally, you will remember this exercise and remember to think in terms of fundamental tools.

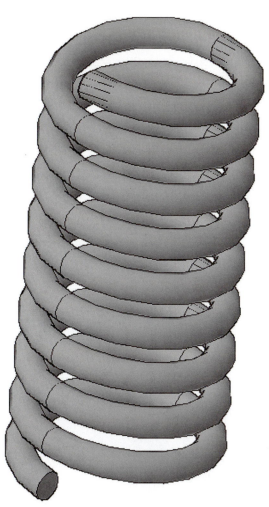

FIGURE 26.30
Completed helical coil.

SUMMARY

You should understand and know how to use the following concepts and commands before
moving on to Chapter 27:

- Polysolid
- Helix
- 3D path array
- 3Dalign
- Faces
  - Move
  - Offset
  - Delete
  - Rotate
  - Copy
  - Color
- Edges
  - Copy
  - Color

REVIEW QUESTIONS

Answer the following based on what you learned in Chapter 26:

1. Describe the polysolid command. Can you turn regular objects into polysolids?
2. Describe the helix command.
3. Describe the 3D path array command.
4. Describe the 3Dalign command.
5. What are the six faces commands?
6. What are the two edges commands?

EXERCISES

1. Color the faces of an extruded box as follows. Create a hole in the middle as shown. (Difficulty level: Easy; Time to completion: 1–2 minutes)

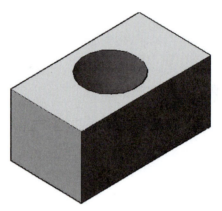

2. Copy faces on an extruded box as follows. Offset the hole to a smaller size as shown. (Difficulty level: Easy; Time to completion: 1–2 minutes)

CHAPTER 27

Surfaces and Meshes

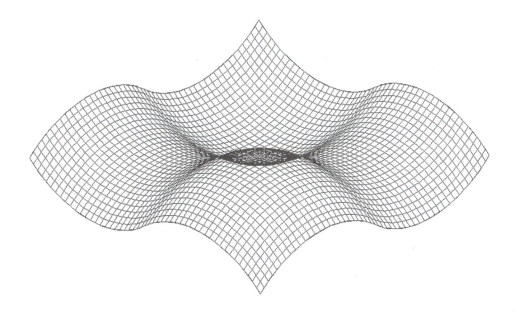

LEARNING OBJECTIVES

In this chapter, we focus on surfaces instead of solid objects. These objects have
no depth the way solids do but possess many desirable properties for design work.
Specifically, we mention

- Planar surface
- Region
- Rulesurf
- Tabsurf
- Revsurf
- Edgesurf
- 3Dface
- 3Dmesh
- Smooth Mesh Primitives
- Mesh smoothing
- Face move and rotate
- Collapse Face or Edge
- Convert to Solid or Surface
- Section Planes

Up and Running with AutoCAD® 2012: 2D and 3D Drawing and Modeling.

By the end of the chapter, you will be able to create a variety of surfaces and meshes.

Estimated time for completion of chapter: 2 hours.

27.1 INTRODUCTION TO SURFACES AND MESHES

In this chapter, we explore some of AutoCAD's surfacing and meshing capabilities. They underwent some significant enhancements in last year's AutoCAD 2011 release. A surface and a mesh are often the same, since in all but a few cases, a surface, by definition, is made up of a mesh. A mesh can then be described as an interconnecting weave of lines representing a surface (as opposed to a solid).

AutoCAD has several general approaches to creating, and working with, these objects. The most recent approach parallels what has been done in advanced 3D software like CATIA for some time now. You create a rough "blob" representing a general shape, then you adjust it one mesh at a time, pushing, pulling, and sculpting it like a piece of clay.

Another method involves creating a "framework" for the surface, a sort of skeleton, if you will. This framework is generally made using splines, which can be shaped easily via control points (grips). You can then create surfaces on top of this framework.

Finally, surfaces can be created via ordinary design methods already covered. Recall how you revolved an "open" profile to create a surface model in Chapter 25. In the same chapter, you also used the loft command to create surfaces between two arcs. Loft is actually quite a useful command for basic surfacing of noncomplex curvature.

We start from the beginning and cover "legacy" meshing capabilities first. These tools, such as planar surface, region, and the "surf" family (rulesurf, tabsurf, revsurf, and edgesurf), were around for quite some time and are useful for basic surfacing and meshing tasks. We then move on to some advanced legacy tools such as 3D faces and meshes and discuss the new family of tools introduced (or enhanced) in AutoCAD 2011. Remember, more than ever, these tools depend on your imagination and creativity to shine. They work best when combined by a skilled user to create some very advanced shapes.

27.2 SURFACING COMMANDS
Planar Surface

Perhaps the simplest thing you can surface is a rectangle, and this is where we start. While loft can accomplish this surfacing in theory, you need cross sections. Here, you just have a mundane shape and want to cover it with a surface, and a simpler command, called *planar surface* or just *planesurf*, is available. To try it out, switch to 3D and create a rectangle of any size, as seen in Figure 27.1.

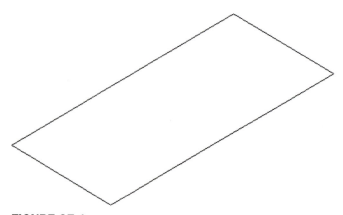

FIGURE 27.1
Basic rectangle in 3D.

We would like to add a simple surface to the rectangle without extruding or making it a true solid.

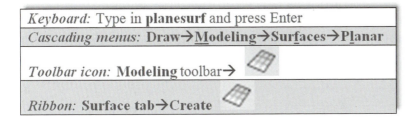

Keyboard: Type in **planesurf** and press Enter	
Cascading menus: **Draw→Modeling→Surfaces→Planar**	
Toolbar icon: **Modeling** toolbar→	
Ribbon: **Surface tab→Create**	

Step 1. Start the planesurf command via any of the preceding methods.
- ○ AutoCAD says: `Specify first corner or [Object] <Object>:`

Step 2. Using OSNAPs, select one of the corners of the rectangle.
- ○ AutoCAD says: `Specify other corner:`

Step 3. Select the diagonally *opposite* corner of the rectangle. A surface is formed, as seen in the wireframe and shade modes in Figure 27.2.

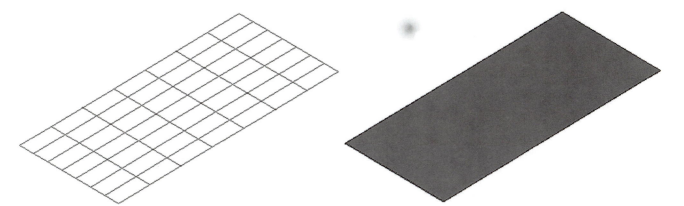

FIGURE 27.2
Rectangle with planesurf.

You can also use planesurf to create surfaces out of objects by typing in o for `Object`. You can then just select the object and turn it into a surface; no clicking of points is needed. This is useful when a curvature, such as an arc, is involved and point-by-point clicking is not practical. Figure 27.3 shows a closed border drawn with a pline, with planesurf then applied in 3D.

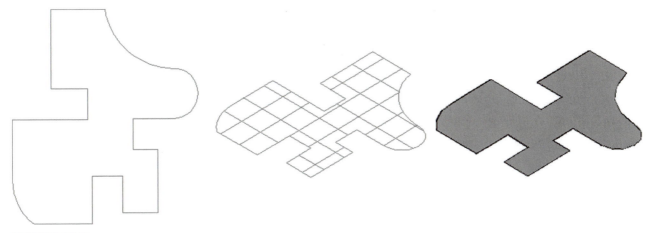

FIGURE 27.3
Shape with planesurf (`Object`).

Region

Region gets a brief mention here as an alternate method to create a simple surface. The difference here is that a region is a surface but *not* a mesh, a rare instance when this is the case. You are not able to edit it in any way, and that may be just fine for simple shapes. To try out region, create another rectangle or circle in 3D with shading on.

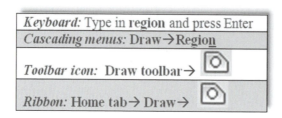

Keyboard: Type in **region** and press Enter
Cascading menus: Draw→Region
Toolbar icon: Draw toolbar→
Ribbon: Home tab→ Draw→

Step 1. Start the region command via any of the preceding methods.
- ○ AutoCAD says: `Select objects:`

Step 2. Pick your shape and press Enter.
- ○ AutoCAD says:
  ```
  1 loop extracted.
  1 Region created.
  ```

A region is created.

Rulesurf

We now go over the family of "surfs." These are all legacy commands that perform different varieties of surfacing, depending on the given starting conditions.

Rulesurf stands for "ruled surface" and is a handy tool to create surfaces between various straight lines or curves (without cross sections) that are not necessarily straight or parallel to each other or even closed. It fills a niche somewhere between the loft and planar surface options. To try it out, create two lines at a random angle to each other, as seen in Figure 27.4.

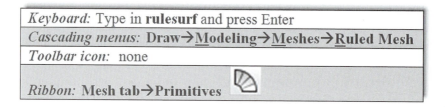

Keyboard: Type in **rulesurf** and press Enter
Cascading menus: Draw→Modeling→Meshes→Ruled Mesh
Toolbar icon: none
Ribbon: Mesh tab→Primitives

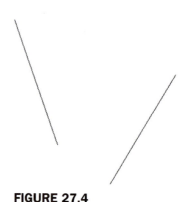

FIGURE 27.4
Random lines for rulesurf.

Step 1. Start the rulesurf command via any of the preceding methods.
- ○ AutoCAD says:
```
_rulesurf
Current wire frame density: SURFTAB1=6
Select first defining curve:
```
Step 2. Select the first line, as seen in Figure 27.5.

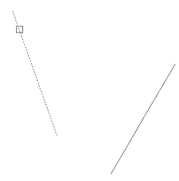

FIGURE 27.5
Select first line for rulesurf.

Step 3. Select the second line at roughly the same upper location (if you select a bottom location the surface self-intersects). The shaded and colored results are shown in Figure 27.6.

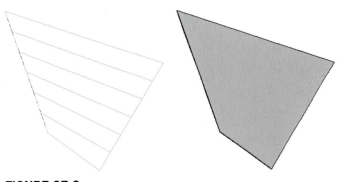

FIGURE 27.6
Final result for rulesurf.

589

As with many of these surfacing techniques, at first, the results may not seem very useful, but with some imagination and creativity, you can use them to fulfill rather complex surfacing needs. For example, in Figure 27.7, we create two arcs, rotate one of them 90°, and connect them end to end.

You can then apply rulesurf to these arcs and get a unique result as seen in Figure 27.8. The arcs became one surface. It was then colored, shaded and rotated for a different view. The density of the mesh can be set by the SURFTAB1 and SURFTAB2 system variables by simply inputting a higher value.

Tabsurf

Tabsurf stands for "tabulated surface" and is a surfaces version of the extrude command, with some elements of path extrusion mixed in. The command requires a path curve and a direction vector. To try it out, create a circle with a perpendicular line, as seen in Figure 27.9.

Keyboard: Type in **tabsurf** and press Enter
Cascading menus: **Draw→Modeling→Meshes→Tabulated Mesh**
Toolbar icon: none
Ribbon: **Mesh tab→Primitives**

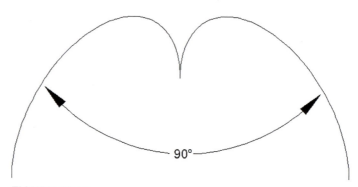

FIGURE 27.7
Perpendicular arcs.

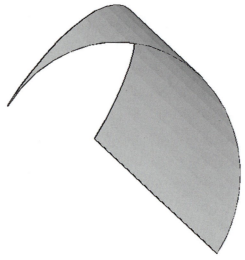

FIGURE 27.8
Final result for rulesurf of arcs.

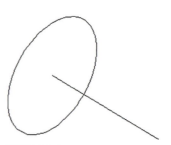

FIGURE 27.9
Circle and line for tabsurf.

590

Step 1. Start the tabsurf command via any of the preceding methods.
- ○ AutoCAD says: `Current wire frame density: SURFTAB1 = 50`
  `Select object for path curve:`

Step 2. Select the circle.
- ○ AutoCAD says: `Select object for direction vector:`

Step 3. Select the line.

The result, shown in Figure 27.10, is a surface projected along the direction vector, not unlike path extrusion or regular extrude, with the difference, however, that the object is not a solid but a surface (be sure to set the SURFTAB1 and SURFTAB2 system variables to a high value for smoothness). A similar example with an I-beam is shown in Figure 27.11.

FIGURE 27.10
Final result for tabsurf.

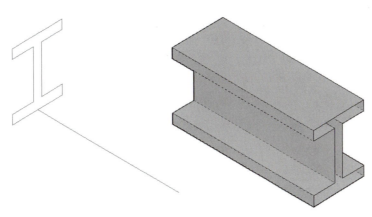

FIGURE 27.11
Tabsurf example for I-beam.

The final tabsurf example is slightly more complex. In 2D, you need to draw a string of connected arcs, all pedited together, as seen in Figure 27.12. Go into 3D and rotate the string of arcs into a straight up position if need be. Then, draw a perpendicular direction vector, attached to one end of the arc string, as seen in Figure 27.13, and finally tabsurf the shape, as seen in Figure 27.14. These exercises should generate some ideas as far as applying the techniques to actual design work.

FIGURE 27.12
Arcs in 2D.

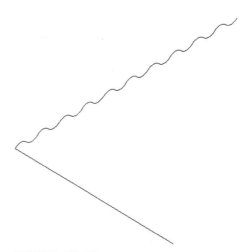

FIGURE 27.13
Arcs with a direction vector in 3D.

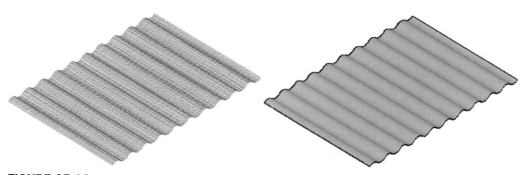

FIGURE 27.14
Tabsurf example completed.

Revsurf

Revsurf stands for "revolved surface" and is a close relative of the revolve command covered in earlier 3D chapters, with the only difference in many cases being that revsurf produces somewhat smoother curves, if the SURFTAB1 and SURFTAB2 system variables are set high enough (around 50 is a good starting point). Generally, you can treat this command as an alternative to revolve. To illustrate it, create the profile seen in Figure 27.15—it is similar to the cup profile used with revolve. Be sure to pedit the profile.

Keyboard: Type in **revsurf** and press Enter
Cascading menus: **Draw→Modeling→Meshes→Revolved Mesh**
Toolbar icon: none
Ribbon: **Mesh tab→Primitives** 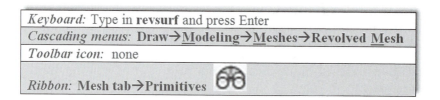

Step 1. Start the revsurf command via any of the preceding methods.
  ○ AutoCAD says: `Current wire frame density: SURFTAB1 = 50 SURFTAB2 = 6`
    `Select object to revolve:`
Step 2. Select the profile.
  ○ AutoCAD says: `Select object that defines the axis of revolution:`

FIGURE 27.15
Profile for revsurf.

Step 3. Select the rotation axis (the vertical line).
- ○ AutoCAD says: `Specify start angle <0>:`

Step 4. Press Enter.
- ○ AutoCAD says: `Specify included angle (+=ccw, −=cw) <360>:`

Step 5. Key in any value or press Enter for the full 360°. The result, shaded and colored, is shown in Figure 27.16.

FIGURE 27.16
Revsurf example
completed.

Edgesurf

The edgesurf command creates a Coons surface patch mesh or a mesh in the M and N directions between a set of edges that are attached to each other with endpoints in a closed loop. The hard part is of course properly defining these edges. Once that is done correctly, AutoCAD fills in the rest and a sophisticated surface is created.

For a good mesh rendering, you need to make it dense using the SURFTAB1 and SURFTAB2 system variables set to 50 or more (too much, though, slows down your system's performance). Now let us try to create a basic mesh as an example. In 2D, draw two arcs and pedit them as seen in Figure 27.17.

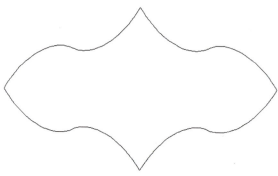

FIGURE 27.17
Two connected arcs for edgesurf.

Now switch to 3D, and using the copy and rotate3D commands, arrange them in a closed loop, with each side 90° perpendicular to its neighbor, as seen in Figure 27.18.

FIGURE 27.18
Four connected arcs for edgesurf.

Keyboard: Type in **edgesurf** and press Enter
Cascading menus: **Draw→Modeling→Meshes→Edge Mesh**
Toolbar icon: none
Ribbon: **Mesh tab→Primitives** 

Step 1. Start the edgesurf command via any of the preceding methods.
- ○ AutoCAD says: `Current wire frame density: SURFTAB1=50 SURFTAB2=50`
  `Select object 1 for surface edge:`
Step 2. Select the four edges, one after the other.
- ○ AutoCAD says:
  `Select object 2 for surface edge:`
  `Select object 3 for surface edge:`
  `Select object 4 for surface edge:`
Step 3. After the final edge is selected, you see the mesh image shown on the left in Figure 27.19. On the right, it is shaded, colored, and slightly rotated for a better view.

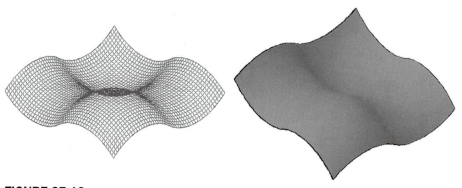

FIGURE 27.19
Final result for edgesurf.

Creating these meshes can be time consuming and occasionally troublesome, but if you follow the basic rules of pediting all the curves (edges) and connecting them together, you should be fine. The sophistication and versatility of this command is rather impressive, and almost any curve can be represented if you can find a clever way to create the necessary edges. The only limitation is that you are limited to four edges at a time, but by connecting separate meshed objects together, you can get around this.

3Dface and 3Dmesh

Our final brief discussion on legacy mesh commands focuses on free-form meshes, ones that require no cross section profile or framework. However, just because the framework is not required, does not mean it is not necessary to a meaningful design. The 3Dface command is actually the one referred to in the chapter introduction as a way to dress up a "framework." The idea is to simply click your way around points in the design, and when AutoCAD senses a closed path, it installs a surface.

Let us give this a try on a random set of surfaces. The design shown in Figure 27.20 can be created by carefully rotating the UCS icon while drawing straight lines to establish a "skeleton" shape. Then, using grips, you can move the intersections of lines together to other intersections. Finally you can add new lines by just drawing from endpoint to endpoint. Try to create a number of faces that can be filled in. Your shape need not look exactly like this one.

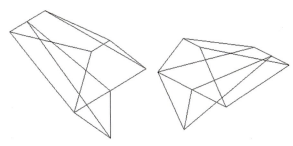

FIGURE 27.20
Angular shapes for 3Dface.

595

The way to activate this command is via typing or cascading menu.

Step 1. In 3D mode, with shading and Ortho on, type in `3dface` and press Enter, or use the cascading menu **Draw→Modeling→Meshes→3D Face.**
- AutoCAD says: `3dface Specify first point or [Invisible]:`
Step 2. Using OSNAP endpoints, click around each facet of the shape. After each click,
- AutoCAD says: `Specify second point or [Invisible]:`
  `Specify third point or [Invisible]`
Step 3. After three or four clicks (depending on which facet you chose to do first) you should be able to get what is seen in Figure 27.21. Continue covering the other faces, and add color, to get what is shown in Figure 27.22 (viewed from several angles).

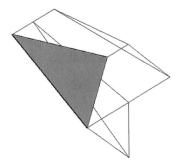

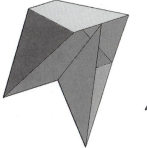

FIGURE 27.21
3Dface.

FIGURE 27.22
3Dface completed.

Little is said of 3Dmesh. This command is a good subject for programming and automation. The 3Dmesh command asks you for the size of the mesh in the M and N directions and location (vertex) points. These can be entered manually or as points on a framework. Using AutoLISP, a programmer can automate meshing of complex surfaces, but this is beyond the scope of the book.

The 3Dface and 3Dmesh commands conclude a basic overview of legacy commands, ones that have been available for years. One of the goals of last year's AutoCAD 2011, and continuing with AutoCAD 2012 is to introduce new mesh tools, which we now look at. These tools involve "sculpting" a shape, which is a popular way to create meshes in advanced 3D solid modeling software, as mentioned at the start of this chapter.

27.3 SMOOTH MESH PRIMITIVES

The way to start "sculpting" a shape is to first create one that has built-in mesh surfaces. This can be easily done via smooth mesh primitives. Primitives, as introduced in Chapter 22, are simple solid shapes, such as a box, sphere, and cone. You again create these shapes, but now they have mesh surfaces that can be manipulated into complex shapes. To create mesh primitives, you can use the Ribbon, the Smooth Mesh Primitives toolbar (Figure 27.23), or cascading menus, as seen in the command matrix.

FIGURE 27.23
Smooth Mesh Primitives toolbar.

Keyboard: none
Cascading menus: Draw→Modeling→Meshes→Primitives→Box
Toolbar icon: **Smooth Mesh Primitives** toolbar
Ribbon: **Mesh tab→Primitives**

The steps to create the box are similar to the regular primitives box (review Chapter 22 if needed), but the results, viewed in wireframe, are definitely not the same, as seen in Figure 27.24.

The primitive is now composed of multiple faces that can be shaped and morphed into the desired final shape.

You can alter the number of faces (technically called *tessellation divisions*) the primitive has via the Mesh Primitive Options dialog box (Figure 27.25). It can be accessed via the small arrow at the bottom right of the Ribbon's Mesh→Primitives tab.

Select the shape with which you want to work and change the tessellation values, that is, length, width, or height, thereby creating more faces. Having more working surfaces on a shape yields smoother shapes and more "control points" with which to alter the meshes. It also allows for more accuracy in shaping the part. The trade-off is more complexity and possible introduction of errors, as well as an extra load on computing resources if the meshes are extremely dense.

How dense to make the mesh grip is a common engineering dilemma in finite element analysis (FEA) for stresses and loads and computational fluid dynamics (CFD) for fluid flow modeling. With modern computers, taxing the resources is less of an issue but can still be a

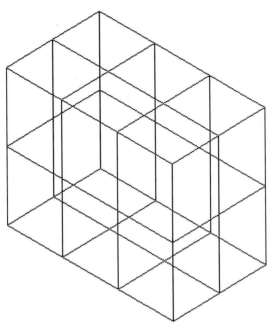

FIGURE 27.24
Smooth Mesh Primitives box.

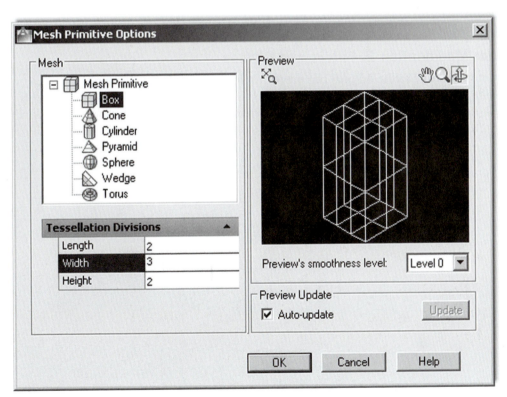

FIGURE 27.25
Mesh Primitive Options dialog box.

problem if you add one too many digits to the density value. Since you may be working with individual mesh facets in this case, make the overall grid only as dense as you, the user, not the computer, can comfortably handle. This can also be automated via the mesh smoothing command as described next.

Mesh Modification 1. Smoothness

You can smooth out the primitive shapes automatically via the mesh smoothing command, best accessed via the Ribbon's Mesh tab. Your choices are Smooth More (increases the number of facets in the mesh), Smooth Less (decreases the number of facets in the mesh), or Refine Mesh (same as Smooth More but works only after that command is employed). Take a few minutes to try out the results on the box. After one Smooth More, followed by a Refine Mesh, you get the modified version of the shape shown in Figure 27.26, as seen in colored wireframe and realistic visual style mode side by side. Notice the marked increase in facets.

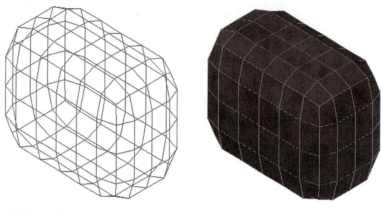

FIGURE 27.26
Smooth and Refined Mesh.

Mesh Modification 2. Filters and Gizmos

Return your shape to the original box via the undo command. Next, we explore the face filter and the move and rotate gizmo (first seen in Chapter 23). From Mesh tab→Subobject, select Move Gizmo followed by Face Filter. Then, click on the top right two facets, as seen in Figure 27.27.

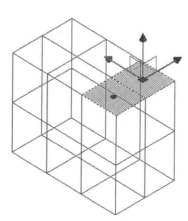

FIGURE 27.27
Facets selected.

The mesh facets are highlighted and the move gizmo appears. With the UCS icon aligned vertically and Ortho on, move the facets up, noticing how the entire shape is "morphed," as seen in Figure 27.28.

In a similar manner, select the rotate gizmo and rotate the facets to a different angle, as seen in Figure 27.29. After a few rounds of adding smoothness and another color change, the shape is presented in Conceptual Visual Style (Figure 27.30).

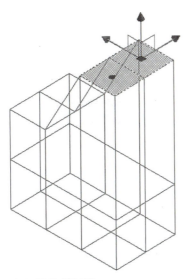

FIGURE 27.28
Facets removed.

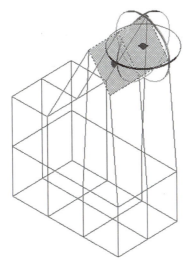

FIGURE 27.29
Facets rotated.

FIGURE 27.30
Final mesh shape.

Mesh Modification 3. Additional Tools

Using smoothness, filters, and gizmos, you can already do quite a bit of shape morphing. Here are some additional tools to consider:

- **Collapse Face or Edge.** This tool, found under the Ribbon's Mesh→Mesh Edit, makes a selected face or edge disappear on a shape, with the remaining ones closing ranks around the void, as seen in Figure 27.31 with a mesh box.

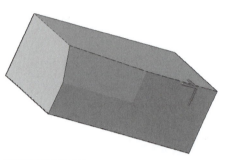

FIGURE 27.31
Collapsed face.

- **Convert to Solid or Surface.** This tool, found under the Ribbon's Mesh→Convert Mesh, converts mesh objects to solids or surfaces. You do not see a tremendous difference in appearance, but the shape is editable in a different way from an organic mesh. For example, individual facets can no longer be selected in a solid.
- **Section Plane.** This tool, found under the Ribbon's Mesh→Section, allows you to easily section a mesh shape to view the interior or just a piece of it. This tool can be used on regular solids as well. Simply press Section Plane and select a method of locating the plane. The result of sectioning a simple box is shown in Figure 27.32. The section itself can be rotated via the gizmo.

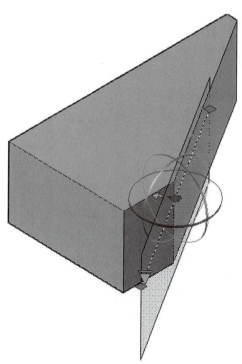

FIGURE 27.32
Section Plane.

SUMMARY

You should understand and know how to use the following concepts and commands before moving on to Chapter 28:

- Planar surface
- Region
- Rulesurf
- Tabsurf
- Revsurf
- Edgesurf
- 3Dface
- 3Dmesh
- Smooth Mesh Primitives
- Mesh smoothing
- Face move and rotate
- Collapse Face or Edge
- Convert to Solid or Surface
- Section Planes

REVIEW QUESTIONS

Answer the following based on what you learned in Chapter 27:

1. Describe the planar surface command.
2. Describe the region command.
3. Describe rulesurf and its applications.
4. Describe tabsurf and its applications.
5. Describe revsurf and its applications.
6. Describe edgesurf and its applications.
7. Describe 3Dface and 3Dmesh.
8. What are the Smooth Mesh Primitives?

601

EXERCISE

1. Based on what you learned of 3D design and the edgesurf command, create the mesh chair that follows (shown from two angles). You need to create a frame in 3D space, attach arcs in strategic locations, and finally execute the edgesurf command for the mesh. A visual step-by-step breakdown is shown for clarity, but you still need to figure out the steps involved. One possible approach is to draw a wireframe 3D box (not a solid) and use that as a reference point to attach the frame. (Difficulty level: Advanced; Time to completion: 30 minutes)

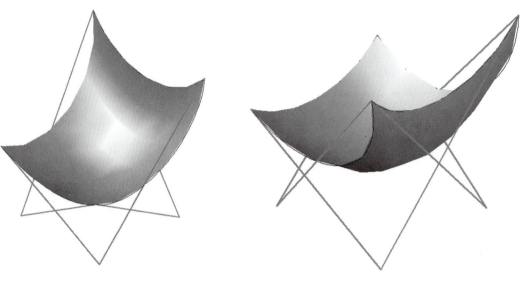

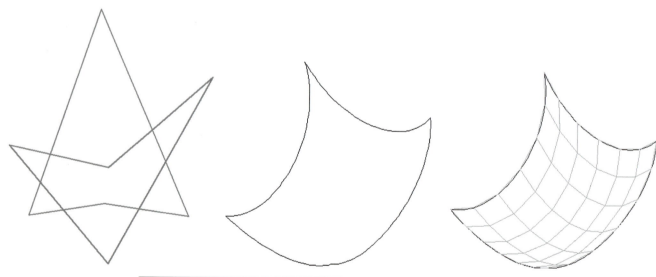

Spotlight On:
Civil Engineering

Civil engineering specializes in design and construction of bridges, roads, buildings, dams, canals, and other infrastructure. Civil engineering is the oldest of the engineering specialties, and some of its principles were already well established during the building of the pyramids in ancient Egypt and aqueducts in Rome. Civil engineering has a number of subspecialties, including environmental engineering, geotechnical engineering, structural engineering, transportation engineering, municipal or urban engineering, construction, and water resources.

FIGURE 1

FIGURE 2

Education for civil engineers starts out similar to that for the other engineering disciplines. In the Unites States, all engineers typically have to attend a four-year ABET-accredited school for their entry level degree, a Bachelor of Science. While there, all students go through a somewhat similar program in their first two years, regardless of future specialization. Classes taken include extensive math, physics, and some chemistry courses, followed by statics, dynamics, mechanics, thermodynamics, fluid dynamics, and material science. In their final two years, engineers specialize by taking courses relevant to their chosen field. For civil engineering, this includes courses in asphalt, concrete, soil, geotechnical, water resources, traffic engineering, seismic, bridge, and advanced structural design.

Upon graduation, civil engineers can immediately enter the workforce or go on to graduate school. Many civil engineers choose to pursue a Professional Engineer (P.E.) license. The process first involves passing a Fundamentals of Engineering (F.E.) exam, followed by several years of work experience under a registered P.E., and finally sitting for the P.E. exam itself. The P.E. license allows the civil engineer to approve and sign off on project plans, and many engineers offer their consulting services to architects on large building projects, where the expertise of a civil engineer is a major requirement.

Civil engineers can generally expect starting salaries (with a bachelor's degree) in the $50,000–$55,000/year range, which is somewhat in the lower to mid-range among engineering specialties. This of course depends highly on market demand and location and is due to the large number of civil engineering graduates in the workforce. The highest starting pay of all ($80,000+) is for petroleum engineering graduates. This specialty is a close cousin of civil engineering that specializes in locating and extracting petroleum. A master's degree is highly desirable and required for many management spots. The job outlook for civil engineers is excellent, as the profession is very much needed as the world's infrastructure ages and population growth creates a need for new development.

How do civil engineers use AutoCAD and what can you expect? Industry wide, AutoCAD enjoys a significant amount of use among civil engineers, as many civil projects are 2D in nature. This is especially true with site plan work and some building design. Several examples of civil site plans are presented here, many of which can be quite complex and detailed in nature. They use xrefs (for the building footprints), plines (for the gradients), and other advanced tools.

As you move more into classic civil engineering projects (bridges, roads, etc.), 2D AutoCAD gives way to dedicated 3D software applications specially geared to this type of work, such as MicroStation, GEOPAK, and InRoads from Bentley. When AutoCAD is used, it is sometimes paired with additional software like Civil 3D, Land Desktop, or various GIS add-ons.

Layering in civil engineering work is highly specialized and depends on the project. Site plan work may include some AIA layers, but mostly they are descriptive and unique to the company; there is no universal standard. Road, Easement Line, Gas Line, and Property Line are all valid layer names. A typical site plan is shown in Figure 3.

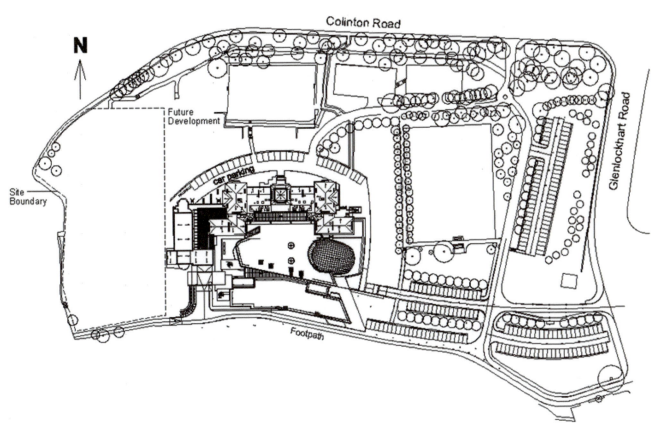

FIGURE 3
An example of a 2D civil engineering site plan.

The figure features the architectural layout (footprint), parking spots, landscaping, and site boundaries among other items. Depending on the plan, utility and power lines may be shown, as well as outdoor lighting, gradient lines, and various legal boundaries, such as easements and offsets. A similar type of plan is shown in Figure 4.

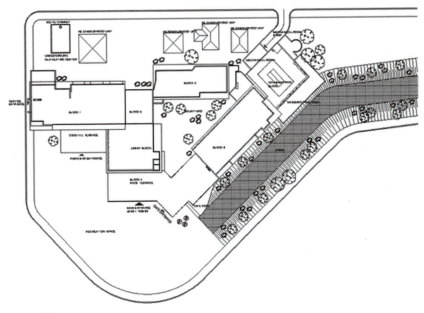

FIGURE 4
Another example of a 2D civil engineering site plan.

CHAPTER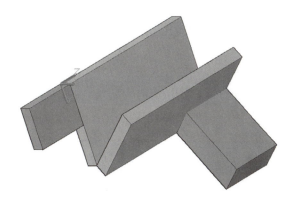

UCS, Vports, Text, and Dimensions in 3D

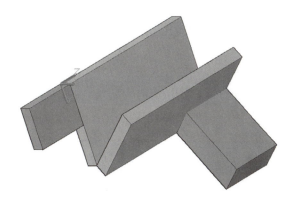

607

LEARNING OBJECTIVES

In this chapter, we complete the design toolbox and expand on your ability to use surfaces of objects to create new shapes. We cover

- Advanced UCS
  - UCS
  - World
  - UCS Previous
  - Face UCS
  - Object
  - View
  - Origin
  - Z-Axis Vector
  - 3 Point
  - X
  - Y
  - Z
  - Apply
- Named UCS
- View
- Vports
- Text and dimensioning in 3D

By the end of the chapter, you will be able to create designs regardless of the surface of orientation.

Estimated time for completion of chapter: 1 hour.

28.1 INTRODUCTION TO UCS, VPORTS, TEXT, AND DIMENSIONS IN 3D

The previous chapter concluded our study of "curved design." As you start this new chapter, you are armed with a formidable set of tools to create virtually any shape or surface required. A summary of your 3D knowledge is as follows.

To operate in 3D space, you learned about

- Axes
- Planes and faces
- UCS rotation
- Visual styles (hide and shade)
- 3D Orbit
- 3D views (Top, SW Isometric, etc.)

To create and modify objects, faces, and surfaces in 3D, you learned about

- Extrude
- Primitives (box, wedge, cone, sphere, cylinder, torus, and pyramid)
- Rotate3D
- Mirror3D
- 3Darray
- 3Dalign
- Fillets and chamfers in 3D
- Boolean operations (union, subtract, and intersect)
- Revolve
- Shell
- Taper
- Loft
- Path extrusion
- Sweep
- Helix
- 3D path array
- Polysolid
- Face operations (delete, move, copy, rotate, offset, and color)
- Edge operations (copy and color)
- Surfaces (planar, region, rulesurf, tabsurf, revsurf, edgesurf, and meshes)

This is it for object creation tools, and in the final three chapters of the book, we move in a different direction. In this chapter, we cover some additional UCS tools that expand the usefulness of that command. We then discuss vports, which is a way to split your screen so your design can be viewed from many angles while working on it and even print it out in such a way. We conclude with some ideas of how to add text and dimensions in 3D.

The next chapter provides an important set of tools under the dynamic view (dview) command that allows you to set perspective views and even cut into designs to reveal internal workings. We then consider how to do basic animations and camera views, and conclude the book with some basics on lighting and rendering.

28.2 ADVANCED UCS

So far the Universal Coordinate System (UCS) icon has been sitting in the background, acting as a reference point when we need to rotate something about an axis or mirror something about a plane. The icon itself is also rotated occasionally when we need to draw into another plane. This is all good, but we can make the icon work even harder for us and add some more usefulness to its repertoire. To investigate it further, bring up UCS's two toolbars: UCS and UCS II, as seen in Figure 28.1. We cover additional icons (and of course their typed, cascading menu, and Ribbon counterparts).

FIGURE 28.1
UCS and UCS II toolbars.

Most of the new commands have something to do with aligning the UCS icon with chosen geometry. This is the central theme in learning UCS's new features. Why is such alignment important? Because this gives you a new "platform" off which to work. Suppose you have an object that is leaning in some direction and you need to draw something else that sits on it. You can draw this new object and then rotate it into position, but a better approach is to right away align it with the tilted surface and use that as a base from which to draw. All this should tie in with the fundamental theory presented way back in Chapter 21 concerning drawing planes and their rotation. Let us jump right in, and you will clearly see what we are talking about very soon.

First of all, though, we need something on which to practice. The various UCS functions are best understood by just trying them out, so we use the shape in Figure 28.2 for all UCS demonstrations. Draw a rectangle of any size and extrude it, then rotate3D it some value (perhaps 20°) and mirror3D it for symmetry. The result in the SW Isometric view is shown in Figure 28.2.

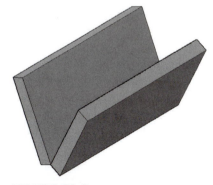

FIGURE 28.2
Basic shapes for UCS practice.

UCS

This first tool presents many options but is primarily used for aligning the UCS via several clicks of the mouse.

609

Keyboard: Type in **ucs** and press Enter
Cascading menus: none
Toolbar icon: **UCS toolbar→**
Ribbon: **Home tab→Coordinates**

Step 1. Start the UCS command via any of the preceding methods.
   ○ AutoCAD says:
     ```
     Specify origin of UCS or
     [Face/NAmed/OBject/Previous/View/World/X/Y/Z/ZAxis] <World>:
     ```
Step 2. Click somewhere on the object to locate the origin (always using OSNAPs).
   ○ AutoCAD says: `Specify point on X-axis or <Accept>:`
Step 3. Here, you can just press Enter, but if you click on another point, the positive X axis of the UCS icon passes through that point.
   ○ AutoCAD says: `Specify point on the XY plane or <Accept>:`
Step 4. Here, you can again press Enter, but if you click on a third and final point somewhere on the object the Y axis passes through it.

Notice how the UCS is now aligned to the points you indicated. We revisit this a bit later with the object and 3 Point options. The basic UCS command can also be used to rotate the UCS around any of the axes, although there are actually dedicated icons, as we explore soon.

World

This next tool should be a familiar one, as it "resets" the UCS icon and was mentioned in earlier chapters. The result is that the UCS is returned to the World view from whatever configuration it may have been in previously. Try it out via the following methods.

Keyboard: Type in **ucs**, press Enter, then Enter again
Cascading menus: **Tools→New UCS→World**
Toolbar icon: **UCS toolbar→**
Ribbon: **Home tab→Coordinates**

UCS Previous

This next tool simply returns you to whatever UCS configuration you had previously. It is quite useful if you are using just two settings and need to toggle quickly back and forth. Try it out via the following methods.

Keyboard: Type in **ucs**, press Enter, then **p** and Enter
Cascading menus: **Tools→New UCS→Previous**
Toolbar icon: **UCS toolbar→**
Ribbon: **Home tab→Coordinates**

Face UCS

This next tool is one of two very important alignment tools (the other being 3 Point) that are the core of this section on advanced UCS. Here, you can align your UCS to any face on the screen as long as it is flat. The major advantage here of course is that you do not need to know the angle of this face.

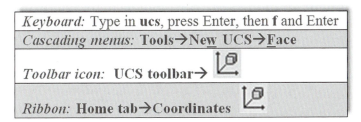

Keyboard: Type in **ucs**, press Enter, then **f** and Enter
Cascading menus: **Tools→New UCS→Face**
Toolbar icon: **UCS toolbar→**
Ribbon: **Home tab→Coordinates**

Step 1. Start the command via any of the preceding methods.
- ○ AutoCAD says:
  ```
  Current ucs name: *NO NAME*
  Specify origin of UCS or
  [Face/NAmed/OBject/Previous/View/World/X/Y/Z/ZAxis]
  <World>: _fa
  Select face of solid object:
  ```
Step 2. Click any face to which you need to align, and the face is highlighted.
- ○ AutoCAD says: `Enter an option [Next/Xflip/Yflip] <accept>:`

You can easily change the direction of the UCS icon while keeping the 0,0,0 origin in one spot by using the `Xflip`, `Yflip`, or `Next` options before pressing Enter to accept.

Object

This next tool is an interesting variation on Face UCS. Here, you can select almost any object and the UCS icon aligns itself to that object in various preset ways. With an arc or circle, it is the center; with a line, it is a point closest to where you clicked; and so forth.

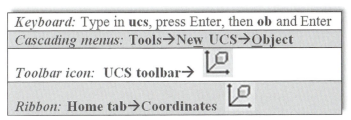

Keyboard: Type in **ucs**, press Enter, then **ob** and Enter
Cascading menus: **Tools→New UCS→Object**
Toolbar icon: **UCS toolbar→**
Ribbon: **Home tab→Coordinates**

A full listing of how it reacts is in the Help files. In Figure 28.3, we see the UCS icon aligned to a few of these objects.

View

This next UCS tool establishes a new coordinate system with the XY plane perpendicular to your viewing direction (parallel to your screen).

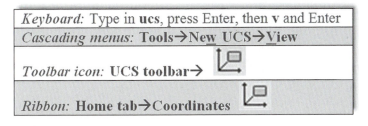

Keyboard: Type in **ucs**, press Enter, then **v** and Enter
Cascading menus: **Tools→New UCS→View**
Toolbar icon: **UCS toolbar→**
Ribbon: **Home tab→Coordinates**

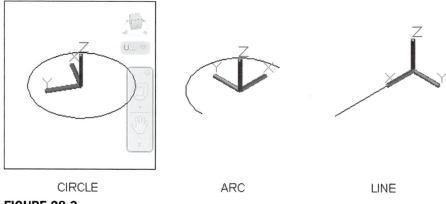

CIRCLE   ARC   LINE

FIGURE 28.3
Object UCS alignment.

This is equivalent to going to the Top View while still in 3D, as seen in the before and after screen shots of Figure 28.4. This technique will probably not be used too often but *is* the only way to accomplish this if needed.

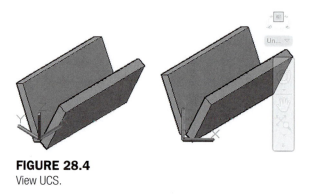

612

FIGURE 28.4
View UCS.

Origin

This next tool simply establishes a new origin for the UCS icon.

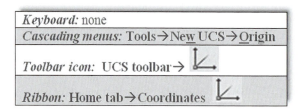

Keyboard: none	
Cascading menus: Tools→New UCS→Origin	
Toolbar icon: UCS toolbar→	
Ribbon: Home tab→Coordinates	

Simply start the command and click anywhere on or near the object; the UCS moves its origin there, as seen in Figure 28.5. Be sure that the UCS icon `ORigin` option is on, or the icon does not move anywhere.

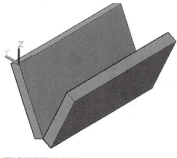

FIGURE 28.5
Origin UCS.

Z-Axis Vector

The next tool simply points the Z axis in a specified direction and is just another way to orient the UCS icon.

Keyboard: Type in **ucs**, press Enter, then **za** and Enter	
Cascading menus: **Tools→New UCS→Z Axis Vector**	
Toolbar icon: **UCS toolbar→**	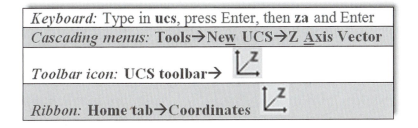
Ribbon: **Home tab→Coordinates**	

3 Point

This next tool is the other extremely useful orientation tool (along with Face UCS). It works on the concept of three points being used to indicate a plane. We present a full demonstration of its use. This demonstration also is relevant to the Face UCS method, as it also orients the UCS to a plane, although in a different way.

Keyboard: Type in **ucs**, press Enter, then **3p** and Enter	
Cascading menus: **Tools→New UCS→3 Point**	
Toolbar icon: **UCS toolbar→**	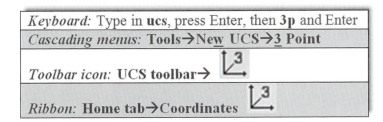
Ribbon: **Home tab→Coordinates**	

Step 1. Start the command via any of the preceding methods.
- AutoCAD says:
  ```
  Specify origin of UCS or
  [Face/NAmed/OBject/Previous/View/World/X/Y/Z/ZAxis] <World>: _3
  Specify new origin point <0,0,0>:
  ```
Step 2. Select the three points in consecutive order to define a plane, as seen in Figure 28.6.
- With each click AutoCAD says:
  ```
  Specify point on positive portion of X-axis <1.00,0.00,0.00>:
  Specify point on positive-Y portion of the UCS XY plane
  0.00,1.00,0.00>
  ```

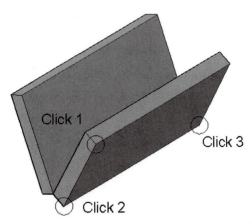

FIGURE 28.6
Refining 3 Point UCS.

When done, the crosshairs and the UCS icon align themselves with the face defined by the three clicks, as seen in Figure 28.7.

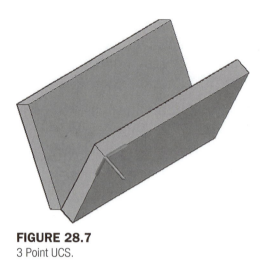

FIGURE 28.7
3 Point UCS.

You can now easily draw something based on that surface, as seen in Figure 28.8. Performing the same 3 point alignment and extrusion on the front surface, we have what is shown in Figure 28.9.

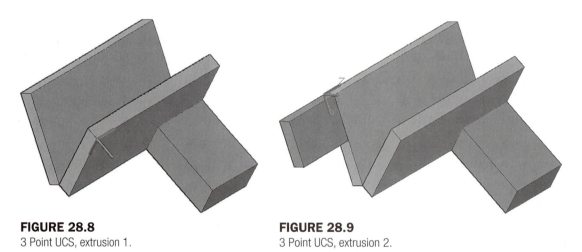

FIGURE 28.8
3 Point UCS, extrusion 1.

FIGURE 28.9
3 Point UCS, extrusion 2.

Note that, as mentioned before, Face UCS works the same way, except that you are just selecting the entire face, not defining it as with 3 Point. But, once you establish the new UCS orientation, proceed in the same manner to create new geometry.

X, Y, and Z

These tools are simply icons for the familiar UCS rotation you learned about in the first few chapters of Level 3. Simply press the icons instead of typing in ucs, then enter the angle for the rotation. That is all you need to do, and the UCS rotates accordingly.

Keyboard: Type in **ucs**, press Enter, then **x**, **y** or **z** and Enter	
Cascading menus: **Tools→New UCS→X, Y or Z**	
Toolbar icon: **UCS toolbar→**	
Ribbon: **Home tab→Coordinates**	

Taking a look next at the UCS II toolbar, we examine two icons, the Named UCS and the drop-down window next to it. The first icon (UCS) is the same as the one on the previous main UCS toolbar, and we do not go over it again.

Named UCS

If you are going to spend some time with a UCS configured in a certain useful way, you may want to save it in case you need it again. AutoCAD allows you to save as many custom UCS presets as you want. To save a view, do the following.

Step 1. Type in UCS and press Enter.
  ○ AutoCAD says: `Specify origin of UCS or [Face/NAmed/OBject/Previous/ View/World/X/Y/Z/ZAxis]<World>:`
Step 2. Type in `save` and press Enter. The Save option does not appear in the preceding menu, but it will work.
  ○ AutoCAD says: `Enter name to save current UCS or [?]`
Step 3. Enter a name for your view, which in this case will just be a generic `MyView`, and press Enter.

The new view is now in AutoCAD's memory, saved for future use. It can be found in the UCS dialog box, which is described next.

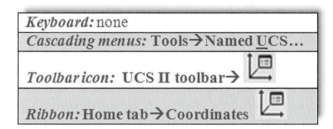

The dialog box (Figure 28.10) can be called up via any of the preceding methods. Notice the presence of the highlighted MyView UCS setting. You can go to other settings and always go back to it later.

The UCS setting also appears in the toolbar drop-down menu, as seen in Figure 28.11. This drop-down menu is a convenient way to quickly access not only your custom UCS settings but also the preset ones, including World and Previous.

28.3 VIEWS AND VIEW MANAGER

In a closely related topic to saving preset UCS views, you can also save general views of your design. This is a very convenient tool if you find a good view with which to work and may want to come back to it later. The views are saved and managed through a View Manager dialog box. Access the command through any of the following methods:

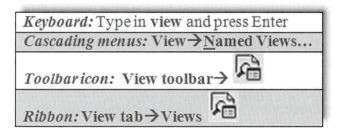

You then see the dialog box in Figure 28.12. Press the New… button and the New View/Shot Properties dialog box appears (Figure 28.13). We are not concerned with the Shot Properties

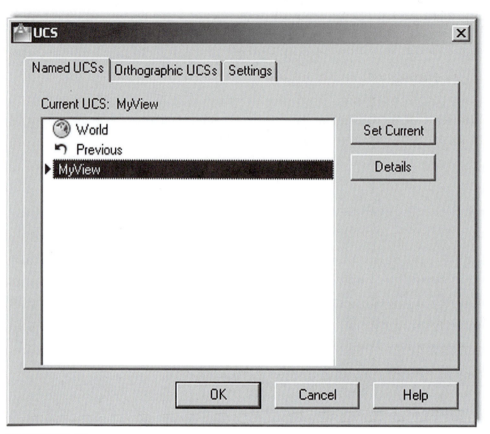

FIGURE 28.10
Named UCS.

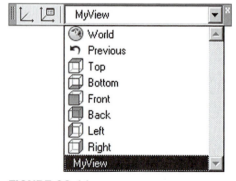

FIGURE 28.11
Named UCS toolbar drop-down menu.

tab for now and just focus on the View Properties tab. Enter a descriptive view name in the first field and press OK. The name then appears in the View Manager. To retrieve the view, just bring up the View Manager again and double-click on the view you want retrieved.

28.4 VPORTS

Vports, which is short for *viewports*, is a unique tool in AutoCAD that allows you to view the same model from multiple angles all at once. It splits the screen in one of the several available ways that best fits the situation. Your design then appears in all screens, and you can assign each screen its own view and properties, such as shade or wireframe mode. To access this tool, let us first bring up a simple 3D model, such as the wastebasket created in Chapter 25, shown again in Figure 28.14.

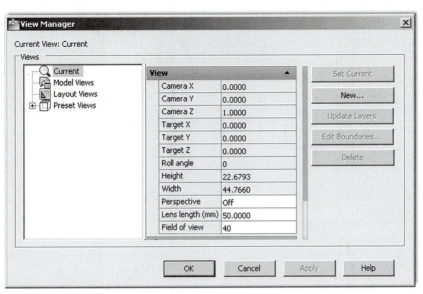

FIGURE 28.12
View Manager.

FIGURE 28.13
New View/Shot Properties.

617

You can access vports via any of the following methods:

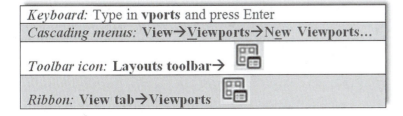

Keyboard: Type in **vports** and press Enter

Cascading menus: View→Viewports→New Viewports...

Toolbar icon: **Layouts toolbar→**

Ribbon: **View tab→Viewports**

FIGURE 28.14
Wastebasket model.

This brings up the dialog box seen in Figure 28.15.

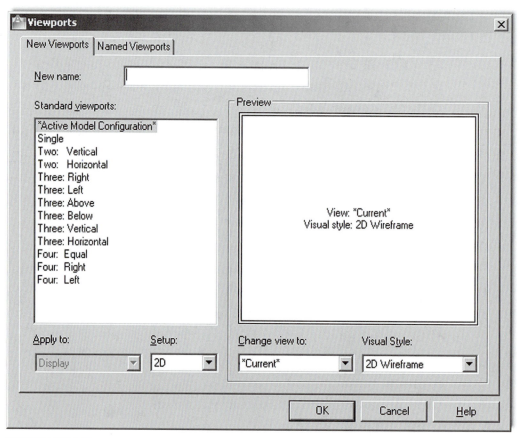

FIGURE 28.15
Viewports.

As you can see from the configuration choices on the left, you have a few ways to arrange the screen. The default view is always Single, and that is the choice you will select later on, when you want to get back to the familiar single screen. For now choose the Four: Equal configuration and press OK. You should see what is shown in Figure 28.16. Notice that the view of the wastepaper basket has been duplicated four times and has not undergone any further changes. Keep in mind that only the view has been duplicated, not the actual 3D model.

You may have to zoom to extents in each viewport to get the model to take up all the available area.

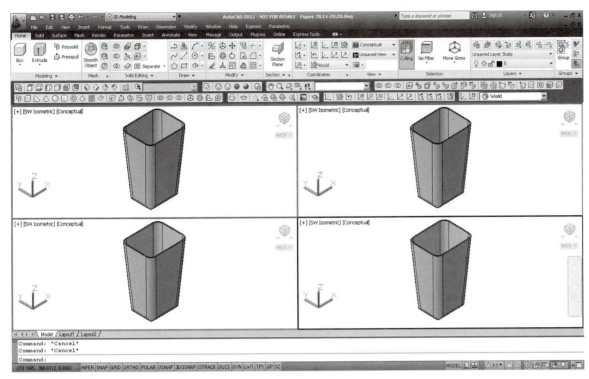

FIGURE 28.16
Four: Equal split.

You can now easily jump from screen to screen by just clicking once in it. Go ahead and visit each one and modify the views to a combination of top, side, and isometric, as seen in Figure 28.17.

619

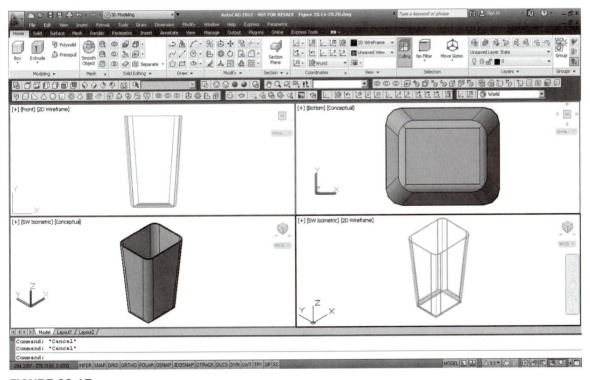

FIGURE 28.17
Four: Equal split with view modifications.

Now, try working in one viewport and notice that anything you do in one is reflected in the others, as it should be, since it is just one model with which you are working. Another popular configuration, Four: Left, is shown in Figure 28.18. Here, you have one primary large view and three supplemental views to see other parts of the model.

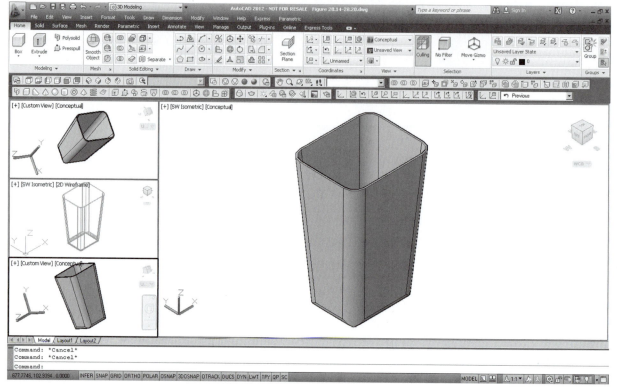

FIGURE 28.18
Four: Left split.

How you use these vports (or whether you use them at all) is up to you and your style of working in 3D. In some cases, they are very useful; in others, it is best to just rotate the model as you need and not maintain a constant view.

Another use for these vports is in plotting out the design for presentations, and it is here that this command gets a chance to really shine. While printing all the vports in Model Space is not possible, if you switch to Paper Space and repeat the vports Four: Equal command, you get what is shown in Figure 28.19. Note that you have to actually draw the area where you

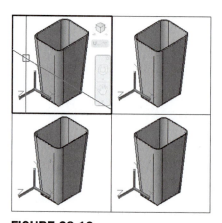

FIGURE 28.19
Four: Equal in Paper Space.

want the vports to go before they appear. Zoom to extents in each one to arrive at what is shown in the figure.

Each view is still active and can be modified, the same as in Model Space, but now also the overall scene with all the viewports can be plotted or printed for an impressive display.

Finally (and this is purely for entertainment purposes), you can split the screen into a maximum of 64 vports, as seen in Figure 28.20. The standard vports dialog box does not let you do this, but type in -vports (a dash means it is command line only) and press Enter.

○ AutoCAD says: Enter an option[Save/Restore/Delete/Join/SIngle/?/2/3/4] <3>:

Enter the value 4 to split the screen into four views. Then, click into each new view and split it again in the exact same manner. The final result is shown in Figure 28.20. In case you are wondering, no, this does not serve any real purpose.

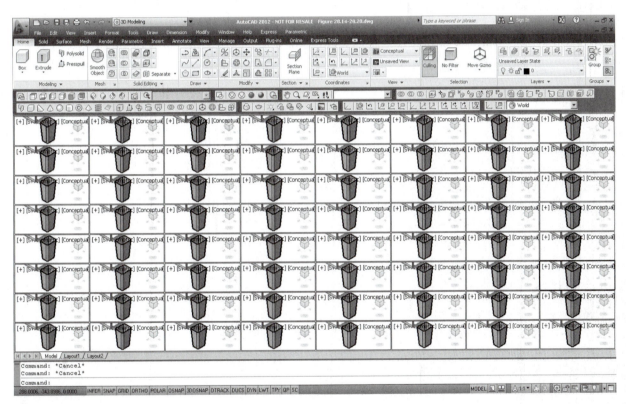

FIGURE 28.20
64 vports.

28.5 TEXT AND DIMENSIONS IN 3D

Creating text and dimensions in 3D is not necessarily any different than in 2D; you just use the standard text, mtext, or ddim commands. However, placing them correctly takes some practice because, in 3D, what is up or down is not always clear. You have to pay very close attention to what plane you are on and which way is "up," otherwise the dimensions and text appear in the wrong place or the text is inverted. Let us run through a few examples. In each of these cases, pay close attention to everything shown, such as the position of the UCS icon (and the associated crosshairs), as well as the positioning of the dimension and text.

The easiest case is to use the standard World UCS view (SW Isometric) that you are accustomed to in this course. If the Z axis is pointed up, then the dimensions and text come out as shown in Figure 28.21.

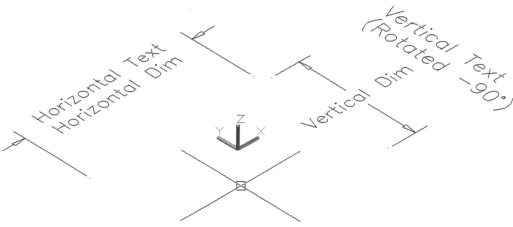

FIGURE 28.21
3D dimensions and text, Example 1.

Now, what would happen if we reversed the orientation of the Z axis and it faced down instead? As Figure 28.22 shows, if we attempt to then add dimensions and text (horizontal or vertical), we are faced with a complete reversal and mirror imaging of the text and dims, which is not the desired outcome.

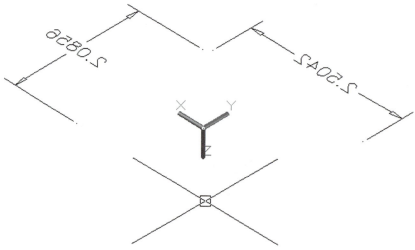

FIGURE 28.22
3D dimensions and text, Example 2.

As this example shows, you have to be very careful about flipping the UCS icon around when dimensioning, as the results may be unpredictable (or at least seem that way). What happens in this example is that you are writing on the other side of the face. So, how do we add dimensions along the Z axis? The next example addresses this.

Let us bring the Z axis back to World view and rotate the UCS icon 90° along the X axis, as seen in Figure 28.23. Notice the effect this has on horizontal and vertical dimensions.

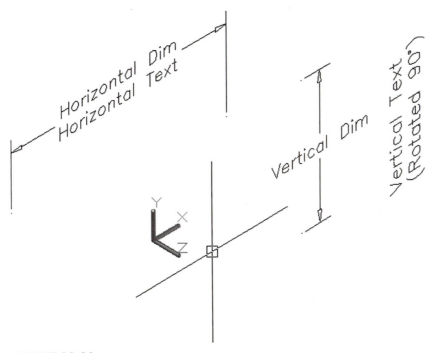

FIGURE 28.23
3D dimensions and text, Example 3.

With these two UCS settings you should be able to dimension just about anything, as the dimensioned cube in Figure 28.24 indicates, although do not be surprised to run into some difficulties if you are dealing with complex rotated surfaces.

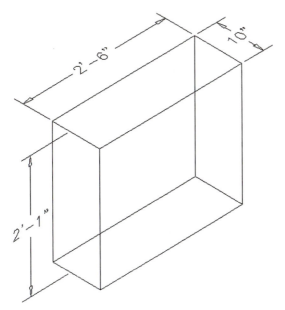

FIGURE 28.24
3D dimensioned cube.

623

SUMMARY

You should understand and know how to use the following concepts and commands before moving on to Chapter 29:

- Advanced UCS
  - UCS
  - World
  - UCS Previous
  - Face UCS
  - Object
  - View
  - Origin
  - Z-Axis Vector
  - 3 Point
  - X
  - Y
  - Z
  - Apply
- Named UCS
- View
- Vports
- Text and dimensioning in 3D

REVIEW QUESTIONS

Answer the following based on what you learned in Chapter 28:

1. List all 13 advanced UCS functions and what they do.
2. What are vports useful for? List two functions.

EXERCISE

1. Create the following fictional suspension and tire 3D model. A breakdown of the parts is shown for clarity, along with hints, but you still need to know how to create each part. Sizing is not specifically given, as the focus is on the commands, but you can base everything on a 2' diameter tire. (Difficulty level: Advanced; Time to completion: 1–2 hours)

Step 1. The tire is an extruded circle with fillets and a subtracted center. The hub is also an extruded circle. The "spokes" are a 3Darray.

Step 2. The struts are all extruded circles. The coil spring is helix and was demonstrated in previous chapters. You can choose which method you want to use.

626

Step 3. The axle and flat panel are extruded shapes. For the axle boot, you need to use revolve.

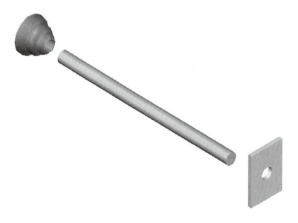

Step 4. The control arms are sweep or path extrusion, your choice. The cylinders are basic extrusions.

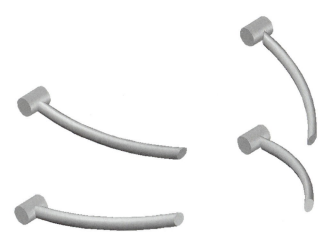

Step 5. The back panel is an easy rectangular extrusion, but the disk brake is more challenging. It requires extrusion, subtraction, 3Darray, and other commands.

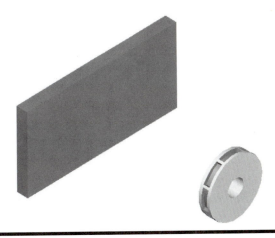

627

CHAPTER 29

Dview, Walk and Fly, Animation, and Action Recording

629

LEARNING OBJECTIVES

In this chapter, we look at advanced presentation views and various animation options. The specific topics are as follows:

- Dview
  - CAmera
  - TArget
  - Distance
  - POints
  - PAn
  - Zoom
  - TWist
  - CLip
  - Hide
  - Off
  - Undo
- Cameras
- Walk and Fly
- Path animation
- Action recording

By the end of the chapter, you will be able to present your model in new and unique ways as well as build basic walk-through animation. You will also be able to build a basic macro to automate repetitive steps by recording your screen actions.

Estimated time for completion of chapter: 2 hours.

29.1 DYNAMIC VIEW

Dynamic view, or dview, is just one command but it has many options. Some of the options are less useful or duplicate existing commands, while a few others are very important and worth a closer look. The full list of dview commands includes `CAmera`, `TArget`, `Distance`, `POints`, `PAn`, `Zoom`, `TWist`, `CLip`, `Hide`, `Off`, and `Undo`.

Out of these options, two are in high demand by AutoCAD designers and worth covering in detail. The first is the `Distance` option, which allows you to set perspective; the other is the `CLip` option, which allows you to set clipping planes and strip away parts of the design to look inside. We of course cover all of the mentioned options, but with a special focus on these two.

Before we try any of these options, you need to get used to something unique to the dview command. In a blank file, go to SW Isometric and turn on shading. Then type in `dview` and press Enter (there is no Ribbon, toolbar, or cascading menu equivalent).

- AutoCAD says: `Select objects or <use DVIEWBLOCK>:`

Here, you are being asked to what you would like to apply dynamic view options. We do not yet have a 3D design so just press Enter; you see the house in Figure 29.1.

630

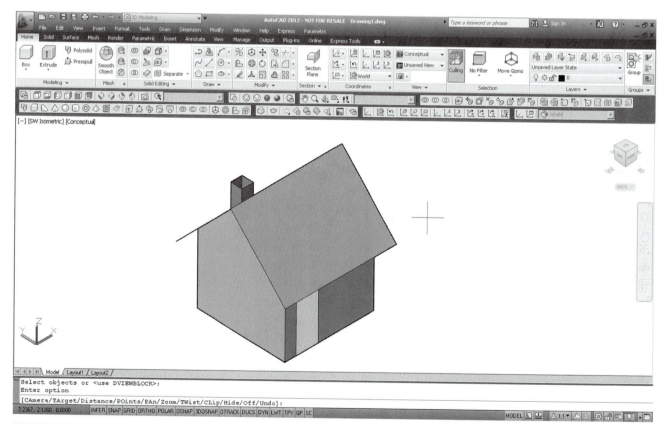

FIGURE 29.1
Dview "house."

This little house appears as a point of reference only if there is nothing else on your screen and is perhaps one of AutoCAD's oddest quirks, prompting many a student to exclaim "what the heck is that" in the middle of class. You can see the effect of whatever you do in dview on this house, but immediately after running the option, the house disappears. In the interest of keeping something permanent on the screen, draw a simple 20" × 5" rectangle and extrude it to 5". Then add a 1" fillet to its sides, as seen in Figure 29.2.

Now we can proceed with learning dview.

FIGURE 29.2
Extruded shape for dview.

CAmera

This command works on the principle of a camera rotating around an object. You are asked to enter the camera angle from the XY plane as well as the angle in the XY plane from the X axis. You are also given the option to use the mouse to just click on a suitable position, which is what most users end up doing.

To try it out, type in dview and press Enter. Select the block when prompted and type in ca for camera. Spin your mouse around and observe the results.

TArget

This command is similar to CAmera, but works on the principle of the target rotating around the camera. It has the same basic options of entering angle values or mouse clicking.

To try it out, type in dview and press Enter. Select the block when prompted and type in ta for target. Spin your mouse around and observe the results.

Distance

We explore this option in more detail. Distance is the method used to set a perspective view. These views of objects are more realistic and feature foreshortened geometry with lines that tend toward a vanishing point, as in reality. For more information on perspective views with detailed descriptions of 0-point, 1-point, 2-point, and 3-point perspectives (AutoCAD uses 3-point), see any graphic design or art textbook. Distance can be applied to any object, but just as in real life, the effect is more pronounced with larger objects.

Also be aware that this view is only for presentations; it cannot be used during construction of the design, as perspective changes dimension values by its very nature. To try out this option type in dview and press Enter as before. Select the object and press Enter. Then type in d for the Distance option. A slider horizontal bar appears at the top of the screen, as seen in Figure 29.3.

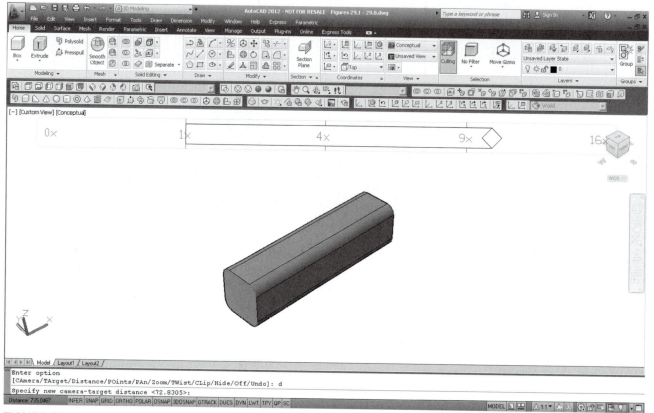

FIGURE 29.3
Distance option.

This yellow and red slider is the camera to target distance meter, with 1" being full size, and increasing to the right (decreasing to the left of the 1"). Slide over to the right until you get to 9" and click. The object remains in perspective mode along with the UCS icon. Press Enter to accept the settings and get out of dview, being careful not to press Esc at this point. You can also set this view as described in an earlier chapter via the menu of the little house icon at the top right of the screen as seen in Figure 29.4.

This perspective view, while not of much use in small part engineering drawings, is quite useful for 3D architectural and civil engineering design presentation. Once in perspective view, print the drawing and, to get out of this view, change to wireframe.

POints

This option allows you to select a specific view based on the locations of three points. The first is the current camera position, and you need to select the second (target point) and third (camera point) points. This tool has some interesting applications when you need to set a specific view, such as looking down a hallway in a building. You can then set your points in strategic locations, such as one end of the hallway for one point and the top of an opposing door for another point. For the most part, however, designers prefer to just use 3D Orbit and zoom or pan to get the right view.

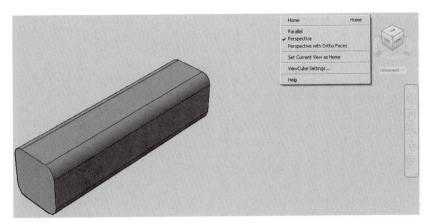

FIGURE 29.4
Perspective view, alternate setting.

PAn

This is just a variation on the basic pan command that shifts the image without changing the level of magnification. Give it a try.

Zoom

This is also just a variation on the zoom command that utilizes the same slider you first saw with the distance option. Give it a try.

TWist

This is another variation of the 3D Orbit command, where the object is rotated or "twisted" around. Give it a try.

CLip

This is the second of two dview commands that deserve a longer discussion. CLip introduces the concept of "clipping planes," which (in various forms) is a common tool in 3D computer-aided design. The idea is to create a flat plane that slices open a design to reveal the internal workings. The plane itself is invisible, only its effect is seen, and can be applied to the front or back of the object. When you reveal as much as you need to, you set the view and print the design. This tool, like most other dview options, is not meant to be permanent and is not for use during design work, only for final presentation.

To illustrate the CLip tool, add some geometry in the form of five 2" diameter cylinders to the inside of the previously used block, as seen in wireframe mode in Figure 29.5. In full shade mode, the inside geometry cannot be seen, as shown in Figure 29.6.

Let us apply a clipping plane to the front and strip away some of the outside material to show the insides. Type in dview, press Enter, and select the object, pressing Enter again. Type in cl for CLip and press Enter.

- AutoCAD says: `Enter clipping option [Back/Front/Off] <Off>:`

Clipping planes can be thought of as flat plates, with one in the front and one in the back. As the object is moved back and forth, the clipping plane cuts in, or clips, the object either from the front or the back. Typically, you use the front clipping plane, so type in f and press Enter.

- AutoCAD says: `Specify distance from target or [set to Eye(camera)/ON/OFF]` `<451.8572>:`

633

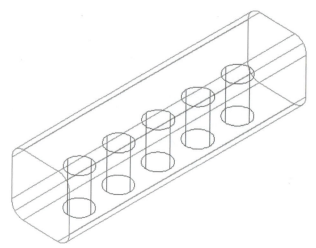

FIGURE 29.5
Internal geometry.

FIGURE 29.6
Shade mode.

634

You also see a yellow slider (similar to the one seen when learning distance), which can be moved back and forth to establish the position of the clipping plane, as seen in Figure 29.7.

FIGURE 29.7
Front clipping plane.

As you can see, the internals of the part can be revealed easily. Figure 29.8 shows a view with the back clipping plane set in a similar manner, though the results are not as useful.

You can actually 3D Orbit through a clipping plane and get a full view of the internal workings of a design from any angle. To remove the planes, you have the Off command in the clipping menu.

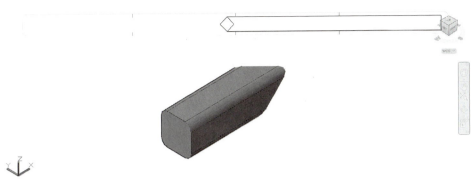

FIGURE 29.8
Back clipping plane.

Hide

This command is essentially the same as the standard hide command learned Chapter 21.

Off

This command turns off perspective viewing, although you can do the same thing by switching to wireframe mode.

Undo

Undo basically undoes the last command in dview and can be used multiple times, just like the regular undo command in 2D AutoCAD.

635

29.2 CAMERAS

The camera tool provides an interesting twist on setting views. The idea is pretty much what it sounds like: You set up a camera at some strategic location and a view is created that looks out through that camera in perspective. This view is saved in the View Manager and you can double-click on it to go into it. You generally stay in that view but are able to move around (rotate, pan, zoom, etc.). These views are used for presenting and plotting a design, and you can easily switch out of them at will. Let us take a look at some of the details.

Step 1. First of all, create a few random blocks of any size, and then shade and color them while in SW Isometric, as seen in Figure 29.9.

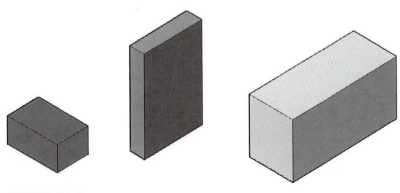

FIGURE 29.9
Blocks for cameras.

Keyboard: Type in **camera** and press Enter	
Cascading menus: **View→Create Camera**	
Toolbar icon: **View toolbar→**	
Ribbon: none	

Step 2. Start the camera command via any of the preceding methods.

    ○ AutoCAD says: `Current camera settings: Height = 4.0000 Lens Length = 100.0000mm` `Specify camera location:`

Your values may of course be different, but they all can be adjusted. You also see a small camera symbol attached to the mouse, as shown in Figure 29.10.

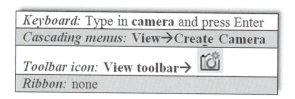

FIGURE 29.10
Camera symbol.

Step 3. Position the camera in a location that is some distance away from the objects and at a good angle to view the blocks. As soon as you click to position the camera, you see the "target view cone" shown in Figure 29.11. This is the camera's field of view.

    ○ AutoCAD also says: `Specify target location:`

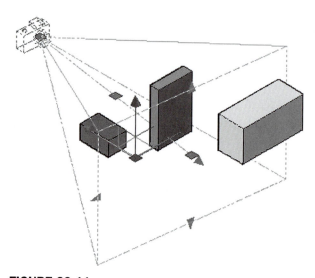

FIGURE 29.11
Camera field of view.

Step 4. Drag out the view cone to cover the blocks and click again.

    ○ AutoCAD says: `Enter an option [?/Name/LOcation/Height/Target/LEns/ Clipping/View/eXit] <eXit>:`

Here, you can change some of the camera values, but for now just press Enter. The little camera symbol sits there at the location you specified, looking in the direction you pointed. If you check the View Manager, you find Camera1 there. Before you go into that view, however, you may want to first check what it looks like by clicking on the camera. You see what is shown in Figure 29.12.

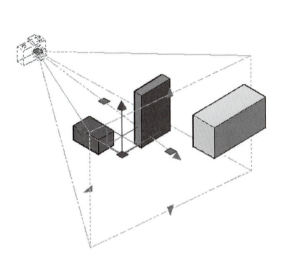 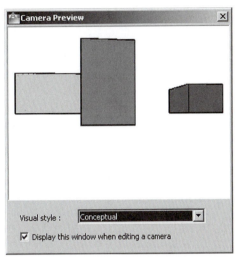

FIGURE 29.12
Camera Preview.

You can now adjust the camera settings via the blue grips (e.g., widen the field of view), and these actions are reflected in the preview window on the right. When done, go to the View Manager and select the Camera1 view, as seen in Figure 29.13. Press Set Current, then OK.

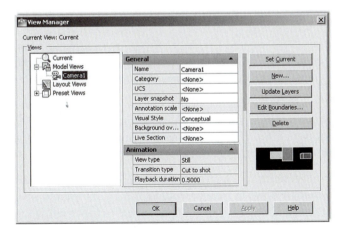

FIGURE 29.13
View Manager with Camera1.

You then see the Camera Perspective view (Figure 29.14), which in this case is a floor level view of the blocks. Notice also how the screen is split between a "ground" and a "sky." This view can be rotated and zoomed or panned. Go ahead and try all of those. To get out of it, go to the little "house" introduced earlier (Figure 29.15), and select Parallel.

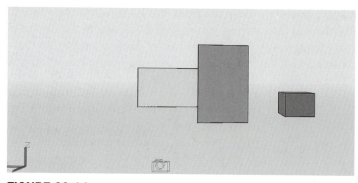

FIGURE 29.14
Camera1 view in perspective.

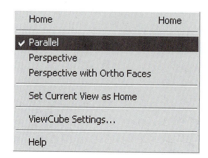

FIGURE 29.15
Exiting Camera1 view.

638

Camera placement is an extensive topic that allows for creativity in where you put the camera, and the resulting generation of some unique views. You should spend some time exploring this further.

29.3 WALK AND FLY

Walk and Fly sound fancier than they really are. All we are talking about here is the ability to move around and through a 3D design using keyboard input (typically the arrow keys). Walking involves forward/reverse and left/right movements and you remain in the XY plane. Flying is the same movements but unconstrained by the XY plane, so you have the appearance of flying through the design as you move forward and in reverse.

The best analogy to this is the experience of playing a first person action video game where you see what the character sees as you move through hallways and rooms while shooting at the bad guys—that last part of course is missing in AutoCAD, though a few well-placed petitions to Autodesk may get designers thinking!

All the Walk and Fly movements can be saved and played back in an animation, and this is usually how things would be done, as a client would typically not want to sit around as you stumble your way through a 3D house model. It takes practice to smoothly toggle between Walk and Fly modes; we are not talking precise joystick control here. During your strolls, you can keep track of where you are in the big picture via a Position Locator palette, which also allows you to fine-tune some aspects of Walk and Fly. Walk and Fly are always conducted in perspective view and AutoCAD sets that view automatically (after a prompt) before anything begins.

To begin exploring this, we need a small architectural floor plan with which to work. Our apartment, first started in Level 1 and worked on throughout some of the chapters, fits the bill perfectly. If not, you can use any 3D model. If you still have the apartment model saved, bring it up in 2D and clean it up a bit by removing everything but the walls, as seen in Figure 29.16.

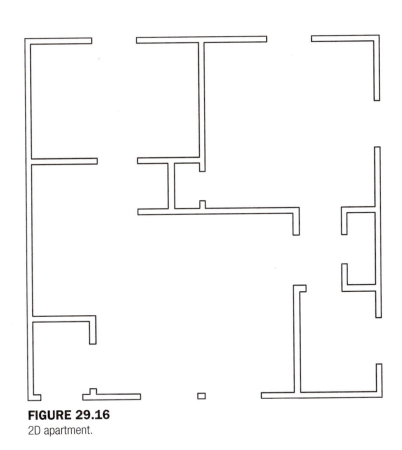

FIGURE 29.16
2D apartment.

Now, go ahead and pedit all the visible linework. Then, extrude all the walls out to a height of 8 feet (96 in.). You should have what is shown in Figure 29.17 after shading, coloring, and switching to the SW Isometric view.

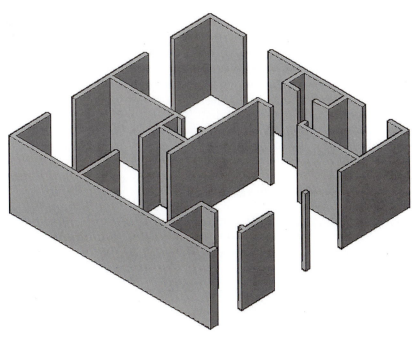

FIGURE 29.17
3D apartment: extruded, colored, and shaded.

To finish the design, let us add a floor and fill in the gaps in the wall where there needs to be windows. For the floor, simply draw an extruded rectangle from corner to corner of the apartment, making sure of course that it is located at the bottom (no roof at this time). Then, draw rectangles above and below the window openings, extruding 24" down and 36" up. That makes for 36" windows.

Do the same for the door openings, extruding 12" down, thereby making the doors 96"– 12" = 84", or 7 feet tall. Finally, add the doors themselves where needed, according to the original floor plan (see Level 1 images). The result is shown in Figure 29.18. Be sure to union the wall sections together and add some color for clarity.

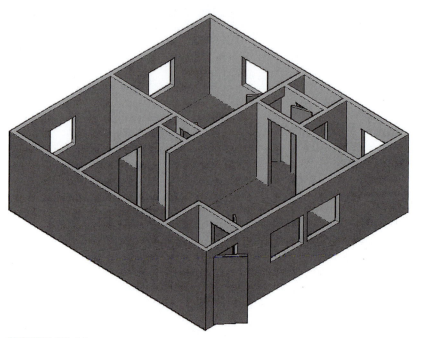

FIGURE 29.18
3D apartment with windows, doors, and floor.

After you are done, bring up the Walk and Fly toolbar, as shown in Figure 29.19. The toolbar consists of Walk (the feet), Fly (the paper airplane), and a third icon for settings, which we explore as well. Let us try the Walk tool. When you first attempt to use it, you likely get the prompt shown in Figure 29.20. After clicking on Change, the design enters perspective view and the Position Locator palette appears on the screen (Figure 29.21).

FIGURE 29.19
Walk and Fly toolbar.

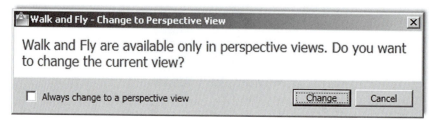

FIGURE 29.20
Walk and Fly warning prompt.

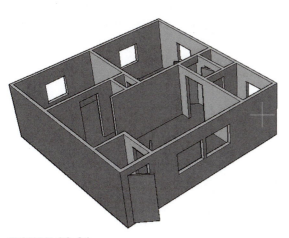
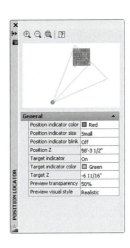

FIGURE 29.21
Position Locator palette.

We look at this palette in detail later on, but for now set it aside and let us focus on navigating the design. You can navigate it (walk) via the tools shown in Figure 29.22.

Up arrow / W key	Move forward
Down arrow / S key	Move backward
Left arrow / A key	Move left
Right arrow / D key	Move right
Drag mouse	Look around & turn
F key	Toggle Fly mode

FIGURE 29.22
Walk and Fly navigation tools.

The information needed to use Walk and Fly is spelled out in the list. While you can certainly use the letter keys (W, S, A, and D), it is probably easier to use the keyboard arrows (Up, Down, Left, and Right). To toggle between Walk and Fly, press the F key. Take some time to practice walking and flying via the four keyboard arrows, with the letter F key as the toggle between the two. Once you have the basic idea of Walk and Fly, let us take a look at the settings icon. When you press that, you see the box shown in Figure 29.23.

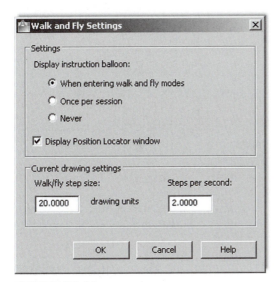

FIGURE 29.23
Walk and Fly Settings.

The main settings have to do with the Display Instruction and Display Position window and palette. You can get rid of both of them if needed; although they are helpful, they are not essential to Walk and Fly operation. Just as important, however, is what is found at the bottom of the dialog box, the drawing units and Steps per second. You can increase both to Walk and Fly through the design faster.

Do not forget also to use the mouse itself to rotate and spin the design around if needed. Walk and Fly are just part of the equation. Let us take another look at the Position Locator. In the top window, you see the design's position relative to you, the observer, as seen in Figure 29.21. Although you are not likely to get lost while traveling through this simple one-bedroom design, you may in a more sophisticated walk-through, and this "bird's eye view" of where you are can be helpful. The rest of the general settings below that preview window are related to how the preview window looks.

29.4 PATH ANIMATION

Path Animation has to do with creating animations of designs. These animations are not the classic type, where we have objects moving and interacting (for example, AutoCAD cannot animate the motions of pistons inside cylinders of a car engine model). Rather, what we refer to is an animated walk-through, a sort of frame-by-frame set of scenes strung together into a smooth motion.

To begin animation type in `anipath` and press Enter. Alternatively, you can use the drop-down menu: **View**→**Motion Paths Animation**…. The dialog box shown in Figure 29.24 appears.

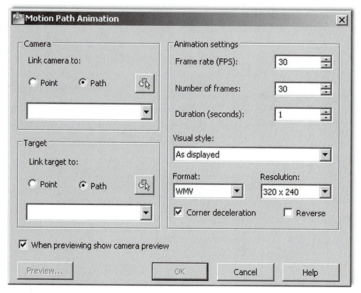

FIGURE 29.24
Motion Path Animation.

This is the combined settings and animation dialog box. Here, you can set your preferences, preview the animation, and export it to the desired format. Some of the options are listed in more detail:

- **Frame rates:** The number of frames per second that play at normal playback speeds. The more frames, the larger the output file. The maximum is 60, and 40 is usually just fine. The number of frames divided by the duration yields the Frame rate (FPS).

- **Resolution:** How sharp the image is. The larger the resolution, the larger the output file. The range is from 160 × 120 to 1024 × 768. The default, 320 × 240, is usually adequate.
- **Format:** The available exporting formats are avi, mov, mpg, and wmv.
- **Camera or Target:** You can specify either one; we use camera.

As the name implies, motion path animation depends on a path. Once the path is defined, along with a number of animation-related parameters, AutoCAD sends a camera down the path and plays back what it sees as it travels. It is smooth and continuous and a good way to present a tour of an entire building.

To try this out, cancel the dialog box and create a path in the building using pline. This is by far the hardest and most time-consuming task, and it is also the most important. A good path ensures a clear, smooth view of the interior. Here are some tips for path placement:

- Place the path in the middle of doorways using temporary guidelines. If you use the edges of the door, the camera literally slams into the door frame as it travels, and the view is briefly obscured.
- Add fillets to round out the turns instead of right angles for smoother motion.
- Create a realistic and steady path, as a real person would walk. Do not jump from floor to ceiling unless there's a good reason.

In Figure 29.25, several doors are removed for clarity, and a sample path through our 3D apartment is carefully drawn in using pline. The path comes in through the front door and walks through the living room and into the kitchen, following a typical path a person would take.

643

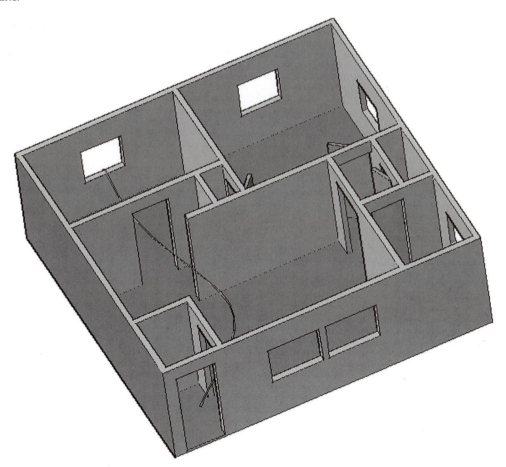

FIGURE 29.25
Motion pathway.

This path may or may not be ideal and may have to be fixed up later, but it is fine for now. When done, select **V**iew→**M**otion Paths Animation… again, and follow these steps:

Step 1. The first step is to link the camera to a point or a path. Typically, the camera is linked to the path, so in the upper right-hand area, use the button to select the path and confirm its name as Path 1 in the field.

Step 2. Next, you need to link the target to a point or a path. If you choose Path, then the Target and Camera settings are the same, meaning the camera follows the path and turns when the path turns. If you choose Point (and select a distant point), then the camera is always fixed on that point as it follows its path.

Step 3. Now, you need to run through the animation settings. Here, everything should be more or less intuitive, as all three fields are linked. So, if you select a frame rate of 10 and specify 120 frames, then 120/10 = 12, so there will be 12 seconds of "footage" or animation. Select values that are appropriate to the task at hand. If you have a large walk-through, you may need to run the animation for a minute or more. Just be aware that the files generated can get big very quickly, and overall performance and quality also depend on your computer (RAM, video card, etc.).

Step 4. Finally, select the Visual Style, Format, and Resolution and press Preview…. A preview window appears and runs through the animation, as seen in Figure 29.26.

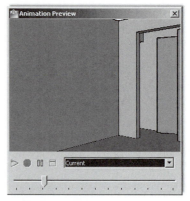

FIGURE 29.26
Animation Preview.

When done, simply close the Animation Preview via the X in the upper right corner; you are taken back to the Motion Path Animation dialog box. Press OK and you are taken to the Save As dialog box and asked where you would like to save the file, as well as to confirm the output format (Figure 29.27). Select a location and the generation sequence commences, as seen in Figure 29.28. The file is generated and you can watch it with whatever player you have on your computer.

29.5 ACTION RECORDING

Action recording is a relatively new tool in AutoCAD that is used to build macros. *Macros* is short for macroinstructions, a computer term that can be defined as a short set of instructions to perform a task. The idea here is to actually record your on-screen actions and save them for future use. Needless to say, these actions should be some sort of useful steps to build a tedious or repetitive object and not just any simple steps. You can then use this macro to make your life easier later on. You can do this with both 2D and 3D modeling, and the action recorder was included here in a 3D chapter only to group it together with the more common path animation just discussed.

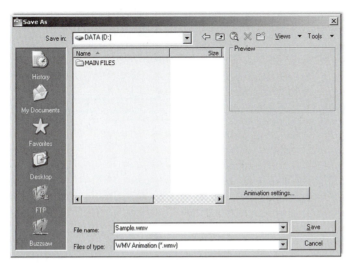

FIGURE 29.27
Saving animation.

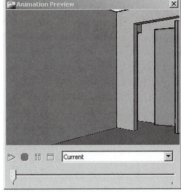

FIGURE 29.28
Animation generation.

To give this a try, open up a brand-new file with the Ribbon present, and start the command via any of the following methods:

Keyboard: Type in **actrecord** and press Enter	
Cascading menus: **Tools→Action Recorder→Record**	
Toolbar icon: none	
Ribbon: **Manage tab→Action recorder**	

You then see what is shown in Figure 29.29. Notice the Ribbon tab on the left has turned light pink and the record button became a stop button. The mouse has also acquired a red "Recording" dot.

Begin drawing anything you like, such as a piece of furniture, as seen in Figure 29.30. Notice how each action of your mouse, keyboard, menu, Ribbon, and so on is recorded in the log on the upper left.

When done, press the Stop button and the Action Macro dialog box appears. Give the macro a descriptive name (no spaces allowed) and press OK, as seen in Figure 29.31.

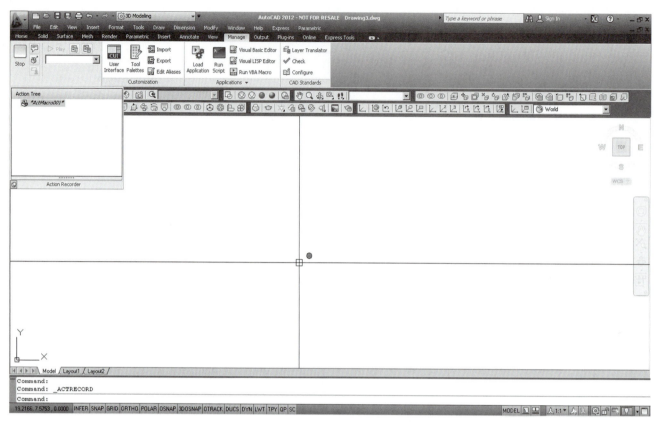

FIGURE 29.29
Action recorder.

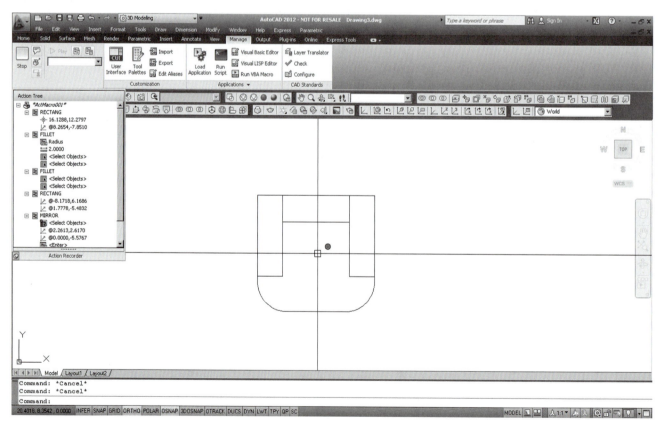

FIGURE 29.30
Recording log.

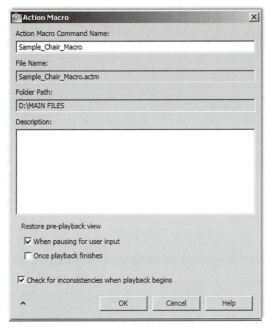

FIGURE 29.31
Action Macro.

FIGURE 29.32
Playback Complete.

647

The macro has now been completed and stored as an *.actm file in a default location (which can be changed via Options...→Files). You can put it to use by erasing the chair and pressing the Play button. The macro executes and the chair is drawn. You then see the message shown in Figure 29.32.

This is a very simple example to illustrate the basics. You can get quite complex with macros. For example, you can insert user input or a message. Macros recognize only a few dialog boxes (like layers), so if you need to use one, then make it a command line dialog box via the "-"(hyphen) prior to typing it in. So hatch would be −hatch, and so on.

Experiment with this on your own; there are many possibilities.

SUMMARY

You should understand and know how to use the following concepts and commands before moving on to Chapter 30:

- Dview
  - CAmera
  - TArget

- ○ Distance
- ○ POints
- ○ PAn
- ○ Zoom
- ○ TWist
- ○ CLip
- ○ Hide
- ○ Off
- ○ Undo
- Cameras
- Walk and Fly
- Path animation
- Action recording

REVIEW QUESTIONS

Answer the following based on what you learned in Chapter 29:

1. List all 11 dview functions and what they do.
2. List any additional views that are covered.
3. What is the purpose of Walk and Fly?
4. How are the Walk and Fly functions controlled?
5. How do you create path animations?
6. What is an action recorder?

EXERCISE

1. Using the suspension design from the previous chapter, set up a title block and viewports as shown in the following illustration. (Difficulty level: Advanced; Time to completion: 20 minutes)

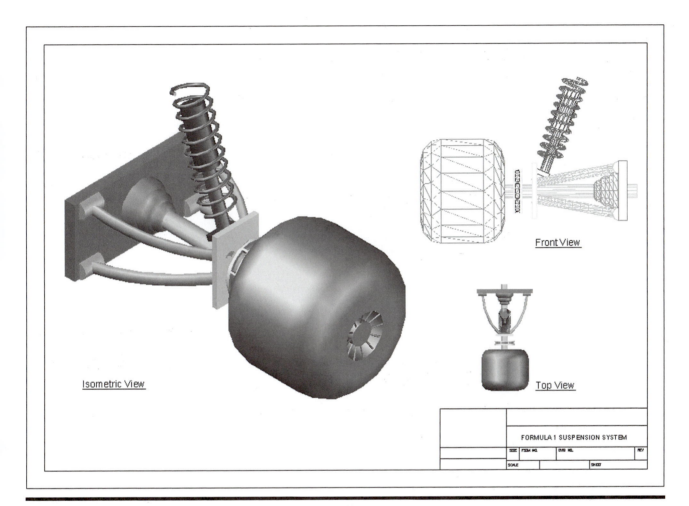

Isometric View

Front View

Top View

FORMULA 1 SUSPENSION SYSTEM

| SIZE | FSCM NO. | DWG NO. | | REV |
| SCALE | | | SHEET | |

CHAPTER 30

Lighting and Rendering

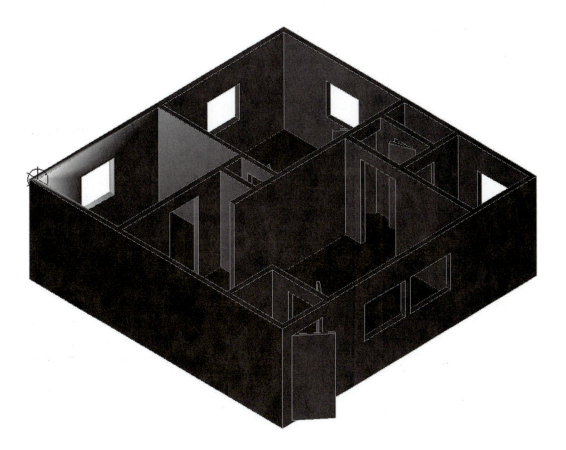

651

LEARNING OBJECTIVES

In this final chapter, we explore lighting and rendering and discuss

- Point light
- Spotlight
- Distant light
- Shadows
- Sun and sky
- Rendering
- Materials

By the end of the chapter, you will be able to add various lighting effects and basic materials to your design.

Estimated time for completion of chapter: 2 hours.

30.1 INTRODUCTION TO LIGHTING AND RENDERING

This final chapter of the textbook is dedicated to "finishing" a design. This means the application of lighting, shadows, environmental effects, backgrounds, materials, and rendering. It is something that is done last, after a design is completed, as these effects do not stand alone but are applied to an existing model. All finishing effects are loosely grouped under two major categories: lighting and rendering.

It should be stated up front that, despite major improvements in AutoCAD's lighting and rendering ability, it is still not quite on par with available dedicated rendering software. Packages such as Form Z, Rhino, and the famous 3ds Max (3D Studio Max) can create stunning photorealistic images and, in some cases, advanced animation. Autodesk has had a bit of a dilemma with AutoCAD. Should it pursue advanced rendering development and risk taking sales away from its own 3ds Max or add to AutoCAD to please those few customers who may need some rendering ability? After all, the vast majority of AutoCAD designs are still plain old 2D schematics or floor plans and it makes little sense to load AutoCAD with rendering power that customers rarely need.

The answer is somewhere in the middle. AutoCAD always had limited lighting and rendering ability, going back to the Advanced Modeling Extension of the early releases, and slowly new features were added. These were just enough to satisfy casual 3D users but never so much that they stopped considering the expensive dedicated software. One way around this of course is to purchase reasonably priced add-ons, like AccuRender (see Appendix I). This software adds dramatically to the existing capabilities without expensive investment.

In the meantime, AutoCAD's development team (always in need of new features to advertise as a new version rolls out) slowly but surely kept improving the built-in lighting and rendering. Today, with the latest version, Release 2012, we have quite a bit to discuss. It is still not 3ds Max, but if you have never worked with advanced dedicated effects software, you may be impressed. It is also a great introduction into that world. Let us take a look first at lighting and related topics (such as shadows and sun and sky), followed by materials and rendering.

30.2 LIGHTING

Lighting in digital art and CAD software is based on several established techniques and principles, and AutoCAD is no exception. A model can be lit in one of several ways, which involve variations on ambient, directional, or spotlight light sources. This of course mimics our everyday experiences, where we have a variety of lighting conditions, from diffused sunlight to the intense headlights of a car and everything in between.

Most of the options that have to do with lighting are either on the Lights toolbar or the Render tab of the Ribbon. Both are shown in Figure 30.1.

The three main options for lighting are

- Point light
- Spotlight
- Distant light

Let us give these various lights a try. Bring up the same 3D apartment model used in the previous chapter and switch to the Realistic Visual Style (RVS), as seen in Figure 30.2.

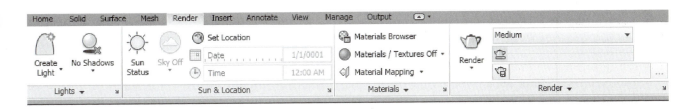

FIGURE 30.1
Ribbon Render tab and Lights toolbar.

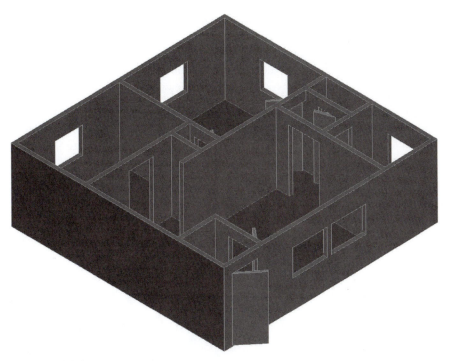

FIGURE 30.2
RVS shading.

Point Light

We can now try to set a lighting condition and look over some of the available options, which are relatively similar from one lighting scenario to another. The one most often used is the New Point Light option, described here.

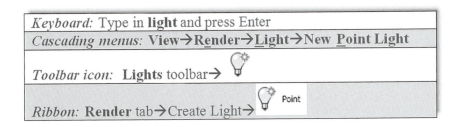

Keyboard: Type in **light** and press Enter	
Cascading menus: **View→Render→Light→New Point Light**	
Toolbar icon: **Lights** toolbar→	
Ribbon: **Render** tab→Create Light→ Point	

Step 1. Begin the New Point Light via any of the preceding methods. You may see a warning dialog box asking if you want to turn off default lighting (not shown here). Go ahead and agree to that.

Step 2a. If you type in `light`,
- ○ AutoCAD says: `Enter light type [Point/Spot/Web/Targetpoint/` `Freespot/freeweB/Distant] <Point>:`
  Go ahead and select `P` for `Point` and press Enter. Go on to Step 2b.
Step 2b. If you used any of the other methods or just did Step 2a,
- ○ AutoCAD says: `Specify source location <0,0,0>:`
Step 3. Select the extreme upper left corner of the building.
- ○ AutoCAD says: `Enter an option to change [Name/Intensity factor/` `Status/Photometry/shadoW/Attenuation/filterColor/eXit] <eXit>:`
Step 4. We go over some of these options in detail momentarily, but for now press `i` for `Intensity`.
- ○ AutoCAD says: `Enter intensity (0.00 — max float) <1.0000>:`
Step 5. Enter the value `20`. Then press `c` for `Color`.
- ○ AutoCAD says: `Enter true color (R,G,B) or enter an option [Index` `color/Hsl/colorBook] <255,255,255>:`
Step 6. Press `i` for `Index color` and enter color number 30 (it is one of the yellows).

The result is a spotlight located at an upper left corner of the 3D model (notice the circular marker), shining yellow light on the entire structure (Figure 30.3). Note that the light goes through the walls, illuminating everything equally, as if the walls are not even there. We change this later with the LIGHTINGUNITS system variable. You can locate as many point lights as you want; we do only one in this example.

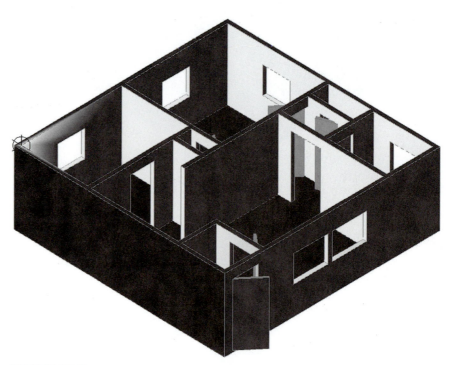

FIGURE 30.3
Point light (LIGHTINGUNITS 0).

So, what options can you change to enhance this very basic example? Let us go over them, as they appear in the other light choices as well—Name, Intensity factor, Status, Photometry (if LIGHTINGUNITS is set to 1 or 2 only), shadoW, Attenuation, filterColor, eXit:

- **Name.** You can give the light setting a name so you know to what you are referring later on. Use logical names like `Bedroom Spotlight 1`.

- **Intensity factor.** This is the brightness of the light; you can enter almost any value. The exact number is dictated by design needs and often arrived at by trial and error.
- **Status.** This simply means on or off. You can position a light and then turn it off, as with a regular light switch, if not needed in a scenario.
- **Photometry.** This option depends on the LIGHTINGUNITS system variable, to be set to 1 or 2 (International or American with Photometry enabled), not 0, which is Generic lighting, as seen in Figure 30.3. Once that is set, *photometry* refers to the perceived power emitted by a light source. The result of setting that system variable is that a more realistic light gradient is shown, where the intensity fades as you move farther from the lamp source, as seen in Figure 30.4. The three options under Photometry are

FIGURE 30.4
Point light (LIGHTINGUNITS 1).

- ○ **Intensity.** This is similar to the previous generic intensity setting, but allows for more options. The default values are in Candelas (Cd), the SI unit of luminous intensity. You can also press f for the Flux option (Lm) or press i for Illuminate, in which case you can use Lux (Lx) or Foot Candle (Fc) values. As a suboption, you can even use Distance. These settings are familiar to any architect or designer who works with lighting. Photometry is a science unto itself; experiment with the various options, and if you have never worked with lighting before, seek out information online or in a library on this interesting topic.
- ○ **Color.** This option assigns color to the light (otherwise it is the default white) but works differently from the color filters in the main menu. Here, color represents a type of lamp or a Kelvin temperature setting. If you type in a question mark and press Enter twice you get this list of standard colors or lamps:
  "D65"
  "Fluorescent"
  "Coolwhite"
  "Whitefluorescent"
  "daylightfluorescent"
  "Incandescent"

```
"Xenon"
"Halogen"
"Quartz"
"Metalhalide"
"mErcury"
"Phosphormercury"
"highpressureSodium"
"Lowpressuresodium"
```

You can choose a lamp type and it is shown. Another interesting option is the ability to use temperature as a light setting. Type in k for `Kelvin` and enter a temperature value. As mentioned before, it takes experience and knowledge to use these tools properly, and there is a science of proper lighting. For those who work in this field, AutoCAD gives some basic tools with which to work, and for those who do not, this is an invitation to learn more.

○ **eXit.** This option exits the Photometry command.

- **shadoW.** This command controls the appearances of shadows when your light source encounters an obstruction (such as a piece of furniture). The options are
  ○ **Off.** This option turns off shadows.
  ○ **Sharp.** This displays shadows with sharp edges (less realistic but better computer performance).
  ○ **soFtmapped.** This entire option has to do with soft shadows and how much computer resources (map size and memory) to allocate to the shadow. As with all such settings, there is a trade-off between quality and system performance.
  ○ **softsAmpled.** This is another set of settings for realistic shadows.

- **Attenuation.** Attenuation (with AutoCAD anyway) is the decay of light as distance from the source increases. There are a variety of settings here, such as
  ○ **attenuation Type.** The two choices are Inverse Linear and Inverse Squared, which have to do with the strength of the light (whether the value is squared or linear) as the distance increases.
  ○ **Use limits.** This option turns limits (see later) on and off.
  ○ **attenuation start Limit.** This option specifies where the light starts.
  ○ **attenuation End limit.** This option specifies where the light ends.
  ○ **eXit.** This option exits the Attenuation command.

- **filterColor.** This is simply the color of the light emitted with no regard to temperature or type of lamp, as seen with Photometry settings. You can enter RGB values, Index colors, or use any of the other color choices AutoCAD provides. Figure 30.2 features Index color 30.

- **eXit.** This option exits the filterColor command.

Spotlight

Now that you have had an introduction to these tools, let us briefly go over the remainder of the lights and a few other associated options. The New Spotlight option is roughly similar in theory to the New Point Light option just covered.

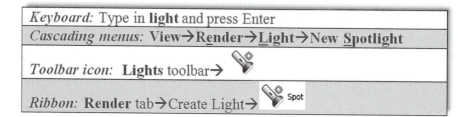

Keyboard: Type in **light** and press Enter
Cascading menus: View→R<u>e</u>nder→Light→New **S**potlight
Toolbar icon: **Lights** toolbar→
Ribbon: **Render** tab→Create Light→ Spot

Step 1. Begin the Spotlight via any of the preceding methods. You may see a warning dialog box asking if you want to turn off default lighting (not shown here). Go ahead and agree to that.

Step 2a. If you type in `light`,

- AutoCAD says: `Enter light type [Point/Spot/Web/Targetpoint/ Freespot/freeweB/Distant] <Point>:`
  Go ahead and select S for `Spot` and press Enter. Go on to Step 2b.

Step 2b. If you used any of the other methods or just did Step 2a,

- AutoCAD says: `Specify source location <0,0,0>:`

Step 3. Select a corner of the building (a flashlight is attached to the crosshairs).

- AutoCAD says: `Specify target location <0,0,-10>:`

Step 4. Select a point in such a way as to indicate a direction for the spotlight to shine. A large menu appears that has essentially the same choices seen previously.

- AutoCAD says: `Enter an option to change [Name/Intensity factor/ Status/ Photometry/Hotspot/Falloff/shadoW/Attenuation/ filterColor/eXit] <eXit>:`

There are however some new commands in this menu, as described next:

Hotspot. This command specifies the angle that defines the brightest cone of light, which is known as the *beam angle*. This value can range from 0° to 160°.

Falloff. This command specifies the angle that defines the full cone of light, which is also known as the *field angle*. This value can range from 0° to 160°.

Step 5. Using values of 100 for the `Intensity factor` and 255,127,0 for the `filterColor`, go ahead and adjust those options.

Figure 30.5 shows the results. The spotlight is attached to the right corner of the design. It is clicked on to show its light cone (called a *glyph*) and grips. These glyphs can be adjusted via the grips to vary the "sweep" of the light source and the area it affects (illuminates).

657

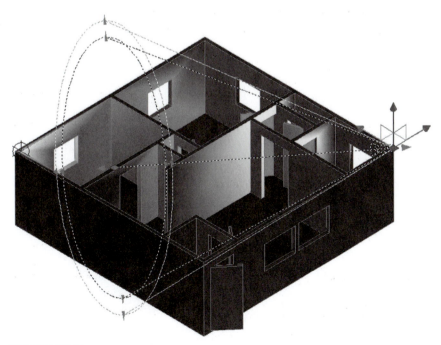

FIGURE 30.5
Spotlight.

Distant Light

The final light type we try out is called New Distant Light. This is more of a floodlight that saturates everything and is not used as often.

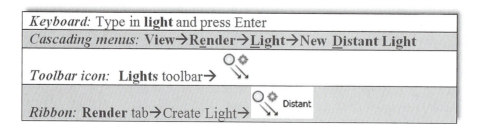

Keyboard: Type in **light** and press Enter	
Cascading menus: **View→Render→Light→New Distant Light**	
Toolbar icon: **Lights** toolbar→	
Ribbon: **Render** tab→Create Light→	

Step 1. Begin the Distant Light via any of the preceding methods. You may see a warning dialog box asking if you want to disable distant lights when the lighting unit is photometric (not shown here). Allow it in this case!

Step 2a. If you type in `light`,

- AutoCAD says: `Enter light type [Point/Spot/Web/Targetpoint/ Freespot/freeweB/Distant] <Point>:`

Go ahead and select `D` for `Distant` and press Enter. Go on to Step 2b.

Step 2b. If you used any of the other methods or just did Step 2a,

- AutoCAD says: `Specify light direction FROM <0,0,0> or [Vector]:`

Step 3. Pick a direction from which the light comes.

- AutoCAD says: `Specify light direction TO <1,1,1>:`

Step 4. Pick a direction into which the light goes. The familiar menu appears (not shown this time), and you can set some options to complete the distant light setting.

To complete our tour of AutoCAD's lighting features, there are a few more items to cover. Perhaps the most important one is the Properties palette as it applies to lighting. Double-click on any light to call it up, as seen with the spotlight in Figure 30.6.

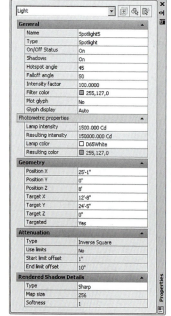

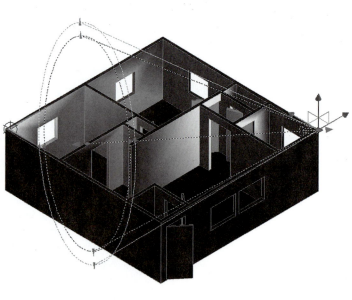

FIGURE 30.6
Properties palette.

Using the Properties palette you can change virtually any aspect of the newly created light, with the added benefit that you can see the effect right away.

To keep track of all your lights, you can activate the Lights in Model palette (Figure 30.7) by clicking on the Lights List icon on the toolbar or the arrow flyout on the Ribbon's Render tab. This palette shows every light created, which is why it is important to give them descriptive names. In our example, we did not, and they are simply referred to as Pointlight4 and Spotlight5.

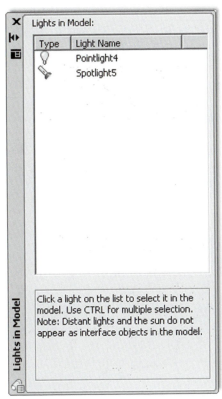

FIGURE 30.7
Lights in Model palette.

Shadows, Sun, and Sky

Shadows are an integral part of the sun and sky settings, which are nothing more than shadows customized to a specific location and time. You can get a simple shadow right away, however. Draw and position some simple 3D shapes, and in the Ribbon's Render tab find the shadow settings and select Ground Shadows (Figure 30.8). The result is seen in Figure 30.9.

Sun and sky, as mentioned already, refer to a background light setting where you can set a particular city location anywhere in the world and its particular latitude and longitude are factored into the lighting conditions. To set the time zone, press the Geographic Location… icon (it looks like a globe) and select the Enter the location values button from the Define Geographic Location box seen in Figure 30.10.

The Geographic Location dialog box then appears (Figure 30.11). Set your particular region, country, or city via the drop-down menus. You can use the provided time zones as an alternative. You can also click anywhere on the map or enter values manually, although not too many people have that kind of information handy. When done, click on OK.

Finally, let us bring up sun properties via the final icon on the Lights toolbar. The Sun Properties palette appears, as seen in Figure 30.12.

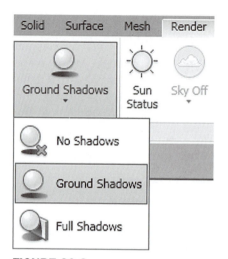

FIGURE 30.8
Ribbon's Ground Shadows.

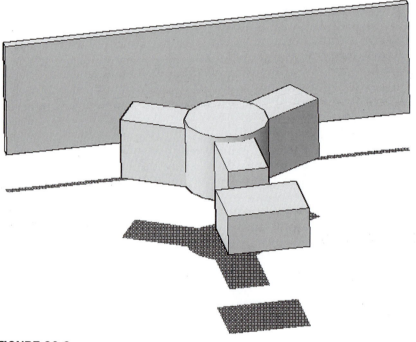

FIGURE 30.9
Basic ground shadows.

Using this palette and a bit of trial and error (at least until you understand what each property does), you can set the ambient sunlight properties. The exact description of each category is found in the Help files, reproduced here for convenience and in edited form for clarity:

General. This is where you set the general properties of the sun.

- **Status.** This option turns the sun on and off. If lighting is not enabled in the drawing, this setting has no effect.
- **Intensity Factor.** This option sets the intensity or brightness of the sun. The range is from 0 (no light) to maximum. The higher is the number, the brighter the light.

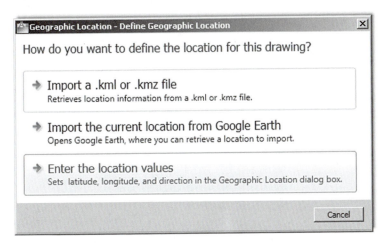

FIGURE 30.10
Define Geographic Location.

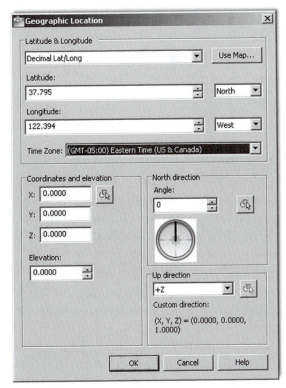

FIGURE 30.11
Geographic Location.

- **Color.** This option controls the color of the light. Enter a color name or number or click on Select Color to open the Select Color dialog box.
- **Shadows.** This option turns on and off the display and calculation of shadows for the sun. Turning shadows off increases performance.

Sky Properties. This is where you set the general properties of the sky.

- **Status.** This option determines if the sky illumination is computed at render time. This has no impact on the viewport illumination or the background. It simply makes the

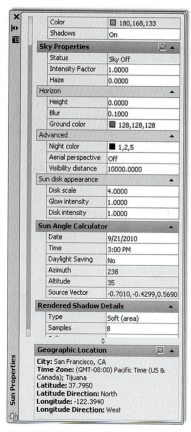

FIGURE 30.12
Sun Properties palette.

sky available as a gathered light source for rendering. Note: This does not control the background. Its values are Sky Off, Sky Background, and Illumination.

- **Intensity Factor.** This option provides a way to magnify the effect of the sky light. Its values are 0.0 to MAX.
- **Haze.** This option determines the magnitude of scattering effects in the atmosphere. Its values are 0.0 to 15.

Horizon. This category of properties pertains to the appearance and location of the ground plane.

- **Height.** This option determines the absolute position of the ground plane relative to world zero. This parameter represents a world-space length and should be formatted in the current length unit. Its values are + to MAX.
- **Blur.** This option determines the amount of blurring between ground plane and sky. Its values are 0 to 10.
- **Ground Color.** This option determines the color of the ground plane by selecting a color from the drop-down list. You can also select the Select Color dialog box to make a color choice.

Advanced. This category of properties adds various artistic effects.

- **Night Color.** This option specifies the color of the night sky by selecting a color from the drop-down list. You can also select the Select Color dialog box to make a color choice.
- **Aerial Perspective.** This option specifies if aerial perspective is applied. Its values are On or Off.
- **Visibility Distance.** This option specifies the distance at which 10% haze occlusion results. Its values are 0.0 to MAX.

Sun Disk Appearance. This category of properties pertains to the background only, and they control the appearance of the sun disk.

- **Disk Scale.** This option specifies the scale of the sun disk (1.0 = correct size).
- **Glow Intensity.** This option specifies the intensity of the sun glow. Its values are 0.0 to 25.0.
- **Disk Intensity.** This option specifies the intensity of the sun disk. Its values are 0.0 to 25.0.

Sun Angle Calculator. This category of properties sets the angle of the sun.

- **Date.** This option displays the current date setting.
- **Time.** This option displays the current time setting.
- **Daylight Saving.** This option displays the current setting for daylight savings time.
- **Azimuth.** This option displays the azimuth, the angle of the sun along the horizon clockwise from due north. This setting is read-only.
- **Altitude.** This option displays the altitude, the angle of the sun vertically from the horizon. The maximum is 90°, or directly overhead. This setting is read-only.
- **Source Vector.** This option displays the coordinates of the source vector, the direction of the sun. This setting is read-only.

Rendered Shadow Details. This category of properties specifies the properties of the shadows.

- **Type.** This option displays the setting for shadow type. This setting is read-only when display of shadows is turned off. The selections are Sharp and Soft (mapped), which display the Map size option, and Soft (area), which displays the Samples option. Soft (area) is the only option for the sun in photometric workflow (LIGHTINGUNITS set to 1 or 2).
- **Map size (Standard lighting workflow only).** This option displays the size of the shadow map. This setting is read-only when display of shadows is turned off. Its values are 0 to 1000.
- **Samples.** This option specifies the number of samples to take on the solar disk. This setting is read-only when display of shadows is turned off. Its values are 0 to 1000.
- **Softness.** This option displays the setting for the appearance of the edges of shadows. This setting is read-only when display of shadows is turned off. Its values are 0 to 50.0.

Geographic Location. This category of properties displays the current geographic location settings. This information is read-only. When a city is not stored with latitude and longitude, the city does not appear in the list. Use the Edit Geographic Location button to open the Geographic Location dialog box.

30.3 MATERIALS AND RENDERING

Rendering in AutoCAD is a two-part story. The first part involves materials and simply adding them to designs. AutoCAD has a sizable built-in library of materials, with most residing in the tool palettes, and adding them is simply a matter of click and drag onto the design. These materials are pictures, usually bitmaps, and they simply cover the design similar to wallpaper. You can also create and modify your own materials, within limits.

The other side to rendering is actual rendering, where more advanced settings are applied and the software makes passes over the design, rendering the surfaces as desired. In regard to both material application and rendering, other dedicated software packages have far more capabilities than AutoCAD, as explained earlier in the chapter, but this is still a good introduction to this fascinating world. And remember that quality of output is often a result of knowledge. An expert using limited software may still achieve better results than a novice on superior software. If you wish to pursue this type of work further, consult Appendix I for some additional information on this topic.

Materials

Let us jump right into materials by introducing the two relevant palettes (materials is very much a palette driven feature): the Materials Browser and the Materials Editor. Both are shown in Figure 30.13 and can be accessed via the cascading menus: **Tools→Palettes→Materials Browser** (or **Materials Editor**).

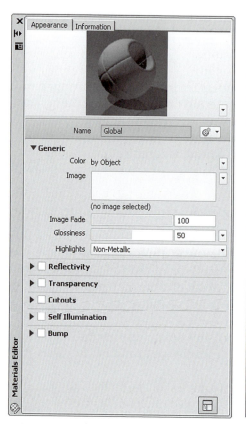
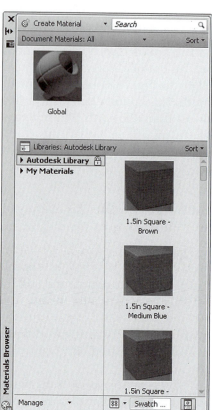

FIGURE 30.13
Materials Editor and Materials Browser.

Let us first take a look at the Materials Browser. It was expanded significantly over the last several releases of AutoCAD and now features over 1200 materials (see Figure 30.14). These materials are bitmap graphics designed in house at Autodesk (or acquired) and are superimposed onto geometry in a simple manner, akin to wrapping paper. They are not "procedural materials" and do not become one with the design (see Appendix I for more on this). The Materials Browser allows you to visually search the materials database and create a library of materials relevant to your design intent.

Applying these materials without further modification is simply a matter of click and drag. Draw a cube of any size, as seen in Figure 30.15, and select a random material. Then, click and drag the material onto the cube, which acquires those properties instantly. Some additional examples are shown in Figure 30.16.

If you right-click on any of the materials in the selection list, you can add it to the Material Browser's My Materials collection for future reuse. These materials can be deleted and renamed if necessary. The main Autodesk library, however, is locked, as indicated by a padlock. You can sort My Materials (or the main library) according to Name, Category, Type, or Color as well as filter what is shown by All, Applied, Selected, or Unused. Spend some time getting acquainted

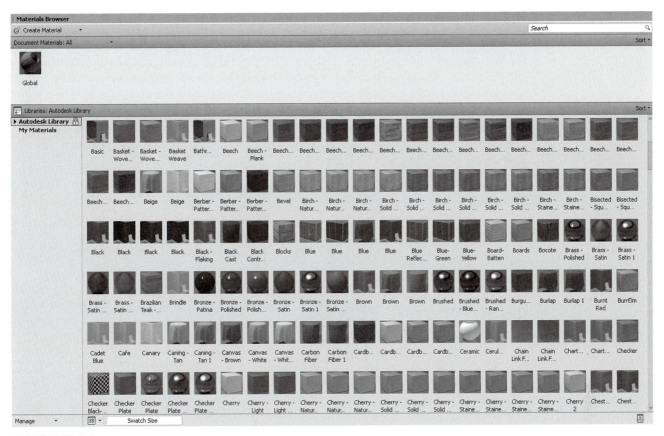

FIGURE 30.14
Expanded Materials Browser.

BASIC CUBE   ASHLAR MATERIAL APPLIED

FIGURE 30.15
Material application.

FIGURE 30.16
Additional material application examples.

with the Material Browser, so you have a good idea of what is available and what you can do with these materials.

The Materials Editor is exactly that: It edits and modifies existing materials. To enter something into the editor, click on a material once in the browser (but do not drag it anywhere). The material appears in the editor's preview window, with associated menus below that. An example is shown with the previously used Ashlar material (Figure 30.17). Some of the more useful settings that can be modified include Reflectivity, Transparency, and Self Illumination. Explore all these settings to see what the effects are. When done, the material appears in the Materials Browser. Be sure to rename it before using.

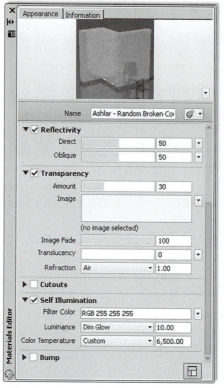

FIGURE 30.17
Materials Editor with a material loaded.

Rendering

Our final discussion involves rendering. Because of AutoCAD's improved materials and excellent visualization and coloring options, rendering in general is not something you often require, and your options here are not exactly extensive. In advanced modeling software, *rendering* is a general catch-all term that refers to advanced colorizing, materials, and lighting effects applied to a design. In AutoCAD, these procedures are separate from a rendering, and we already covered them in this chapter or elsewhere. What *rendering* means in our discussion is simply a processed output that can be saved to a file. You still have to add colors, materials, and lighting prior to a render. Then, the process of rendering an image produces a tangible output in the form of a jpg, or another format such as a bmp (Bitmap) or a tiff, that can be printed on a LaserJet (for example). These renderings can then be inserted into a presentation or given to a client.

The gateway to rendering is the Render toolbar (Figure 30.18) or the Ribbon's Render tab. Let us give this a try and render our familiar apartment (Figure 30.19), introducing features as needed. Bring up the design and add a new material for some variety, as shown in the figure.

FIGURE 30.18
Render toolbar.

FIGURE 30.19
Rendering model.

Let us jump right in and render the design without changing any settings. We will return and tweak things right afterward. Click the Render icon (the teapot); you can also just type in render and press Enter. The render panel appears and the apartment is rendered in several seconds. When done, information about the rendering process is shown to the right and below the rendered model (Figure 30.20).

The next step is to output the design to one of several formats. Select File→Save… and the Render Output File dialog box appears (Figure 30.21). Select the output format, jpeg in this case, and give it a name and location. Once you press Save, the JPEG Image Options dialog box appears (Figure 30.22). Adjust settings if desired and press OK. The image is saved to the previously selected location, ready for printing or insertion into a presentation.

We conclude rendering with a brief survey of rendering settings, in case you want to deviate from the basic procedure just outlined. The Settings palette is accessed via the Advanced Render Settings… icon or the flyout arrow on the bottom right of the Ribbon's Render tab. It is shown in Figure 30.23.

Review the settings on your own or with the Help files, as there is quite a bit here. Some of the more relevant ones (reading top to bottom) are the options that allow you to render only a small section of the image if needed via Procedure, control resolution via Output Size, and Apply Materials to determine what type of view gets rendered. You can also do all this via the Ribbon; just drop down the Render arrow on the Render tab.

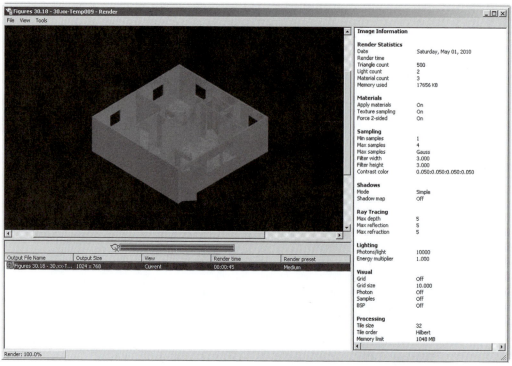

FIGURE 30.20
Rendering panel with output.

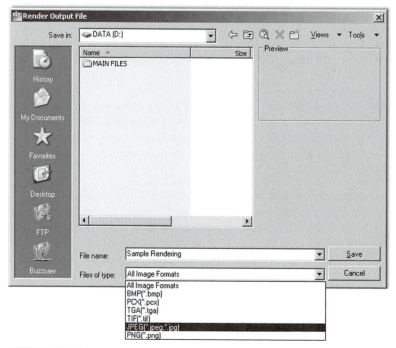

FIGURE 30.21
Render Output File.

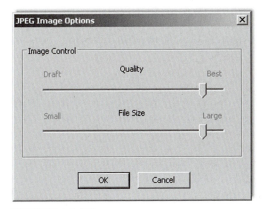

FIGURE 30.22
JPEG Image Options.

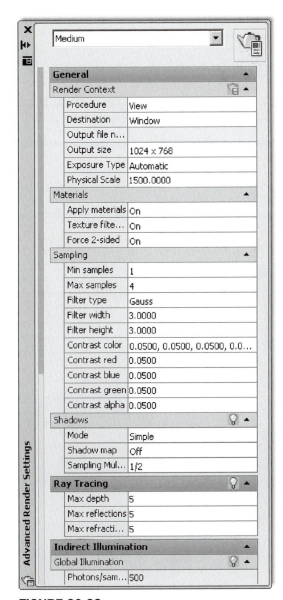

FIGURE 30.23
Advanced Render Settings.

669

SUMMARY

You should understand and know how to use the following concepts and commands at the conclusion of this chapter:

- Point light
- Spotlight
- Distant light
- Sun and sky
- Materials
- Rendering

REVIEW QUESTIONS

Answer the following based on what you learned in Chapter 30:

1. Describe point light.
2. Describe spotlight.
3. Describe distant light.
4. Describe sun and sky.
5. Describe how to bring up materials.
6. Describe basic rendering procedures.

EXERCISE

1. Create the following 20" × 50" box, extruded to 10" and shelled out to 1". Attach a point light to both corners. Set one point light to an intensity of 100 and a blue color. Set the other to an intensity of 75 and a red color. Then render the figure to a jpg file. (Difficulty level: Easy; Time to completion: 5–10 minutes)

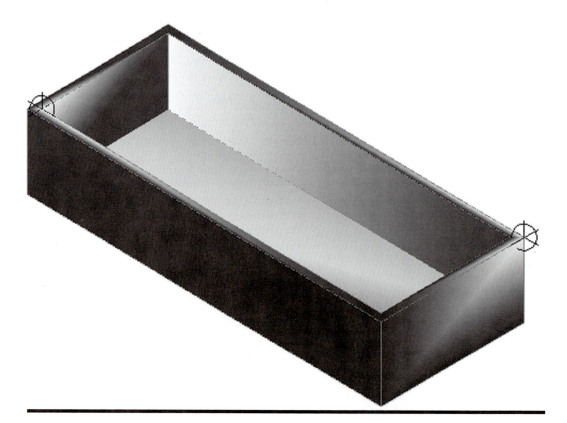

Answers to Review Questions

CHAPTER 21 REVIEW QUESTIONS

1. What three axes are discussed?
   X,Y, and Z.

2. What three planes are discussed?
   XY, ZX, and YZ.

3. What 3D toolbars were introduced in this chapter?
   View, 3D Navigation, Visual Styles, UCS, Modeling, and Render.

4. What command is used most often to create 3D objects?
   Extrude.

5. How do you rotate the UCS icon?
   Via the UCS command or UCS toolbar icons.

6. What does hide do?
   Hide turns of lines that would not be visible if the shape weres a solid.

7. What does shade do? What are the two types of shade?
   Shade adds colored shading, with light effects, to your design. The two types are conceptual and realistic.

8. What is the purpose of the Navigation bar and ViewCube?
   Both present additional 3D navigation and viewing view options.

CHAPTER 22 REVIEW QUESTIONS

1. Name the seven primitives covered in this chapter.
> Box, Wedge, Cone, Sphere, Cylinder, Torus, and Pyramid.

2. What four methods are used to create primitives?
> Typing, cascading menus, Modeling toolbar, and the Ribbon's Solid tab.

3. What two primitives cannot be created any other way?
> Sphere and Torus.

CHAPTER 23 REVIEW QUESTIONS

1. Describe the Rotate3D procedure. Do you need axes or planes for this command?
> Rotate3D rotates objects around axes. You need axes.

2. How is 3Drotate different from Rotate3D?
> 3Drotate features a graphical gizmo to assist in the rotation process. The net effect is the same as Rotate3D.

3. Describe the Mirror3D procedure. Do you need axes or planes for this command?
> The mirror3D command mirrors objects around planes. You need planes.

4. Describe the 3Darray procedure, both rectangular and polar.
> Rectangular 3Darray is command line driven and asks for rows, columns, and levels as well as the distances between all of these. Polar 3Darray is also command line driven and asks for number of items, angle to fill, and two points that define the axis of rotation for the array.

5. Describe 3Dscale and 3Dmove.
> Both commands are similar to their 2D versions; however, 3Dmove features movements constrained by axes.

6. Describe fillets and chamfers in 3D.
> A 3D fillet must have a radius and can be performed instantly on any edge, as well as dynamically in real time. A 3D chamfer is similar.

CHAPTER 24 REVIEW QUESTIONS

1. What are the three Boolean operations?
> Union, subtract, and intersect.

2. In what toolbar are they found?
> Modeling toolbar.

3. Is union a permanent setting?
> Yes

4. Why is it a good idea to make a copy of something that is being subtracted?
> Because you may need that part for later design work and the subtract operation deletes it.

CHAPTER 25 REVIEW QUESTIONS

1. Describe the revolve command.
> This command creates a revolved shape around a chosen axis of revolution.

2. Does the profile need to be open or closed? Describe the effect of each.
> Either way works; however, a closed profile yields a solid, and an open one creates a surface of revolution.

3. Describe the shell command.

   This command creates a hollow cavity inside a solid based on a given wall thickness.

4. How many surfaces can you select with shell?

   You can select a minimum of one and a maximum of all of them. Typically, only one or two are selected.

5. Describe the taper command.

   This command tapers (adds slope to) a solid.

6. Describe the loft command.

   This command creates a solid out of a sequence of profiles.

7. Describe the path extrusion command.

   This command creates a solid extrusion along an indicated pathway.

8. Describe the sweep command.

   The sweep command is similar to a path extrusion but more general in nature and creates a solid out of virtually any profile and along virtually any pathway.

9. Why is sweep better than path extrusion?

   It is more tolerant of geometric variations and inconsistencies.

CHAPTER 26 REVIEW QUESTIONS

1. Describe the polysolid command. Can you turn regular objects into polysolids?

   This command creates a solid shape out of linework or basic geometric objects. Yes.

2. Describe the helix command.

   This command creates a wire helix based on indicated radiuses of the top and bottom along with the number and spacing of the coils.

3. Describe the 3D path array command.

   The 3D path array works in a similar manner to the 2D path array. A solid is drawn, a path is selected (in our example – a helix), and the solid is duplicated along the path.

4. Describe the 3Dalign command.

   This command aligns objects in 3D space via a three-point tangency.

5. What are the six faces commands?

   Move face, offset face, delete face, rotate face, copy face, and color face.

6. What are the two edges commands?

   Copy edge and color edge.

CHAPTER 27 REVIEW QUESTIONS

1. Describe the planar surface command.

   A planar surface is a very simple mesh surface shading tool for basic geometry and surfaces.

2. Describe the region command.

   A region is similar to a planar surface but is not a mesh.

3. Describe rulesurf and its applications.

   Rulesurf (ruled surface) is used to create surfaces between straight lines or curves.

4. Describe tabsurf and its applications.

   Tabsurf (tabulated surface) is used to create surfaces along a path.

5. Describe revsurf and its applications.

Revsurf (revolved surface) is similar to the revolve command but produces surface revolutions.

6. Describe edgesurf and its applications.

The edgesurf command is used to create mesh (Coon) surfaces.

7. Describe 3Dface and 3Dmesh.

3Dface is a free form mesh creating tool, usually used to cover surfaces. 3Dmesh is used more for automation and programming, as it is too tedious to work with by hand.

8. What are the Smooth Mesh Primitives?

These are basic primitive shapes that are made up of mesh faces, which in turn can be worked with to "sculpt" an organic form.

CHAPTER 28 REVIEW QUESTIONS

1. List all 13 advanced UCS functions and what they do.

UCS, World, UCS Previous, Face UCS, Object, View, Origin, Z-Axis Vector, 3 Point, X, Y, Z, and Apply. See Chapter 28 for detailed descriptions of each.

2. What are vports useful for? List two functions.

Vports allow you to view and print a model from multiple angles all at once.

CHAPTER 29 REVIEW QUESTIONS

1. List all 11 dview functions and what they do.

Camera, Target, Distance Points, Pan, Zoom, Twist, Clip, Hide, Off, and Undo. See Chapter 29 for detailed descriptions of each.

2. List any additional views that are covered.

The camera tool provides for additional viewing tools.

3. What is the purpose of walk and fly?

To create a walking tour through a 3D model.

4. How are walk and fly functions controlled?

Via the keyboard arrows or hot keys.

5. How do you create path animations?

You first create a pathway, followed by setting up a camera and fine-tuning the animation by specifying the frame rate, resolution, and format.

6. What is an action recorder?

This tool allows you to record a set of commands and play back the animation to automate certain tasks.

CHAPTER 30 REVIEW QUESTIONS

1. Describe point light.

The point light is a direct lighting effect, and the most often used of the lighting options. It shines evenly from a point in space indicated by the user.

2. Describe spotlight.

The spotlight is somewhat similar to the point light, but more concentrated, akin to a flashlight, with less reach through walls, casting softer shadows.

3. Describe distant light.

The distant light is similar to a floodlight, saturating everything in its cone.

4. Describe sun and sky.

Sun and sky refers to the background lighting and the ability to set a particular worldwide location, which in turn accounts for latitude and longitude.

5. Describe how to bring up materials.

Materials are accessed via a palette called the Materials Browser (also called the Materials Editor) and can be clicked and dragged onto shapes.

6. Describe basic rendering procedures.

Rendering involves the render command and saving the output to a file such as in jpg format.

Appendices

Additional Information on AutoCAD

WHO MAKES AUTOCAD?

As you may already know, AutoCAD is designed and marketed by Autodesk®, Inc. (NASDAQ: ADSK), a software giant headquartered in San Rafael, California. Its website (a good resource center) is www.autodesk.com. The company was founded in 1982 by programmer John Walker and his associates and went public in 1985 (see history). Walker was one of AutoCAD's early developers, building on original source code purchased from another programmer, Michael Riddle. Autodesk today is headed by president and CEO Carl Bass; has over 7300 employees, down after some layoffs in early 2009; and posted revenues of $1.71 billion in FY 2010, down from $2.31 billion in FY 2009. Although the company markets nearly 100 software applications, AutoCAD remains its flagship product and biggest source of revenue.

It is difficult to pin down the exact size of AutoCAD's worldwide "installed base," meaning the number of copies of the software that have been purchased and used over the years. Some estimates from 2008 show this base to be 2.8 million units of AutoCAD and 3.6 million units of AutoCAD LT. An Autodesk press release from the same year claimed 4.16 million seats, but did not specify if that was only AutoCAD or LT as well. Industry analysts have estimated that Autodesk currently sells about 40,000 copies of AutoCAD per quarter, or 160,000 seats per year. Due to piracy, the actual installed base is probably significantly higher. Also, the number of seats can vary depending on how you count educational versions and those "add-on" vertical software purchases that need AutoCAD to run but are counted separately.

Any way you look at this, though, these numbers indicate that AutoCAD is indisputably the world's number one drafting software, with a market share often estimated to be somewhere in the 55–60% range in the 2D world and 30–35% among all CAD software (both 2D and 3D) worldwide. Considering that there are over a dozen major competitors and hundreds of smaller ones, this market dominance is likely to assure AutoCAD's relevance for years to come.

WHAT IS AUTOCAD LT?

Contrary to popular belief, LT does not stand for "lite" (or as we joke in class, "AutoCAD with fewer calories"). It also does not mean "less trouble" or "learning and training," or even "limited," although Autodesk likes to throw that last term around during conferences. Instead, it means "laptop" and was originally meant for slower and weaker early laptop computers, a sort of portable AutoCAD, if you will. Today laptops are just as powerful as desktops and LT exists to compete in the lower-end CAD market with similarly priced

679

software. It is essentially stripped-down AutoCAD. What is missing is detailed exhaustively on the Autodesk website, but it essentially boils down to no 3D ability, no advanced customization, and no network licensing. In addition to these, some other minor missing features are scattered throughout the software, but the differences are *not* that great, and aside from the lack of 3D, it takes a trained eye to spot LT. It really looks and acts like AutoCAD, and with good reason. It was created by taking regular AutoCAD and "commenting out" (thereby deactivating) portions of the C++ code that Autodesk wanted to reserve for its pricier main product.

Having said this, however, AutoCAD LT is an excellent product (its estimated sales outnumber AutoCAD) and is recommended for companies that do not have ambitious CAD requirements. The simple truth and something Autodesk does not always promote is that some users never need the full power of AutoCAD and do just fine with LT, while paying a fraction of the cost of full AutoCAD. Electrical engineers especially need nothing more than LT for basic linework and diagrams, as do architects and mechanical engineers who deal with schematics. The size of the files you anticipate working on, by the way, is of no consequence. LT lacks certain features, not the ability to work with multistory skyscraper layouts if needed. It truly is a hidden secret in CAD. Autodesk must have been taken by surprise by the strong LT sales since 2002 and dramatically raised the price of the software from $700 to $1200 in 2008.

HOW IS AUTOCAD PURCHASED AND HOW MUCH DOES IT COST?

A number of years ago, you could not have purchased AutoCAD at a retail store or from Autodesk itself. You had to contact an authorized dealer (or reseller) and the sale was then usually arranged by phone, paid for with a credit card, and AutoCAD was shipped via registered mail. You can still do this, and some customers continue to; however, since Release 2000, Autodesk's website is configured for direct purchase. Look under Products, then AutoCAD, in the main section of the site. AutoCAD is still not available at retail outlets, although AutoCAD LT always has been, and continues to be, sold in some retail computer stores.

How much AutoCAD costs can be a more convoluted question. If purchasing from a dealer, called an *authorized reseller*, prices may vary but are generally the lower "street prices." If purchasing from the Autodesk site, however, the prices are fixed. Currently, AutoCAD 2012, the latest release, is shipping for $3995. A download is $3975 (you do not have to pay for shipping and the cardboard box). An upgrade from 2011, 2010, and 2009 is $1995 and higher from older versions (AutoCAD 2008 and earlier). This is an incentive, and not a well-liked one, to force customers to upgrade now for fear of paying more later on. AutoCAD LT, as mentioned, is now $1200 retail. In general, use these prices only as guidelines; they may be lower with dealers, and of course this is all subject to change. AutoCAD LT can be found for much less at various online retailers such as Amazon.com.

There are also variations in cost if you are buying only AutoCAD versus buying AutoCAD with vertical add-ons. The cost also varies depending on how many copies you are buying (for large volume purchases). With a dealer, the purchase may come bundled with a telephone or an on-site support contract for any anticipated problems or training needs. Sometimes, these are sold separately. Autodesk in general, like other software firms, has moved toward a subscription model where you pay a yearly fee to own AutoCAD but upgrades are free. The subscription price for AutoCAD 2012 is $4445, shipped or downloaded. This makes more sense to larger firms that anticipate upgrading often. If you are tasked with purchasing AutoCAD for your company, explore all your options and shop around. You also need not upgrade as often as Autodesk suggests.

ARE THERE SIGNIFICANT DIFFERENCES BETWEEN AUTOCAD RELEASES?

This is a common question at the start of training sessions. Many schools are authorized training centers and, as such, have the latest version of AutoCAD installed. That may or may not be the case with the students and their respective companies. No issue seems to cause as much confusion and anxiety as this one. To put your fears to rest, if you are learning AutoCAD 2012 (this book) and have 2011, 2010, 2009, 2008, or earlier versions at home or at work, do not worry. The differences, while real, are for the most part relatively minor and cosmetic (such as the introduction of the Ribbon in 2009). While AutoCAD has certainly been modernized, it has *not* fundamentally changed since it moved to Windows and away from DOS over 17 years ago and finally stabilized with R14 in 1997. If you were to open up that release, you would feel right at home with 90% of the features. Purists and AutoCAD fanatics may scoff, but although there have been many great enhancements, the fact remains that the vast majority of what is truly important has already been added to AutoCAD and new features are incremental improvements. Most of what you learn in one version is applicable to several versions backward and forward, especially at a beginner's level.

IS THERE AN AUTOCAD FOR THE MAC?

The short answer is "Yes." After a nearly 20 year absence, Autodesk announced in August 2010 that the Mac platform is once again supported, and AutoCAD for the Mac was shipped in November of that year. Why was AutoCAD offered for the Mac up until 1992, then removed from the market, only to be reintroduced now? The reasons are all purely financial ones. In the late 1980s and early 1990s, for those old enough to remember, the PC was not yet the dominant force it is today. Mac and other manufacturers were all strong competitors and there were many choices on the market. Autodesk decided to go with the leading two—PC and Mac—and invested in both. As the PC became the de facto world computer standard and Apple went into a decline, resources had to be focused on the PC, as it takes a lot of effort to reprogram and support a major piece of software for another configuration.

We all of course know what happened to Apple Inc. Under leadership form Steve Jobs, it roared back strong with a line of high-demand consumer products, such as the iPod and iPhone. Its computers also regained some of its former market share with superb desktop and laptop offerings.

Autodesk took notice of these developments! Such is the nature of the software business: Your fortunes are closely tied in with the ups and downs of hardware, operating systems, and chip makers. The Mac once again began to look like an attractive platform and Autodesk invested in reprogramming AutoCAD to support it. When exactly this effort started is not known outside of the company, but it was likely authorized a few years ago, as it takes a lot of lead time to rework code of such complexity. This move should net Autodesk quite a few new sales of AutoCAD and spells trouble for other Mac-based CAD software like VectorWorks, which became popular precisely *because* AutoCAD was not available for the Mac and catered to those designers who refused to switch to a PC. Figure A.1 is a screen shot of the new software from Autodesk's promotional literature.

It of course remains to be seen how successful AutoCAD for the Mac will be or even how good the software is. It has been rewritten from the ground up as a native Mac application, not simply a port from Windows. Upon first look, the release seems to be a bit of a "work in progress," as evidenced by the fact that some standard AutoCAD features are not yet supported. The full list is available on the Autodesk website, but it includes the following. If you completed Levels 1, 2, and 3 of this textbook, you should recognize all these tools.

681

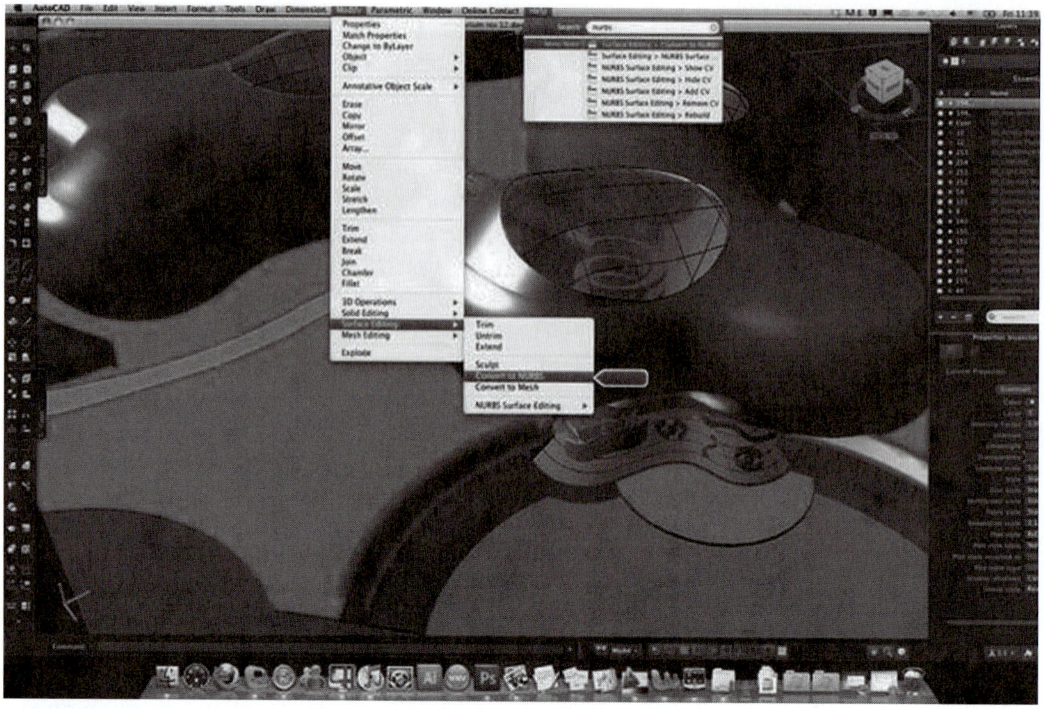

FIGURE A.1
AutoCAD for the Mac screen shot.

Partial list of AutoCAD for the Mac missing features (as of January 2011):

- 3D Walk and Fly
- Action Recorder
- Standards Checker
- "Classic" dialogs for xref, and layers
- Data Extraction Wizard—EATTEXT
- Design Center
- Dynamic block authoring
- eTransmit
- Express Tools
- Filters
- Hyperlinks
- In-place editing of blocks or xrefs
- Layer filters
- Layer State Manager
- Layer Walk
- Navigation bar and Steering Wheel
- OLE objects
- Plot style configuration
- Plotting to file
- Purge dialog box
- Sheet Set Manager
- Table style setup
- Tool palettes

Some of these are reasonably important, some are more obscure, and it is likely many of these issues will be resolved in the coming releases. Other than that, AutoCAD for the Mac is not entirely different from the PC version; after all, it still has to behave like AutoCAD. Although the Ribbon is removed, the cascading menus and toolbars are still present, and the software can accept files from its PC-based cousins. Much more will be written about this version in the next edition of this textbook as the new software matures.

A BRIEF HISTORY OF AUTODESK AND AUTOCAD

The father of Autodesk and AutoCAD is John Walker. A programmer by training, he formed Marinchip Systems in 1977 and, by 1982, seeing an opportunity with the emergence of the personal computer, decided to create a new "software only" company with business partner Dan Drake. The plan was to market a variety of software, including a simple drafting application they acquired earlier. Based on code written in 1981 by Mike Riddle and called MicroCAD (later on changed briefly to Interact), the software would sell for under $1000 and would run on an IBM PC. It was just what they were looking for to add to their other products.

Curiously, the still unnamed company did not see anything special in the future of only MicroCAD and focused its energy on developing and improving other products as well. Among them were Optical (a spreadsheet), Window (a screen editor), and another application called Autodesk or Automated Desktop, which was an office automation and filing system meant to compete with a popular, at the time, application called Visidex. This Automated Desktop was seen as the most promising product; and when it came time to finally incorporate and select a name, Walker and company (after rejecting such gems as Coders of the Lost Spark) settled on Autodesk Inc. as the official and legal name.

Written originally in SPL, MicroCAD was ported by Walker onto an IBM machine and recoded in C. As it was practically a new product now, it was renamed AutoCAD, which stood for Automated Computer Aided Design. AutoCAD-80 made its formal debut in November 1982 at the Las Vegas COMDEX show. As it was new and had little competition at the time, the new application drew much attention. The development team knew a potential hit was on their hands and a major corner was turned in AutoCAD's future. AutoCAD began to be shipped to customers, slowly at first, then increasing dramatically. A wish list of features given to the development team by customers was actively incorporated into each succeeding version—they were not yet called *releases*.

In June 1985, Autodesk went public and was valued at $70 million, in 1989 it was at $500 million, and in 1991 at $1.4 billion. With each release, milestones, large and small, such as introduction of 3D in Version 2.1 and porting to Windows with Releases 12 and 13, were achieved. In a story not unlike Microsoft, Autodesk, through persistence, innovation, firm belief in the future of personal computers, and a little bit of luck, came to dominate the CAD industry.

Figure A.2 shows perhaps the most famous early AutoCAD drawing of all, a nozzle drawn in 1984 by customer Don Strimbu on a very early AutoCAD version. This was the most

683

FIGURE A.2
A nozzle drawn by Don Strimbu in 1984 on a very early version of AutoCAD.

complex drawing on AutoCAD up to that point, used by the development team as a timing test for machines, and featured on the covers of many publications, including *Scientific American* in September 1984.

AUTOCAD RELEASES

Autodesk first released AutoCAD in December 1982, with Version 1.0 (Release 1) on DOS, and continued a trend of releasing a new version sporadically every one to three years until R2004, when they switched to a more regular once a year schedule. A major change occurred with R12 and R13, when AutoCAD was ported to Windows and support for UNIX, DOS, and Macintosh was dropped. R12 and R13 were actually transitional versions (R13 is better left forgotten), with both DOS and Windows platforms supported. It was not until R14 that AutoCAD regained its footing as sales dramatically increased. Figure A.3 is AutoCAD's release timeline. A release history follows:

    Version 1.0 (Release 1), December 1982
    Version 1.2 (Release 2), April 1983
    Version 1.3 (Release 3), August 1983
    Version 1.4 (Release 4), October 1983
    Version 2.0 (Release 5), October 1984
    Version 2.1 (Release 6), May 1985
    Version 2.5 (Release 7), June 1986
    Version 2.6 (Release 8), April 1987
    Release 9, September 1987
    Release 10, October 1988
    Release 11, October 1990
    Release 12, June 1992 (last release for Apple Macintosh until 2011)
    Release 13, November 1994 (last release for UNIX, MS-DOS, and Windows 3.11)
    Release 14, February 1997
    AutoCAD 2000 (R15.0), March 1999
    AutoCAD 2000i (R15.1), July 2000
    AutoCAD 2002 (R15.6), June 2001
    AutoCAD 2004 (R16.0), March 2003
    AutoCAD 2005 (R16.1), March 2004
    AutoCAD 2006 (R16.2), March 2005

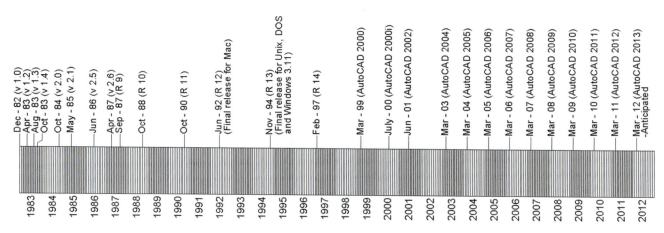

AutoCAD History Timeline

FIGURE A.3
AutoCAD timeline, 1982 through 2012.

AutoCAD 2007 (R17.0), March 2006
AutoCAD 2008 (R17.1), March 2007
AutoCAD 2009 (R17.2), March 2008
AutoCAD 2010 (R17.3), March 2009
AutoCAD 2011 (R17.4), March 2010
AutoCAD 2012 (R17.5), March 2011
AutoCAD 2013 (R17.6), March 2012 (anticipated)

MAJOR AUTODESK PRODUCTS

Autodesk markets almost 100 software products, some major and some obscure. Only one application sold today is the company's original product, which is AutoCAD. The rest have been developed in-house from scratch over the years, purchased outright from the original developers, or as is often the case, acquired in takeovers and mergers. Many products are designed to "hitch a ride" on the back of AutoCAD and extend its power to serve a niche market with additional functionality and menus not found in the base product. Those are often homegrown applications and are referred to as *verticals*. In other cases, Autodesk markets radically different software, such as one used for high-end animations and cartoons. Those have been acquired to extend the company's reach into new markets.

The following lists Autodesk's flagship products that you should know about, as you may come across them in the industry; also listed are the popular add-on products purchased with AutoCAD.

Flagship products:

- **AutoCAD.** Autodesk's best-selling and biggest product for computer-aided design.
- **AutoCAD LT.** A limited functionality version of AutoCAD for an entry-level market.
- **3ds Max.** A powerful high-end rendering and animation software product.
- **Autodesk AliasStudio.** A high-end surface modeling, sketching, and illustration product.
- **Inventor.** A mid-range parametric solid modeling and design package for engineering.
- **Revit.** A high-end parametric software for architectural and building design.
- **Maya.** A very high-end animation, effects, and rendering software.

Add-ons to AutoCAD (verticals):

- **AutoCAD Architecture.** Additional functionality for architecture design applications.
- **AutoCAD Civil.** Additional functionality for civil engineering applications.
- **AutoCAD Electrical.** Additional functionality for electrical engineering applications.
- **AutoCAD Land Desktop.** Additional functionality for land, civil, and environmental design.
- **AutoCAD Mechanical.** Additional functionality for mechanical engineering.
- **AutoCAD P&ID.** Additional functionality for creating piping and instrumentation diagrams.

These are just some of the major software applications in Autodesk's inventory. A significant number of products, including all the less known or obscure ones, have been omitted. A full and comprehensive list of all of them can be found on the Autodesk website. You are encouraged to look them over to stay ahead of the curve and know what other design professionals may be using.

AUTOCAD RELATED WEBSITES

Next is a listing of important and valuable websites for AutoCAD and general CAD related information. Online content has grown from occasionally useful to indispensable, and experienced AutoCAD users must have at least the following at their fingertips when the need arises. Note that link fidelity is not guaranteed, and except for the few major ones at the start of the list, the sites may and do change.

- **Autodesk Inc.** (www.autodesk.com). Mentioned previously, it is home base for the makers of AutoCAD. Here you find locations of AutoCAD dealers and training centers as well as endless information on AutoCAD and other software products and services. Well worth visiting occasionally.

- **Autodesk User Group International** (www.augi.com). The Holy Grail for all things related to AutoCAD. This site is the officially sanctioned user community for Autodesk and its products. AUGI puts out numerous publications and runs conferences and training seminars all over the United States and worldwide. Its website is packed full of tips, tricks, articles, and forums where you can ask the community your hardest AutoCAD questions. Claiming over 100,000 members, AUGI has the distinction of being the only organization that can officially submit "wish lists" directly to Autodesk, meaning that if you are a member and have an idea for a new feature or how to improve AutoCAD, the company just may listen to you.

- *Cadalyst* **magazine** (www.cadalyst.com). A great magazine dedicated to all things CAD, including solid modeling and rendering, but with a large amount of coverage dedicated specifically to AutoCAD. A must read for AutoCAD managers, the magazine features informative articles on all aspects of the job, including programming, industry trends, and advanced topics. Subscription was free up until 2010. You can still access free archived issues from as far back as 2004, but the magazine now charges $9.95 for the new issues.

- **CAD Block Exchange Network** (http://cben.net). Many sites post helpful articles; this site posts actual AutoCAD files. Blocks, blocks, and more blocks. If you need it and really do not want to draw it, check here first. Great resource for people, trees, and vehicle blocks for civil and landscape engineering, in plan, elevation, and 3D, though just about everything else can be found there, too.

- **Lynn Allen's blog** (http://lynn.blogs.com). You may have never heard of Lynn Allen if you are just starting out with AutoCAD, but hang around long enough and you probably will. She is what the AutoCAD community refers to as an Autodesk Technical Evangelist, an expert on all things AutoCAD. Although there are certainly many experts around, what sets Lynn apart is that she actually works for Autodesk and gets to learn AutoCAD directly from the developers. Her blog is the place to go to when a new release is about to hit the market. She will likely have already analyzed it and written a book by then. Lynn is also a prolific speaker, appearing at just about every AutoCAD conference, and has been a columnist for *Cadalyst* for many years as well. She writes and speaks well and employs humor in her approach to thorny technical issues.

- **AutoCAD Lessons on the Web** (www.cadtutor.net). A well organized and extensive site for learning AutoCAD that includes some tutorials on 3ds Max, VIZ, Photoshop, and web design.

- **The Business of CAD, Part I** (www.upfrontezine.com). The website describes itself thus: "Launched in 1985, the upFront.eZine e-newsletter is your prime independent source of weekly business news and opinion for the computer-aided design (CAD) industry." The site is a no-nonsense, cut through the hype source for CAD (mostly AutoCAD, but others, too) information, latest releases, and analysis of Autodesk business decisions; a great resource for a CAD manager.

- **The Business of CAD, Part II** (http://worldcadaccess.typepad.com). A spin-off blog from the preceding site, it also addresses news in the media about all things CAD and how they will affect you; recommended for the serious CAD manager or user.

- **John Walker's site** (www.fourmilab.ch). This site, while not essential AutoCAD reading, is still interesting, fun, and often quite informative. As co-creator of AutoCAD, former programmer John Walker has put together *the* definitive early history of how it all began. Fascinating reading from a fascinating man who is now "retired" in Switzerland and pursuing artificial intelligence research. His site is chock full of readings on numerous subjects, as you would expect from a restless and inquisitive mind; truly someone who changed computing history in a significant way.

As with many topics in this brief appendix, what is listed here only scratches the surface. Remember that you are not the only one out there using AutoCAD, so feel free to reach out and see what else is out there if you have questions or would just like to learn more. At a minimum, it is recommended to join AUGI and get a subscription to *Cadalyst*.

Other CAD Software, Design and Analysis Tools, and Concepts

While it is true AutoCAD is the market leader for drafting software, it certainly does not operate in a vacuum and is not the only game in town. The list that follows summarizes other major players in the drafting field. The reason you should know about them is because you may run into them in use and may have to exchange files if collaborating. Saying "I've never heard of that" is usually not a good way to start a technical conversation in your area of expertise, so take note of the following products.

689

MICROSTATION

A major (and really the only serious) competitor to AutoCAD, MicroStation is produced and marketed by Bentley Systems (*www.Bentley.com*). For historical reasons, the software is widely used in government agencies and the civil engineering community. It comes in full and PowerDraft variations (akin to AutoCAD and LT). The current version is MicroStation V8*i*. Its file extension is dgn. AutoCAD does not open these files directly, although that may soon change. MicroStation, on the other hand, does open AutoCAD's dwg files. Of course, many techniques and even third party software are available to quickly and easily convert one file format to another.

MicroStation is more icon driven but otherwise similar to AutoCAD in many ways. Its 3D capabilities are considered excellent, but its interface can be a bit cumbersome to use for a novice. Its main disadvantage is that it is not an industry standard; it is otherwise excellent software, and its ability to handle xrefs (multiple external file attachments) is unparalleled. Its users sometimes are a fanatical bunch, promoting its virtues against AutoCAD at every opportunity. An objective view of both applications yields an opinion that both have their pros and cons, and those who disparage one over the other usually do not know what they are talking about.

ARCHICAD

ArchiCAD is a popular application among architects, geared toward the Macintosh but also available for Windows. Developed by a Hungarian firm, Graphisoft (*www.graphisoft.com*), at around the same time as AutoCAD, ArchiCAD took a different approach, allowing for a parametric relationship between objects, as well as easier transitions to 3D, as all objects are created inherently with depth. The software has a loyal installed user base of about 100,000 architects worldwide and can interoperate fully with AutoCAD files. The current release is ArchiCAD 14.

TURBOCAD

TurboCAD (*www.turbocad.com*) is a mid-range 2D/3D drafting software, first developed and marketed in South Africa in 1985 and brought to the United States the following year. It was marketed as an entry level basic drafting tool (it once sold for as low as $49), but with constant upgrading and development, TurboCAD has grown over the years to a $1200–$1500 software product for the architectural and mechanical engineering community. It operates on both Windows and Macintosh. The current release is TurboCAD 16.

OTHER DESIGN SOFTWARE

Some other notable CAD applications include *RealCAD*, *VectorWorks* (formerly MiniCAD), *IntelliCAD*, *Form Z*, and *Rhino*, although the last two are often used only for modeling and visualization. This list is not exhaustive by any means, as hundreds of other CAD programs are available.

Many of my students are interested in other aspects of design, and questions often come up as to how AutoCAD fits into the "big picture" of a much larger engineering or design world. To address some of these questions, the following information, culled from a chapter written in 2007 as part of an undergraduate machine design theory workbook, is included. AutoCAD students, especially engineers, may find this informative. The subject matter is extensive and this brief overview only scratches the surface. Ideally, if the reader was not aware of these concepts and the engineering design and analysis software described, this serves as a good introduction and a jumping off point to further inquiry.

- **CAD (computer-aided design).** This is a broad generic term for using a computer to design something. It can also be used in reference to a family of software that allows you to design, engineer, and test a product on the computer prior to manufacturing it. CAD software can be used for primarily 2D drafting (such as with AutoCAD) or full 3D parametric solid modeling. The latter is the software that aerospace, mechanical, automotive, and naval engineers commonly use. Examples include CATIA, NX, Pro-Engineer, IronCAD, and SolidWorks, among others. Included under the broad category of CAD is testing and analysis software.
- **FEA (finite element analysis).** This is used for structural and stress analysis, testing thermal properties, and much more as related to frames, beams, columns, and other structural components. Examples of FEA software include NASTRAN, ALGOR, and ANSYS.
- **CFD (computational fluid dynamics).** This is software used to model fluid flow around and through objects to predict aerodynamic and hydrodynamic performance. Examples of CFD software include Fluent and Flow 3-D.
- **CAM (computer-aided manufacturing).** The basic idea is this: OK, we designed and optimized our product, now what? How do we get it from the computer to the machine shop to build a prototype? The solution is to use a CAM software package that takes your design (or parts that make up your design) and converts all geometry to information useable by the manufacturing machines, so the drills, lathes, and routing tools can get to work and create the part out of a chunk of metal. Examples of dedicated CAM software include Mastercam and FastCAM.
- **CAE (computer-aided engineering).** Strictly defined, CAE is the "use of information technology for supporting engineers in tasks such as analysis, simulation, design, manufacture, planning, diagnosis, and repair." We use this term to define an ability to do this all from one platform, as explained next.

As you read the CAD and CAM descriptions you may be wondering why CAD companies have not provided a complete solution to design, test, and manufacture a product using one software package. They have, to varying degrees of success. For a number of years, an engineering team designing an aircraft, for example, had to typically create the design (and subsystems) in a 3D program like CATIA, then import the airframe for FEA testing into NASTRAN, followed by airflow testing in CFD software (and a wind tunnel), and finally send each component through CAM software like Mastercam to manufacture it (a greatly simplified version of events, of course).

Obviously data could be lost or corrupted in the export/import process, and a lot of effort was expended in making sure the design transitioned smoothly from one software application to another. While the final product was technically a result of CAE, software companies had something better in mind: "one stop shopping." As a result, CAE today means one source for most design and manufacturing needs. CATIA for example has "workbenches" for FEA analysis and CAM. Code generated by CATIA is imported into the CNC machines that (typically after some corrections and editing) produce virtually anything you can design, with some limitations, of course.

As you may imagine, not everything is rosy with this picture. As the specialized companies got better and better at their respective specialties (CAM, FEA, or CFD), it became harder and harder for the CAD companies to catch up and offer equally good solutions embedded in their respective products. Mastercam is still the undisputed leader in CAM worldwide, and NASTRAN is still the king of FEA. CATIA's version of these has gotten better but is still a notch behind. The situation is the same with Pro/E, NX, and other CAD software.

These high-end 3D packages are also referred to as *product lifecycle management* (PLM) software, a term you will hear a lot of in the engineering community. As described already, these software tools aim to follow a product from design and conception all the way through testing and manufacturing in one integrated package.

691

- **CNC/G-Code (computer numerical code).** This is the language (referred to as *G-Code*), created by the CAM software and read by the CNC machines. G-Code existed for decades. It was originally written by skilled machinists based on the geometry of the component to be manufactured (and sound manufacturing practices). This code can still be written by hand. It is not hard to understand. Most lines start with G, hence its name; and they all describe some function that the CNC machine does as related to the process. After some practice, the lines start to make sense and you can roughly visualize what is being manufactured, just by reading them.

Here are some more specific details on the CAD software mentioned in the preceding section. You may run into many of these throughout your engineering career. It is well worth being familiar with these products. The information is up to date as of press time.

CATIA

Computer Aided Three-dimensional Interactive Application (CATIA) is the world's leading CAD/CAM/CAE software. Originally developed in France by Dassault (*www.3ds.com*) for its Mirage fighter jet project (although originally based on Lockheed's CADAM software), it is now a well-known staple at Boeing and many automotive, aerospace, and naval design companies. It has even been famously adopted by architect Frank Geary for certain projects. CATIA has grown into a full PLM solution and is considered the industry standard, although complex and with a not always user-friendly reputation. The latest version is CATIA V6 R2011.

NX

Another major CAD/CAM/CAE PLM software solution is from a company formerly called UGS (*www.ugs.com*). This package is a blend of the company's Unigraphics software and SDRC's popular I-DEAS software, which UGS purchased and took over. Siemens AG now owns UGS and the NX software. It is used by GM, Rolls Royce, Pratt & Whitney, and many others. The current version is NX 7.5.

PRO/ENGINEER

The third major high-end CAD/CAM/CAE PLM software package, Pro/E (*www.ptc.com*) is historically significant. It caused a major change in the CAD industry when first released (in 1987) by introducing the concept of parametric modeling. Rather than models being constructed like a mound of clay with pieces added or removed to make changes, the user constructs the model as a list of features, which are stored by the program and can be used to change the model by modifying, reordering, or removing them. CATIA and NX also now work this way. The current version is no longer called Pro/Engineer; it was changed to Creo Elements/Pro 5.0 in October 2010.

SOLIDWORKS

SolidWorks (*www.solidworks.com*) is a mid-range CAD/CAM/CAE software product, first introduced in 1995 by a Massachusetts company of the same name. Dassault, the makers of CATIA, acquired the company in 1997. The SolidWorks installed base is estimated at 1.3 million users, with FY 2009 revenues of over $350 million, making the product a major player in the 3D CAD field. SolidWorks is marketed as a lower-cost competitor to major CAD packages, but the software still has extensive capabilities and caught on quickly in smaller design companies. The current version is SolidWorks 2011 SP1.

INVENTOR

Inventor (*www.autodesk.com*) is another mid-range product, from the makers of AutoCAD, for solid modeling and design. Considered a low- to mid-range product when first introduced, Inventor matured to become a major competitor to SolidWorks and even Pro/Engineer. The current version is Inventor 2011.

IRONCAD

IronCAD (*www.ironcad.com*) is another vendor in the mid-range market. The software gained ground in recent years due to a user-friendly reputation and a more intuitive approach to 3D design. The current version is IronCAD XG 2011.

SOLID EDGE

Solid Edge (*www.solidedge.com*) is also a mid-range product from UGS (now owned by Siemens AG), the same company that makes the high-end NX package. It is comparable to SolidWorks in cost and functionality. The current release is Solid Edge ST3.

Listed next are some of the major software products in FEA and CFD, as mentioned earlier.

NASTRAN

Originally developed for NASA as open source code for the aerospace community to perform structural analysis, NASA structural analysis, or NASTRAN, was acquired by several corporations and marketed in numerous versions. NEiNastran (*www.NEiNastran.com*) is just one (commonly used) flavor of it. NASTRAN in its pure form is a solver for finite element analysis and cannot create its own models or meshes. Developers, however, have added pre- and postprocessors to allow for this. The current release for NEiNastran is V9, though other companies have different designations for their products.

ANSYS

ANSYS (*www.ansys.com*) is a major FEA product for structural, thermal, CFD, acoustic, and electromagnetic simulations.

ALGOR

ALGOR (*www.algor.com*) is another major FEA product for structural, thermal, CFD, acoustic, and electromagnetic simulations.

FLUENT

Fluent (*www.Fluent.com*) is the industry standard for CFD software, holding about 40% of the market. It imports geometry, creates meshes (using GAMBIT software) and boundary conditions, and solves for a variety of fluid flows, using the Navier-Stokes equations as theoretical underpinnings. The company was recently acquired by ANSYS. The current release is Fluent 6.3.

File Extensions

File extensions are the bane of the computer user's existence. They number in the many hundreds (AutoCAD alone claims nearly 200) and even if you pare the list down to a useful few, it is still quite a collection. Here, we focus on those extensions that are important to an AutoCAD user, as chances are you will encounter most, if not all, of them while learning the software top to bottom. They are listed alphabetically (with explanations and descriptions) but grouped in categories to bring some order to the chaos and set priorities for what you really need to know and what you can just glance over. Following AutoCAD's list is a further listing of extensions from other popular, often-used software, which you may find useful if you are not already familiar with most of them.

AUTOCAD PRIMARY EXTENSIONS

- `.bak` **(backup file).** A `bak` is an AutoCAD backup file created and updated every time you save a drawing, provided that setting is turned on in the Options dialog box. The `bak` file has the same name as the main `dwg` file, generally sits right next to it, and can be easily renamed to a valid `dwg` if the original `dwg` is lost.
- `.dwf` **(design web format file).** The `dwf` is a format for viewing drawings online or with a `dwf` viewer. The idea here is to share files with others who do not have AutoCAD in a way that enables them to look but not modify. If this sounds like Adobe `pdf`, you are correct; `dwf` is a competing format.
- `.dwg` **(drawing file format).** The `dwg` is AutoCAD's famous file extension for all drawing files. This format can be read by many of Autodesk's other software products as well as by some competitors. The format itself changes somewhat from release to release, but these are internal programming changes and invisible to most users.
- `.dwt` **(drawing template).** The `dwt` is AutoCAD's template format and is essentially a `dwg` file. The idea is to not repeat setup steps from project to project but use this template. It is equivalent to just saving a completed project as a new job and erasing the contents.
- `.dxf` **(drawing exchange format).** The `dxf` is a universal file format developed to allow for smooth data exchange between competing CAD packages. If an AutoCAD drawing is saved to a `dxf` file, it can be read by most other design software (in theory anyway). This of course is not always the case, and some CAD packages open `dwg` files anyway without this step, but the `dxf` remains an important tool for collaboration. The `dxf` is similar in principle and intent to IGES and STEP for those readers who may have worked with solid modeling software and understand what those acronyms mean.

AUTOCAD SECONDARY EXTENSIONS

- `.ac$` A temporary file generated if AutoCAD crashes; it can be deleted if AutoCAD is not running.

- `.ctb` This color table file is created when pen settings and thicknesses are set (Chapter 19 topic).
- `.cui` This customizable user interface file is for changes to menus and toolbars (Chapter 14 topic).
- `.err` This error file is generated upon an AutoCAD crash. You can delete it.
- `.las` This layer states file can be imported and exported (Chapter 12 topic).
- `.lin` This linetype definition file can be modified and expanded (see Appendix D).
- `.lsp` This LISP file is a relatively simple language used to modify, customize, and automate AutoCAD to effectively expand its abilities.
- `.pat` This hatch definition file can be modified and expanded (see Appendix D).
- `.pgp` This program parameters file contains the customizable command shortcuts (Chapter 14 topic).
- `.scr` This script file is used for automation (mentioned in Chapter 12).
- `.sv$` This is another temporary file for the AutoSave feature. It appears if AutoSave is on and remains in the event of a crash. It can be then renamed to `dwg` or deleted.

MISCELLANEOUS SOFTWARE EXTENSIONS

These extensions are for software and formats likely to be encountered daily or at least occasionally by AutoCAD users and are always good to know. Some extensions can be found in AutoCAD as export/import choices.

- `.ai` Adobe Illustrator files.
- `.bmp` Bitmap, a type of image file format that, like `jpg`, uses pixels.
- `.dgn` The MicroStation file extension. This competitor to AutoCAD can open native `dwg` files. Appendix B discusses this software.
- `.doc` MS Word files.
- `.eps` Encapsulated PostScript file.
- `.jpg` The jpeg (joint photographic experts group) is a commonly used method (algorithm) of compression for photographic images. It is also a file format (discussed in Chapter 16).
- `.pdf` The famous Adobe Acrobat file extension. You can print any AutoCAD drawing to a `pdf` assuming Acrobat is installed.
- `.ppt` MS PowerPoint file extension.
- `.psd` The famous Photoshop file extension.
- `.vsd` MS Visio drawing file extension. Visio is a simple and popular "drag and drop style" CAD package for electrical engineering schematics and flowcharts, among other applications. It accepts AutoCAD files.
- `.xls` MS Excel file extension.

This list is by no means complete, but it covers many of the applications discussed in Chapter 16.

Custom Linetypes and Hatch Patterns

This appendix outlines the basic procedures for creating custom linetypes and hatch patterns if you need something that is not found in AutoCAD itself or with any third party vendor. All linetypes and hatch patterns were originally "coded" by hand and there is nothing inherently special or magical about making them; they just take some time and patience to create. Though it is admittedly rare for a designer to need something that is not already available, creating these entities is a good skill to acquire. We just present the basics here and you can certainly do additional research and come up with some fancy designs and patterns.

LINETYPE DEFINITIONS (BASIC)

All linetypes in AutoCAD (about 38 standard ones, to be precise) reside in the `acad.lin` or the `acadiso.lin` file. These are linetype definition files that AutoCAD accesses when you tell it to load linetypes into a drawing. In AutoCAD 2012, these files can be found in the Program Files\AutoCAD 2012\backup folder. Anything you do to these files, including adding to them, immediately shows up the next time you use linetypes. Our goal here is to open up the `acad.lin` file, analyze how linetypes are defined, then make our own.

Locate and open (usually with Notepad) the `acad.lin` file and take a look at it closely. Here is a reproduction of the first six linetype definitions: Border (Standard, 2, and X2) and Center (Standard, 2, and X2).

```
*BORDER,Border __ __ . __ __ . __ __ . __ __ . __ __ .
A,.5,2.25,.5,2.25,0,2.25
*BORDER2,Border (.5x) __.__.__.__.__.__.__.__.__.
A,.25,2.125,.25,2.125,0,2.125
*BORDERX2,Border (2x) ____ ____ . ____ ____ . ____
A,1.0,2.5,1.0,2.5,0,2.5
*CENTER,Center ____ _ ____ _ ____ _ ____ _ ____ _ ____
A,1.25,2.25,.25,2.25
*CENTER2,Center (.5x) ___ _ ___ _ ___ _ ___ _ ___ _ ___
A,.75,2.125,.125,2.125
*CENTERX2,Center (2x) _____ __ _____ __ _____
A,2.5,2.5,.5,2.5
```

Let us take just one linetype and look at it closer. Here is the standard size center line:

```
*CENTER,Center ____ _ ____ _ ____ _ ____ _ ____ _ ____
A,1.25,2.25,.25,2.25
```

The two lines that you see are the header line and the pattern line; both are needed to properly define a linetype. The header line consists of an asterisk, followed by the name of the linetype (CENTER), then a comma, and finally the description of the linetype (Center). In this case, they are the same.

The pattern line is where the linetype is actually described, and the cryptic-looking "numerical code" (A,1.25,2.25,.25,2.25) is the method used to describe it and needs to be discussed for it to make sense.

The A is the alignment field specification and is of little concern (AutoCAD accepts only this type so you always see A). Next are the linetype specification numbers, and they actually describe the linetype using dashes, dots, and spaces. Let us go over them, as really this is the key concept.

The basic elements are

- Dash (pen down, positive length specified, such as 0.5).
- Dot (pen down, length of 0).
- Space (pen up, negative length specified, such as –0.25).

Now, you just have to mix and match and arrange these in the order you want them to be in to define a linetype that you envisioned.

What would this look like?

    (A,.75,2.25,0, 2.25, .75)

Well you know that dashes are positive, so everywhere you see .75, there will be a dash of that length. Spaces are negative, so everywhere you see 2.25, there will be a space of that length, and finally a 0 is a dot. This is the result:

    ___ . ___

These are the essentials of creating basic linetypes. AutoCAD, however, also allows for two other types to be created, and they are part of the complex linetype family: string complex and shape complex.

LINETYPES (STRING COMPLEX AND SHAPE COMPLEX)

The problem with the previously mentioned linetypes is obvious; you have only dashes, dots, and spaces with which to work. While this is actually enough for many applications, you may occasionally (especially in civil engineering and architecture) need to draw boundary or utility lines that feature text (Fence line, Gas line, etc.) or you may want to put small symbols between the dashes. It is here that strings and shapes come in.

String complex linetypes allow you to insert text into the line. Here is a gas line example:

    *GAS_LINE,Gas line ----GAS----GAS----GAS----GAS----GAS----GAS--
    A,.5,2.2,["GAS",STANDARD,S=.1,R=0.0,X=20.1,Y=2.05],2.25

The header line is as discussed before, and the pattern line still starts with an A. After the A, we see a dash of length .5 followed by a space of length .2; so far it is familiar. Next is information inside a set of brackets, defined as follows:

    ["String", Text Style, Text Height (S), Rotation (R), X-Offset (X),
    Y-Offset (Y)].

Finally there is a .25 space and the string repeats. The meaning of the information inside the brackets is as follows:

- String. This is the actual text, in quotation marks (GAS, in our example).
- Text Style. The predefined text style (STANDARD, in our example).

- `Text Height`. The actual text height (`.1`, in our example).
- `Rotation`. The rotation of the text relative to the line (`0.0`, in our example).
- `X-Offset`. The distance between the end of the dash and the beginning of the text as well as the end of the text and the beginning of the dash (a space of `.1`, in our example).
- `Y-Offset`. The distance between the bottom of the text and the line itself. This function centers the text on the line.

With these tools you can create any text string linetype.

Shape complex linetypes are basically similar, aside from one difference; instead of a text string, we have a shape definition inserted. Shape definitions are a separate topic in their own right and are not covered in detail in this textbook. Briefly, however, shape files are descriptions of geometric shapes (such as rectangles or circles but can be more complex) and are written using a Text Editor and compiled using the compile command. They are saved under the `*.shp` extension. AutoCAD accesses these when it reads the linetype definitions. If you are familiar with shapes, then you can proceed in defining shape complex linetypes as with string complex ones.

HATCH PATTERN DEFINITIONS (BASIC)

Knowing the basics of linetype definitions puts you on familiar ground when learning hatch pattern definitions. Even though they are more complex, the basics and styles of description are similar.

Hatch pattern definitions are found in the `acad.pat` and `acadiso.pat` files, and they reside in the same folder as the `.lin` files. Open the `acad.pat` file and take a look at a few definitions. Some of them can get long and involved due to the complexity of the pattern (Gravel, for example), but Figure D.1 shows a simpler one to use as a first example: Honeycomb.

The pattern definition file that describes it looks like this:

```
*HONEY, Honeycomb pattern
0, 0,0, .1875,.108253175, .125,2.25
120, 0,0, .1875,.108253175, .125,2.25
60, 0,0, .1875,.108253175, 2.25,.125
```

The first line (header line) is as discussed previously with linetypes. The next three lines are the hatch descriptors, and they have somewhat different definitions as shown next:

```
Angle, X-Origin, Y-Origin, D1, D2, Dash Length
```

where (in this particular example),

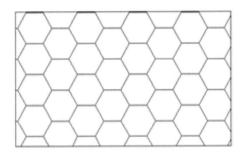

FIGURE D.1
Honeycomb pattern.

- `Angle`. Angle of the pattern's first hatch line from the X axis (0°, in our example).
- `X-Origin`. X coordinate of hatch line (0,0, in our example).
- `Y-Origin`. Y coordinate of hatch line (.1875, in our example).
- `D1`. The displacement of second line (.1083, in our example).
- `D2`. The distance between adjoining lines (.125, in our example).
- `Length`. Length of dashes or spaces.

The pattern is repeated two more times to form the overall shape and repeated many more times in the overall hatch pattern.

Principles of CAD Management

So, you have been hired on as a CAD manager or maybe forced into this due to your substantial AutoCAD skills—or was it because no one else wanted the job? Either way, you are now the top dog, head honcho, and have to take the bull by the horns, well many bulls actually, coworkers who run around and draft whatever they please just to get the job out. Then again, maybe you are simply a student finishing your AutoCAD education. The bottom line is that you could use some ideas and a short discussion on this elusive science of CAD management. Well, you came to the right appendix.

All jokes aside, this is an important topic. You must take everything you learned in Levels 1 and 2 and apply it, while keeping others in line. You must be the keeper and the enforcer of standards and the technical expert who can resolve any issue. A true CAD manager who does nothing else is not all that common. You may have to multitask and do your own work while doing the management, especially in smaller companies.

Here are a few rules, ideas, and thoughts gathered from my 15+ years of experience as a CAD consultant, educator, and CAD manager. They outline some of the challenges and issues, and provide some possible answers, to those in charge of the software used to design a lot of what is around us.

PART 1. KNOW THE SEVEN GOLDEN RULES OF AUTOCAD

1. Always draw one-to-one (1:1).
2. Never draw the same item twice.
3. Avoid drawing standard parts.
4. Use layers correctly.
5. Do not explode anything.
6. Use accuracy: Ortho and OSNAPs.
7. Save often.

Notice that most of those rules were discussed in some way in the first few chapters of Level 1. It is the basics people screw up, not some fancy new feature. Let us briefly go over these one by one; they bear repeating.

1. **Always draw one-to-one (1:1).** This is the most important one. Always draw items to real life sizes. Endless problems arise if users decide to scale items as they draw them. Leave scales and scaling to the pencil-and-paper folks. A drawing must pass a simple test. Measure anything that is known to be a certain value, for example, a door opening that is 3 feet wide, by using the list command. AutoCAD should tell you its value as 36 inches or 3 feet, and nothing else. You must enforce this as a CAD manager.
2. **Never draw the same item twice.** In hand drafting, just about everything needed to be drawn one item at a time. Some inexperienced CAD users do the same in AutoCAD. That

is not correct. AutoCAD and computer drafting in general require a different mindset, philosophy, and approach. If you have already drawn something once, then copy it, do not create it again. While this one may be obvious, another overlooked tool is the block or wblock command. If you anticipate using the new item again (even if there is a small chance), then make a block out of it. As a CAD manager, you need to sometimes remind users of this. Remember: Time = $$.

3. **Avoid drawing standard parts.** This, like the preceding rule, aims to prevent duplicate effort. Before starting a project, as a CAD manager, you need to see to it that any standard parts (meaning ones that are not unique but constantly reused in the industry) are available for use and are *not* drawn and redrawn. Chances are good that you either already have these parts or can get the "cads," as they say, from the part supplier. Another resource is the web. Sites such as www.cben.net and many others have pre-drawn blocks available for use. Buy them, borrow them, or download them, but do not redraw items that someone already has done before. Again, Time = $$.

4. **Use layers correctly.** Layers are there to make life easier, and they are your friends. While overdoing it is always a danger (500+ layers is pushing reason), you need to use layers and place items on the appropriate ones. As a CAD manager, you need to develop a layering standard (more on that later) and enforce it. Users in a hurry do not always appreciate what having all items organized in layers does ("it all comes out the same on paper!" they may say), so you must stand firm and enforce the standard. And do not use layer zero. It cannot be renamed with a descriptive name, so why use it?

5. **Do not explode anything.** How do you destroy an AutoCAD drawing and render weeks of work and thousands of dollars in work hours useless? Why, explode everything, of course. Short of deleting the files, this is the best way to ruin them. While this is generally a far-fetched scenario, remind users to not ruin a drawing in small ways, like once in a while exploding an mtext, dimension, or hatch pattern. They are liable to do so, if these items do not behave as they are supposed to and users explode them to manually force them into submission. Not a good idea; lack of skill is not an excuse. Remind them to ask you for help if a dimension just does not look right. There is always a way to fix something. Leave the exploding for the occasional block or polygon.

6. **Use accuracy: Ortho and OSNAPs.** Drawing without accuracy is not drafting, it is doodling. Most users learn the basics, such as how to draw straight lines and connect them, but you may still see the occasional project manager (who gets to draft on average only once a month) fix a designer's linework and forget to turn on OSNAPs. When zoomed out, the lines look OK and print just fine (reason enough to throw accuracy out the window for the occasional user), but a closer inspection reveals they are not joined together, causing a world of problems for a variety of functions such as hatch, area, distance, and other commands. Besides, it is just plain sloppy. How accurate is AutoCAD? Enough to zoom from a scale model of the earth to a grapefruit (more on that later), so be sure to use all the available accuracy.

7. **Save often.** Let us state the obvious. There is no crash-proof software. AutoCAD does eventually lock up and crash. Remind users to save often, and do daily and weekly file backups. But you already knew this, right?

PART 2. KNOW THE CAPABILITIES AND LIMITATIONS OF AUTOCAD

As a CAD manager, you have to know what you have to work with. All computer-aided design software, including AutoCAD, has an operating range for which the software is designed and optimized. Some tasks it excels at, and you need to leverage that advantage, and some tasks are outside the bounds of expectation. AutoCAD cannot be all things to all people (although you would be hard-pressed to tell based on its marketing). What follows is a short discussion on the capabilities and limitations and how you need to best take advantage of what you have.

Capabilities

Often, as a CAD manager, you are asked to provide an estimate on how long it will take to complete the drafting portion of a job or just recreate something in CAD. Sometimes, these requests come from individuals (senior management) who have had extensive experience in hand drafting, and they expect an answer in the form of a significantly smaller time frame. AutoCAD has an aura of magic about it, as if a few buttons just need to be pushed and out pops the design. Where *is* that "Finish Project" icon?

The surprising truth is that a professional hand drafter can give an average AutoCAD designer a run for the money if the request is a simple floor plan. After all, the drafter is not burdened with setting up layers, styles, dims, and other CAD prep work. But, this is not where AutoCAD's power lies. Once you begin to repeat patterns and objects, AutoCAD takes a commanding lead. A layout that took one hour to draft by hand or via computer can be copied in seconds on the computer, but adds another hour to the hand drafter's task.

Do not be surprised by the seeming obviousness of this. I have met many users and students who do not always see where exactly CAD excels; they think work is always faster on a computer. It may not be and may lead to erroneous job estimates. But, in all cases with patterns, copying, and use of stored libraries, CAD is much quicker. So when estimating a job, look for

- Instances where the offset command can be used.
- Instances where there are identical objects.
- Minimal cases of unique, uneven, and non-repetitive linework.
- Minimal cases of abstract curves.
- Abundance of standard items.

Knowing what to look for enables you to make more accurate estimates. Remember, the more repetition, the greater is the CAD advantage.

This of course is just one example. Other advantages include accuracy (discussed next) and ease of collaboration. Another interesting advantage noticed by me is that non-artistic people or ones who would never have been able to draft with pencil, paper, and T-square often excel at CAD, and skills such as neatness and a steady hand become less critical.

In support of Golden Rule 6, a brief discussion on accuracy follows. This is one often brought up in class. Search the web for a drawing called SOLAR.DWG. Download and open the file. It is a drawing of the solar system created in the early days of AutoCAD by founder John Walker and has become quite legendary over the years (Figure E.1). You see, the drawing is to scale. This means that the orbits of all the planets are drawn to full size as well as correct relative to each other. Zoom in to find the earth. Then zoom in to see the moon around the earth. Zoom in further and find the lunar landing module on the moon. Zoom in again and find a small plaque on one of the legs of the landing module. On it is a statement from humankind to anyone else who may read it. What does it say? And what is the date?

Think about what you just witnessed. You have zoomed in from the orbit of Pluto to look at a 5" × 4" plaque. That is quite an amazing magnification. To put things into perspective, this is the equivalent of looking at the earth and then at a grapefruit or looking at a grapefruit then an atom. Can you design microchips in AutoCAD? Safe to say, yes. Can you connect walls together in a floor plan? Absolutely.

Limitations

One of the chief limitations of AutoCAD is its 3D capabilities. Not that they are shabby, but sometimes they are pushed to do what they were not created for. The essence is that AutoCAD is not modeling and analysis software, rather, it is visualization software when it comes to 3D. It was not designed to compete with CATIA, NX, Pro/Engineer,

703

The Solar System
To Scale

Units: Km

FIGURE E.1
The solar system and plaque to scale.

and SolidWorks. Parametric design, dimension driven design, FEA, and CFD analysis are not what AutoCAD is about. These packages are dedicated engineering tools. AutoCAD was intended to be powerful 2D software with tools for presenting a design in 3D when necessary. As such, it is good at pretty pictures, to sell or present an idea.

Another limitation of the software, an obvious one about which a CAD manager needs to be aware, is that AutoCAD does not do the thinking for you. It contains little to no error checking, interference checking, or design fidelity tools, so garbage in will definitely give you garbage out. Be aware as a CAD manager that AutoCAD lets users do almost anything, and the word *anything* can swing in either a positive or a terribly negative direction. And finally understand the learning curve of this software. It *will* take a few months for a novice user to get up to speed. Keep an eye on that person until then.

PART 3. MAINTAIN AN OFFICE CAD STANDARD

A standard is a necessity for a smooth-running office, especially for one with more than a few AutoCAD users. A standard ensures uniformity and a professional look to the company's output, minimizes CAD errors, and makes it easy for new users to get up to speed. A manual needs to be created, published, and given to all users who may come in contact with AutoCAD. This manual should have, at a minimum,

- The exact location of all key files and the overall office computer file structure, indicating the naming convention for all jobs and the locations of libraries, templates, and standard symbol folders.
- A listing of the office layering convention. This may follow the AIA standard in an architecture office or just an accepted internal company standard.
- A listing of acceptable fonts, dimension styles, and hatch patterns.
- A listing of print and plot procedures, including `*.ctb` files.
- A listing of proper xref naming procedures.
- A description of how Paper Space is implemented. List title blocks, logos, and necessary title block information.
- A policy on purging files, deleting backup files, and AutoSave.

Keep the manual as short as possible while including all the necessary information. Do not attempt to cover every scenario that could possibly be encountered, as the longer the manual, the less likely it will be looked at cover to cover.

PART 4. BE AN EFFECTIVE TEACHER AND HIRING MANAGER

Part of being a CAD manager is sometimes explaining the finer points of AutoCAD to new hires or updating the occasional user as well as promotion of better techniques and habits. If you studied Levels 1 and 2 cover to cover, you may have noticed a concerted effort to simplify as much as possible. You should do the same with any potential students. While this is not meant to be a teaching guide, some basic points apply. If you are tasked with training new hires, then try to observe the following:

- Cover basic theory first—creating, editing, and viewing objects.
- Cover accuracy next, such as Ortho and OSNAP.
- Move on to the fundamental topics in the order of layers, text, hatch, blocks, arrays, dimensions, and print and plot, similar to Level 1.
- If advanced training is needed, cover advanced linework, followed by xref, attributes, and Paper Space, in that order. The rest of Level 2 can wait, as some of it is geared mainly to CAD managers anyway.
- Stress to users the secrets to speed: the pgp file, right-click customization, and intense practice of the basics; remember the 95/5 rule from the second paragraph of Chapter 1.
- Accurately assess the skill level of users and do not assign advanced tasks to those not yet ready, unless the drawings are not critical for a particular project. I have personally seen the subtle damage to a drawing that can be caused by an inexperienced drafter.

Also, be ready to test potential hires with an AutoCAD test. This test does not have to be long; 45 minutes to an hour should be the absolute maximum, although truthfully, an experienced user can tell if another person knows AutoCAD within 3 minutes of watching him or her draft. The trick is to maximize every minute and squeeze the most amount of demonstrated skill out of a potential recruit, which means do not have that person engage in lengthy drawing of basic geometry but rather perform broad tasks.

For example, the floor plan of Level 1 is a good test for an architectural drafting recruit (the drawing can be simplified further if necessary by eliminating one of the closets or even a room). It should take an experienced user about 15 minutes total to set up a brand-new drawing (fonts, layers, etc.) and draft the basic floor layout (most students can do it in an hour after graduating Level 1—not too bad). Then, it should take another 15 minutes to add text, dims, and maybe some furniture, followed by another 5–10 minutes of hatch and touchups, followed by printing. So it is roughly a 45-minute test and runs the recruit through just about every needed drafting skill. When looking over the drawing, note the layers chosen, accuracy demonstrated, and how much was completed in what time frame.

Note the following. It is *not* critical if a new hire does not know xrefs or Paper Space. These can be explained in a few minutes and learned to perfection "on the job." However, nothing but hard time spent drafting teaches someone how to draft quickly and efficiently with minimal errors. This *is* a requirement, and do not hire anyone who does not have that, as it will take time climbing that learning curve. Exceptions should be granted, of course, for individuals hired for other skills, with AutoCAD being of only secondary importance.

The bottom line (in my hiring experience) is that a fast, accurate drafter is far more desirable than a more knowledgeable one with a sloppy, slow manner of drafting. Realize also that extensive knowledge is not always a good substitute for experience, and while many drafters are fine workers and can come in and be productive on an entry level, be sure you understand what you are getting. I have for a long time encouraged constant practice in class, even at the cost of occasionally not covering an advanced topic or two, knowing full well what these students will face in a saturated market full of experienced AutoCAD professionals.

As a final word, be sure the AutoCAD test is fair but challenging. Leave units in decimal form as the student needs to demonstrate a grasp of nuances and situational awareness and the ability to change units to architectural if the foot/inch input does not work. Include in the test

- Basic linework to be drawn.
- A handful of layers to set (colors, linetypes, etc.).
- Some text to write (set fonts, sizes, etc.).
- Some hatches to create.
- Some arrays to create.
- Some blocks to create.
- Some dimensions to put in.
- Plot of output.

PART 5. STAY CURRENT AND COMPETENT

While we are on the subject of technical competency, *you*, as a recent advanced class graduate or CAD manager, are not off the hook for continuing to add to your skill set. It is much too easy to settle into a comfortable routine of drafting or designing and not learn anything outside the comfort zone (and not just with AutoCAD either). Some thoughts on this topic follow.

First of all, realize that it is virtually impossible for one individual to know everything about AutoCAD, although some claim to come close. It is truly an outrageous piece of software, developed and updated constantly by hundreds of software programmers and designers. This statement should make you realize that there is always more to learn (do not throw your hands up in despair, however), and you will not run out of new, easier, and creative ways to do a task. The main trait shared by all successful AutoCAD experts is that they are genuinely interested in the software and in the concept of computer-aided design in general.

To maintain your skills or advance further, look through the following list and determine what applies to you, then take the steps needed to acquire that knowledge. Some of this you will learn as part of studying all levels of the book; others you have to initiate on your own:

- Learn the 3D features. Many users do not know 3D well, and it is a good way to stand out from the crowd. It is fun to learn and is an important side of AutoCAD.
- Learn basic AutoLISP. This relatively simple programming language allows you to customize AutoCAD and write automation routines, among other uses.
- Explore AutoCAD's advanced features that are sometimes overlooked, such as sheet sets, dynamic blocks, eTransmit, security features, CUI, and much more. Keep notes of key features and effects (a "cookbook" in software lingo).
- Read up on the latest and greatest from AutoCAD, new tips and tricks, and what the industry is talking about. Cadalyst.com and Augi.com are two good sites for this. You should also get a *Cadalyst* subscription. It is no longer free but still worth the cost. At the very least, check up on the latest in the AutoCAD world on all of the relevant websites.
- Do not work in isolation; talk to others in your field and see how they do things. Also, be sure to take an AutoCAD update class if upgrading to a new release.
- Be curious about other design software, including AutoCAD add-on programs (verticals) such as AutoCAD Mechanical, Electrical, Civil, and MEP. Be knowledgeable about as much software as possible. In the world of CAD, Revit and SolidWorks are two software packages to keep an eye on in the future.
- Enjoy what you are doing. If you are not, you should seek other work duties in your profession. AutoCAD is very miserable to deal with if you hate it. Occasionally, you may think it is just one big software virus out to get you; just do not let it be a daily thought.

AutoLISP Basics and Advanced Customization Tools

You may have heard of AutoLISP, Visual LISP, VBA, .NET Framework, Active X, and ObjectARX. What exactly are they? These are all programming languages, environments, and tools used, among other things, to customize and automate AutoCAD beyond what can be done by simpler methods like shortcuts, macros, and the CUI, all explored earlier in this book. To properly discuss all these methods would take up quite a few volumes of text, and indeed much has been written concerning all of them. Our goal here is to introduce the very basics of AutoLISP and only gloss over the rest, so you at least have an idea of what they are. If advanced customization or software development work interests you, you can pursue it much further with other widely available publications.

OVERVIEW 1. AUTOLISP

This is really the very beginning of the discussion. AutoLISP is a built-in programming language that comes with AutoCAD and is itself a dialect of a family of historic programming languages, collectively referred to as LISP (list processing language). LISP was invented by an MIT student in 1958 and found wide acceptance in artificial intelligence among other research circles. The language and its developer, J. McCarthy, pioneered many expressions and concepts now found in modern programming.

AutoLISP is a somewhat scaled back version of LISP uniquely geared toward AutoCAD. It allows for interaction between the user and the software, and indeed one of the primary uses of AutoLISP is to automate routines or complex processes. Once the code is written, it is saved and recalled as needed. The code is then executed as a script (no compiling needed) and performs its operation.

One advantage of AutoLISP over other methods, discussed shortly, is that it requires no special training, except some familiarity with syntax, and is thus quite well suited for nonprogrammers. While it is not a compiled, object oriented language like C++ or Java, and therefore not suitable for writing actual software applications, it also does not need to be for its intended use.

OVERVIEW 2. VISUAL LISP

Over the years, AutoLISP has morphed into a significantly more enhanced version, called *Visual LISP*. More specifically, Visual LISP was purchased by Autodesk from another developer and became part of AutoCAD since release 2000. It is a major improvement and features a graphical interface and a debugger. Visual LISP also has Active X functionality (more on that later). However, Visual LISP still reads and uses regular AutoLISP, and that is still where a beginner needs to start before moving up. We briefly touch upon Visual LISP and its interface (the vlide command) later on. The subject is an entire book all in itself.

OVERVIEW 3. VBA, .NET, ACTIVE X, AND OBJECTARX

There is evidence to suggest Autodesk is moving away from using Visual LISP toward other, more advanced tools for customization and programming.

VBA (Visual Basic for Applications) is a language based on Visual BASIC, a proprietary Microsoft product. VBA allows you to write routines that can run in a host application (AutoCAD) but not as stand-alone programs. It has some additional power and functionality over Visual LISP.

.NET Framework is another Microsoft software technology that was intended as a replacement for VBA. It features pre-coded solutions to common programming problems and a virtual machine to execute the programs.

Active X, also developed by Microsoft, is used to create software components that perform particular functions, many of which are found in Windows applications. Active X controls are small program building blocks and are often used on the internet, such as for animations and unfortunately also for some viruses and spyware applications.

ObjectARX is the highest level of customization and programming you can do with AutoCAD, short of working on developing AutoCAD itself. Unlike the scripting-type languages and methods discussed so far, ObjectARX (ARX stands for AutoCAD runtime extension) is an application programming interface (API). Using C++, programmers can create Windows DLLs (Data Link Libraries) that interact directly with AutoCAD itself.

Everything needed to do this is part of the ObjectARX Software Development Kit, freely distributed by Autodesk. This is not easy to master, as you would expect from professional-level software coding tools. Typical users are third party developers who need a way to write complex software that interacts seamlessly with AutoCAD. ObjectARX is specific to each release of AutoCAD and even requires the use of the same compiler Autodesk uses for AutoCAD itself.

We hope this discussion gives you some insight into what is out there for AutoCAD customization. This is of course only a cursory overview, and you can pursue these topics much further. Any journey to learn advanced AutoCAD customization, however, begins with AutoLISP, and this is where we go next to introduce some basics of this programming language.

AUTOLISP FUNDAMENTALS

As mentioned before, despite the proliferation of new tools, AutoLISP is still very much in use, thousands of chunks of code are still being written annually, and many are easily found online for just about any desired function. The main reason for AutoLISP's popularity remains the fact that AutoCAD designers are not programmers and AutoLISP is accessible enough for almost anyone to jump in and learn. The value it adds to your skill set is substantial. Let us cover some basics and get you writing some code.

AutoLISP code can be implemented in two distinct ways:

- You can type and execute a few lines of code right on the command line of AutoCAD. This is good for a simple routine that is useful only in that drawing.
- You can load saved AutoLISP code and execute it, usually the case with more complex and often-used routines.

We generally use the second method of writing and saving our routines. At this stage, it is helpful to know how (and where) to save AutoLISP code and, by extension, how to load what is written.

All AutoLISP code is saved under the `*.lsp` extension. Once the code is written, it can theoretically be stored just about anywhere (AutoCAD Express folder contains a sizable number of built-in routines). To load the code, go to Tools→Load Application… or Tool→AutoLISP→Load Application… (or just type in `appload`) and the dialog box in Figure F.1 appears.

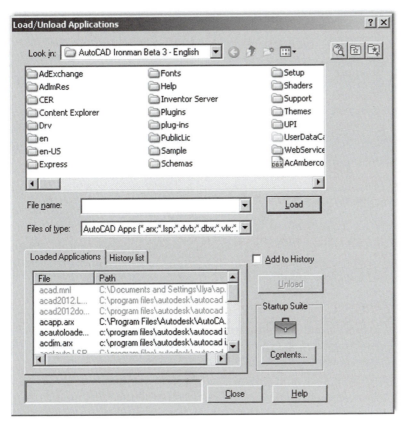

FIGURE F.1
Load/Unload Applications.

Then, you just browse around, find the AutoLISP code you are interested in, click on it, and press Load. Once loaded, you can close the dialog box; the code is ready to execute with a command input. There is a way to automatically load all existing code upon starting AutoCAD by adding it to the Startup Suite (lower right corner, the briefcase icon).

Now that you know how to save new code and where to find it, we should try writing a simple example that illustrates some AutoLISP basics. Here are two critical points that need to be understood before you begin:

- AutoLISP is closely tied to AutoCAD commands, so to write something useful, you need to know the commands well. If you do not know how to do it in AutoCAD manually, AutoLISP is of no use.
- Approach AutoLISP for an answer to a specific problem. State what it is you want to solve or automate, then write out the steps to do it manually. Only then, using proper syntax, write and test the AutoLISP code.

Following this philosophy, let us state a problem we would like to solve. We would like to modify the zoom to extents command by having it zoom to only 90% of the screen, not 100%. That way we can see everything and nothing gets lost at the edges of the screen. How do we do this by hand, first of all?

You of course type in zoom, press Enter, select e for Extents, and press Enter again. The drawing is zoomed all the way. To bring it back a bit, you repeat the zoom command but type in .90x instead of e. The drawing then zooms in a bit to fill only 90% of the screen.

Let us now automate these steps so all you have to do is type in zz (a random choice for a command name; you could pick anything) instead of the two-step process just described.

We need to

1. Open up the Notepad editor (we use the built-in Visual LISP editor later).
2. Save the blank file as ZoomSample.lsp in any folder you wish.
3. Define the function zz by typing in exactly the following:
   (defun C:zz ()
4. List the relevant commands in order by typing in exactly the following:
   (command "zoom" "e" "zoom" ".90x")
5. Finally add (princ) and close the expression with another).

So, here is the final result:

```
(defun C:zz ()
(command "zoom" "e" "zoom" ".90x")
(princ))
```

Save the file again and let us run it. Open up any drawing, load the AutoLISP routine using appload or the drop-down menu and run it by typing in zz and pressing Enter. The drawing should zoom out to 90% as expected.

This simple routine has some essential ingredients of all AutoLISP code:

- It starts out with a parenthesis, indicating that a function follows.
- The defun or define function (an AutoLISP expression) is next followed by C, indicating it is an AutoCAD command, and finally the actual command, zz.
- The set of opening and closing parentheses can hold arguments or variables.
- The next line features command, an AutoLISP expression meaning AutoCAD commands (in quotation marks, meaning it is a string) follow.
- The (princ) function is added so the routine does not constantly return a "nil" value.
- Finally, another, matching parenthesis closes out the program.

Here are two simple routines that I use often to rotate the UCS crosshairs to align them to some geometry (this general task was covered in Chapter 15).

The first routine does the rotation and is called perp:

```
(defun C:perp ()
(command "snap" "ro" "" "per" pause "snap" "off")
(princ))
```

The second one brings the crosshairs back to normal and is called flat:

```
(defun C:flat ()
(command "snap" "ro" "" "0" pause "snap" "off")
(princ))
```

The perp and flat AutoLISP routines feature two new concepts. The first one is the pause function, which allows for user input. In this case, it is for the user to select the line to which the crosshairs are to be aligned. Then, the function resumes and turns off `snap`. The other concept is the two quotation marks back to back with nothing between them. They simply force an Enter.

Study these three examples closely. They are simple; many AutoLISP routines are, as that is all that is needed to get the job done. Here is what we learned so far:

- **String**. Anything that appears in quotes (often an AutoCAD command).
- **Expression**. AutoLISP built-in commands such as `defun`.
- **Functions**. Can be user defined (`perp`, `flat`) or built-in (`pause`, `princ`, `command`).
- Functions are sometimes followed by arguments in parentheses.

VARIABLES AND COMMENTS

Let us now move on to other concepts. To take a step up in sophistication from what we did so far, we need to introduce program variables. Variables can be numbers, letters, or just about anything else that can change. The reason they are needed is because they allow for user interaction. Because user input is not known when the code is first written, a variable is used. When that information is acquired, the program proceeds accordingly. Variables are a fundamental concept in all programming.

Variables in AutoLISP can be global (defined once for the entire routine) or local (defined as needed for one-time use). Local is usually how variables get defined in AutoLISP. The function that creates variables is `setq`. To set a variable equal to whatever the user inputs (such as the radius of a circle), you write: `setq RadCircle`. The next logical step with variables is to pick up on whatever the user inputs. For this we need the `getstring` function that returns anything the user types in.

Here is how these concepts look in practice, with the \n simply meaning that a new line is started.

```
(setq RadCircle (getstring "\nCircle Radius: "))
```

In addition to `getstring` are many other `get` functions, such as `getfiled` (asks user to select a file using a dialog box), `getangle`, and `getdist`.

As in any programming language, you are allowed to comment on your work, so you or someone else can understand your intent when examining the code at a later time. In AutoLISP, comments are preceded by a semicolon. You can of course add more than one semicolon to indicate sections or headers:

```
;This is a comment
;;So is this
```

ADVANCED FEATURES AND THE VLIDE COMMAND

We have only scratched the surface of AutoLISP, and if you enjoyed writing and running these simple routines, then consider taking a full class on this subject or reading specialized books. There is much more to learn. For example,

- **The IF→THEN statements.** This is a common feature in all of programming. If certain conditions are met, then the program does something; if not, it does something else.
- **The WHILE function.** Another important concept in programming. This function continues to do something until a condition is met.

Finally, we come to the vlide command and the Visual LISP Editor, as shown in Figure F.2.

Everything you typed earlier into Notepad can be typed into the Visual LISP Editor, which has many additional useful features, such as debugging, error tracing, and checks for closed

711

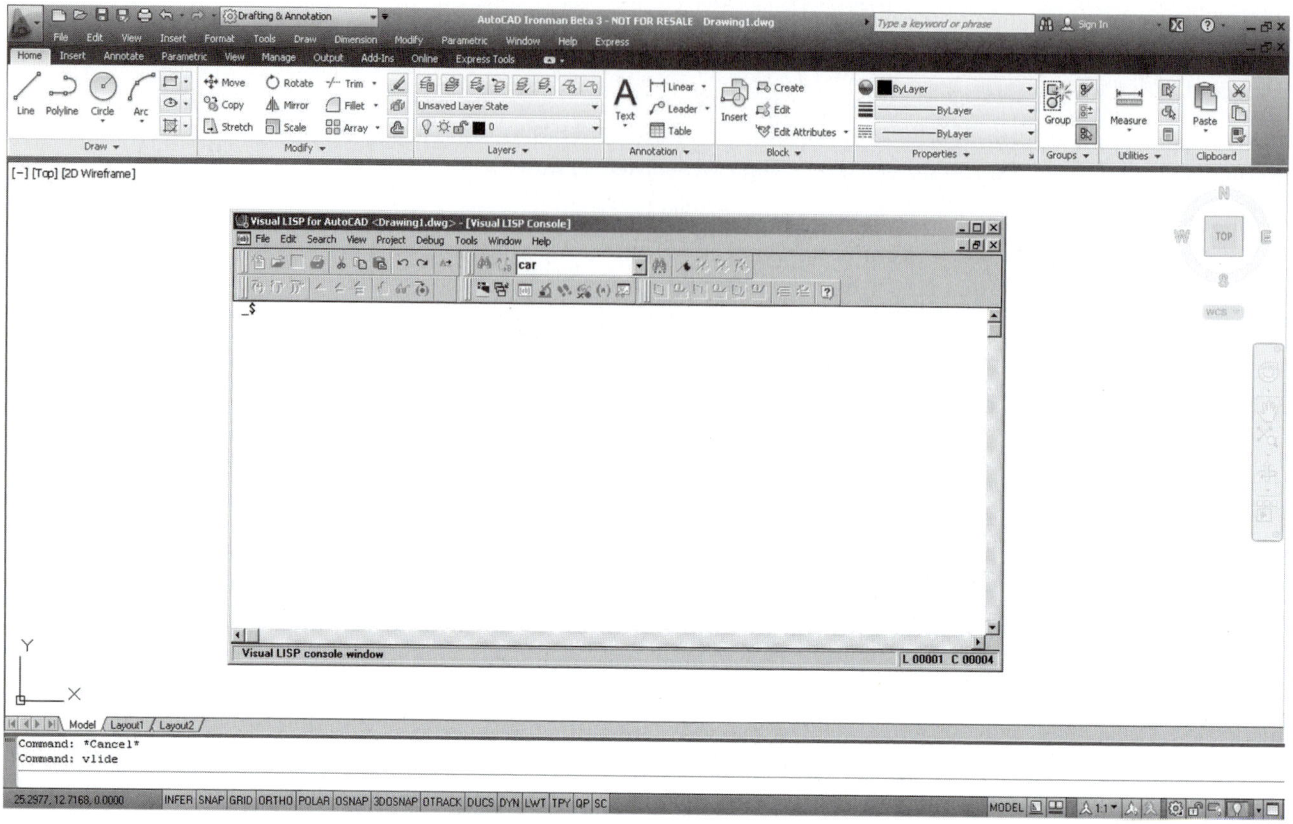

FIGURE F.2
Visual LISP Editor.

quotations and parentheses (formatting tools). Everything you type is also color coded for easy identification.

In-depth coverage of Visual LISP is beyond the scope of the book, but you are encouraged to pick up a specialized textbook and try to progress with this invaluable skill.

PC Hardware, Printers and Plotters, and Networks

This is another collection of topics that can easily span volumes, and indeed much has been written about all of them. We try to focus on hardware from an AutoCAD user's point of view exclusively, which may be a less covered topic (if only slightly).

Software generally goes hand in hand with the hardware. And, while computers, like cars, have become so reliable that a "hands-off" approach works fine, remember that CAD in general, and AutoCAD specifically, is not ordinary software. It is a high-end, powerful application that places heavy demands on the associated hardware and does not run at its peak on a clunky PC. Much like the case where the owner of a Ferrari is more likely to be interested in (and look under the hood of) the car versus the owner of a Toyota, so should AutoCAD designers express interest in what is "under the hood" of their computer. After all, they are manipulating and creating multi-megabyte sophisticated drawings or 3D models, not merely typing up a memo on Word, and it is in their best interest to use a PC that meets or exceeds AutoCAD's demands.

713

PC HARDWARE

By its nature, AutoCAD is not as fussy as some other CAD software out there (3D solid modeling applications come to mind), but it does need a high-end PC to operate optimally. Here is a rundown of a recommended setup with comments.

- **Processor.** The "brain" of the computer, also referred to as a *chip*, or *processor*, is where all the calculations and operations are performed. Intel and AMD are the major players in the PC market. Which is better is a moot point; while Intel was once the only game in town, these days the two companies constantly one-up each other with performance benchmarks. At this point, both manufacturers produce high-quality products, and the Intel versus AMD comparison is essentially meaningless to the average user. The top-of-the-line Intel processor (as of this writing) is the Intel Core i7 Extreme Edition with six cores, a 3.33 GHz clock speed, and a 12 MB cache. AMD has a comparable Phenom II X6 1090T Black Edition series chip. It also has six cores and a 3.2 GHz clock speed. Autodesk recommends a tamer 3.0 GHz Pentium 4 or AMD Athlon dual core processor as a minimum, so you have some room to shop on price between these two extremes.
- **RAM.** Random access memory is the next consideration, and the more the merrier. RAM has a direct effect on performance as it is there that software resides while in operation. More available RAM allows for more functions to execute faster. Autodesk recommends 2 GB, but go as high as you can afford. Most motherboards will accept 4 GB and some up to 12 GB of memory.

- **Hard drive.** Also known as *storage disk, main drive*, and the like, it needs to be at least 1.8 GB to install AutoCAD. Fortunately, this is rarely a problem, as most machines have drives on the order of 100–250 GB. It is not unusual to find a terabyte-sized drive as well. Prices have come down quite a bit.
- **Video graphics card.** Perhaps the most overlooked part of the PC, the video or graphics card is similar to a processor but is specially designed to control the screen images or graphics. It needs to be beefed up to deliver the kind of graphics AutoCAD is capable of producing (essential in 3D rendering). Autodesk recommends only a "workstation-class" graphics card, but get the best one possible under the allowed budget. High-quality graphic cards are not cheap; a high-end card tops out at over $2500. One in the $500–$700 range should be just fine for AutoCAD applications.
- **OS.** Until late 2010, operating systems that AutoCAD could run on numbered in the single digits (actually precisely one, Windows). In November of that year, after a 20-year break, Autodesk announced AutoCAD would once again be available for the Mac. AutoCAD still does not run on UNIX or Linux. Windows XP (with Service Pack 2), Vista, and Windows 7 are the main Microsoft operating systems. Older NT may run AutoCAD 2012, but few companies still have that OS. Although AutoCAD is officially certified for Vista, students have reported sporadic operational problems. Maybe as of this writing things have improved; if not, Windows 7 is finally established and seems to be doing just fine.
- **Other PC components.** These include the motherboard (must be picked according to processor requirements), the sound card and speakers (if needed), the CD/DVD, and power, cooling, and wiring systems. The DVD capability is needed as AutoCAD now comes on DVD when you buy it (unless you chose the download option, of course). The program has gotten to be too big for a reasonable number of CDs to hold it.
- **Monitor.** Get the best and largest flat screen monitor you can afford; your eyes will thank you many times over.
- **Mouse and keyboard.** Get a laser precision mouse (forget roller ball) and the best soft-touch keyboard you can find. Do not try to cut corners here, as you will be in physical contact and using these devices constantly while working on the PC, so get the good stuff.

Whether building your own PC or ordering an assembled one, take the PC purchase seriously and try to set realistic budget expectations. CAD workstations are in the same family as gaming stations and you need to budget over $2000 (or more) for a serious machine, not including the monitor. Some well-known vendors include Dell, but if you are inclined to assemble your own, check out online stores such as TigerDirect.com and buy parts on sale. You will get phenomenal performance for a fraction of the cost, as long as you know what you are doing when assembling the machine.

PRINTERS AND PLOTTERS

There are essentially two ways to output something from AutoCAD, a printer or a plotter. Printers are generally Inkjet or LaserJet and black and white or color capable. Prices can range from a $49 desktop Inkjet to a "sky's the limit" corporate high-speed color LaserJet. Which to get depends entirely on budget and desired level of quality for output, as AutoCAD does not care. Once AutoCAD is installed, it automatically senses any printer attached to the PC (assuming it works and has drivers in the first place). Be aware that most small desktop printers cannot handle 11" × 17" paper, but larger LaserJet printers can. The 11" × 17" format is very important in AutoCAD, as many jobs are handed in on this size paper, so be sure your printer can handle it.

Plotters (Figure G.1) are a familiar sight in architecture or engineering offices. Anything sized above an 11" × 17" sheet of paper needs to be routed to a plotter, all the way up to a 48" × 36" or larger in some cases. Plotters as a rule are more expensive than printers, and buying one should be a carefully researched decision. Many plotters are color capable and

FIGURE G.1
A typical HP professional plotter.

715

accept almost any type of paper, although bond or vellum is the norm. Hewlett-Packard (HP) is the premier manufacturer of plotters, with some models on sale well over $20,000, though their 500, 800, and 1055 series plotters are much more affordable. Other companies, such as OCE, Canon, and Encad, round out the field. The typical average cost of a professional grade plotter hovers around $5000–$6000. Do your research carefully when selecting one and be aware of recurring costs such as ink and paper.

NETWORKS

This topic of course is an entire profession all to itself. A computer network is defined as a group of interconnected computers. Networks can range from local area network (LAN) to city, country, or worldwide (WAN, or wide area network). Networks allow for collaboration and sharing of information from computer to computer and have their own set of hardware familiar to any network engineer, such as routers, hubs, switches, repeaters, and other gear. Networks are operated via a protocol or set of instructions, such as the IEEE 802.3 protocol running on Ethernet technology for a typical LAN. The medium of data transfer between computers is usually a standard Cat5e cable but can be wireless or fiber optic for larger networks.

As an AutoCAD designer, you will likely encounter a local area network consisting of a server (host) with some other computers (clients) connected to it at your company. All the data are stored on the server, and once you log on to the network, your computer can access what it needs.

In regard to networks, AutoCAD can be installed on a per seat or network license method. Per seat is cost effective to a certain limit (usually ten seats). If your company requires more, a site license is purchased, which is cheaper and allows more computers to run the software, which now resides on the server only.

What Are Kernels?

If you have taken an active interest in 3D CAD software and are learning what is "under the hood," then you may have read or heard about kernels. What are they and what do farming and planting terms have to do with software? Well, as you may have already deduced, the word *kernel* is a metaphor for describing the heart of the software, a starting point from which everything else grows, the same as a kernel of corn. This analogy is quite accurate: A 3D modeling kernel is the computer code that depicts the "brains" of any 3D software package, such as CATIA, NX, Pro/Engineer, SolidWorks, Inventor, and of course AutoCAD.

As it turns out, kernels are quite rare. This is because they are extremely difficult, expensive, and labor intensive to develop from scratch. As a result, while hundreds of 3D capable CAD software applications are available, there are only a small handful of kernels. If you wanted to start a new CAD company and develop a competitor to any of the established products, then chances are good that you would simply purchase a kernel license and design your software around the kernel, not try to create your own. In many cases, this is indeed what happens in the industry.

Why would kernel developers license their kernels, as opposed to using them for just their own CAD programs? Well, for starters, they do use them in such a way, and most kernel developers have a popular 3D CAD program for sale. But the licensing fee to "rent out" kernel code is substantial, so they do this as well to increase the revenue stream.

What customers get when they purchase the kernel is a large chunk of code made up of what are called *classes* and *components*—essentially mathematical functions that perform specific modeling tasks. The CAD company then writes its own code to interface with the kernel via numerous APIs (application programming interfaces) and essentially creates the new software around the kernel.

This new CAD software is unique, according to the design philosophy and talent of the CAD company, and may be different in how it operates compared to another company's software that uses the same kernel. In essence, however, most modelers are essentially the same, and one can switch relatively easily from, say, CATIA to NX or Pro/Engineer. All three of these software applications feature parametric modeling, part trees, dimension driven design, and other hallmarks of solid modeling. The differences are chiefly in the interfaces, ease of use, extra features offered with each package, and of course price.

One notable exception is AutoCAD. The kernel it uses, called ACIS, is made by Spatial Technologies. Its modeling methods are somewhat different than the other kernels on the market, giving AutoCAD a unique (and not necessarily the best) approach to 3D. Other, more high-end kernels include Parasolid (used by NX, SolidWorks, and SolidEdge), Granite (used by Pro/Engineer), and CAA (used by CATIA). Autodesk set out to develop its own kernel in recent years, and the fruit of that labor is in its Inventor software.

717

Lighting, Rendering, Effects, and Animation

The world of digital effects is a very expansive one, and under that umbrella falls everything from simple shaded models (extensively used in this book), all the way up to full-blown Hollywood-style effects. Architects and engineers use of some of these digital effects and techniques, which are roughly in the middle of the pack as far as sophistication goes.

What architects look for includes visualization renderings that present a building or an interior scene in the most realistic way possible, so the clients see what they are getting and can offer input. They may also want some basic animation such as a walk-through. Some high-end requests may even include photorealistic scenes that have weather effects (wind, precipitation, etc.) or realistic images of people interacting with the design.

Engineers have the same basic goal, to present a design to a client, although their renderings must often account for metals, plastics, and other materials, including material properties. Engineering animation needs are also more sophisticated. An architectural animation of a walk-through is essentially a collection of snapshots played back in succession. With engineering animations, complex part interaction must be accounted for and a full motion path must be shown. The entire animation must also be looped for constant playback. An example of this is a cutaway of an engine, with all the components (pistons, connecting rods, crankshaft, etc.) working in unison to demonstrate the design in operation.

As far as engineering applications go, AutoCAD plays a relatively minor role. From the beginning, the intent of AutoCAD was to be a superb 2D drafting tool, not a 3D solid modeling and testing software. Other software have evolved over the years (with some of these applications predating AutoCAD), for use in such cases. Examples, quoted elsewhere in this book, include CATIA, Pro/Engineer, NX, SolidWorks, SolidEdge, IronCAD, Autodesk's own Inventor, and many others. These software applications often take care of the drafting, modeling, rendering, testing, presentation, animation, and manufacturing in one complete package. Therefore, AutoCAD rarely finds its way into truly high-end situations; it is simply not built for that.

In the world of architecture and related fields, however, AutoCAD 3D models often find themselves in the middle of the rendering world. These models are imported into a variety of software and are the "skeletons" on which advanced rendering and even animation techniques are applied. Studio VIZ was the affordable choice for many professionals until it was discontinued a few years back. Its replacement was the top of the line, more expensive 3ds Max Design. This software can create stunning photorealistic renderings that feature advanced use of lights, shadows, and materials. This 3ds Max software can also do sophisticated animations, even to the point of full feature cartoon films, but it is not

719

commonly used for that, as even more advanced software is available for that purpose. There are of course many other software solutions for creating renderings. Two popular ones include Rhino and Form Z. Both software packages can generate their own shapes and solids but are often used as pure rendering engines with imported models.

Another excellent (and cost effective) solution is to purchase add-on software that simply extends AutoCAD's built-in abilities. One well-known choice is AccuRender. It installs and integrates seamlessly with any AutoCAD, going back to AutoCAD 2002. The software creates sophisticated renderings and light effects and has a huge collection of materials, plants, and light fixtures to use. It also has a desirable feature called *Procedural Materials*. This allows materials to become part of the design, not just wrapped around it like wrapping paper. The material property penetrates through the part and it becomes that material. So a piece of wood actually has grains and jagged edges if viewed up close.

The world of animation is tied in with rendering, as the animator simply animates an already complete rendering and adjusts for changes in lights and shadows. A major player in that field is Maya, formerly of Alias Wavefront but now owned by Autodesk. Maya has extensive customizing capabilities and studios take its core (the kernel) and write custom code around it, tuned to their specific production needs. Maya in various forms was one of the software applications behind movies such as *Jurassic Park, The Abyss,* and *Terminator 2,* although the amount of customization needed for such high-end productions practically turned Maya into whole new software.

The software listed thus far are just some of the high-end market products. Many other companies in the field produce software in many other price ranges. If you are interested in rendering and animation, it is a great field to get into, and many such students passed through my classroom, as they wanted to start with being able to model 3D designs (in plain old AutoCAD) and move on from there.

AutoCAD Certification Exams

It seems like there is an exam or certification for just about every profession or occupation. Although AutoCAD and drafting in general are not really part of the greater information technology (IT) world, where certification exams are the benchmarks of competence and almost as valuable as college degrees, there still exists a series of exams for AutoCAD designers to take if they wish to "formalize" their knowledge and hang up a piece of paper in their office.

These exams have changed somewhat over the years and have varied in the number of questions, difficulty level, and the amount of time given to complete them. The typical method of delivery has been online, with users logging in and taking the exam in a timed fashion—although in years past there have been exams with no time limit. You are able to use the Help files and any manuals or texts you can find.

Are these exams important? This question does not have an easy answer. The overriding belief among established AutoCAD professionals is that the exams are not well organized and test for trivial knowledge while not accurately assessing an individual's true mastery of the software, which can really be tested only by having the person draw a complex design. It is partially Autodesk's fault for not developing these exams into the powerhouse institutions that Microsoft, Cisco, Novell, and others have done with theirs.

Then again, however, there was little demand for the exams in the past. As AutoCAD was exploding in popularity, anyone who was reasonably good was hired and allowed to mature on the job. In recent years, as the supply of knowledgeable AutoCAD designers finally caught up with the demand, employers wanted ready-made experts and needed a way to separate the "wheat from the chaff." This would explain the resurgence in popularity of testing for AutoCAD competence and the renewed interest from students as to how to study for and pass these exams. So, yes, these exams may be important to get a foot in the door with certain companies, but having passed the exam does not necessarily guarantee a knowledgeable AutoCAD expert; and I never used them as the sole reason to hire someone. Let us go over the exams as they exist now for AutoCAD 2012. There are two exams:

1. AutoCAD Certified Associate Exam.
2. AutoCAD Certified Professional Exam.

The Certified Associate Exam (CAE), which is taken first, consists of 30 online questions, with the following breakdown by topic:

- Introduction to AutoCAD, one question.
- Creating drawings, four questions.
- Manipulating objects, six questions.

- Drawing organization and inquiry commands, three questions.
- Altering objects, two questions.
- Working with layouts, one question.
- Annotating the drawing, three questions.
- Dimensioning, two questions.
- Hatching objects, two questions.
- Working with reusable content, three questions.
- Creating additional drawing objects, two questions.
- Plotting, one question.

The passing score is 70% and you have a 60-minute time limit.

The Certified Professional Exam (CPE) is more focused on doing tasks, not just answering questions. It is taken after the CAE, and consists of 20 questions/tasks, with the following breakdown by topic:

- Manipulating objects, five questions.
- Drawing organization and inquiry commands, one question.
- Altering objects, two questions.
- Working with layouts, one question.
- Annotating the drawing, three questions.
- Dimensioning, three questions.
- Hatching objects, one question.
- Working with reusable content, one question.
- Creating additional drawing objects, two questions.
- Plotting, one question.

The passing score is 80% and you have a 90-minute time limit.

As mentioned before, the CAE focuses more on multiple choice and matching questions and basic knowledge, while the CPE is more performance based and focuses on drawing (you need to download, open, and work on actual AutoCAD files while taking the exam; you then are asked to draw something and answer a question based on this drawing). The intent here is to have an individual take the CAE shortly after completing a training course, and then the CPE a few months later, after some practical drawing skills are gained.

Practice sample tests are available, and it is highly recommended to go over these at least to see what types of questions to expect. The cost for taking these exams is $50 per exam, but that is of course subject to change, so check the latest information on the Autodesk website. You may also update your certification from AutoCAD 2011 by taking only the CPE. If you have an older certification, you need to take both exams again.

Information on purchasing and scheduling your exam can be found at www.autodesk .starttest.com.

AutoCAD Employment

AutoCAD students vary widely in their backgrounds. Some are already in mid-career and are registered architects or engineers looking to pick up new skills. Others are recent college graduates entering the job market with architecture or engineering degrees. These students need AutoCAD as a small part of their job, or maybe even the main part of their job, but are ultimately hired for their engineering or design knowledge and may draft in only the first few years of their career as they learn the ropes.

Another very large group of students, however, are attracted to CAD drafting in and of itself and attend one of the many programs at local community colleges or tech/vocational schools that ultimately lead to a certificate or even a two-year associate's degree in architectural or engineering drafting. These students intend to enter the workforce as professional AutoCAD drafters, a career track in itself. This can lead to CAD management positions as well as serve as a springboard for a career in engineering or architecture if the students decide to return to school to complete the bachelor's or master's degree. Even if they do not, a professional draftsperson is a well-paid professional who is a valuable asset and, as CAD software increases in sophistication, is needed more than ever. So the question is then, how do you get your "foot in the door" and begin and AutoCAD career?

Once you graduate and are ready to look for work, there are a number of things to consider. First of all, decide if you would like to concentrate on architectural, mechanical, electrical, or some other drafting field. Although I moved around a lot in my early days of consulting work, this approach does not work in today's saturated market. It is easier to succeed if you focus your interests instead of jumping around from company to company and field to field. What you ultimately decide to pursue should be something you are interested in (houses vs. mechanical devices, for example) as well as something you are good at. Generally, while learning AutoCAD in school, you are drawn to one type of work over another; go with your gut feeling.

Then consider your options to actually find work. Two major sets of resources are presented next:

- **Staffing firms.** This is one of the best ways to find a CAD job, though sometimes staffing firms prefer experienced CAD designers, as those command higher salaries and therefore higher commissions and fees for the staffing agencies. These staffing firms can either place you outright with a client (collecting a onetime finder's fee) or place you on an hourly rate as one of their own workers "on loan" to a client (collecting a premium over your pay). It may or may not matter which route you go, although being hired on directly with benefits may be more desirable, even if the initial salary is lower. Search for staffing firms (also informally called *job shops*) on the internet in your area. Some well-known, large, nationwide firms include Kelly Staffing, Manpower Technical, and Aerotek.

- **Job boards.** These include Monster.com, Dice.com, and Craigslist.com, among others. Often these sites are populated with postings from staffing firms (see the preceding), but independent companies can be found as well. Simply enter AutoCAD and your city or town into the search engine and listings appear. Of special note is Craigslist.com. You may have already used it for everything from apartment hunting to selling something, and its job boards are excellent and very informal and direct. The only downside is that Craigslist.com is most utilized by residents and companies in large metropolitan centers, and if you live in a small town far from major cities, you may have slim pickings (even if the actual job scene is pretty good).

Other places to look include your school's employment office and classified newspaper ads (more likely to be online, not in print these days). Also never forget the value of word of mouth and personal contacts. Let everyone know you are looking for AutoCAD work, and just maybe someone's cousin's friend knows something.

AutoCAD Humor, Oddities, Quirks, and Easter Eggs

REAL AUTOCAD USERS

- Have been doing AutoCAD for six or more years.
- Never bother to keep .bak files.
- Do not do drawings on a dot matrix printer.
- Have a graphics card that can support two monitors.
- Have two monitors.
- Have games on their system.
- Get *real* pissed when someone else uses their machine just to "look around."
- Know AutoLISP.
- Know how to edit an AutoLISP file.
- Wonder why Autodesk gives you so much worthless stuff with a software upgrade.
- Wonder why Autodesk charges so much for an upgrade, then AutoCAD gets pirated.
- Maintain at least three older versions of AutoCAD on their system.
- Never ask for the newest copy of AutoCAD—they already have it.
- Have a subscription to *Cadalyst*.
- Do not really bother to read *Cadalyst*.
- Could write an article that could go in *Cadalyst*.
- Do not run AutoCAD on a Mac.
- Do not have friends that run AutoCAD on a Mac.
- Despise Macs.
- Are not concerned with losing a drawing.
- Use the purge command about 60 times in a drawing.
- Use the spacebar, not return.

725

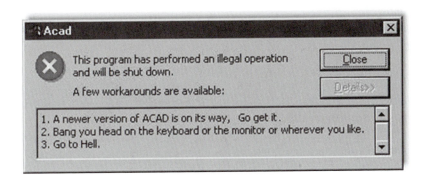

Up and Running with AutoCAD® 2012: 2D and 3D Drawing and Modeling.

- Have drawn a picture of at least one of the following:
  - Their car.
  - Their house.
  - Their dream house.
  - Their dream house with their dream car parked in the drive.
- Like to waste time, such as like typing ZOOM, E twice in a large drawing.
- Can talk on the phone and draw at the same time.
- Drink lots of liquid while drawing.
- Go to the bathroom a lot.
- Get some exercise while at work.
- Are not afraid to eat while drawing.
- Can leave the room while plotting.
- Do not care that the work system is on 24 hours a day; it is not their machine.
- Do not really like the system they run AutoCAD on; they are sure there is a better one.
- Would be damn lucky if they got the system they have at work, for their home.
- Always complain that AutoCAD on their system is too slow.
- Work in the dark with their AutoCAD screen color set as black.
- Get pissed when someone turns on the light.
- Get *real* pissed when people say, "How can you work in the dark like this?"
- Listen to free downloaded MP3s while working.
- Are reading this while they should be working.

THE AUTOCAD MONKEY JOKE THAT NEVER GOES AWAY

A tourist walked into a pet shop and was looking at the animals on display. While he was there, another customer walked in and said to the shopkeeper, "I'll have an AutoCAD monkey, please." The shopkeeper nodded, went over to a cage at the side of the shop and took out a monkey. He fitted a collar and leash, and handed it to the customer, saying, "That'll be $5000." The customer paid and walked out with his monkey.

Startled, the tourist went over to the shopkeeper and said, "That was a very expensive monkey. Most of them are only a few hundred dollars. Why did that one cost so much?"

The shopkeeper answered, "Ah, that monkey can draw in AutoCAD—very fast, clear layouts, no mistakes, well worth the money."

The tourist looked at a monkey in another cage. "That one's even more expensive! $10,000! What does it do?"

"Oh, that one's a design monkey; it can design systems, layout projects, mark up drawings, write specifications, some even calculate. All the really useful stuff," said the shopkeeper.

The tourist looked around for a little longer and saw a third monkey in its own cage. The price tag around its neck read $50,000. He gasped to the shopkeeper, "That one costs more than all the others put together! What on earth does it do?" The shopkeeper replied, "Well, I haven't actually seen it do anything, but it says it's an engineer."

ODDITIES AND QUIRKS

- AutoCAD has an *oops* command; try it to see what it does.
- AutoCAD had an *end* command. It used to shut down your drawing, even if you meant that as an OSNAP command and just forgot to type in line and press Enter first. End has been permanently discontinued for obvious reasons.
- AutoCAD has a *battman* command. It has nothing to do with the caped crusader and everything to do with the block attribute manager (similar to the enhanced attribute manager of Chapter 18).

EASTER EGGS

Easter eggs are surprises that one can find in software, hence the analogy to the Easter egg hunt. These surprises can be anything from a silly message to an entire video game buried in the code. To find and activate Easter eggs you really just have to be told how, as the complex combination of keys that need to be pressed in correct sequence is just too much for a random guess. Why are they in there? Just for fun is the simple answer. Also because programmers can get away with it; even a medium-size application has millions of lines of code in which to hide the much smaller Easter eggs. AutoCAD has its own share of them that are well documented.

Here is a known AutoCAD 2008 Easter egg, which reveals a credit roll of the design team's names:

- On the AutoCAD command line start the Sun Properties by typing in `sunproperties`.
- Click on the magnifying glass in Sky Properties.
- Change the date to 3/23/2007 (the date AutoCAD 2008 was released) via the pop-up calendar. Press Enter to set it.
- Click on the time.
- Press the Home key on your keyboard (this sets you to midnight).
- With the Ctrl key down, press the down arrow twice on your keyboard.
- Now you should see the credits roll with music.

There are other Easter eggs in virtually all releases of AutoCAD. The more you look, the more you find.

INDEX

729